Lecture Notes in Computer Science

Commenced Publication in 1973
Founding and Former Series Editors:
Gerhard Goos, Juris Hartmanis, and Jan van Leeuwen

Carlos A. Coello Coello Vincenzo Cutello
Kalyanmoy Deb Stephanie Forrest
Giuseppe Nicosia Mario Pavone (Eds.)

Parallel Problem Solving from Nature - PPSN XII

12th International Conference
Taormina, Italy, September 1-5, 2012
Proceedings, Part II

 Springer

Volume Editors

Carlos A. Coello Coello
CINVESTAV-IPN, Mexico City, Mexico
E-mail: ccoello@cs.cinvestav.mx

Vincenzo Cutello
Giuseppe Nicosia
Mario Pavone
University of Catania, Italy
E-mail:{cutello, nicosia, mpavone}@dmi.unict.it

Kalyanmoy Deb
Indian Institute of Technology, Kanpur, India
E-mail: deb@iitk.ac.in

Stephanie Forrest
University of New Mexico, Albuquerque, NM, USA
E-mail: forrest@cs.unm.edu

ISSN 0302-9743   e-ISSN 1611-3349
ISBN 978-3-642-32963-0   e-ISBN 978-3-642-32964-7
DOI 10.1007/978-3-642-32964-7
Springer Heidelberg Dordrecht London New York

Library of Congress Control Number: 2012944753

CR Subject Classification (1998): J.3, I.2, F.1, F.2, I.4-5, G.2

LNCS Sublibrary: SL 1 – Theoretical Computer Science and General Issues

Typesetting: Camera-ready by author, data conversion by Scientific Publishing Services, Chennai, India

Printed on acid-free paper

Springer is part of Springer Science+Business Media (www.springer.com)

Preface

This LNCS volume contains the proceedings of the 12th International Conference on Parallel Problem Solving from Nature (PPSN 2012). This biennial event constitutes one of the most important and highly regarded international conferences in evolutionary computation and bio-inspired metaheuristics. Continuing with a tradition that started in Dortmund, in 1990, PPSN 2012 was held during September 1–5, 2012 in Taormina, Sicily, Italy.

PPSN 2012 received 226 submissions from 44 countries. After an extensive peer-review process involving more than 230 reviewers, the Program Committee Chairs went through all the reports and ranked the papers according to the reviewers' comments. Each paper was evaluated by at least four reviewers. The top 105 manuscripts were finally selected for inclusion in this LNCS volume and for presentation at the conference. This represents an acceptance rate of 46%, which guarantees that PPSN will continue to be one of the most respected conferences for researchers working in natural computing around the world.

PPSN 2012 featured four distinguished keynote speakers: Angelo Cangelosi (University of Plymouth, UK), Natalio Krasnogor (University of Nottingham, UK), Panos M. Pardalos (University of Florida, USA), and Leslie G. Valiant (Harvard University, USA).

The meeting began with six workshops: "Evolving Predictive Systems" (Bogdan Gabrys and Athanasios Tsakonas), "Joint Workshop on Automated Selection and Tuning of Algorithms" Part A: Continuous Search Spaces—Focus on Algorithm Selection (Heike Trautmann, Mike Preuss, Olaf Mersmann, and Bernd Bischl), Part B: Discrete Search Spaces – Focus on Parameter Selection (Andrew Parkes and Ender Özcan), "Theoretical Aspects of Evolutionary Multiobjective Optimization: Interactive Problem Solving Sessions and New Results" (Dimo Brockhoff and Günter Rudolph), "Modeling Biological Systems" (Julia Handl, Joshua Knowles, and Yaochu Jin), and "Parallel Techniques in Search, Optimization, and Learning" (Enrique Alba and Francisco Luna). The workshops offered and ideal opportunity for the conference members to explore specific topics in evolutionary computation, bio-inspired computing, and metaheuristics in an informal and friendly setting.

PPSN 2012 also included eight tutorials: "Introduction to Bioinformatics" (Jaume Bacardit, University of Nottingham, UK), "Evolutionary Multi-Objective Optimization" (Jürgen Branke, University of Warwick, UK), "Implementing Artificial Evolution on GPGPU-Based Computing Eco-Systems with the EASEA-CLOUD Massively Parallel Platform" (Pierre Collet, Strasbourg University, France), "Programming by Optimization—A New Paradigm for Developing High-Performance Software" (Holger H. Hoos, University of British Columbia, Canada), "Computational Intelligence and Games" (Pier Luca Lanzi, Polytechnic of Milan, Italy), "Ant Colony Optimization" (Vittorio Maniezzo, University of Bologna,

Italy), "Complex Systems Science in Its Thirties" (Roberto Serra, University of Modena and Reggio Emilia, Italy), and "Expressive Genetic Programming" (Lee Spector, Hampshire College, USA).

We wish to express our gratitude to the authors who submitted their papers to PPSN 2012 and to the Program Committee members and external reviewers who provided thorough evaluations of all these submissions. We also express our profound thanks to Marisa Lappano Anile, Claudio Angione, Jole Costanza, Giovanni Carapezza, Giovanni Murabito, and all the members of the Organizing Committee for their substantial efforts in preparing for and running the meeting. Thanks to all the keynote and tutorial speakers for their participation, which greatly enhanced the quality of this conference. Finally, we also express our gratitude to all the organizations that provided financial support for this event.

September 2012

Carlos Coello Coello
Vincenzo Cutello
Kalyanmoy Deb
Stephanie Forrest
Giuseppe Nicosia
Mario Pavone

Organization

PPSN 2012 was organized and hosted by the Optimization and BioComputing Group of the Department of Mathematics and Computer Science, University of Catania, Italy. The University of Catania is the 29th oldest university in the world. Its establishment dates back to 1434.

Conference Committee

General Chairs

Vincenzo Cutello   University of Catania, Italy
Mario Pavone   University of Catania, Italy

Honorary Chair

Hans-Paul Schwefel   Technische Universität Dortmund, Germany

Program Chairs

Carlos A. Coello Coello   CINVESTAV-IPN, Mexico
Kalyanmoy Deb   Indian Institute of Technology, India
Stephanie Forrest   University of New Mexico, USA
Giuseppe Nicosia   University of Catania, Italy

Tutorial Chairs

Giuseppe Narzisi   Cold Spring Harbor Laboratory, USA
Germán Terrazas Angulo   University of Nottingham, UK

Workshop Chair

Alberto Moraglio   University of Birmingham, UK

E-Publicity Chairs

Heder Bernardino Soares   LNCC, Brazil
Fernando Esponda   Instituto Tecnologico Autonomo de Mexico, Mexico

Financial Manager

Marisa Lappano Anile   University of Catania, Italy

Local Organization

Giovanni Carapezza   University of Catania, Italy
Piero Consoli   University of Catania, Italy
Jole Costanza   University of Catania, Italy

Matteo De Felice	ENEA, Italy
Luigi Malagó	Politecnico di Milano, Italy
Giovanni Murabito	University of Catania, Italy
Annalisa Occhipinti	University of Catania, Italy
Elisa Pappalardo	University of Catania, Italy
Giovanni Stracquadanio	Johns Hopkins University, USA
Renato Umeton	University of Rome "La Sapienza", Italy

Steering Committee

Carlos Cotta	Universidad de Málaga, Spain
David W. Corne	Heriot-Watt University Edinburgh, UK
Kenneth A. De Jong	George Mason University, USA
Agoston E. Eiben	Vrije Universiteit Amsterdam, The Netherlands
Juan Julián Merelo Guervós	Universidad de Granada, Spain
Günter Rudolph	Technische Universität Dortmund, Germany
Thomas P. Runarsson	University of Iceland, Iceland
Robert Schaefer	University of Krakow, Poland
Marc Schoenauer	Université Paris Sud, France
Xin Yao	University of Birmingham, UK

Workshops

Evolving Predictive Systems
   Bogdan Gabrys and Athanasios Tsakonas

Workshop on Automated Selection and Tuning of Algorithms
Part A: Continuous Search Spaces – Focus on Algorithm Selection
   Heike Trautmann, Mike Preuss, Olaf Mersmann, and Bernd Bischl

Workshop on Automated Selection and Tuning of Algorithms
Part B: Discrete Search Spaces – Focus on Parameter Selection
   Andrew Parkes and Ender Özcan

Theoretical Aspects of Evolutionary Multiobjective Optimization:
Interactive Problem Solving Sessions and New Results
   Dimo Brockhoff and Günter Rudolph

Modeling Biological Systems Workshop
   Julia Handl, Joshua Knowles, and Yaochu Jin

Parallel Techniques in Search, Optimization, and Learning
   Enrique Alba and Francisco Luna

Tutorials

Introduction to Bioinformatics
Jaume Bacardit

Evolutionary Multi-Objective Optimization
Jürgen Branke

Implementing Artificial Evolution on GPGPU-Based Computing Eco-Systems with the EASEA-CLOUD Massively Parallel Platform
Pierre Collet

Programming by Optimization—A New Paradigm for Developing High-Performance Software
Holger H. Hoos

Computational Intelligence and Games
Pier Luca Lanzi

Ant Colony Optimization
Vittorio Maniezzo

Complex Systems Science in Its Thirties
Roberto Serra

Expressive Genetic Programming
Lee Spector

Keynote Speakers

Angelo Cangelosi	University of Plymouth, UK
Natalio Krasnogor	University of Nottingham, UK
Panos M. Pardalos	University of Florida, USA
Leslie G. Valiant	Harvard University, USA

Program Committee

Enrique Alba	Wolfgang Banzhaf	Hans-Georg Beyer
Youhei Akimoto	Helio Jose Barbosa	Mauro Birattari
Jaroslaw Arabas	Thomas Bartz-Beielstein	Christian Blum
Paolo Arena	Simone Bassis	Yossi Borenstein
Dirk Arnold	Roberto Battiti	Peter Bosman
Anne Auger	Gerardo Beni	Pascal Bouvry
Dogan Aydin	Heder S. Bernandino	Anthony Brabazon
Jaume Bacardit	Adam Berry	Jürgen Branke

Petr Posík
Mike Preuss
Christian Prins
Adam Pruegel-Bennett
Günther Raidl
Vitorino Ramos
William Rand
Khaled Rasheed
Mauricio Resende
Katya Rodriguez
Eduardo A. Rodríguez
  Tello
Philipp Rohlfshagen
Andrea Roli
Günter Rudolph
Thomas Runarsson
Thomas A. Runkler
Conor Ryan
Erol Sahin
Michael Sampels
Ivo Sbalzarini
Robert Schaefer
Andrea Schaerf
Marc Schoenauer
Oliver Schütze
Michele Sebag

Bernhard Sendhoff
Roberto Serra
Marc Sevaux
Jonathan Shapiro
Moshe Sipper
Roman Slowinski
Christine Solnon
Terence Soule
Dipti Srinivasan
Catalin Stoean
Giovanni Stracquadanio
Thomas Stützle
Dirk Sudholt
Ponnuthurai Suganthan
Jerry Swan
Daniel Tauritz
Jorge Tavares
Andrea G.B. Tettamanzi
Madeleine Theile
Lothar Thiele
Dirk Thierens
Jon Timmis
Jerzy Tiuryn
Julian Togelius
Marco Tomassini
Heike Trautmann

Vito Trianni
Bianca Truthe
Elio Tuci
Andrew M. Tyrrell
Renato Umeton
Leonardo Vanneschi
Sebastien Verel
Carlos Martín Vide
Verel Carlos
Markus Wagner
Lipo Wang
Darrel Whitley
R. Paul Wiegand
Carola Winzen
Carsten Witt
Man-Leung Wong
John Woodward
Ning Xiong
Xin Yao
Gary Yen
Tina Yu
Yang Yu
Christine Zarges
Ivan Zelinka
Qingfu Zhang
Eckart Zitzler

Sponsor

ESTECO
IBM Italy
SolveIT Software Pty Ltd

Patronage

Angelo Marcello Anile Association
IET - The Institute of Engineering and Technology
Tao Science Research Center, Italy
UNINFO
University of Catania, Italy

Table of Contents – Part II

Multiobjective Optimization

Swarm Intelligence, Collective Behaviour, Coevolution and Robotics

Memetic Algorithms, Hybridized Techniques, Meta and Hyperheuristics

Applications (II)

Table of Contents – Part I

Theory of Evolutionary Computation

Machine Learning, Classifier Systems, Image Processing

Experimental Analysis, Encoding, EDA, GP

Applications (I)

Temporal Evolution of Design Principles in Engineering Systems: Analogies with Human Evolution

Kalyanmoy Deb[1], Sunith Bandaru[1], and Cem Celal Tutum[2]

[1] Indian Institute of Technology Kanpur, Kanpur, UP 208016, India
{deb,sunithb}@iitk.ac.in
[2] Denmark Technical University, Lyngby, Denmark
cctu@mek.dtu.dk

Abstract. Optimization of an engineering system or component makes a series of changes in the initial random solution(s) iteratively to form the final optimal shape. When multiple conflicting objectives are considered, recent studies on *innovization* revealed the fact that the set of Pareto-optimal solutions portray certain common design principles. In this paper, we consider a 14-variable bi-objective design optimization of a MEMS device and identify a number of such common design principles through a recently proposed automated innovization procedure. Although these design principles are found to exist among near-Pareto-optimal solutions, the main crux of this paper lies in a demonstration of temporal evolution of these principles during the course of optimization. The results reveal that certain important design principles start to evolve early on, whereas some detailed design principles get constructed later during optimization. Interestingly, there exists a simile between evolution of design principles with that of human evolution. Such information about the hierarchy of key design principles should enable designers to have a deeper understanding of their problems.

Keywords: multi-objective optimization, automated innovization, MEMS design, evolution, design principles.

1 Introduction

Gathering better and richer knowledge about a problem always fascinated man. In the context of engineering design, this amounts to discovering and understanding a number of aspects related to the design problem at hand. First and foremost, the designer is interested in knowing what shape, parameters, materials etc. would make a solution *optimal* with respect to one or many objectives of design. Optimality is an important consideration, as the designers are aware that an optimal design is always competitive and can never be bettered by any other solution. With the classical mathematics-oriented [11] and non-traditional optimization tools, such as evolutionary algorithms, simulated annealing, etc. that are available today, finding a near-optimal solution to a complex engineering problem involving non-linear objectives and constraints, mixed nature of

C.A. Coello Coello et al. (Eds.): PPSN 2012, Part II, LNCS 7492, pp. 1–10, 2012.

variables, computationally expensive evaluation procedures, and stochasticities in evaluation process can all be achieved reasonably well.

Secondly, with the machine learning and data mining tools available today, designers can hope to know more beyond just finding the optimal solutions of a problem. They can provide a deeper understanding about the properties of optimal solutions and gather valuable knowledge for their future use. A recent study on *innovization* proposed the use of two or more conflicting objectives to find a set of trade-off near Pareto-optimal solutions and then an analysis of the solutions to unveil hidden properties common to them [1,2,7]. These properties are referred to as design principles. They convey useful information about 'what makes a solution optimal?'.

Optimization is an iterative process in which the task is started with one or more random solutions. Solutions are then modified by the algorithm's operators to hopefully find better solutions. The solution update procedure is continued iteratively till one or more satisfactory solutions are found. The process, if thought carefully, is an evolutionary process, in which a set of random naive solutions (most likely not resembling at all with the final optimal solutions) get modified to take shape of optimal solutions. Ignoring a number of complex effects associated with natural evolution (such as environmental changes, sexual reproduction, dominance-diploidy etc.), the above-described optimization process can be viewed similar to the human evolution, a process that started from the creation of prokaryotes cells (around 4,000 million years ago (Ma)) to eukaryotes (around 2,100 Ma) to sponges (around 600 Ma) to vertebrates (around 500 Ma) to tetrapods (around 390 Ma) to synapsida (around 256 Ma) to reptiles (around 250 Ma) to placental mammal (around 160 Ma) to primates (around 75 Ma) to Hominidae (around 15 Ma) to Australopithecus Afarensis (around 3.6 Ma) to Homo erectus (around 1.8 Ma) to Homo Sapiens (160 thousand years ago) and to the ancestors of modern day Homo Sapiens (around 12,000 years ago) [9]. We concentrate on the fact that several milestone developments made the evolution of modern human possible and the information about these key developments are important for the evolutionists to have a better understanding of how we have come and where we may go from here. The development of back-bone (vertebrate) as early as around 500 Ma was the first major event in the human evolution. Thereafter, the development of legs around 390 Ma was another major breakthrough that allowed the creatures to leave water and come to land. Many other significant anthropological developments took place along the way, which eventually helped create high-performing living creatures like humans.

In this paper, we consider a specific engineering design task for our study and first find a set of trade-off, near-Pareto-optimal solutions using an EMO procedure. These high-performing solutions can be viewed somewhat similar to the human population of today who can be considered better and high-performing compared to all of their ancestors since the beginning of life formation about four billion years ago. Thereafter, we perform an automated innovization task on these high-performing design solutions to reveal a set of design principles common to them. These principles may be thought of as similar to the features that are common to the present human population, such as presence of a backbone, legs, skull etc. of certain type. As the history of human evolution reveals a chronology of developments (such as being a vertebrate first, then developing legs, and so on), in this paper, we are particularly interested in investigating the evolutionary history of the key design principles. For this purpose, we suggest a computational procedure and reveal interesting time-line of formation of design ideas along an optimization process. Such information about a problem provides valuable insight about the importance of various design principles and should help designers to better understand their designs and eventually create better designs.

1.1 Multi-objective Optimization and Automated Innovization

Multi-objective optimization considers multiple conflicting objectives and theoretically gives rise to a set of Pareto-optimal solutions, each of which is optimal corresponding to a trade-off among the objectives. Since the outcome are multiple solutions, multi-objective optimization is ideal for finding a set of alternate solutions either for finally choosing a single preferred solution or to launch a future analysis. Due to the population approach and ability to introduce artificial niches within a population, evolutionary algorithms (EAs) are ideal for solving multi-objective optimization problems. For the purpose of future analysis of Pareto-optimal solutions, as mentioned above, recent studies have proposed an *innovization* task for discovering innovative solution principles [7]. Since Pareto-optimal solutions are all optimal, they are likely to possess some common properties related to design variables, objectives and constraints that remain as 'signatures' to Pareto-optimal solutions. A few recent studies have also attempted to discover common design principles automatically using sophisticated machine learning procedures [2,1], which we discuss here in brief.

Automated innovization, proposed in 2010 [2], uses a grid-based clustering technique to identify correlations in any multi-dimensional space whose dimensions are provided by the user. The procedure was later extended [1] so that design principles hidden in all possible Euclidean spaces formed by the variables and objectives (and any other user-defined functions) can be obtained simultaneously without any human interaction. This is achieved at the cost of restricting the mathematical structure of the design principles to a multiplicative form given by, $\prod_{j=1}^{N} \phi_j(\mathbf{x})^{a_j b_j} = c$, where ϕ_j's are basis functions. A total of N basis functions need to be provided by the user. A basis function can be any function of the problem variables. The usual choices are the objectives and the variables themselves. a_j is a Boolean exponent determining the presence (=1) or absence (=0) of j-th basis function and b_j is a real-valued exponent. Automated innovization is capable of extracting multiple design principles of multiplicative form by optimizing a_j's and b_j's. It is argued that since many natural, physical, biological and man-made processes are governed by formulae with the same structure (power laws [10]), most correlations are expected to be mathematically captured by it. By definition, the expression on the left side of above equation is a design principle (DP) if it evaluates to approximately the same value for a majority of the Pareto-optimal solutions. Thus the c value on the right is a measure of commonality and the extent of this commonality is obtained by clustering the evaluated c values. For further details readers are referred to [1].

2 MEMS Design Study

MEMS (Microelectromechanical systems) are tiny mechanical devices that are built upon semiconductor chips. They usually integrate across different physical domains a number of functions, including fluidics, optics, mechanics and electronics and are used to make numerous devices such as pressure sensors, gyroscopes, engines and accelerometers. The present design problem concerns a comb-drive micro-resonator shown in Figure 1. There are 14 design variables as shown in Figure 1, V is the voltage amplitude and N_c is number of rotor comb fingers. The variable bounds are: $2\mu m \leq L_b, L_t, L_{sy}, L_{sa} \leq 400\mu m$, $2\mu m \leq w_b, w_t, w_c \leq 20\mu m$, $10\mu m \leq w_{sy}, w_{sa}, w_{cy} \leq 400\mu m$, $4\mu m \leq x_0 \leq 400\mu m$, 7 Volts $\leq V \leq 50$ Volts and $3 \leq N_c \leq 66$. The design is subject to 10 linear constraints and 14 non-linear constraints. The objectives of design are (i) minimization of the power consumption (applied voltage), and (ii) minimization of the total area of MEMS device. Further details about the problem can be found in [8].

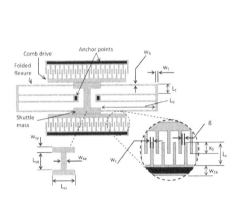

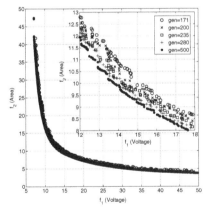

Fig. 1. MEMS model adapted from [8]   **Fig. 2.** Progress of EMO solutions towards Pareto-optimal front

2.1 Generation of Pareto-optimal Front

This highly non-linear design optimization problem was previously solved using NSGA-II [6] with an external archive for collecting the non-dominated solutions [12]. In this study, the original implementation of NSGA-II is used instead.

To have statistical invariance, 10 different runs are performed each with $P = 500$ population members for $t_{max} = 500$ generations. Each NSGA-II run uses the same parameters: SBX (simulated binary crossover [5]) operator with $p_c = 0.9$ and $\eta_c = 15$, and polynomial mutation operator [4] with $p_m = 0.033$ and $\eta_m = 20$. All variables except N_c are real-valued. A binary string of length six bits is used to represent N_c, for which p_c and p_m are 0.9 and 0.0125, respectively. The non-dominated solutions from each run are accumulated and sorted using the dominance criterion. This process gives rise to $1,198$ high-performing trade-off solutions.

To ensure a proper convergence, a local search procedure (the nonlinear gradient-based minimization algorithm `fmincon` from MATLAB) is applied to the ϵ-constrained MEMS design problem [3] on each of $1,198$ solutions. However, since gradient-based algorithms cannot efficiently handle discrete variables, for every solution, we keep N_c fixed to its current value and the other 13 variables are modified during the local search procedure. It is observed that the difference between NSGA-II solutions and the local-searched solutions are quite small. The improved non-dominated front is shown in Figure 2.

3   Results

The previous study [12] attempted to visually decipher design trends among these solutions. In the following section, we apply the automated innovization algorithm [2] to unveil design knowledge in a quantitative way.

3.1   Design Principles Using Automated Innovization

The $1,198$ non-dominated solutions obtained above are used for the innovization study. All 14 design variables and the two objective functions are chosen as the basis functions needed for the automated innovization study. The optimization formulation of the automated innovization problem is solved using a single-objective NSGA-II which uses the following settings: (i) population Size = 500, (ii) number of generations = 500, (iii) niched tournament selection operator, (iv) single-point binary crossover with $p_c = 0.85$ and SBX with $p_c = 0.90$, and $\eta_c = 10$, (v) bitwise mutation with $p_m = 0.15$ and polynomial mutation with $p_m = 0.05$ and $\eta_m = 50$. Table 1 lists all 15 design principles found by the automated innovization study. The last column is a measure of the extent of commonality among the $1,198$ non-dominated solutions. It is referred to as the significance and is simply the percentage of the trade-off solutions whose c values are clustered. Minimum allowable significance is a user parameter in innovization and has been set to 70% in this case. A few interesting aspects of the comb driven micro-resonator design problem obtained from the automated innovization study are as follows:

1. Six design principles, namely DP1, DP2, DP4, DP7, DP8 and DP9 are all more or less constant for at least 80% of the data. The third column shows

Table 1. Automated innovization results for the MEMS design problem

Notation	Design principle	Cluster average ($\mu_{largest}$)	Significance
DP1	$w_c^{1.0000} = c$	2.000231E-06	98.50 %
DP2	$w_{sy}^{1.0000} = c$	1.000441E-05	97.16 %
DP3	$L_{sa}^{1.0000} = c$	1.169490E-05	88.23 %
DP4	$w_t^{1.0000} = c$	2.001497E-06	87.65 %
DP5	$L_t^{1.0000} = c$	6.873649E-06	87.40 %
DP6	$L_{sy}^{1.0000} = c$	3.605399E-05	86.56 %
DP7	$w_{sa}^{1.0000} = c$	1.000482E-05	86.06 %
DP8	$w_b^{1.0000} = c$	2.000028E-06	84.72 %
DP9	$w_{cy}^{1.0000} = c$	1.000088e-05	79.63 %
DP10	$f_1^{1.0000} L_b^{0.6470} = c$	1.078929E-01	78.46 %
DP11	$f_2^{1.0000} L_b^{-0.4888} = c$	3.671301E+02	74.12 %
DP12	$f_1^{0.2546} f_2^{1.0000} L_b^{-0.3563} = c$	2.812855E+02	73.79 %
DP13	$f_2^{1.0000} L_b^{-0.4800} L_c^{-0.1160} = c$	1.258088E+03	72.70 %
DP14	$f_1^{1.0000} L_b^{0.6490} L_c^{0.1429} = c$	2.112050E-02	72.70 %
DP15	$f_1^{0.7737} f_2^{1.0000} = c$	7.301285E+01	70.45 %

that all these design principles tend to their lower bounds. It is interesting to note that all the associated variables are widths, indicating that for this MEMS component, the overall width should be as low as physically possible (provided they satisfy the constraints) for (near) Pareto-optimality.

2. Each of DP3, DP5 and DP6 are also approximately constant on the front. However they surprisingly take a value intermediate in their variable ranges. This indicate that the corresponding length variables are very important and will determine Pareto-optimality for this problem.

3. The flexure beam length L_b is important in an indirect way. DP10 signifies that it is inversely proportional to the voltage f_1. The instantaneous voltage applied across the comb drive is associated with the force created to move the shuttle mass and the flexure beams are designed to compensate this movement. It becomes clear that more the force (which in turn is due to higher voltage), the stiffer the structure should be and hence a shorter beam length (L_b) is required.

The design principles and their implications mentioned above are interesting and provide a designer interesting insights about the particular MEMS design. However, in the following section, we discuss a further post-optimality analysis procedure that is more revealing and also has a deeper connection to the time-line developments of human evolution.

3.2   Evolution of Design Principles

Consider the various anthropological features that homo-sapiens acquired during the process of human evolution. There is sufficient documented evidence which tells us that these features evolved *gradually* over millions of years, rather than appearing out as a single event, driven by the natural mechanisms of reproduction, genetic mutation and natural selection. Despite the simplicity in our design environment (being static, deterministic, asexual, non-cooperative, etc.), the design principles obtained in Table 1 can be thought as somewhat analogous to these features, since they are common to most of the solutions (at least 70%), just like the anthropological features that distinguish humans from other living beings. We are interested here in investigating if there exists a gradual and chronological evolution of the above design principles over iterations just like the chronology of anthropological feature development over millions of years. If such a gradual development of key design features is observed, the information would be valuable to the designers for a better understanding and further their future use. Similarity between natural and artificial evolutions can help both fields with cross-breeding of their key concepts.

We propose the following procedure for recording the evolutionary time-line of design principles. The non-dominated solutions from each of the 10 runs at each generation t is stored. Next, each of the 13 identified design principles (DPi, $i = 1, \ldots, 13$) is checked for their appearance in the combined data in each generation. The significance of DPi ($S_t^{\text{DP}i}$) at generation t is calculated as the proportion of points satisfying the DP to the total non-dominated points at the final generation. Thereafter, a plot of the significance value of each DP with generation will reveal the relative appearance of the DP during the optimization process. Here, we provide the algorithm in step by step format with the following input: (i) design principles (DPi, $i = 1, 2, \ldots$) obtained after the automated innovization task, (ii) cluster information associated with each DPi, and (iii) generation-wise population members for each run:

Step 0: Set $t \leftarrow 0$.
Step 1: Collect non-dominated solution set $\mathbf{P}_t$ at generation t from all runs. Thereafter, remove the dominated points from $\mathbf{P}_t$.
Step 2: Evaluate DPi at all solutions in $\mathbf{P}_t$ to compute the c values and collect them in set $\mathbf{C}_t$.
Step 3: Every element $c_j \in \mathbf{C}_t$ is checked for its existence in any of the K clusters of DPi using the criterion, $c_j \in$ cluster $k \Leftrightarrow \mu_k - d\,\sigma_k \leq c_j \leq \mu_k + d\,\sigma_k$, where μ_k and σ_k are the mean and standard deviation of c-values of the k-th cluster, respectively. The number of elements E_t in $\mathbf{C}_t$ that do belong any one of the K clusters is recorded. $d = 4$ is used here and also recommended.
Step 4: Calculate the significance of DPi in the current generation t as follows: $S_t^{\text{DP}i} = (E_t/|\mathbf{P}_{t_{max}}|) \times 100\%$, where $|.|$ represents the set size. For the MEMS design case, we have $|\mathbf{P}_{t_{max}}| = 1,198$.
Step 5: If $t = t_{\max}$ **Stop** else $t \leftarrow t + 1$ and **Goto Step 1**. Here $t_{\max}$ is the number of generations used for solving the original multi-objective problem.

We apply the above procedure to the MEMS design problem for the first 13 of the 15 design principles obtained by automated innovization. DP14 is a combination of DP10 and DP13. DP15 does not involve any decision variables. Hence, we do not consider them for the evolution analysis. Figure 3 shows $S_t^{\text{DP}i}$ for each of the 13 DPs at various generations. The evolution history shown in the figure reveals the time at which each of DPs started to evolve during the optimization process.

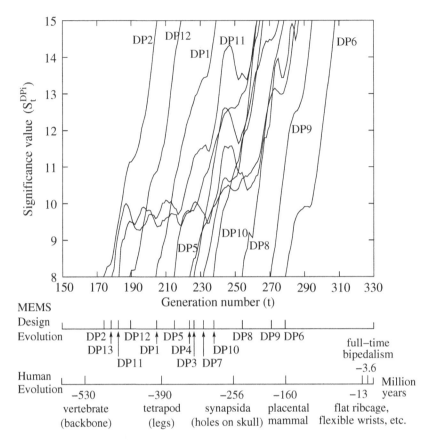

Fig. 3. Evolution of 13 design principles show a gradual development. Similarity with events in human evolution is also shown for a comparison.

We show the evolution when there is around 10% existence of the particular DP in the combined population. Clearly, a gradual evolution pattern of DPs can be seen: (DP2, DP13, DP11, DP12, DP1, DP5, DP4, DP3, DP7, DP10, DP8, DP9, DP6). This information of some DPs which form later due to existence of some other DPs which formed early on during the optimization process provide valuable knowledge to the designers.

For the first 10 generations, no feasible solution was found. Thereafter, when some feasible solutions were created, it took another 160 generations for the first design principle DP2 to emerge among the non-dominated solutions. At around 171 generations, about 10% of the non-dominated points of all 10 runs have the DP2 property: w_{sy} is constant. This variable denotes the thickness of the web of the I-shaped element. The fact that this property has started to evolve first means that this design principle is a fundamental requirement for a design to take shape of an optimal solution. This is equivalent to the development of the 'backbone' as early as around 530 million years ago for the eventual evolution of a human.

After the emergence of DP2, the next few generations created DP13 which is a relationship between the area of MEMS device, L_b and L_c. The principle states that for a MEMS with smaller area, smaller values of L_b and L_c are needed. Third and fourth DPs (DP11, DP12) emerge after a while. These DPs enables a more direct relationship between the area and L_b to be created in the form of DP11 and DP12. As an analogy, the emergence of DP11 and DP12 may be compared with further anthropological developments, such as formation of legs, that made a significant leap towards human evolution. In this sense, fixation of w_{sy}, L_b and L_c early on during the optimization process remain as fundamental developments towards becoming optimal.

Thereafter, after a gap of 15 generations, a new DP emerges. This is DP1 denoting that the variable w_c must be constant. The variable w_c is the thickness of the comb tooth. When the MEMS with a previously evolved feature (DP12) fixed a direct relationship between L_b and the area, the thickness of each comb was the next parameter to get fixed. DP1 dictates that the optimal design requires a fixed tooth size. From this generation onwards, detail design principles involving a few other variables (L_t from DP5, w_t from DP4, L_{sa} from DP3, w_{sa} from DP7, w_b from DP8, w_{cy} from DP9) evolved. As the solutions approach the Pareto-optimal front, DP15 relating two objective values starts to get formed and around 235 generation a direct relationship (DP10) between the first objective (applied voltage) and L_b forms.

Finally, DP6 that requires the variable L_{sy} to be constant evolves at around 280 generations. This variable relates to the width of the web of the I-shaped element. When more characteristic variables get settled with evolution, this was the final fixation needed for the solutions to become close being Pareto-optimal.

It is interesting to note from Figure 2 that evolution of all DPs take place when the non-dominated points are close to the Pareto-optimal front. This observation is similar to the fact that most of the major anthropological developments in human evolution took place in a relatively short time span since the creation of life forms. The chronology of evolution of design principles discovered above from multiple EMO runs clearly puts forward a hierarchy of importance of them and highlights their inter-relationships. Such important information are difficult to obtain in any other ways.

4 Conclusions

In this paper, we have extended the use of automated innovization principles to make a deeper understanding of an engineering design problem. The key design principles found by the *innovization* procedure have been investigated for their chronological evolution during the optimization process. A computational procedure has been suggested for this purpose. It is observed that certain design principles get created early on during the optimization process, while some other detail design principles form later. We have argued that the evolution of design principles during the course of optimization has, barring some details, a remarkable similarity to the time-line history of significant anthropological developments for human evolution over many millions of years. The connection is interesting and puts natural and artificial design of systems on a similar platform, thereby allowing cross-breeding of ideas between two areas. The evolutionary information thus obtained may provide a clear hierarchy of important design features needed to constitute an optimal design. Such knowledge is vital for designers in having a clear understanding of key features and their interrelationships and also to make use of them in their future design scenarios.

Acknowledgments. Authors wish to thank Dr. Zhun Fan for introducing them to the MEMS design problem.

References

1. Bandaru, S., Deb, K.: Automated Innovization for Simultaneous Discovery of Multiple Rules in Bi-objective Problems. In: Takahashi, R.H.C., Deb, K., Wanner, E.F., Greco, S. (eds.) EMO 2011. LNCS, vol. 6576, pp. 1–15. Springer, Heidelberg (2011)
2. Bandaru, S., Deb, K.: Towards automating the discovery of certain innovative design principles through a clustering-based optimization technique. Engineering Optimization 43(9), 911–941 (2011)
3. Chankong, V., Haimes, Y.Y.: Multiobjective Decision Making Theory and Methodology. North-Holland, New York (1983)
4. Deb, K.: Multi-objective optimization using evolutionary algorithms. Wiley, Chichester (2001)
5. Deb, K., Agrawal, R.: Simulated binary crossover for continuous search space. Complex Systems 9(2), 115–148 (1995)
6. Deb, K., Pratap, A., Agarwal, S., Meyarivan, T.: A fast and elitist multi-objective genetic algorithm: NSGA-II. IEEE Transactions on Evolutionary Computation 6(2), 182–197 (2002)
7. Deb, K., Srinivasan, A.: Innovization: Innovating design principles through optimization. In: Proceedings of the 8th Annual Conference on Genetic and Evolutionary Computation, GECCO 2006, pp. 1629–1636. ACM, New York (2006)
8. Fedder, G., Mukherjee, T.: Physical design for surface-micromachined MEMS. In: Proceedings of the Fifth ACM SIGDA Physical Design Workshop, Virginia, USA (April 1996)
9. Haeckel, E.: The evolution of man, vol. 1. Kessinger Publishing (1879)
10. Newman, M.: Power laws, pareto distributions and zipf's law. Contemporary Physics 46(5), 323–351 (2005)
11. Reklaitis, G., Ravindran, A., Ragsdell, K.: Engineering optimization: Methods and applications. Wiley, New York (1983)
12. Tutum, C.C., Fan, Z.: Multi-criteria layout synthesis of mems devices using memetic computing. In: IEEE Congress on Evolutionary Computation, pp. 902–908 (2011)

Exploiting Prior Information in Multi-objective Route Planning

Antony Waldock[1] and David W. Corne[2]

[1] Advanced Technology Centre, BAE Systems,
Bristol, UK
antony.waldock@baesystems.com
[2] School of MACS, Heriot-Watt University,
Edinburgh, UK
dwcorne@macs.hw.ac.uk

Abstract. Planning a satisfactory route for an autonomous vehicle over a complex unstructured environment with respect to multiple objectives is a time consuming task. However, there are abundant opportunities to speed up the process by exploiting prior information, either from approximations of the current problem instance or from previously solved instances of similar problems. We examine these two approaches for a set of test instances in which the competing objectives are the time taken and likelihood of detection, using real-world data sources (Digital Terrain Elevation Data and Hyperspectral data) to estimate the objectives. Five different instances of the problem are used, and initially we compare three multi-objective optimisation evolutionary algorithms (MOEA) on these instances, without involving prior information. Using the best-performing MOEA, we then evaluate two approaches that exploit prior information; a graph-based approximation method that pre-computes a collection of generic 'coarse-grained' routes between randomly selected zones in the terrain, and a memory-based approach that uses the solutions to previous instances. In both cases the prior information is queried to find previously solved instances (or pseudo-instances, in the graph based approach) that are similar to the instance in hand, and these are then used to seed the optimisation. We find that the memory based approach is most effective, however this is only usable when prior instances are available.

1 Introduction

Route-planning is one of the increasingly many application domains in which a multi-objective optimisation (MOO) approach [1] has been found to have significant advantages over single-objective approaches [2]. In this paper, we further explore multi-objective optimisation algorithms for route-planning of manned and unmanned vehicles in a hostile and unstructured environment, and focus on the question of accelerating the process by exploiting prior information. Speed of optimization can be particularly vital in the application environments of interest to us – broadly speaking, this is because there will often be a need to have a

C.A. Coello Coello et al. (Eds.): PPSN 2012, Part II, LNCS 7492, pp. 11–21, 2012.

viable route plan within seconds of the decision being made to start to move the vehicle to a given new location. Meanwhile, in the route planning domain, as well as a large number of other interesting application domains, a range of prior information is available and could be used in various ways to bootstrap the optimisation process.

In the case of route planning, the broad geographic area and terrain characteristics where routes are to be planned are known in advance. The solutions to previously solved route planning instances within the same terrain may also be known but it would be infeasible to pre-compute all instances from every possible start and end location (some geographic features in the environment may move between instances). However, it is very appealing to utilise prior information whenever possible. In this paper, we examine two approaches to integrating prior information into a multi-objective optimisation algorithm; (i) solving approximations of the current problem instance, or (ii) information derived from previously solved problems with sufficient similarity.

In the remainder, we briefly cover background material in Section 2 and then introduce multi-objective route-planning. Section 3 then evaluates three MOEAs on our five test instances, without exploiting prior information. In Sections 4 and 5 we then respectively explore two different approaches to include prior information. Section 6 concludes and discusses future work.

2   Background

A multi-objective optimization (MOO) problem is posed as $\arg\min_{x \in X} G_m(x)$, where $G_m(x)$ is a set of m objective functions and x is defined as a vector of decision variables (or a solution) in the form $x = (x_1, x_2.., x_N)$ from the set of solutions X. The aim is to find the *Pareto* set which contains all the solutions that are not dominated by any other solution. A solution x is said to be non-dominated by y, if and only if, x is as good as y in all objectives and x is strictly better then y in at least one objective.

The most effective MOO approaches to date are generally regarded to be multi-objective evolutionary algorithms (MOEAs). Typically, MOEAs (as well as most optimization algorithms) make little or no use of prior information that may be available about the problem at hand. The concept of exploiting prior information is pervasive in artificial intelligence, appearing in several different guises (e.g. case based reasoning (CBR) [3], or, more recently, per-instance tuning [4]), however it is infrequent in the optimization literature, perhaps because appropriate approaches are highly domain-specific. Nevertheless some examples include work in the Genetic Programming community [5,6] in which populations were seeded with solutions to previous instances, while an approach was recently proposed in [7] which exploits extensive pre-computation of solutions to *potential* instances that may be faced in a given domain. Meanwhile, [8] explores the re-use of the probability models built by an estimation of distribution algorithm (EDA) on previous instances, while seeding with previous solutions is occasionally explored, especially for dynamic optimization [9,10].

We examine two approaches with which to integrate prior information in route planning, in the scenario that instances will occur with previously unknown start and end locations, but within a known geographic area (e.g. a 5km by 5km square). The first approach is to prepare in advance Pareto optimal but coarse grained route solutions for a large collection of potential start and end locations within the region. For any such start/end pair, we abstract the search space as a directed graph, and then use a multi-objective extension to traditional A* called NAMOA* [11] to find Pareto optimal coarse-grained routes. which, in turn, seed the population of a MOEA solving the instance at hand. The second approach uses solutions to previously solved similar route planning instances to seed the initial population for the new instance.

We consider a route planning scenario where a route is required that minimises a set of competing objectives such as the fuel used, the time taken, the distance travelled, or the likelihood of being detected by observers. We build on the route planning problem defined in [2] and are informed by previous studies of motion planning for autonomous vehicles [12,13]. Route planning is the construction of a route that navigates between two geographic locations. The start and end location are defined by a latitude, longitude and heading from true north. For convenience, we encode a route in relative polar coordinates where α_i is the heading relative to the next way point and r is the distance to travel in this direction. To evaluate a route, the objective functions used here are the time taken and likelihood of detection, as defined in [2], and the route is divided into 30 segments. The objective functions are calculated using Digital Terrain Elevation Data (DTED) [1] and the NASA LandSat Multispectral data. The Multispectral data is combined with a classifier to infer the terrain type and hence the maximum speed allowed on that portion of the route segment. To evaluate the performance of different MOEAs, five instances, P_1 to P_5, were generated. The definition of the routes and java code for the objective functions is available at http://code.google.com/p/multi-objective-route-planning/.

3   Comparison of Multi Objective Optimisation Algorithms

First, we compare three different MOEAs, MOEA/D, SMPSO, and NSGA-II, on our five problem instances, without using prior information. Multi Objective Evolutionary Algorithms Based on Decomposition (MOEA/D)[14] was selected to represent the current state of the art MOEA. Speed-constrained Multi-objective PSO (SMPSO)[15] is used to provide a baseline algorithm from the Particle Swarm Optimisation (PSO) community, while the Non-dominated Sorting Genetic Algorithm II (NSGAII)[16] is also tested as a commonly used effective benchmark MOEA.

The implementations of MOEA/D, SMPSO and NSGAII have been taken from JMetal[2] and the results presented are averaged over 50 independent runs

---

[1] EarthExplorer (http://earthexplorer.usgs.gov)
[2] http://jmetal.sourceforge.net/

which were limited to 200,000 evaluations per run. NSGAII and SMPSO had a population size of 100 and MOEA/D a population size of 600. Comparisons are quantified using the Inverted Generational Distance (IGD), as defined in [17]. The 'known' Pareto optimal front is calculated by combining the solutions generated from all experiments presented in addition to one million randomly generated samples. Results are summarised in Table 1.

Table 1. The IGD values (M. = mean and Std = standard deviation) for (M)=MOEA/D, (S)=SMPSO and (N)=NSGAII on Problems 1 to 5

Prob	P_1			P_2			P_3			P_4			P_5		
Alg	S	N	M	S	N	M	S	N	M	S	N	M	S	N	M
Mean	61.0	464.7	33.8	152.4	817.9	47.9	130.8	345.5	29.8	107.2	407.5	47.7	54.1	205.7	5.8
Std	12.0	150.9	28.2	41.6	193.4	36.1	25.5	141.2	19.2	33.6	76.3	25.5	9.8	56.9	14.8

The results in Table 1 follow preliminary experiments which hand-optimised the parameters of each of the algorithms. For these five problem instances, MOEA/D clearly produces solutions closest, in terms of IGD, to the reference Pareto optimal front. Hence, MOEA/D is used in the remainder of this paper.

4   Graph-Based Approximation

In the route planning scenario of interest, before an instance of the problem arises, we know the broad geographical region in which the start and end locations will be. We describe here a way to exploit that prior information, based on *a priori* finding coarse-grained solutions to many potential instances, based on all possible pairs of start and end locations over a 50m by 50m mesh. For each such pair, a solution tree is generated by using the encoding outlined in Section 2, but with each bearing (α_i) restricted to a discrete set, e.g. $[-30, 0, +30]$. Once the maximum number of segments has been reached, the current location is joined to the end location using a single straight line segment. The Pareto set of solutions on this tree is then extracted by using NAMOA* (of which more below). Given a new instance of the route planning problem, we then find the pre-solved coarse-grained instance whose start and end locations best match the new instance, and use the Pareto set found by NAMOA* to seed the MOEA/D population.

Figure 1 (a) shows the final IGD value for the five different route planning problems when MOEA/D is initialised randomly and with the non-dominated solutions found by solving two configurations of the graph-based approximation. The results clearly shown that initialising the MOEA/D population with solutions generated from an initial graph-based approximation has an improvement in the final IGD value. With 3 segments and 3 bearings (3,3) at each node, the final IGD is statistically different (according to a two tailed, paired T test with confidence level 0.99) for 1 of the problems (P_4) and with 5 segments and 5 bearings (5,5), the final IGD value is statistically different for all 5 problems.

Figure 1 (b) shows the number of route evaluations required to reach the 110% of the maximum final IGD value when using MOEA/D with a randomly

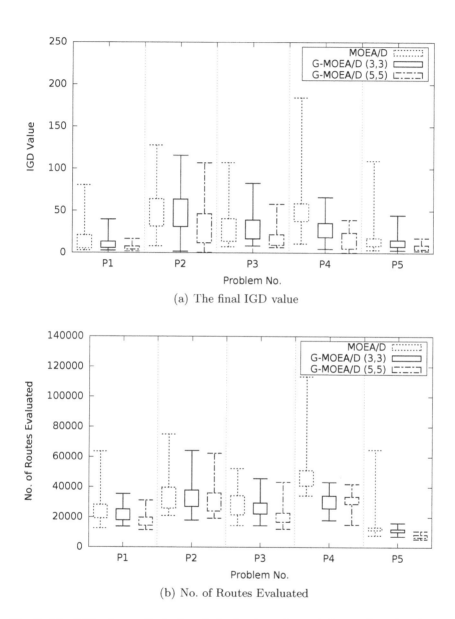

(a) The final IGD value

(b) No. of Routes Evaluated

Fig. 1. The IGD value on the last optimisation iteration and the number of evaluations to reach 110% of the maximum IGD value (as defined by MOEA/D) for MOEA/D and Graph-Based MOEA/D with (3,3) and (5,5) segments and bearings for Problems 1 to 5

initialised population. The results show that the number of evaluations required is statistically reduced on all route planning problems except for problem 2 (with 5 segments and 5 bearings). The results show that, on average, using a graph-based approximation can reduce the number of routes evaluated by 5,318 for 3 segments and 3 bearings and 8,582 routes for 5 segments and 5 bearings where NAMOA* only evaluates, on average, 11 and 143 routes for each of these configurations.

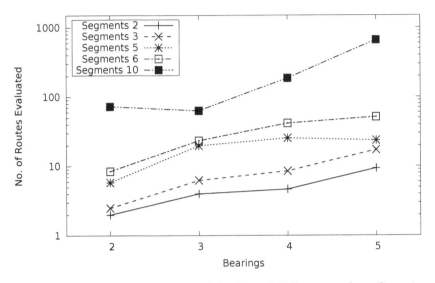

Fig. 2. The No. of Routes Evaluated for P_1 with different graph configurations

Although, the exponential complexity of NAMOA* is clearly shown in Figure 2 (Note the log scale) where the number of routes evaluated significantly increases (for a small reduction in the IGD). The number of routes evaluated are calculated by totalling the number of edges evaluated during the operation of the NAMOA* algorithm. One reason for this is that a suitable tree-pruning heuristic is unavailable for the likelihood of detection and hence a lower bound must be assumed (likelihood of detection is zero) therefore reducing the removal of dominated branches (or partially explored routes) in the graph.

This section has shown that seeding MOEA/D with solutions generated from an approximation of the problem, solved using NAMOA*, has a significant impact on the IGD value but NAMOA* can only be run for very crude approximations of the problem before the number of evaluations required quickly becomes infeasible. The next section evaluates a different seeding strategy that is based on storing solutions between problem instances.

5   Memory-Based

In this section, a memory-based approach is examined, where previous solutions are reused from previously solved instances of similar problems. The approach is only appropriate when the problems are not evolving rapidly and when information from previously solved problems have some bearing on the current problem being solved i.e. the shape and distribution of the routes. The non-dominated solutions generated by MOEA/D are stored in a k-d tree [18] which is referenced by the latitude and longitude of the start and end locations. All of the solutions from previous problems are currently stored in the k-d tree but only a subset of the K closest neighbours, up to half the total population, are used to initialise the population of MOEA/D when solving a new instance.

To evaluate the approach, it is necessary to generate a sequence of problem instances. A sequence of ten instances of route planning problems, $P_{i,j}$, where i is the problem number and j is the index in the sequence, are generated by randomly selecting a start and end location in the area of the route planning problems P_1 to P_5. Once a route planning problem in the sequence has been solved, the non-dominated solutions are added to the k-d tree and the next route planning problem is tackled. At the beginning of the next route planning problem, the closest neighbours to this route planning problem are extracted from the k-d tree and the non-dominated solutions for these problems added to the initial population of MOEA/D. The remaining population used in MOEA/D is randomly initialised within the input parameter range.

Figure 3 (a) shows the final IGD value for MOEA/D and the Memory-Based MOEA/D (neighbours=3) on P_1 over the sequence of ten route planning instances. The IGD value for each iteration is generated over 50 runs of MOEA/D and the same sequence is used for each run. The IGD results show that initialising MOEA/D with solutions from previously similar instances results in a statistically better (using a two tailed, paired with 0.99 probability) set of solutions for five of the ten instances (4,6, 8,9 and 10). Figure 3 (b) shows the number of evaluations required to reach 110% of the maximum final IGD value as found by MOEA/D. The results show that for some instances the number of evaluations is significantly lower (4,6,8,9 and 10) with on average a reduction in the number of routes evaluated for these four instances is 13,369. The results show that MOEA/D initialised using solutions from previously solved problems has the potential, on some instances, to reduce the number of evaluations required to produce a reasonable approximation of the Pareto optimal front.

For example, the reduction in the number of evaluations can be graphically seen in Figure 4 (a) where the memory-based MOEA/D (shown in dashed) is compared with a randomly initialised MOEA/D (shown in black). The reason for this earlier reduction in IGD value can be seen by comparing the solutions extracted from the memory at the start of the optimisation. The solutions extracted from the memory provide a reasonable approximation of the Pareto optimal front in the first few iterations of the algorithm. A comparison of the initial and optimised routes can be seen in Figure 4 (b) where the initial routes provide a broad spectrum of possible routes with previously successful shapes without having to exhaustively search all possible combinations as with the previous graph-based approach.

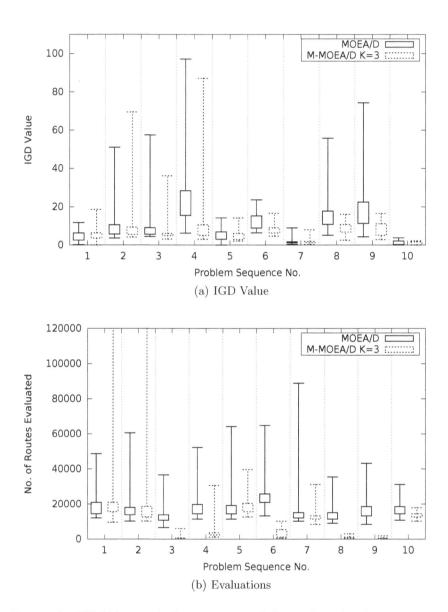

(a) IGD Value

(b) Evaluations

Fig. 3. The IGD Value on the last iteration and the number of evaluations to reach 110% of the maximum final IGD value (as defined by MOEA/D) for MOEA/D and Memory-Based MOEA/D for Problem Sequence 1

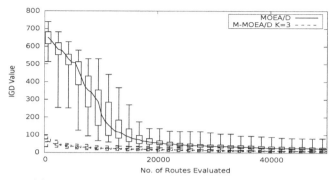

(a) Comparison of IGD Value during the optimisation

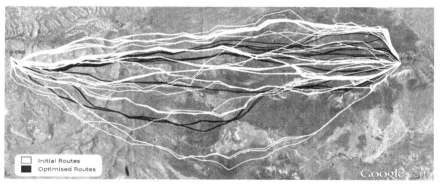

(b) Geo-spatial representation of the routes for Problem $P_{1,9}$

Fig. 4. Analysis of Problem $P_{1,9}$ - ©Google and GeoEye

6   Conclusions and Future Work

This paper has presented an examination of two approaches to using prior information, graph-based and memory-based, for MOEA/D when applied to a route planning problem over an unstructured environment. The experimental results have shown that both approaches enable MOEA/D to generate a set of solutions closer to the known Pareto optimal front in fewer iterations than a traditional random initialisation. Solving a graph-based approximation has been shown to produce routes closer to the known Pareto optimal front for the five route planning problems but is only feasible for small graphs because the number of evaluations required increases exponentially. Using a memory-based approach has also been shown to generate routes closer to the known Pareto optimal front for a sequence of problem instances but this is highly dependent on whether a sufficiently similar problem has been solved previously and whether the environment has evolved since solving that problem. The approaches examined in this paper are applicable to a wide range of multi-objective optimisation applications.

Future work will concentrate on how to apply these techniques to dynamic environments where the problem is evolving either during or between route planning problems and to analyse methods of storing and extracting subsets of solutions based on the similarity of the instances.

Acknowledgements. This project was funded by the Systems Eng. for Autonomous Systems (SEAS) Defence Tech. Centre (DTC).

References

1. Deb, K.: Multi-Objective Optimization Using EAs. John Wiley and Sons Inc. (2002)
2. Waldock, A., Corne, D.: Multi-objective optimisation applied to route planning. In: Genetic and Evolutionary Computation Conference, GECCO 2011, vol. 1, pp. 1827–1834 (2011)
3. Leake, D. (ed.): Case-based Reasoning: Experiences, Lessons and Future Directions. AAAI Press, Menlo Park (1996)
4. Hutter, F., Hoos, H., Stutzle, T.: Automatic algorithm configuration based on local search. In: Proc. AAAI, pp. 1152–1157. MIT Press (2007)
5. Arcuri, A., White, D.R., Clark, J., Yao, X.: Multi-objective Improvement of Software Using Co-evolution and Smart Seeding. In: Li, X., Kirley, M., Zhang, M., Green, D., Ciesielski, V., Abbass, H.A., Michalewicz, Z., Hendtlass, T., Deb, K., Tan, K.C., Branke, J., Shi, Y. (eds.) SEAL 2008. LNCS, vol. 5361, pp. 61–70. Springer, Heidelberg (2008)
6. Westerberg, C.H., Levine, J.: Investigation of Different Seeding Strategies in a Genetic Planner. In: Boers, E.J.W., Gottlieb, J., Lanzi, P.L., Smith, R.E., Cagnoni, S., Hart, E., Raidl, G.R., Tijink, H. (eds.) EvoWorkshop 2001. LNCS, vol. 2037, pp. 505–514. Springer, Heidelberg (2001)
7. Corne, D., Reynolds, A.: Optimization by precomputation. In: Proc. NABIC 2011. IEEE CIS Press (2011)
8. Hauschild, M., Pelikan, M.: Intelligent bias of network structures in the hierarchical boa. In: Proc. GECCO 2009. ACM Press (2009)
9. Branke, J.: Memory enhanced evolutionary algorithms for changing optimization problems. In: Proc. CEC 2009, pp. 1875–1882 (1999)
10. Potter, M., Wiegand, R., Blumenthal, H., Sofge, D.: Effects of experience bias when seeding with prior results. In: Proc. CEC 2005, vol. 3, pp. 2730–2737 (2005)
11. Mandow, L., de-la Cruz, J.L.P.: Multiobjective a[*] search with consistent heuristics. J. ACM 57 (2010)
12. Pivtoraiko, M., Kelly, A.: Efficient constrained path planning via search in state lattices. In: The 8th International Symposium on Artificial Intelligence, Robotics and Automation in Space (2005)
13. Lavalle, S.M.: Rapidly-exploring random trees: A new tool for path planning. Technical report (1998)
14. Li, H., Zhang, Q.: Multiobjective optimization problems with complicated pareto sets, moea/d and nsgaii. IEEE Trans. Evolutionary Computation (2008)

15. Nebro, A., Durillo, J., García-Nieto, J., Coello Coello, C., Luna, F., Alba, E.: Smpso: A new pso-based metaheuristic for multi-objective optimization. In: 2009 IEEE Symp. on MCDM, pp. 66–73. IEEE (2009)
16. Deb, K., Pratap, A., Agarwal, S., Meyarivan, T.: A fast elitist multi-objective genetic algorithm: Nsga-ii. IEEE Trans on Evol. Comp. 6, 182–197 (2000)
17. Zhang, Q., Liu, W., Li, H.: The performance of a new version of moea/d on cec09 unconstrained mop test instances. In: Proc CEC 2009, pp. 203–208 (2009)
18. Bentley, J.L.: Multidimensional binary search trees used for associative searching. Commun. ACM 18, 509–517 (1975)

Analysis on Population Size and Neighborhood Recombination on Many-Objective Optimization

Naoya Kowatari[1], Akira Oyama[2], Hernán Aguirre[1], and Kiyoshi Tanaka[1]

[1] Faculty of Engineering, Shinshu University
4-17-1 Wakasato, Nagano, 380-8553 Japan
[2] Institute of Space and Astronautical Science, Japan Aerospace Exploration Agency
{kowatari@iplab,ahernan,ktanaka}@shinshu-u.ac.jp,
oyama@flab.isas.jaxa.jp

Abstract. This work analyzes population size and neighborhood recombination in the context of many-objective optimization. Large populations might support better the evolutionary search to deal with the increased complexity inherent to high dimensional spaces, whereas neighborhood recombination can reduce dissimilarity between crossing individuals and would allow us to understand better the implications of a large number of solutions that are Pareto-optimal from the perspective of decision space and the operator of variation. Our aim is to understand why and how they improve the effectiveness of a dominance-based many-objective optimizer. To do that, we vary population size and analyze in detail convergence, front distribution, the distance between individuals that undergo crossover, and the distribution of solutions in objective space. We use DTLZ2 problem with $m = 5$ objectives in our study, revealing important properties of large populations, recombination in general, and neighborhood recombination in particular, related to convergence and distribution of solutions.

1 Introduction

Recently, there is a growing interest on applying multi-objective evolutionary algorithms (MOEAs) to solve many-objective optimization problems [1], where the number of objective functions to optimize simultaneously is considered to be more than three. It is well known that conventional MOEAs [2,3] scale up poorly with the number of objectives of the problem, which is often attributed to the large number of non-dominated solutions and the lack of effective selection and diversity estimation operators to discriminate appropriately among them, particularly in dominance-based algorithms. Selection, indeed, is a fundamental part of the algorithm and has been the subject of several studies, leading to the development of evolutionary algorithms that improve the performance of conventional MOEAs on many-objective problems [1]. However, finding trade-off solutions that satisfy simultaneously the three properties of convergence to the Pareto front, well spread, and well distributed along the front is especially difficult to achieve in many-objective problems and most search strategies for many-objective optimization proposed recently compromise one in favor of the other [1].

C.A. Coello Coello et al. (Eds.): PPSN 2012, Part II, LNCS 7492, pp. 22–31, 2012.
© Springer-Verlag Berlin Heidelberg 2012

In addition to selection, detailed studies on the characteristics of many-objective landscapes, the effectiveness of operators of variation, and the effects of large populations are important to move forward in our understanding of evolutionary many-objective optimization in order to develop effective and efficient algorithms. From this standpoint, we have presented initial evidence that MOEAs can improve their performance on many-objective problems by using large populations and neighborhood recombination [4].

In this work, our aim is to understand why and how population size and neighborhood recombination increase the effectiveness of the algorithm. To study that, we choose NSGA-II [5] as our base algorithm and incorporate neighborhood recombination into it. We vary population size and analyze in detail convergence, front distribution, the distance between individuals that undergo crossover, and the distribution of solutions in objective space. We use DTLZ2 problem [6] with $m = 5$ objectives in our study.

The motivation to look into large populations is that they might support better the evolutionary search to deal with the increased complexity inherent to high dimensional spaces. On the other hand, the motivation to study recombination is to understand better the implications of a large number of solutions that are Pareto-optimal from the perspective of decision space and the operator that make moves on it. A large number of non-dominated solutions could cause a large diversity of individuals in the instantaneous population and recombination of very dissimilar individuals could be too disruptive. Neighborhood recombination aims to reduce dissimilarity between crossing individuals.

Our study reveals important properties of large populations, recombination in general, and neighborhood recombination in particular, related to convergence and distribution of solutions.

2   Method

In many-objective problems the number of non-dominated solutions grows substantially with the number of objectives of the problem [7,8]. A side effect of this is that non-dominated solutions tend to cover a larger portion of objective and variable space compared to problems with fewer objectives [9]. The implications of a large number of non-dominated solutions have been studied in objective space, where selection operates. However, little is known about the implications in decision space, where recombination and mutation operate. It is expected that the large number of non-dominated solutions in many objective problems induce a large diversity of individuals in the instantaneous population. In which case recombination of very dissimilar individuals could be too disruptive affecting its effectiveness.

Neighborhood Recombination encourages mating between individuals located close to each other, aiming to reduce dissimilarity between crossing individuals and improve the effectiveness of recombination in high dimensional objective spaces. We choose NSGA-II as our base algorithm and incorporate neighborhood recombination into it. In this work, we leave untouched selection of NSGA-II,

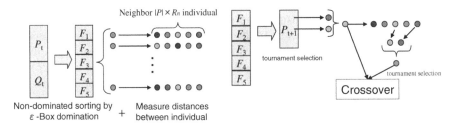

(a) Neighborhood creation   (b) Mating for recombination

Fig. 1. Neighborhood Recombination

which uses a primary ranking based on dominance and a secondary ranking based on crowding distance. This would allow us to show and explain the effects of population size and recombination with a well known selection.

The main steps of Neighborhood Recombination are as follows. During the calculation of dominance relationships, the proposed method calculates the distance between individuals in objective space and keeps a record of the $|P| \times R_n$ closest neighbors of each individual. Note that when the ranked population of size $|P| + |Q|$ is truncated to form the new population of size $|P|$, some individuals would be deleted from the neighborhood of each individual. When individuals are selected for recombination, the first parent p_A is chosen from the parent population P using a binary tournament, while the second parent p_B is chosen from the neighborhood of p_A using another binary tournament. Then, recombination is performed between p_A and p_B. That is, between p_A and one of its neighbors in objective space. If all neighbors of individual p_A were eliminated during truncation, the second parent p_B is selected from the population P similar to p_A. Fig.1 illustrates the neighborhood creation and mating for recombination. In this work, we set the parameter that defines the size of the neighborhood of each individual to $R_n = 0.1$ ($10\%|P|$).

3   Test Problem and Analysis Indicators

3.1   Test Problem

We study the behavior of the algorithms using the continuous function DTLZ2 [5], setting the number of objectives to $m = 5$ and the total number of variables to $n = m + 9$. Problem DTLZ2 is designed in such a way that the Pareto-optimal front corresponds to a non-convex surface in objective space, which lies in the positive quadrants of the unit hyper-sphere, with Pareto-local fronts constructed parallel to it. To achieve this, the n variables of a solution $x = (x_1, x_2, \cdots, x_n)$ are classified in two subsets. The first $m - 1$ variables $x_1, x_2, \cdots, x_{m-1}$, denoted $x_{1:m-1}$, determine the position of solutions within the Pareto-local/optimal front, whereas the remaining $n - m + 1$ variables $x_m, x_{m+1}, \cdots, x_n$, denoted $x_{m:n}$, determine the distance of the Pareto-local front to the Pareto-optimal front. When $x_m = x_{m+1} = \cdots = x_n = 0.5$, the solution is located in the Pareto-optimal front. The m objective functions used in DTLZ2 are as follows

$$
\left.
\begin{aligned}
f_1(\boldsymbol{x}) &= (1 + g(\boldsymbol{x}_{m:n})) \prod_{i=1}^{m-1} \cos(\tfrac{\pi}{2} x_i) \\
f_2(\boldsymbol{x}) &= (1 + g(\boldsymbol{x}_{m:n})) \left(\prod_{i=1}^{m-2} \cos(\tfrac{\pi}{2} x_i) \right) \sin(\tfrac{\pi}{2} x_{m-1}) \\
f_3(\boldsymbol{x}) &= (1 + g(\boldsymbol{x}_{m:n})) \left(\prod_{i=1}^{m-3} \cos(\tfrac{\pi}{2} x_i) \right) \sin(\tfrac{\pi}{2} x_{m-2}) \\
&\;\;\vdots \\
f_{m-1}(\boldsymbol{x}) &= (1 + g(\boldsymbol{x}_{m:n})) \cos(\tfrac{\pi}{2} x_1) \sin(\tfrac{\pi}{2} x_2) \\
f_m(\boldsymbol{x}) &= (1 + g(\boldsymbol{x}_{m:n})) \sin(\tfrac{\pi}{2} x_1)
\end{aligned}
\right\}
\tag{1}
$$

$$
g(\boldsymbol{x}_{m:n}) = \sum_{i=m}^{n} (x_i - 0.5)^2
\tag{2}
$$

3.2 Analysis Indicators

Proximity Indicator(I_p) [10]: Measures convergence of solutions by

$$
I_p = \operatorname*{median}_{\boldsymbol{x} \in P} \left\{ \left[\sum_{i=1}^{m} (f_i(\boldsymbol{x}))^2 \right]^{\frac{1}{2}} - 1 \right\},
\tag{3}
$$

where $\boldsymbol{x}$ denotes a solution in the population P. Smaller values of I_p indicate that the population P is closer to the Pareto-optimal front and therefore mean better convergence of solutions.

Mates Distance(D_c): Euclidian distances in variable space between pairs of solutions that undergo crossover are computed separately for the subsets of variables $\boldsymbol{x}_{1:m-1}$ and $\boldsymbol{x}_{m:n}$ that determine the position of solutions within the front and their distance to the Pareto-optimal front, respectively. Here we report the average distances $D_c(\boldsymbol{x}_{1:m-1})$ and $D_c(\boldsymbol{x}_{m:n})$ taken over all pairs of solutions that undergo crossover at a given generation.

Distribution of Solution in Objective Space (ψ_k): In order to observe where solutions of the parent population P are located in objective space, we classify each solution according to the number $k = 0, 1, \cdots, m-1$ of its objective values that are very small compared to the maximum objective value of the solution. More formally, The class k a solution belongs to is determined by

$$
k = \sum_{i=1}^{m} \theta_i
\tag{4}
$$

$$
\theta_i = \begin{cases} 1 & \text{if } f_i(\boldsymbol{x}) < f_{max}(\boldsymbol{x})/100 \\ 0 & \text{otherwise} \end{cases}
\tag{5}
$$

where $f_{max}(\boldsymbol{x}) = max\{f_1(\boldsymbol{x}), f_2(\boldsymbol{x}), \cdots, f_m(\boldsymbol{x})\}$. Roughly, a solution belonging to class $k = 0$ is considered to be in the central region of objective space, whereas solutions belonging to class $k \geq 1$ are gradually considered to be in the edges of objective space and identify dominant resistant solutions. We report ψ_k, the number of solutions in population P belonging to class k.

Front Distribution: Shows the number of solutions per front obtained after applying non-dominated sorting to the combined population of parents P and offspring Q. Here, we report results for fronts $F_1 \sim F_5$.

4   Simulation Results and Discussion

4.1   Preparation

In this work we use NSGA-II[6] as a base algorithm and include in its framework Neighborhood Recombination. We observe the behavior of conventional NSGA-II and NSGA-II with Neighborhood Recombination varying the population size from $|P| = 100$ to 5000 individuals. As genetic operators we use SBX crossover and Polynomial Mutation, setting their distribution exponents to $\eta_c = 15$ and $\eta_m = 20$, respectively. The parameter for the operators are crossover rate $p_c = 1.0$, crossover rate per variable $p_v = 0.5$, and mutation rate $p_m = 1/n$, where n is the number of variables. The maximum number of generations is fixed to $T = 100$. Here we report average results obtained running the algorithms 30 times.

4.2   Analysis Varying Population Size in NSGA-II

In this section we analysis the behavior of NSGA-II. Results are shown in Fig.2. First, we look at the convergence of the algorithm. Fig.2(a) shows I_p of Pareto-optimal solutions obtained in the final generation ($T = 100$) increasing the population size from $|P| = 100$ to 5000. It can be seen that I_p gets smaller by increasing population size $|P|$. That is, a larger population improves convergence of the algorithm. In order to investigate these results with more detail, Fig.2(b) shows the transition of I_p over the generations when the algorithm is set to population sizes $|P| = 100, 1000, 2000, 5000$. In the case of $|P| = 100$, it can be seen that I_p increases substantially. This indicates that the algorithm diverges from the Pareto-optimal front, rather than converge to it, as evolution proceeds. However, signs of convergence gradually appear by increasing population size $|P|$. Eventually, for $|P| = 5000$ no divergence is observed and I_p reduces to 0.05 with very small dispersion. Here an important conclusion is that population size is strongly correlated to the convergence ability of the algorithm.

Then, we analyze the front distribution over the generations. Results for the first five fronts $F_1, \cdots, F_5$ are shown in Fig.2(c)~(e) for population sizes $|P| = 100, 1000, 5000$, respectively. Note that the number of solutions in the first front $|F_1|$, obtained after applying non-dominated sorting to the combined population of parents and offspring of size $2|P|$, is larger than the size of the parent

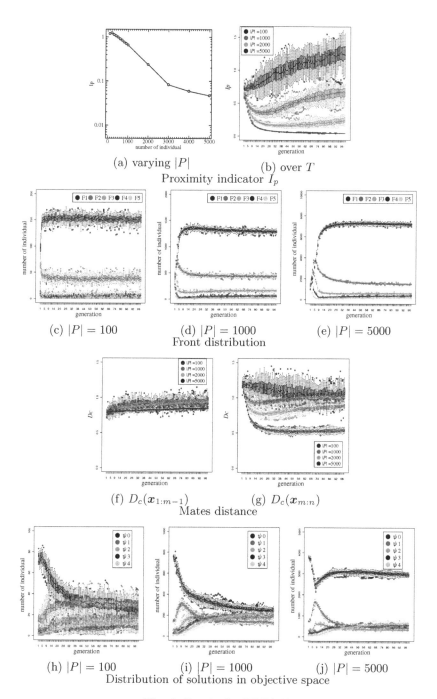

(a) varying $|P|$ (b) over T

Proximity indicator I_p

(c) $|P| = 100$ (d) $|P| = 1000$ (e) $|P| = 5000$

Front distribution

(f) $D_c(\boldsymbol{x}_{1:m-1})$ (g) $D_c(\boldsymbol{x}_{m:n})$

Mates distance

(h) $|P| = 100$ (i) $|P| = 1000$ (j) $|P| = 5000$

Distribution of solutions in objective space

Fig. 2. Results by NSGA-II

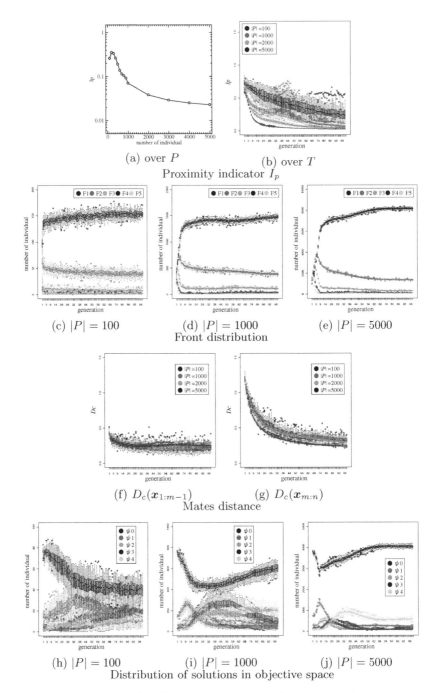

(a) over P (b) over T

Proximity indicator I_p

(c) $|P| = 100$ (d) $|P| = 1000$ (e) $|P| = 5000$

Front distribution

(f) $D_c(\boldsymbol{x}_{1:m-1})$ (g) $D_c(\boldsymbol{x}_{m:n})$

Mates distance

(h) $|P| = 100$ (i) $|P| = 1000$ (j) $|P| = 5000$

Distribution of solutions in objective space

Fig. 3. Results by NSGA-II with Neighborhood Recombination

population $|P|$ for most of the evolution. Especially when $|P| = 100$, the ratio $|F_1|/|P|$ is the highest and $|F_1|$ exceeds $|P|$ at a very early generation. When the population increases to $|P|=1000$ and 5000, the ratio $|F_1|/|P|$ reduces and it takes few more generations until $|F_1|$ exceeds the size of the parent population. Looking closely at Fig.2 (e) that shows results for $|P| = 5000$, we can see that the first 10 generations when $|F_1| < |P|$ is precisely the period where I_p reduces significantly, as shown in Fig.2 (b) $|P| = 5000$. If $|F_1| < |P|$ then the population is composed by solutions coming from two or more fronts, which means that parent selection can discriminate based on dominance ranking and not only on crowding distance as it is the case when $|F_1| > |P|$. These results suggest that a large enough random initial population is able to include lateral diversity (solutions from different fronts), which allows dominance-based selection to pull the population in the direction of the Pareto-optimal front.

Next, we look at $D_c(\boldsymbol{x}_{1:m-1})$ and $D_c(\boldsymbol{x}_{m:n})$, the average distances in decision space between individuals that undergo crossover. Note from Fig.2(f) that $D_c(\boldsymbol{x}_{1:m-1})$, computed on the subset of variables that determine the position within the front, is large at generation 0 and it tends to increase as the evolution proceeds. Note also that this trend becomes less pronounced as we increase the population size $|P|$. This shows the diversity of solutions and it is evidence that recombination takes place among very dissimilar solutions, raising questions about its effectiveness to help convergence. On the other hand, from Fig.2 (g) note that $D_c(\boldsymbol{x}_{m:n})$, computed on the subset of variables that determine the distance to the Pareto-optimal front, becomes smaller with the generations as we increase $|P|$. This reduction of $D_c(\boldsymbol{x}_{m:n})$ is expected if the population converge towards the Pareto-optimal front, which is located at $\boldsymbol{x}_m = \boldsymbol{x}_{m+1} = \cdots = \boldsymbol{x}_n = 0.5$.

Finally, we analyze the distributions of solutions in objective space $\psi_k (k = 0, 1, \cdots, 4)$ shown in Fig.2(h)$\sim$(j) for population sizes $|P| = 100, 1000, 5000$, respectively. Note that in the case of $|P| = 100, 1000$, the number of solutions ψ_0 are initially around 75% of the population size $|P|$, but after few generations it reduces to around 30% of $|P|$, showing that the number of solutions ψ_k, $k \geq 1$, close to the axis of the objectives functions increase significantly as evolution proceeds. On the other hand, when a population size $|P| = 5000$ is used, in few generations the number of solutions ψ_0 reduce from around 75% of $|P|$ to 50% of $|P|$, but then it rises and remains around 60% of $|P|$ until the last generation. This shows that larger populations are not easily pulled towards the extreme regions of objective space, caused by selection based on crowding distance that is at work for most generations since the size of the first front surpasses the population size ($|F_1| > |P|$).

4.3   Analysis of Neighborhood Recombination

In this section we analyze the behavior of NSGA-II with Neighborhood Recombination. Results are shown in Fig.3. Similar to the previous section, first we look at convergence. From Fig.3(a) note that the inclusion of neighborhood recombination reduces I_p drastically compared to NSGA-II for any population size, as shown in Fig.2(a). From Fig.3(b) it is important to note that, contrary to NSGA-II, no divergence of solutions is observed over the generations when neighborhood recombination is used, even for very small populations.

Next, we look at the front distribution shown in Fig.3(c)~(e). Note that a similar trend to NSGA-II can be observed. However, when neighborhood recombination is used $|F_1|$ gradually increases with the number of generations, while in NSGA-II $|F_1|$ remained high but relatively constant.

Then, we analyze the distances among solutions that undergo crossover shown in Fig.3(f),(g). Looking at Fig.3(f), note that $D_c(x_{1:m-1})$ reduces substantially compared to NSGA-II. This is an effect of recombining individuals with one of its neighbors. By selecting a partner close in objective space we are increasing the likelihood of selecting one that is also close in variable space, unless the functions are highly non-linear. Most importantly, a short $D_c(x_{1:m-1})$ indicates that recombination takes place in less dissimilar individuals, which increases significantly the effectiveness of recombination for any population size as corroborated by the reduction of I_p shown above. Fig.3(g) shows that $D_c(x_{m:n})$ shortens as the algorithm approaches better fronts, resembling the reduction of I_p.

Finally, looking at distribution of solutions ψ_k in objective space shown in Fig.3(h)~(j), comparing with NSGA-II it can be seen that the number of solutions ψ_0 increases when neighborhood recombination is used, whereas the number of solutions ψ_k, $k \geq 1$, reduces. This shows that there are more solutions in the central region and fewer in the edges of objective space, even for very small populations. That is, a more effective recombination helps convergence and resists the pull of selection towards extreme regions of objective space.

5   Conclusions

In this work, we have studied the effects of population size and neighborhood recombination on the search ability of a dominance-based MOEA applied to many-objective optimization. We chose NSGA-II as our base algorithm and included in it an operator that keeps track of neighbors and recombine individuals that are close to each other in objective space. We varied population size and analyzed in detail convergence, front distribution, the distance between individuals that undergo crossover, and the distribution of solutions in objective space using problem DTLZ2 with $m = 5$ objectives.

Our results showed that population size is strongly correlated to the convergence ability of the algorithm. A large enough random initial population is able to include lateral diversity (solutions belonging to different fronts), which allows dominance-based selection to pull the population in the direction of the Pareto-optimal front. We also showed that small populations are easily pulled towards extreme regions of objective space by selection based on crowding distance and that larger populations gradually become more resistant to this effect.

We argued and presented evidence that recombination in many-objective optimization takes place on highly dissimilar individuals if no restriction is put to partner selection. Also, we verified that neighborhood recombination takes place in less dissimilar individuals and showed that this increases significantly the effectiveness of recombination for any population size, helping convergence and resisting the pull of selection towards extreme regions of objective space. In this work we have focused mostly on population size and recombination. Selection is a fundamental part of the algorithm and there is ongoing work analyzing the effects of population size and recombination using improved selection mechanisms for many-objective optimization. However, due to space limitations, we shall report our finding elsewhere.

In the future, we would like to extend our analysis to other problems, increase the number of objectives and variables, and look at other ways to perform effective recombination in many-objective spaces. Also, determining an appropriate population size according to the number of objectives is an important area of research.

References

1. Ishibuchi, H., Tsukamoto, N., Nojima, Y.: Evolutionary Many-Objective Optimization: A Short Review. In: Proc. 2008 IEEE Congress on Evolutionary Computation, pp. 2424–2431. IEEE Press (2008)
2. Deb, K.: Multi-Objective Optimization using Evolutionary Algorithms. John Wiley & Sons, Chichester (2001)
3. Coello, C., Van Veldhuizen, D., Lamont, G.: Evolutionary Algorithms for Solving Multi-Objective Problems. Kluwer Academic Publishers, Boston (2002)
4. Kowatari, N., Oyama, A., Aguirre, H., Tanaka, K.: A Study on Large Population MOEA Using Adaptive Epsilon-Box Dominance and Neighborhood Recombination for Many-objective Optimization. In: Proc. Learning and Intelligent Optimization Conference, LION 6. LNCS. Springer (January 2012) (to appear)
5. Deb, K., Thiele, L., Laumanns, M., Zitzler, E.: Scalable Multi-Objective Optimization Test Problems. In: Proc. 2002 Congress on Evolutionary Computation, pp. 825–830. IEEE Service Center (2002)
6. Deb, K., Agrawal, S., Pratap, A., Meyarivan, T.: A Fast Elitist Non-Dominated Sorting Genetic Algorithm for Multi-Objective Optimization: NSGA-II, KanGAL report 200001 (2000)
7. Aguirre, H., Tanaka, K.: Insights on Properties of Multi-objective MNK-Landscapes. In: Proc. 2004 IEEE Congress on Evolutionary Computation, pp. 196–203. IEEE Service Center (2004)
8. Aguirre, H., Tanaka, K.: Working Principles, Behavior, and Performance of MOEAs on MNK-Landscapes. European Journal of Operational Research 181(3), 1670–1690 (2007)
9. Sato, H., Aguirre, H.E., Tanaka, K.: Genetic Diversity and Effective Crossover in Evolutionary Many-objective Optimization. In: Coello Coello, C.A. (ed.) LION 5 2011. LNCS, vol. 6683, pp. 91–105. Springer, Heidelberg (2011)
10. Purshouse, R.C., Fleming, P.J.: Evolutionary Many-Objective Optimisation: An Exploratory Analysis. In: Proc. IEEE CEC 2003, pp. 2066–2073 (2003)

Clustering Criteria in Multiobjective Data Clustering

Julia Handl[1] and Joshua Knowles[2]

[1] Manchester Business School, University of Manchester, UK
julia.handl@mbs.ac.uk
[2] School of Computer Science, University of Manchester, UK
j.knowles@manchester.ac.uk

Abstract. We consider the choice of clustering criteria for use in multiobjective data clustering. We evaluate four different pairs of criteria, three employed in recent evolutionary algorithms for multiobjective clustering, and one from Delattre and Hansen's seminal exact bicriterion method. The criteria pairs are tested here within a single multiobjective evolutionary algorithm and representation scheme to isolate their effects from other considerations. Results on a range of data sets reveal significant performance differences, which can be understood in relation to certain types of challenging cluster structure, and the mathematical form of the criteria. A performance advantage is generally found for those methods that make limited use of cluster centroids and assess partitionings based on aggregate measures of the location of all data points.

1 Introduction

Multiobjective clustering algorithms frame the data clustering problem as a multiobjective optimization problem in which a partitioning is optimized with respect to a number of conflicting criteria. This can be seen as a step beyond traditional clustering techniques, which commonly optimize a single criterion only [11]. It is also a step beyond techniques for internal cluster validation [9], which typically consider combinations of criteria, but usually do so by combining criteria in a linear or non-linear form.

The use of multiple objectives in data clustering has two key advantages. First, the framework of multiobjective optimization provides a natural way of defining a good partitioning: An exact definition of the clustering problem is elusive, but, loosely, a good partitioning can be described as one that meets at least the following two criteria: (i) data points within the same cluster are similar; while (ii) data points within different clusters are dissimilar. Second, single criteria for clustering are biased with respect to the number of clusters (i.e., the criteria naturally increase or decrease for partitionings with a larger number of clusters). One of the consequences of this is that the large majority of single-objective algorithms require the number of clusters to be specified as an input parameter. Multiobjective approaches to data clustering can tackle this issue in a novel way: by selecting two criteria that have opposite biases with respect to the number of clusters, these techniques are able to counter-balance this bias. In principle, multiobjective algorithms are therefore capable of exploring a range of solutions

C.A. Coello Coello et al. (Eds.): PPSN 2012, Part II, LNCS 7492, pp. 32–41, 2012.

with different number of clusters, which can support the user in identifying the most appropriate number of clusters [7].

Previous research on multiobjective clustering [4,7,8] has shown that bicriterion clustering methods often outperform their single-objective counterparts: an algorithm that optimizes two objectives, X and Y, simultaneously, will usually generate certain solutions that are better than the solutions generated by an algorithm that optimizes X or Y only. However, little research (if any) has been done to compare the different choices of (pairs of) criteria in terms of their conceptual aims, or their empirical performance, in bicriterion clustering. In this manuscript, we investigate this issue by comparing four pairs of clustering criteria that have been proposed in previous work on multiobjective clustering. We discuss the conceptual similarities and differences between these choices, and provide empirical results on the use of the criteria in an existing multiobjective evolutionary algorithm for data clustering.

2   Background and Methods

The principle of multiobjective data clustering was first introduced in 1980, when Delattre and Hansen described an exact algorithm for bicriterion clustering [4]. This algorithm was able to identify the set of partitionings corresponding to an optimal trade-off between two objectives, namely the split and the diameter of a partitioning. Given computational resources at the time (as well as the algorithm's reliance on graph colouring [10]), the method was evaluated on small data sets with tens of data items only. More recently, the idea of multiobjective clustering has been extended to a wider set of clustering criteria [1,2,7,8,9,12], which have been optimized using heuristic approaches to multiobjective optimization, principally evolutionary algorithms (EMO).

Here, based on this previous work, four versions of multiobjective clustering were implemented that differed solely in the clustering criteria used. An existing multiobjective clustering algorithm was used as the basis of the implementation, that is the underlying multiobjective evolutionary algorithm (PESA-II, [3]), as well as the encoding, variation operators, initialization and parameter settings are consistent with those described in [8]. The pairs of objectives used within the different versions are as follows.

MOCK [8]: The first method uses the objectives employed in the multiobjective evolutionary clustering algorithm MOCK (Multiobjective clustering with automatic k-determination). The first of these, overall deviation, measures the compactness of the clusters in the partitioning. It is given as:

$$\text{(Min.)} \quad \sum_{c_k \in C} \sum_{i \in c_k} d(i, \mu_k),$$

where C is the given set of clusters, μ_k is the centroid of cluster c_k and $d(,)$ is a distance measure defined between data points. The second objective, connectivity, assesses to what extent data points that are close neighbours are found in the same cluster. It is given as:

$$\text{(Min.)} \quad \sum_{c_k \in C} \sum_{i \in c_k} \sum_{l \in 1..L} \delta(i, l),$$

where L is a parameter specifying the number of neighbours to use (here, the default $L = 20$ is used), and $\delta(i, l)$ is a function which is 0 when data item i and its lth nearest neighbour are in the same cluster and $1/l$ otherwise.

DH [10]: The second method employs the clustering criteria used in Delattre and Hansen's seminal biclustering algorithm. The first objective is the complete link clustering criterion, which minimizes the largest cluster diameter observed in a partitioning. The objective is formally given as

$$(\text{Min.}) \quad \max_{c_k \in C} \max_{i,j \in c_k} d(i, j),$$

where C is the given partitioning of the data. The second objective is the single link clustering criterion, which maximizes the minimum split (distance) between clusters present in a partitioning. This is given as

$$(\text{Max.}) \quad \min_{c_k \in C, c_l \in C, l \neq k} \min_{i \in c_k, j \in c_l} d(i, j).$$

BMM1 [1]: The third pair of objectives is taken from a multiobjective evolutionary algorithm originally designed for fuzzy clustering. For the case of crisp partitioning (considered here), the clustering objectives used simplify to the within-cluster sum of squares, and the minimum distance observed between cluster centroids. Formally, the within-cluster sum of squares is given as

$$(\text{Min.}) \quad \sum_{c_k \in C} \sum_{i \in c_k} d(i, \mu_k)^2,$$

where μ_k is the cluster centroid of cluster c_k. Evidently, this is very similar to the measure of overall deviation defined above with the difference that the distance values are here squared. The minimum distance between cluster centroids is given as

$$(\text{Max.}) \quad \min_{c_k \in C, c_l \in C, l \neq k} d(\mu_k, \mu_l),$$

where μ_k and μ_l are the cluster centroids of cluster c_k and c_l, respectively.

BMM2 [12]: The fourth method also uses the intra-cluster sum of squares (see above) as its first objective. The second objective is the summed pairwise distance between cluster centroids. Formally, this is given as

$$(\text{Max.}) \quad \sum_{c_k \in C, c_l \in C, l \neq k} d(\mu_k, \mu_l),$$

where μ_k and μ_l are the cluster centroids of cluster c_k and c_l, respectively.

3   Conceptual Characteristics

Key similarities and differences between MOCK, DH, BMM1 and BMM2 are summarized in Table 1 and are discussed in this section.

Table 1. Characteristics of the different clustering criteria: (i) Computational complexity associated with evaluating a partitioning of N data points in D dimensions into K clusters; L gives the number of neighbours used in MOCK's connectivity measure; (ii) Resolution of the criteria (the extent to which information about all data points is taken into account); (iii) Use of cluster centroids

	Complexity	Resolution	Centroids
Overall deviation (MOCK)	$\Theta(DN)$	Complete	Yes
Maximum diameter (DH)	$\Theta(N^2)$	Partial	No
Within-cluster sum of squares (BMM1, BMM2)	$\Theta(DN)$	Complete	Yes
Connectivity (MOCK)	$\Theta(LN)$	Complete	No
Minimum split (DH)	$\Theta(N^2)$	Partial	No
Minimum centroid distance (BMM1)	$\Theta(DK^2)$	Partial	Yes
Sum of centroid distances (BMM2)	$\Theta(DK^2)$	Complete	Yes

3.1   Similarities between the Objectives

There are some clear similarities in the way clustering objectives have been combined in the techniques considered. In all four cases, the pair of objectives has been selected to assess both of the key properties of a good partitioning (see Introduction): that (i) data points within the same cluster are similar; while (ii) data points within different clusters are dissimilar.

In MOCK, homogeneity of clusters is assessed using the measure of overall deviation. A similar role is played by the maximum diameter criterion in Delattre and Hansen's method and by the within-cluster sum of squares in Bandyopadhyay et al.'s methods (methods BMM1 and BMM2).

In MOCK, separation between clusters is considered implicitly through the measure of connectivity, which penalizes data points whose nearest neighbours do not reside in the same cluster. In Delattre and Hansen's technique the distance between clusters is assessed using the criterion of minimum split, which identifies the closest pair of data points that are not in the same cluster. Finally, Bandyopadhyay et al. measure cluster distance based on the distance of cluster representatives, either considering the entire set of cluster centres (method BMM2 [12]) or the minimum distance only (method BMM1 [1]).

3.2   Differences between the Objectives

Despite these clear similarities, there are also some fundamental differences between the criteria considered.

One defining characteristic of a clustering criterion is the extent to which its calculation takes into account the cluster assignment of all data points within a data set. This can be most easily understood using the examples of the within-cluster sum of squares and the maximum diameter criterion. The within-cluster sum of squares is calculated as the sum of the distances of all data items to their cluster centre. A change in the cluster assignment of any single data item will therefore usually result in a change to the value of the criterion. In contrast, the maximum diameter of a partitioning is defined as the largest dissimilarity observed between data items that reside in the same cluster. This means that changes in the cluster assignment of individual data items will often have no effect on the value of the criterion, provided that the maximum diameter remains unchanged.

A second defining characteristic is the presence or absence of the concept of cluster centroids in the calculation. Methods that use a cluster centroid make certain implicit assumptions on the shape of the surrounding clusters: it is clear that the definition of a cluster centroid makes relatively little sense for a nonconvex cluster. Out of the objectives discussed, overall deviation, within-cluster sum of squares and the measures of cluster dissimilarity in BMM1 and BMM2 all rely on the definition of a cluster centre. On the other hand, MOCK's connectivity measure, as well as both of Delattre and Hansen's clustering criteria, make no such assumptions on the presence of a centroid and the shape of the underlying cluster.

3.3   Computational Complexity

A further significant difference between the clustering criteria is their computational complexity.

As discussed above, Delattre and Hansen's measure of cluster homogeneity does not make use of a cluster centroid. This comes at the expense of quadratic complexity, as all pairwise dissimilarities between data items need to be considered. In contrast, methods of cluster homogeneity that do utilize a centroid (i.e., overall deviation in MOCK, within-cluster sum of squares in BMM1 and BMM2) have linear complexity.

For measures of cluster separation, the differences in complexity are even more significant. Again, Delattre and Hansen's is the computationally most expensive: the identification of the minimum split requires the pairwise comparison of all data items, resulting in quadratic complexity. MOCK's objective (connectivity) ranks second in complexity: it requires the one-off calculation of N sorted lists of length N (complexity $\Theta(N \times N \log N)$), but has linear complexity for all further evaluations. The objectives in BMM1 and BMM2 have a complexity of only $\Theta(DK^2)$, where K is the number of clusters in the partitioning.

4   Empirical Performance Analysis

4.1   Experimental Setup

For the empirical comparison of the four methods, a benchmark set of Gaussian Clusters in 2, 10 and 100 dimensions was used. This benchmark set has been described previously [8], and the data and results are summarized as supplementary material [6]. In addition, eight two-dimensional data sets available at http://cs.joensuu.fi/sipu/datasets/, were used; these are summarized in Table 2. These feature a variety of challenging cluster properties which we discuss in the next section. The Euclidean distance measure was used for all data sets.

The results returned by each method were evaluated by monitoring the size of the non-dominated set, their quality with respect to the set of eight clustering criteria, as

Table 2. Two-dimensional data sets (also see http://cs.joensuu.fi/sipu/datasets/)

Name	N	D	K	Name	N	D	K
Jain	373	2	2	Compound	399	2	6
Aggregation	788	2	8	Path-based	300	2	3
Flame	244	2	2	r15	600	2	15
Spiral	312	2	3	d31	3100	2	31

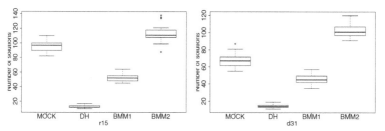

Fig. 1. Size of the solution sets of non-dominated solutions. Representative results over 21 runs for data sets r15 and d31. The data shows a general trend with an ordering of the solutions sets S returned by the methods as $|S_{BMM2}| > |S_{MOCK}| > |S_{BMM1}| > |S_{DH}|$.

well as their accuracy with respect to the known class labels for the data. The latter was assessed using the Adjusted Rand Index (AR, [5]), which is an established external technique of cluster validation that can be used to compare a clustering to a set of known true class labels. It is normalized with respect to the number of clusters in the partitioning, and is therefore well suited for the comparison of partitionings with different numbers of clusters as done in this work [9]. It takes values between 0 and 1, with 1 indicating a partitioning that accurately matches all known class labels.

4.2 Results

A set of 21 runs was obtained for each combination of data set and pair of objectives. Each run generated a set of non-dominated solutions, which was then analyzed with respect to the performance measures discussed above. Full results are available as supplementary material [6]. In the following, we will show selected results obtained by the methods, with the aim of highlighting key strengths and limitations of the four combinations of objectives used.

In terms of the size of the solution sets returned by the methods, we find that BMM2 and MOCK return respectively the most and second most non-dominated partitionings. DH returns the least solutions, followed by BMM1. We postulate that this ordering is due to the different levels of resolution of the objectives used. As discussed in the previous section, measures with partial resolution (such as minimum split) are calculated based on the position of a few, extreme data points only. Consequently, many different

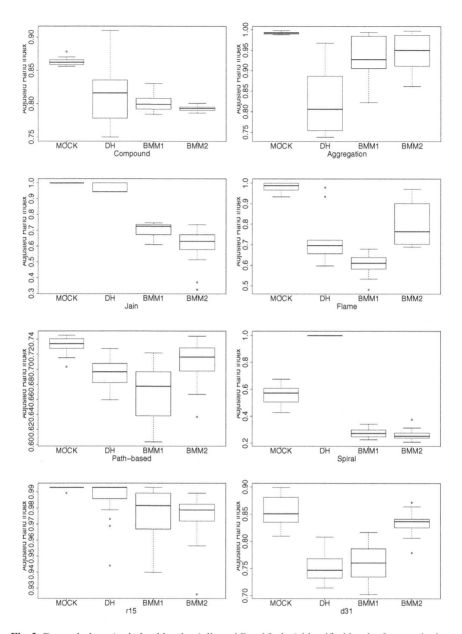

Fig. 2. Best solutions (as judged by the Adjusted Rand Index) identified by the four methods on the two-dimensional data sets. Results are over 21 runs.

partitionings will result in the same objective value such that plateaus are introduced into the search space. Our results indicate that this also reduces the number of Pareto optimal solutions. Figure 1 shows representative results for data sets r15 and d31.

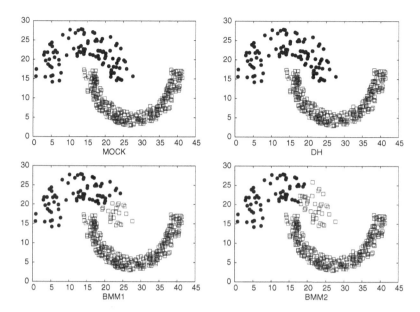

Fig. 3. Illustrative example: Non-spherical clusters. Best two cluster solution returned on the Jain data set by the first run of each method. The nonconvex shape of the clusters introduces problems for methods BMM1 and BMM2, which implicitly assume a convex shape through the use of cluster centroids in both objectives.

Internal validation of clustering results, based on the eight different criteria of clustering quality, indicates that, as expected, all of the four methods outperform their contestant techniques at optimizing their individual pair of objectives (results not shown).

Figure 2 summarizes the results of external cluster validation (based on the Adjusted Rand Index) for the two-dimensional data sets. From these data, it is clear that the choice of objectives has significant impact on the quality of the best solutions returned. This is further confirmed by the results obtained for the Gaussian data sets (see supplementary material [6]). Out of the four methods tested, MOCK shows the best peak performance for the majority of the data sets. The performance differences observed can be understood in more detail by considering the objectives' performance with respect to cluster structures that pose challenges. Using the clustering results obtained for the Jain, Flame and Spiral data, Figures 3 to 5 highlight the effects of nonconvex clusters, chaining between clusters [4] and highly elongated clusters. Key observations from this analysis are limitations of Bandyopadhyay et al.'s techniques with respect to unequally sized and nonconvex clusters (a direct consequence of the use of cluster centroids in both objectives), limitations of Delattre and Hansen's technique with respect to chaining / overlap between cluster, and limitations of MOCK's connectivity measure for extremely elongated clusters (which may be overcome through adjustment of the parameter L in the connectivity measure).

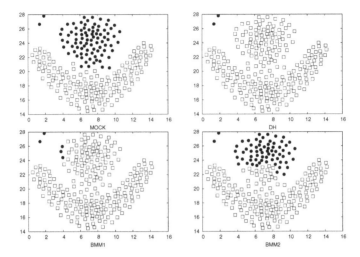

Fig. 4. Illustrative example: Non-spherical clusters with chaining. Best two cluster solution returned on the Flame data set by the first run of each method. The chaining between clusters poses problems for method DH. As the objectives used in DH do not consider the location of all data points, they are more sensitive to this type of noise. The presence of non-spherical clusters makes this data set problematic for methods BMM1 and BMM2.

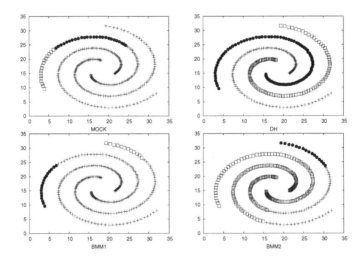

Fig. 5. Illustrative example: Highly elongated clusters. Best three-cluster solution returned on the Spiral data set by the first run of each method. Again, the non-spherical shape of the clusters is problematic for methods BMM1 and BMM2. MOCK also shows a poor performance on this data, as the clusters are so elongated that the connectivity measure ceases to work (some of the L nearest neighbours of each data point are located in another cluster).

5 Conclusion

This manuscript has focused on the comparison of four pairs of criteria for multiobjective clustering. One pair of criteria — from Delattre and Hansen's early bicriterion clustering algorithm — has not previously been evaluated except on very small data sets. The results show that, despite some conceptual similarities in the clustering criteria compared here, significant performance differences can be observed when they are employed within a multiobjective evolutionary algorithm for clustering. Overall, the pair of objectives employed in the multiobjective clustering algorithm MOCK emerges as the strongest combination. We offer two explanations for this result: (i) the limited use of cluster centroids in MOCK's objectives (use in one rather than both objectives) and (ii) the consideration of all data points in the calculation of both of MOCK's objectives. Here, results were all generated using PESA-II; future work may seek to generalize our findings to alternative metaheuristic or exact methods.

References

1. Bandyopadhyay, S., Mukhopadhyay, A., Maulik, U.: An improved algorithm for clustering gene expression data. Bioinformatics 23(21), 2859–2865 (2007)
2. Caballero, R., Laguna, M., Martí, R., Molina, J.: Scatter tabu search for multiobjective clustering problems. Journal of the Operational Research Society 62(11), 2034–2046 (2010)
3. Corne, D., Jerram, N., Knowles, J., Oates, M.: PESA-II: Region-based selection in evolutionary multiobjective optimization. In: Proceedings of the Genetic and Evolutionary Computation Conference, GECCO 2001, pp. 283–290. Morgan Kaufmann Publishers (2001)
4. Delattre, M., Hansen, P.: Bicriterion cluster analysis. IEEE Transactions on Pattern Analysis and Machine Intelligence 2(4), 277–291 (1980)
5. Halkidi, M., Batistakis, Y., Vazirgiannis, M.: On clustering validation techniques. Journal of Intelligent Information Systems 17(2), 107–145 (2001)
6. Handl, J., Knowles, J.: http://personalpages.manchester.ac.uk/mbs/julia.handl/moc.html
7. Handl, J., Knowles, J.: Exploiting the Trade-off — The Benefits of Multiple Objectives in Data Clustering. In: Coello Coello, C.A., Hernández Aguirre, A., Zitzler, E. (eds.) EMO 2005. LNCS, vol. 3410, pp. 547–560. Springer, Heidelberg (2005)
8. Handl, J., Knowles, J.: An evolutionary approach to multiobjective clustering. IEEE Transactions on Evolutionary Computation 11(1), 56–76 (2007)
9. Handl, J., Knowles, J., Kell, D.B.: Computational cluster validation for post-genomic data analysis. Bioinformatics 21(15), 3201–3212 (2005)
10. Hansen, P., Delattre, M.: Complete-link cluster analysis by graph coloring. Journal of the American Statistical Association, 397–403 (1978)
11. Jain, A., Murty, M., Flynn, P.: Data clustering: a review. ACM Computing Surveys 31(3), 264–323 (1999)
12. Mukhopadhyay, A., Maulik, U., Bandyopadhyay, S.: Multiobjective genetic algorithm-based fuzzy clustering of categorical attributes. IEEE Transactions on Evolutionary Computation 13(5), 991–1005 (2009)

Enhancing Profitability through Interpretability in Algorithmic Trading with a Multiobjective Evolutionary Fuzzy System

Adam Ghandar[1], Zbigniew Michalewicz[1,2], and Ralf Zurbruegg[3]

[1] School of Computer Science, University of Adelaide, Adelaide, SA 5005, Australia
[2] Institute of Computer Science, Polish Academy of Sciences, ul. Ordona 21, 01-237 Warsaw, Poland, and Polish-Japanese Institute of Information Technology, ul. Koszykowa 86, 02-008 Warsaw, Poland
[3] Business School, University of Adelaide, Adelaide, SA 5005, Australia

Abstract. This paper examines the interaction of decision model complexity and utility in a computational intelligence system for algorithmic trading. An empirical analysis is undertaken which makes use of recent developments in multiobjective evolutionary fuzzy systems (MOEFS) to produce and evaluate a Pareto set of rulebases that balance conflicting criteria. This results in strong evidence that controlling portfolio risk and return in this and other similar methodologies by selecting for interpretability is feasible. Furthermore, while investigating these properties we contribute to a growing body of evidence that stochastic systems based on natural computing techniques can deliver results that outperform the market.

1 Introduction

Algorithmic trading is an important part of the global financial services industry. In 2008 over 40% of executed market orders were attributed to algorithmic trading methods in major developed stock markets. Growth in the volume of trades generated by automatic signals has risen 30-40% per annum since, with the financial crisis over the period having little effect on the uptake of technology [1]. Financial portfolio management is a complex task that takes place in a highly dynamic and competitive environment with arguably immeasurable uncertainty. These factors make the problem quite different from other applications of computational intelligence in control, pattern recognition, etc, even though the tools and methods used are the same. It is therefore of value to examine the relationships between system design and parameters and specific and extensive performance markers and tools in financial applications distinctly.

This paper applies an evolving fuzzy system based on the representation described in [2] and from a technical viewpoint significantly extends that earlier research to make the models which are learned even closer to those used by financial practitioners. This is done by adding fundamental data (accounting and macro economic information) and by making use of a multiobjective EA to implement criteria for model parsimony. Here we make a contribution to answering

C.A. Coello Coello et al. (Eds.): PPSN 2012, Part II, LNCS 7492, pp. 42–51, 2012.
© Springer-Verlag Berlin Heidelberg 2012

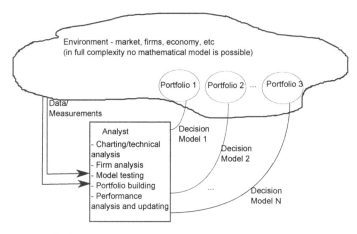

Fig. 1. Practical problem: design a control system to consolidate the tasks of the human portfolio analyst

questions about whether subjective criteria (such as human interpretability) can produce value in the application of heuristic, and specifically computational intelligence, approaches to problem solving in financial trading - a complex and dynamic activity performed on the basis of incomplete information. We find that by controlling rule intelligibility we are able to strongly influence the risk and return profile of portfolios managed by the algorithm. This angle is simply not able to be considered in classical modeling as it is currently espoused in finance because a trading model is viewed in simple terms essentially as a formula rather than an intelligent agent.

The particular methodology used to produce interpretable models is as justified as other techniques capable of representing expressive models computationally because it too has been found to perform at a standard equivalent to other state of the art machine learning approaches on standard benchmarks [3]. Many other multiobjective approaches are described in the literature [4]. The conclusions made here are also to some extent applicable to other methods. In addition, certain aspects of the particular approach make it particularly suitable for this analysis, notably the the structure that is imposed on the decision model representation to cause the optimzation process to in some respects imitate human reasoning in the application domain.

Rule based approximate reasoning systems can be applied to approximate human thinking when combined with a self-learning methodology. Efforts to automate human thinking in this way has been found to lead to improved performance in real world applications: for example in emulating a skilled human operator who controls complex machinery "without a formal quantitative model in mind" [5]. Figure 1 illustrates the rationale of the system presented here in adapting approximate reasoning principles to financial modeling. A financial analyst generally performs a sequence of specific activities: generating or selecting/tuning a range of indicators (these are often called rules in financial parlance and should not be confused with the term rulebase as it is used here) from raw data measurements or feeds; testing resultant models using historic data; building a portfolio using the model and lastly there is constant process of updating the approach based on portfolio management performance.

In the remainder of the paper we describe the design of the system in performing the tasks illustrated in Figure 1, provide empirical results on performance of different portfolios while varying model complexity, and make conclusions applicable more generally regarding the complexity of decision models in the financial problem domain.

2   Methodology

The approach mimics a real financial analyst (see Figure 1). The key tasks are: data measurement; transformation to generate unit strategies based on historic price and volume data (technical analysis) and information about the firms underlying the stocks (fundamental analysis); selecting and combining the numerous models in the previous step into a genotype comprising a fuzzy rulebase and a vector of parameters for the strategies; and lastly the implementation of decisions on portfolio contents. A multiobjective EA facilitates the study of model complexity.

2.1   Measurement, Information Set, and Initial Models

The information set defines the universe of discourse that can be represented from the environment. Technical (price/volume) and fundamental information on underlying firms is incorporated.

Fundamental Strategies. A set of fundamental variables found to be useful in other emprical studies was selected. The relationship between the variables considered and price changes has been found to be farily transient further accentuating the need for an adaptive methodology able to be achieved with a heuristic approach. Dividend yield measures the cash flow benefit with regard to the share investment. The power of dividend yield to forecast stock return has been noted in [6] as being "temporary" component of prices. Price to Book Value has been used in fundamental factor models for some time. The price Earnings ratio (PE) divides the share price, over the total earnings per share, in some periods and markets the PE ratio is a predictor of higher return with reduced risk, see [7]. Earnings per share (EPS) is calculated by taking the ratio of the profit of the firm over its market share. Stocks that have a higher earnings per share generate more income relative to the stock price and thus places upward pressure on the share price [8]. The debt to equity ratio looks at the liabilities per share; it has been shown to be positively correlated with stock price [9].

In a falling or volatile market liabilities may be of more importance. The last three fundamental variables (earnings before interest and tax, return on assets, and return on equity) divide firm income into classifications provide more fine grained picture.

Each variable is processed to obtain a rate of change measurement in the form of an oscillator (O). The oscillator measures the change relative to an earlier point $O = (v_t - v_{t-m})/v_{t-m}$. The parameter m measures this period and belongs to the set $10, 20, 30, \ldots, 260$.

Technical Strategies. Technical strategies use price and volume data. They are widely used in industry and theoretical justifications postulate the importance of behavioural factors and their detection using technical rules [10]. Technical trading rules may also be able to pick up institutional trading activity [11].

The technical inputs is given in table 1. The abbreviations have meaning: SMA, single moving average; DMA, double moving average; PPO, price oscillator; OBV, on balance volume indicator; RSI, relative strength index; MFI, money flow index; Vol. DMA, volume double moving average; PVO, percentage volume oscillator; DMI, directional movement index; %R, percent R. For the OBV indicator the value obv_t for each day t is calculated by initially at $t = 0$ $obv_0 = v_0$, then for each subsequent day t: if $p_t > p_{t-20}$ then $obv_t = obv_{t-1} + v_t$; else if $p_t < p_{t-20}$ then $obv_t = obv_{t-1} - v_t$, else if if $p_t > p_{t-20}$ then $obv_t = obv_{t-1}$.

2.2   Decision Model Representation

A solution aggregates the inputs from the processing described in the previous subsections and is represented using a set of fuzzy rules and an integer vector of parameters (time horizons - see restrictions in Table 1).

A rulebase is a mapping

$$D : \Re^n \to \Omega,$$

from vector of observations $\mathbf{x} = \{x_1, \ldots, x_n\} \in \Re^n$ to a signals $\{\omega_1, \ldots, \omega_c\} \in \Omega$ to buy or sell. The form of the rules is based on [13], the output is interpreted as a degree of certainty the buy or sell signal is correct, given antecedent and training data. A rule r_k in a ruleset M has the format: R_k : if x_1 is $A_1 \wedge \ldots \wedge x_n$ is A_n; then $(z_{k,1}, \ldots, z_{k,c})$ where $x_1 \ldots x_n$ are feature observations that are described by linguistic labels $A_1 \ldots A_n$, these are common in the different rules and precalculated as in [2].

$$z_{k,i} = \frac{\text{Sum of matching degrees of rule with } \omega_i}{\text{Sum of total matching degrees for all rules}}.$$

The mapping D uses the max operation to aggregate rules and the product t-norm to aggregate the antecedent conjunctions. The degree of certainty for i-th signal is $D : eval_i(\mathbf{x}) = \max_{k=1}^{M} \left\{ z_{k,i} \prod_{j=1}^{n} \{\mu_j(x_j)\} \right\}$, In this way, we specify a search space of possible rules to correspond to trading rules that a human expert trader could construct using the same information. A rule is

Table 1. Technical indicators and restrictions on parameters

Name	Formula	Restrictions
Price Change 1		$\delta = 20$
Price Change 2	$\ln\left(\frac{p_t}{p_{t-\delta}}\right)$	$\delta = 50$
Price Change 3		$\delta = 100$
SMA Buy	$\frac{p_t}{ma_t}$	$len_{ma} \in \{10, 20, 30\}$
SMA Sell	$\frac{ma_t}{p_t}$	$len_{ma} \in \{10, 20, 30\}$
DMA Buy 1		$len_{ma2} \in \{10, 20, 30\}$ $len_{ma2} \in \{40, 50, 60\}$
DMA Buy 2	$\frac{ma1_t}{ma2_t}$	$len_{ma1} \in \{60, 70, \ldots, 120\}$ $len_{ma2} \in \{130, 140, \ldots, 240\}$
DMA Buy 3		$len_{ma1} \in \{60, 70, \ldots, 120\}$ $len_{ma2} \in \{130, 140, \ldots, 240\}$
DMA Sell 1		$len_{ma1} \in \{10, 20, 30\}$ $len_{ma2} \in \{40, 50, 60\}$
DMA Sell 2	$\frac{ma2_t}{ma1_t}$	$len_{ma1} \in \{60, 70, \ldots, 120\}$ $len_{ma2} \in \{130, 140, \ldots, 240\}$
DMA Sell 3		$len_{ma1} \in \{60, 70, \ldots, 120\}$ $len_{ma2} \in \{130, 140, \ldots, 240\}$
PPO 1		$len_{ma2} \in \{10, 20, 30\}$ $len_{ma2} \in \{40, 50, 60\}$
PPO 2	$\frac{ma1_t - ma2_t}{ma1_t} \times 100$	$len_{ma1} \in \{60, 70, \ldots, 120\}$ $len_{ma2} \in \{130, 140, \ldots, 240\}$
PPO 3		$len_{ma1} \in \{60, 70, \ldots, 120\}$ $len_{ma2} \in \{130, 140, \ldots, 240\}$
DMI	see [12]	
%R	$\%R = \frac{p_t - min[p_{t-1}, \ldots, p_{t}-10]}{max[p_{t-1}, \ldots, p_t-10] - min[p_{t-1}, \ldots, p_t-10]}$	
RSI	$RSI = 100 - \frac{100}{1+RS}$	$RS = \frac{totalgains \div n}{totallosses \div n}$
MFI	$MFI = 100 - \frac{100}{1+MR}$	$MR = \sum \frac{MF^+}{MF^-}$ $MF^+ = p_i \times v_t, where p_i > p_{i-1},$ and $MF^- = p_i \times v_t, where p_i < p_{i-1}$
Vol. DMA Buy 1	$\frac{vma1_t}{vma2_t}$	$len_{vma1} = 5, len_{vma2} = 20$
Vol. DMA Buy 2		$len_{vma1} = 20, len_{vma2} = 100$
Vol. DMA Sell 1	$\frac{vma2_t}{vma1_t}$	$len_{vma1} = 5, len_{vma2} = 20$
Vol. DMA Sell 2		$len_{vma1} = 20, len_{vma2} = 100$
OBV Buy	$\frac{\left(p_t - max[p_{t-1}, \ldots, p_{t-n}]\right)}{p_t} + \frac{\left(max[obv_{t-1} \ldots obv_{t-n}] - obv_t\right)}{obv_t}$	
OBV Sell	$\frac{\left(min[p_{t-1}, \ldots, p_{t-n}] - p_t\right)}{p_t} + \frac{\left(obv_t - min[obv_{t-1}, \ldots, obv_{t-n}]\right)}{obv_t}$	
PVO 1	$\frac{ma1_t - ma2_t}{SM_t} \times 100$	$len_{ma1} = 5, len_{ma1} = 20$
PVO 2		$len_{ma1} = 20, len_{ma1} = 100$
Bol 1	$Bol = \frac{p_t - ma_t}{2 \times sd(P_t, \ldots, P_{t-\delta})}$	$\delta = 20$
Bol 2		$\delta = 50$

considered in this paper to be a statement in structured language that specifies that if some condition(s) hold, then a particular action ought to be taken. IF [Conditions], THEN do [buy/sell a particular stock] Each rulebase comprises several such statements. The genotype representation and operators are provided in [2].

2.3 Learning Process

The learning model uses a pareto based algorithm (SPEA2, [14]) to obtain a set of solutions balancing objectives of solution simplicity and in sample prediction accuracy. We can express the task to learn rules as an optimization problem in which there are two main criteria. These are to minimize the number of false signals produced by the strategies (accuracy), and reduce the model complexity:

$$\text{minimize } \mathbf{z} = f_{error}(\mathbf{x}), f_{complexity}(\mathbf{x}).$$

The accuracy objective performance is the error in determining buy and sell signals from a set of examples in training data T:

$$f_{error}(\mathbf{x}) = 1 - \frac{\#\text{ correct signals}}{\#\text{ false signals} + \#\text{ correct signals}}.$$

The number of correct signals is the count of the number of times the rulebase correctly anticipated a rise or fall in the share price, a false signal is the number of times the rulebase falsly predicted a rise or fall in the share price. The complexity of a strategy specified by a rulebase is defined by the number of rules and, within each rule, by the number of clauses.

A rulebase has two main sources of complexity which are the number of rules, and inputs quantified per rule. Therefore, $f_{complexity}$ may be decomposed into these components (which are modeled as separate objectives using the multiobjective evolutionary algorithm). In this paper we consider two complexity objectives, the *number of rules*, and the *average number of inputs used per rule (#inputs/#rules)* in the rulebase. Given two fuzzy rulebase solutions, $\mathbf{x_1}$ and $\mathbf{x_2}$, we can say that $\mathbf{x_1}$ dominates $\mathbf{x_2}$ if it is less than or equal to the other in all objectives being minimized or otherwise that it dominates in particular one of the objectives. The final source of complexity is from the definition of the linguistic variables - we set this deterministically prior to running the optimizer so it is not an objective here. The approach is verified using classical benchmarks such as Iris in [3].

3 Financial Portfolio Management

These experiments used historic data from the Standard and Poor ASX 200 oil and gas stocks between April 2000 and November 2009. All data was sourced from Data International. Instead of raw price data we used a total return index adjusted for stock splits, mergers and dividend payments. The oil and gas sector is very volatile in this period (global financial crisis and reversal as a result of the resources trade with China). There was a period of growth followed by fall and a subsequent recovery, this allowed for an evaluation of the system in three different epochs. Transaction costs were 0.25% to buy or sell, population size for SPEA was 100 and the archive size was 10. In the business model, transactions inspired by the decision model were applied to re-balance portfolios to sell and buy recommended stocks at fixed intervals of 20 trading days, for adaptation the optimization process was run before each re-balance using the previous year of data for "training".

Two benchmarks are used for comparison the first is a a buy and hold approach (BH) and the second is a standard active alpha strategy [1]. The alpha is based on the single-factor pricing model which relates a stocks excess return, $r_{i,t} - r_{f,t}$ to market return as follows

$$r_{i,t} - r_{f,t} = \alpha_i + \beta_i \left[r_{m,t} - r_{f,t} \right] + e_{i,t},$$

Table 2. Metrics of performance for the portfolios managed by solutions with varying structural complexity (SC) and linguistic complexity (LC) along with the 2 benchmark portfolios. Confidence bounds are at the 90% level based on 30 test runs.

	HP Ret	An. Ret	σ	IR	Sel.	Net Sel.
Benchmarks						
BH	6.1325	0.2899	0.3654	0.5836	0.2956	0.1703
Alpha	12.7924	0.3873	0.3988	0.7931	0.4095	0.2697
Low LC						
Low SC	16.0 [± 4.74]	0.366 [± 0.0375]	0.501 [± 0.0279]	0.699 [± 0.0694]	0.438 [± 0.0405]	0.262 [± 0.0369]
Medium SC	10.4 [± 2.94]	0.314 [± 0.0351]	0.418 [± 0.0218]	0.621 [± 0.0773]	0.345 [± 0.0362]	0.199 [± 0.0351]
High SC	12.7 [± 4.55]	0.318 [± 0.0421]	0.453 [± 0.0297]	0.617 [± 0.0811]	0.368 [± 0.0450]	0.208 [± 0.0407]
Medium LC						
Low SC	16.5 [± 5.99]	0.368 [± 0.0372]	0.490 [± 0.0587]	0.667 [± 0.0621]	0.419 [± 0.0522]	0.246 [± 0.0366]
Medium SC	13.3 [± 3.04]	0.355 [± 0.0330]	0.526 [± 0.0581]	0.647 [± 0.0617]	0.421 [± 0.0408]	0.235 [± 0.0311]
High SC	10.8 [± 2.32]	0.337 [± 0.0287]	0.491 [± 0.0396]	0.615 [± 0.0563]	0.388 [± 0.0359]	0.215 [± 0.0292]
High LC						
Low SC	12.7 [± 1.80]	0.375 [± 0.0179]	0.394 [± 0.0152]	0.764 [± 0.0377]	0.394 [± 0.0204]	0.256 [± 0.0178]
Medium SC	31.6 [± 7.65]	0.482 [± 0.0279]	0.490 [± 0.0179]	0.912 [± 0.0456]	0.543 [± 0.0332]	0.370 [± 0.0290]
High SC	30.8 [± 7.67]	0.477 [± 0.0292]	0.491 [± 0.0214]	0.903 [± 0.0504]	0.537 [± 0.0341]	0.364 [± 0.0302]

index i indicates the stock and t refers to a day, e is an error term and $r_{i,t}$ is the stocks return on day t, $r_{f,t}$ is the risk free rate, $r_{m,t}$ is the market return. In an efficient market it would be possible to price stocks based solely on their risk components, here the excess return above the market. In the ideal theoretic case returns of any stock above the risk-free rate can be fully explained by the risk component, meaning that β_i would be one and α_i zero. If the alpha is actually positive the stock is outperforming relative to its level of risk and should be bought (conversely if it is negative the stock should be sold). In testing, the alpha portfolio produced an annual return of 38% compared with the buy and hold strategy which resulted in 29.56%. The information ratio of the alpha portfolio was 0.78 and for the buy and hold it was 0.58. Other metrics are given in table 2.

With some levels of complexity, the system was able to out perform both the active and passive benchmarks. Table 2 shows performance metrics achieved while varying linguistic and structural complexity parameters. Linguistic complexity refers to the granularity of the membership functions, and structural complexity refers to the relative number of rules and inputs per rule (this was controlled by selecting solutions from the Pareto front with different trade-offs between performance and complexity objectives).

As well as volatility, we consider risk using two more refined metrics. The first of these, the information ratio (IR), is superior to the commonly used Sharpe ratio (return/volatility) as it considers excess return. It is calculated as the portfolios average return in excess of the benchmark portfolio over the standard deviation of this excess return. It is used to evaluate the active stock-picking ability of the rulebase. The IR is calculated as: $IR = \frac{\sqrt{260}\alpha}{\sigma_e}$, where σ_e is the standard error of alpha in the capital asset pricing model expression of portfolio return given in the previous section and used to manage the alpha benchmark. Selectivity and Net Selectivity [15] provide further refinement of overall performance adjusted for risk. Any returns that a portfolio earns above the risk free rate are adjusted for both the returns that a market benchmark portfolio would earn if it had the same level of systematic risk and the same level of total risk.

Figure 3 shows differences in portfolio return and volatility due to complexity. Linguistic complexity is indicated by the number of membership functions

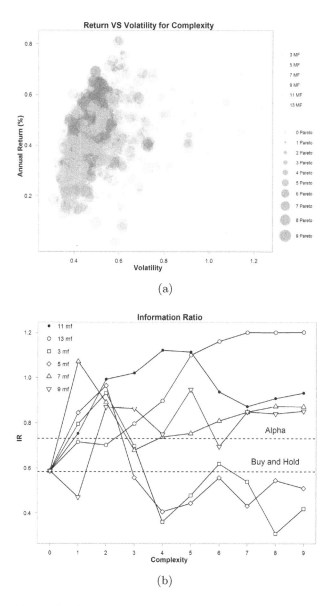

Fig. 2. The risk profile of portfolios managed with solutions of differing complexity. (a) shows Return vs. Volatility by complexity. (b) shows the Information Ratio of the portfolios and the benchmarks.

(i.e. 3 to 13 MF) and structural complexity by numbers (0 to 9 Pareto). Simpler models resulted in of lower return and (often) higher volatility, there was also lower stability between the different runs. The stability between different runs is an important risk as the method is stochastic. For more complex models higher return was observed at the cost of higher volatility (though here the information ratio shows the increase in volatility is justified by the return). In all cases there is a crux where increasing complexity above a certain level does not lead to gain and performance deteriorates. But only the average to higher linguistic complexity models provide the potential to out perform the benchmarks with appropriate levels of structural complexity.

4   Conclusion

In this paper we have described an evolutionary fuzzy system for portfolio management that first of all makes novel contributions by significantly building on earlier work in [2]. The experiments show the system performs well. There was considerable variation however for different levels of complexity and this can be used to hone performance. Almost all related research in finance is focused on relatively simple rules, we find such rules did not result in excess return above the benchmarks while the more complex models could. Therefore, it seems likely that controversy in some circles regarding the possibility of finding profitable rules somewhat misses the point, recent systems based on machine learning methods probably are able to do well by harnessing complexity. Another observation is that as the system approximates human reasoning, additional complexity may indicate to some extent why it is a fact that industry practitioners commonly attempt to generate profits by trading based on past data, despite there being in academia almost a consensus that there is limited possibility to do so.

When computational intelligence is used in algorithmic trading, it can lead to novel ways of controlling performance. For instance, as solution complexity is found to be a strong driver of risk and return, performance can be reliably shaped through identifying the locus where additional return starts to generate higher risk. It is also possible to use model complexity parameters improve stability and thus limit problems associated with the stochastic nature of the learning process which are often viewed as a drawback compared to other static modeling approaches in algorithmic trading.

References

1. Narang, K.: Inside the Black Box The Simple Truth About Algorithmic Trading. John Wiley & Sons, Inc., New York (2009)
2. Ghandar, A., Michalewicz, Z., Schmidt, M., To, T.-D., Zurbruegg, R.: Computational intelligence for evolving trading rules. IEEE Trans. Evolutionary Computation 13(1), 71–86 (2009)
3. Ghandar, A., Michalewicz, Z., Zurbruegg, R.: A case for learning simpler rule sets with multiobjective evolutionary algorithms. In: RuleML Europe, pp. 297–304 (2011)

4. Jin, Y., Sendhoff, B.: Pareto-Based Multiobjective Machine Learning: An Overview and Case Studies. IEEE Transactions on Systems, Man, and Cybernetics, Part C: Applications and Reviews 38(3), 397–415 (2008)
5. Lee, C.-C.: A self-learning rule-based controller employing approximate reasoning and neural net concepts. International Journal of Intelligent Systems 6(1), 71–93 (1991)
6. Ray, Ball: Anomalies in relationships between securities' yields and yield-surrogates. Journal of Financial Economics 6(2-3), 103–126 (1978)
7. Basu, S.: Investment performance of common stocks in relation to their price-earnings ratios: A test of the efficient market hypothesis. The Journal of Finance 32(3), 663–682 (1977)
8. Beaver, W., Lambert, R., Morse, D.: The information content of security prices. Journal of Accounting and Economics 2(1), 3–28 (1980)
9. Bhandari, L.C.: Debt/equity ratio and expected common stock returns: Empirical evidence. The Journal of Finance 43(2), 507–528 (1988)
10. Barberis, N., Shleifer, A., Vishny, R.: A model of investor sentiment. Journal of Financial Economics 49(3), 307–343 (1998)
11. Kavajecz, K., Odders-White, E.: Technical Analysis and Liquidity Provision. Rev. Financ. Stud. 17(4), 1043–1071 (2004)
12. Wilder, J.: New Concepts in Technical Trading Systems. Trend Research (1978)
13. Cordón, O., Gomide, F.A.C., Herrera, F., Hoffmann, F., Magdalena, L.: Genetic fuzzy systems. new developments. Fuzzy Sets and Systems 141(1), 1–3 (2004)
14. Zitzler, E., Laumanns, M., Thiele, L.: Spea2: Improving the strength pareto evolutionary algorithm. Tech. Rep. (2001)
15. Fama, E.F.: Components of investment performance. Journal of Finance 27(3), 551–567 (1972)

Bootstrapping Aggregate Fitness Selection with Evolutionary Multi-Objective Optimization

Shlomo Israel and Amiram Moshaiov

Faculty of Engineering, Tel-Aviv University, Israel
shlomois@post.tau.ac.il, moshaiov@eng.tau.ac.il

Abstract. Aggregate fitness selection is known to suffer from the bootstrap problem, which is often viewed as the main inhibitor of the widespread application of aggregate fitness selection in evolutionary robotics. There remains a need to identify methods that overcome it, while requiring the minimum amount of a priori task knowledge from the designer.

We suggest a novel two-phase method. In the first phase, it exploits multi objective optimization to develop a population of controllers that exhibit several desirable behaviors. In the second phase, it applies aggregate selection using the previously obtained population as the seed. The method is assessed by two non-traditional comparison procedures. The proposed approach is demonstrated using simulated coevolution of two robotic soccer players. The multi objective phase is based on adaptation of the well-known NSGA-II algorithm for coevolution. The results demonstrate the potential advantage of the suggested two-phase approach over the conventional one.

1 Introduction

A major goal of Evolutionary Robotics (ER) is to develop methods for automatically synthesizing autonomous robot systems. However, the majority of ER research has used fitness functions which incorporate moderate to high levels a priori knowledge about the task [1]. Solving generally complex problems without the incorporation of a priori knowledge is still an open problem. Specifically, aggregate fitness, which bases selection only on success or failure to complete the task (a very low level of incorporated a priori knowledge [1]), is known to suffer from the bootstrap problem, in which randomly initialized populations have no detectable level of fitness and thus cannot be evolved [1]. Overcoming this problem is often considered as one of the main challenges of ER [2].

Our proposed approach is associated with two existing methods: (a) one that initially relies on a bootstrapping component and later gives way to aggregate selection [3], and (b) one that uses Multi Objective Optimization (MOO) for incremental evolution, relying on a pre-defined decomposition of the task into sub-tasks [4].

We suggest a novel method building upon these concepts. First, it exploits MOO to evolve a population of controllers which exhibit several useful, non-task specific, behaviors (explore environment, avoid obstacles, etc.). Secondly, it applies aggregate

C.A. Coello Coello et al. (Eds.): PPSN 2012, Part II, LNCS 7492, pp. 52–61, 2012.

selection using the previously obtained population as the seed population. This reduces considerably the probability of the population from being sub-minimally competent. As MOO supports diverse search, the overall proposed method remains non-tailored in nature, and thus should scale better to complex problems.

The contributions of the current work are:

1. Proposing and analyzing a novel method for dealing with the bootstrap problem in ER, which offers a more generic approach than common practices.
2. Expanding a recently suggested procedure (equal-effort comparison), and utilizing a unique one (end-game comparison), for comparing co-evolutionary methods.
3. Introducing a slightly modified version of the well established NSGA-II algorithm that makes it suitable for use in co-evolution.

To demonstrate the proposed method, this work employs a competitive coevolution task, namely a soccer game, for which aggregate selection is a natural choice.

The reminder of this paper is organized as follows: Section 2 provides some relevant background. Section 3 presents the comparison procedures and the proposed method. Section 4 provides details on the simulation study and the outcomes of comparisons. Finally, section 5 provides conclusions and future research suggestions.

2   Background

2.1   Bootstrap Problem

When trying to solve a complex task using a high-level reward function, practitioners of evolutionary algorithms, and other search and optimization methods, often encounter a situation where no initial search pressure exists. Particularly in ER studies, such a bootstrap problem is said to occur if all individual controllers in the initial population are scored with null fitness [5], or with no detectable level of fitness [1], prohibiting the onset of evolution. Overcoming this problem is one of the main challenges of ER [2]. Mouret & Doncieux describe the attempts to overcome the bootstrap problem as different schemes of incremental evolution [2]. Nelson et al. [1] and Mouret & Doncieux [2], independently, suggested categories of the available schemes. Here we refer to that of [2] including four categories: staged evolution, environmental complexification, fitness shaping and behavioral decomposition.

In [3] the bootstrap problem is dealt with by adding a bootstrapping component that is active only when the population is sub-minimally competent. This strategy fits well into the fitness shaping category. In [2] and [4] the bootstrap problem is addressed by adding different objectives to the original fitness function. This approach may be categorized as a fifth category that we hereby call multi-objectivization [6]. In [6] multi-objectivization denotes solving difficult single objective problems with local optima by elevating fitness dimensionality. Here we expand this notion to generally supporting solution of numerical difficulties by adding objectives.

2.2   Multi-objective Evolution

ER problems may involve trade-offs between different and possibly conflicting objectives. In such cases the designers may set the goals as part of their problem definition and obtain a Pareto-optimal set of controllers (e.g. [7]). Here the interest in contradicting objectives, and in the corresponding non-dominated set, is different. We explore their potential to serve as a booster for a single objective evolution. Existing Multi-Objective Evolutionary Algorithms (MOEAs) should support the initial phase of the suggested search method. For this purpose, we slightly modified the well established NSGA-II algorithm [8] to suit it for use in co-evolution (see section 3.1).

2.3   Khepera Robot Soccer

Aggregate selection is a natural choice for competitive co-evolution. Thus, we chose to experiment with a domain from this category. The experimental work reported herein is conducted on a robotic soccer-like game following [9]. In [9], two simulated Khepera robots are competing to scoring more goals than each other. In Fig. 1. the robots are depicted as pacman-like symbols chasing the light colored circle (ball).

The black sector within each robot's symbol shows the Field Of View (FOV) of its ball sensor. The inputs to the controller are sensor primitives including: proximity, ball direction and width (when in FOV), "stuck", and "goal direction". The outputs are motor primitives: translation speed, rotation speed, and rotation direction. The controller is a fixed-structured tree of behavior modules. The top behavior modules act as arbitrators, propagating control resolutions downwards, and the bottom modules activate primitive behaviors (motor primitives).

In [9], the fitness for one round was the sum of three components. The first component is meant to dominate the fitness if a goal is scored (as in aggregate selection). The second component favors earlier goals, and the last component favors the robots that stay close to the ball (a bootstrapping component). The authors indicate that the design of the function was a guess based on intuition and it "turned out to serve its purpose". This type of selection employs fitness shaping for bootstrapping, leaving us with an opportunity to change the method and examine if it is beneficial.

A practical factor in choosing this domain was that its source code has been available through a website provided in [9]. It made it possible for us to test alternative evolutionary schemes against the original one. We did encounter some problems with the simulator and introduced corrections, but nothing on the high-level was changed.

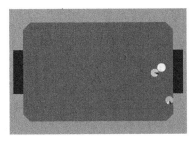

Fig. 1. A snapshot of the domain (as obtained using the website provided in [9])

3   Methodology

3.1   Comparison Procedures

In section 4 of this study several methods are compared, including random search vs. co-evolution, and the proposed two-phase approach vs. co-evolution. In both cases co-evolution is done as in [9]. To compare results the following techniques are used. Their implementation is available at: http://bit.ly/LC0LYA.

Equal-Effort Comparison. Several methods have been utilized to measure progress in coevolution, of which most popular are Masters Tournament [11] and Dominance Tournament [12]. But lately these methods have been criticized to measure historical progress while implying to measure global progress [13]. According to [13], the common fallacy of these methods is that they measure performance against previous opponents – that is, the sequence of successive opponents against which they have evolved. In this case, the training set is being used as a test set.

We decided to turn to alternative methods that do not involve a comparison of individuals with the opponents with which they have coevolved. The underlying assumption here is that separation of co-evolved populations, during a posteriori analysis, will bring us closer to assessing the global progress instead of the historical one.

In this context we utilize the equal-effort comparison technique [13]. This technique is designated to compare the performance of two methods by opposing the champions of one method (its hall of fame) against the champions of the other, and tracking the respective number of victories over the course of coevolution, using several independent runs. The competition is conducted between individuals that were obtained after an equal number of evaluations, thus the name equal-effort comparison.

On top of the basic graph that results from this technique we added some aspects of our own. These include the addition of 95% confidence intervals to the graph, and running a moving average for 5 generations to smoothen the curve. Also, we used a stop condition, which is introduced in the following sub-section, to decide how many independent runs are enough.

End-Game Comparison. Reaching statistical significance is a justifiable requirement; yet, it is not always customary in ER studies. With this respect, we introduce a technique for statistically comparing the performance of two methods by opposing their best performing representatives. The proposed technique follows the principal not to compare individuals with the opponents with which they have coevolved, similarly to the previous technique. As seen in Fig. 2. , for each method we gather the best 5 individuals out of the final population of each run, in a set, which is denoted S^i_j, where i is the method index (i=A or B) and j is the run index (j=1,...,N). We wish to ensure that N is a statistically significant number of runs but to keep it to a minimum. Thus, N is subject to a lower limit. An upper limit for N is also used to manage the computational efforts.

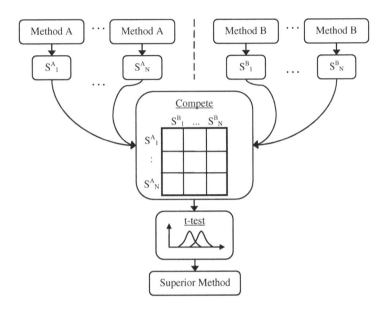

Fig. 2. The sequence of end-game comparison

At i=1, 5x5 competitions are carried out between all pairs from the two sets S^A_1 and S^B_1. At i=2 additional competitions are carried out to complete all possible pairings with the newly introduced individuals and with the existing ones, from i=1. The accumulated results are depicted as a matrix in the 'Compete' block in Fig. 2. . At each change of the run index we calculate an averaged fitness within each best-5 set. To reach statistical significance we run a t-test using the obtained averaged fitness values and cease if the p-value reaches 0.05. If the procedure has ceased before N reached its upper limit then the winning method is the one with the higher mean fitness. Otherwise, the comparison is declared as indecisive.

3.2 Two-Phase Method

Motivated by the idea that MOO supports diverse search, as demonstrated in [2] and [4], we propose a different use of MOO as a novel approach for bootstrapping. Some a priori knowledge is still required, as in any bootstrapping method. However, we aim to make our method more generic and less sensitive to disruptive assumptions.

Our approach is comprised of two phases. First, it exploits MOO to develop a population of controllers evolved to exhibit several desirable behaviors. Secondly, it applies aggregate selection using the previously obtained population as the seed population. We propose to choose the objectives for the first phase from a set of generally desirable behaviors in ER. These may be 'explore environment', 'avoid obstacles' and 'approach target'. Of course these behavioral categories should be realized differently for specific domains, but we claim it is not difficult to design such objectives following a generic guideline - encourage useful behaviors which are not task specific. In Table 1, for instance, we suggested realizations for the domain used in this study.

In contrast to [2] and [4], we propose to multiobjectivize the problem only in early stages of evolution, disregarding the original objective in principal, and then proceed with the original objective alone. This approach is applicable in a wide variety of situations because it doesn't alter the whole algorithm but rather adds a preliminary phase.

For the MOO phase we used an open-source java-based framework, namely jMetal [10]. We slightly modified the well established NSGA-II algorithm to make it suitable for use in co-evolution. We call this variation ceNSGA-II, which means co-evolutionary NSGA-II. The difference lies within how the offspring are evaluated. We let the offspring play three matches (to reduce the influence of randomness): two against themselves and one against each other, such that their objective-values are determined according to the worst score of these three trials. A short examination suggested that in this manner the population is driven towards desirable behaviors.

4   Simulation Study

4.1   Random Search vs. Co-evolution

An initial confidence examination has been conducted to illustrate that co-evolution offers a clear advantage over random search. For this purpose two populations have been co-evolved for a fixed amount of generations, using the procedure of [9]. Then we generated a random population using the same amount of evaluations as was required for co-evolution. The evaluations of the randomly generated individuals served the purpose of establishing the inner ranking of the random population (about 5-10 evaluations per individual to extract the fittest out of a population of 10-20 thousand individuals).

Next, we ran the end-game comparison technique between co-evolution and random search. The comparison was done after 65,000 evaluations (equivalent to 50 co-evolution generations). The number of runs, N, was set in the range of [10, 30]. Meaning that the stop condition was not checked if there were less than 10 entries and that at most 30 runs were made.

As expected, the co-evolutionary method has proven to be superior (after 10 runs with a p-value reaching 10^{-6}). This unequivocal result is not surprising when considering the size of the search space, which makes it hard for randomly generated individuals to succeed.

4.2   Objective Combinations

For the main purpose of this study we explore which objective combinations may potentially lead to better results. We begin by listing, in Table 1, several broad categories of desirable behaviors in ER, including: (a) explore environment, (b) avoid obstacles, and (c) approach target. Then, also in Table 1, we list some possible realizations of these behaviors, as applied to the current particular domain. The realizations are formulated as objectives to be minimized. In the rest of this paper we denote formula (1) by "centerMove" or obj. 1, formula (2) by "proxiSensor" or obj. 2, etc..

Table 1. Summary of suggested behavior categories and corresponding realizations for our particular domain. The realization formulas are objectives to be minimized and are negative by convention; therefore some formulas are normalized by offest and factors.

Behavior Categories	Domain Realizations	Realization Formulas (Objectives)	
Explore environment	Maximize linear speed	$-\sum_{t=1}^{T}\lvert centerMove(t)\rvert$	(1)
Avoid obstacles	Minimize proximity sensors;	$\sum_{t=1}^{T}\sum_{i} proxiSensor_i(t)$	(2)
	Minimize "stuck" sensor	$\sum_{t=1}^{T} stuck(t)-1$	(3)
Approach target	Minimize diversion from goal;	$\sum_{t=1}^{T}\lvert direction(t)\rvert -180\Big/180$	(4)
	Maximize ball size in FOV;	$-\sum_{t=1}^{T} ballSize(t)\Big/64$	(5)
	Minimize diversion from goal when ball in FOV	$\sum_{t=1}^{T}\dfrac{ballSize(t)}{64}\cdot\dfrac{\lvert direction(t)\rvert-180}{180}$	(6)

Next, we employed ceNSGA-II with several bi-objective combinations. The bi-objective approach is used to make the problem easier to handle and analyze, yet the method is not restricted to such an approach. The realization formulas rely on sensory data available to the controller and are mostly self explanatory, except of formula (6). The later formula is the result of an amalgamation of two objectives that seemed reasonable to combine.

The process of evaluating the bi-objective combinations was carried out by examining the non-dominated fronts in objective space and determining which combinations lead to well diversified fronts. We have found "proxiSensor" to be unsuccessful since the ball also activates the proximity sensors. We also found that the objectives "direction" and "ballSize" work better together than separately. This is because it is favorable to head towards the goal when the ball is in close sight.

In Fig. 3., typical performance distributions are shown for two different bi-objective combinations. We should note that the combination {obj. 1, obj. 6} had better diversification not only among the shown two cases but also when compared with other combinations; thus we proceeded with it.

The following search parameters are relevant for the study in this section and the following one: for selection we used the same binary selection operator as in NSGA-II (jMetal's 'BinaryTournament2'); for mutation we used a uniform mutation operator (jMetal's 'BitFlipMutation' operator for integer representation) with probability of 1/638; for crossover we used jMetal's 'SinglePointCrossover' operator with probability 0.9. The operator types and values were adopted from jMetal's examples.

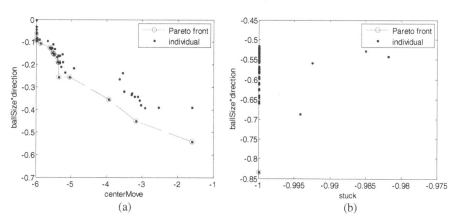

Fig. 3. Typical performance distribution for two bi-objective combinations: (a) {obj.1, obj.6 } and (b) {obj.3, obj.6 }. The instantaneous (local) Pareto front is marked with a dashed line and with circles around the performance vectors which comprise it.

4.3 Two-Phase vs. Co-evolution

Once reaching a seemingly promising objective combination, {obj. 1, obj. 6}, we set to examine its success in the full context of the two-phase method. First we utilized for this purpose the end-game comparison technique, using as competing methods the two-phase method and the original co-evolutionary method from [9], and setting N in the range [10,30].

Table 2 shows the results for the chosen bi-objective combination, for two independent comparisons which differ only in the total number of evaluations. The difference in fitness is due to the 5,000 evaluations used in the 1st phase. It is emphasized that the total number of evaluations in the co-evolution method is equal to the sum of evaluations made by the two-phase method, in both phases.

Table 2. End-game comparisons between the co-evolution and the two-phase methods. Two comparisons appear in the table separated by a dashed line. The 1st row, in each comparison, refers to the co-evolutionary method, and the 2nd to the two-phase method. The 3rd till the 6th columns show how many generations and evaluations were used in the MOO phase and the co-evolution phase respectively. The last two columns under 'comparison' show the average fitness and its difference as obtained in each method. The symbol * that appears in the last column indicates significance at the 5% level; the corresponding t-statistic appear in parentheses.

Comp. №	Method	MOO phase		co-evo. phase		comparison	
		gen.	eval.	gen.	eval.	Avg. fitness	Fitness diff.
1	co-evo.	0	0	50	65,000	1594.0	1247.8*
	two-phase	50	5,000	46	60,000	2841.8	(2.27)
2	co-evo.	0	0	200	250,000	2438.5	143.8
	two-phase	50	5,000	196	245,000	2582.3	(0.45)

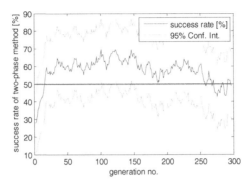

Fig. 4. Equal-effort comparison results showing the success rate of the two phase method and the corresponding 95% confidence intervals versus generations. Generations are counted in terms of co-evolution (not to confuse with the notion of generations in the MOO phase). Success in a match is counted when a champion of the two-phase method defeats the corresponding champion of the co-evolutionary method; disregarding the fitness difference that led to the victory. Here the success rate was calculated based on N=30 independent runs.

It is evident from the table that initially (after 65,000 evaluations – equivalent to 50 co-evolution generations) the two-phase method is superior. However, there is no such clear conviction in the longer case (after 250,000 evaluations – equivalent to 200 generations). To get an idea of the required computational effort we note that one generation is evaluated at approximately 7 seconds on a Core i5 CPU @ 2.66GHz.

An equal-effort comparison is carried out to better understand when the advantage of the two-phase method fades. The results for a 375,000 evaluations long run (equivalent to 300 co-evolution generations) are presented in Fig. 4.

As expected when there is no clear winner, the number of runs was 30. From the graph it is evident that the two-phase method rises to about 70% success rate in 100 generations, but then in about 80 generations it regresses to the 50% equilibrium. This analysis is consistent with the results present in Table 2. The results indicate that the two-phase method has an advantage over the original method in the early stages of co-evolution, but this advantage fades as co-evolution progresses.

5   Conclusions and Future Research

A novel bootstrapping approach is proposed based on multi-objectivization and a two-phase evolution scheme. The suggested approach is compared with a conventional one using co-evolution of robot controllers for a soccer game. The conventional method employs an intuitive domain-specific bootstrapping component, whereas the proposed approach aims to be generic.

In this study we put emphasis on carefully designed and statistically significant comparisons. Using such comparisons, our results confirm that the two-phase method is better in bootstrapping the evolutionary process. We also demonstrated that it is possible to encourage the acquisition of useful behaviors without losing valuable portions of the search space. However, from the erosion in the advantage of the two-phase method we conclude that, at least in our simulation study, it did not succeed in expressing unique behaviors unreachable by conventional means after enough generations.

In regard to ceNSGA-II, it is recommended that in future research various alternatives to evaluate and compare offspring performances will be examined, and the co-evolution algorithm will be optimized. Regarding the MOO phase, it is conceivable that in more complex domains its positive effect will last longer and even persist. Further research is needed, with various domains and fitness aggregation, to further assess the proposed technique, based also on CPU time comparisons. Such studies are also expected to enable automatic selection of satisficing objective combinations in a generic manner.

References

1. Nelson, A.L., Barlow, G.J., Doitsidis, L.: Fitness functions in evolutionary robotics: A survey and analysis. Robotics and Autonomous Systems 57, 345–370 (2009)
2. Mouret, J.B., Doncieux, S.: Overcoming the bootstrap problem in evolutionary robotics using behavioral diversity. In: 2009 IEEE Congress on Evolutionary Computation, CEC 2009, pp. 1161–1168 (2009)
3. Nelson, A.L., Grant, E.: Using direct competition to select for competent controllers in evolutionary robotics. Robotics and Autonomous Systems 54, 840–857 (2006)
4. Mouret, J.B., Doncieux, S.: Incremental evolution of animats' behaviors as a multi-objective optimization. From Animals to Animats 10, 210–219 (2008)
5. Nolfi, S.: Evolutionary robotics: Exploiting the full power of self-organization. Connect. Sci. 10, 167–184 (1998)
6. Knowles, J.D., Watson, R.A., Corne, D.W.: Reducing Local Optima in Single-Objective Problems by Multi-objectivization. In: Zitzler, E., Deb, K., Thiele, L., Coello Coello, C.A., Corne, D.W. (eds.) EMO 2001. LNCS, vol. 1993, pp. 269–283. Springer, Heidelberg (2001)
7. Moshaiov, A., Ashram, A.: Multi-objective evolution of robot neuro-controllers. In: 2009 IEEE Congress on Evolutionary Computation, CEC 2009, pp. 1093–1100 (2009)
8. Deb, K., Pratap, A., Agarwal, S., Meyarivan, T.: A fast and elitist multiobjective genetic algorithm: NSGA-II. IEEE Transactions on Evolutionary Computation 6, 182–197 (2002)
9. Østergaard, E.H., Hautop Lund, H.: Co-evolving Complex Robot Behavior. In: Tyrrell, A.M., Haddow, P.C., Torresen, J. (eds.) ICES 2003. LNCS, vol. 2606, pp. 308–319. Springer, Heidelberg (2003)
10. Durillo, J.J., Nebro, A.J., Alba, E.: The jmetal framework for multi-objective optimization: Design and architecture. In: 2010 IEEE Congress on Evolutionary Computation, CEC 2010, pp. 1–8 (2010)
11. Nolfi, S., Floreano, D.: Coevolving Predator and Prey Robots: Do "Arms Races" Arise in Artificial Evolution? Artif. Life 4, 311–335 (1998)
12. Stanley, K.O., Miikkulainen, R.: The dominance tournament method of monitoring progress in coevolution. In: GECCO 2002, pp. 242–248 (2002)
13. Miconi, T.: Why Coevolution Doesn't "Work": Superiority and Progress in Coevolution. In: Vanneschi, L., Gustafson, S., Moraglio, A., De Falco, I., Ebner, M. (eds.) EuroGP 2009. LNCS, vol. 5481, pp. 49–60. Springer, Heidelberg (2009)

Network Topology Planning Using MOEA/D with Objective-Guided Operators

Wei Peng[1,*] and Qingfu Zhang[2]

[1] School of Computer, National University of Defense Technology,
Changsha, Hunan, 410073, China
`wpeng@nudt.edu.cn`
[2] School of Computer Science & Electronic Engineering, University of Essex
Colchester, Essex, CO4 3SQ, United Kingdom
`qzhang@essex.ac.uk`

Abstract. Multiobjective evolutionary algorithms (MOEAs) have attracted growing attention recently. Problem-specific operators have been successfully used in single objective evolutionary algorithms and it is widely believed that the performance of MOEAs can be improved by using problem-specific knowledge. However, not much work have been done along this direction. Taking a network topology planning problem as an example, we study how to incorporate problem-specific knowledge into the multiobjective evolutionary algorithm based on decomposition (MOEA/D). We propose *objective-guided operators* for the network topology planning problem and use them in MOEA/D. Experiments are conducted on two test networks and the experimental results show that the MOEA/D algorithm using the proposed operators works very well. The idea in this paper can be generalized to other multiobjective optimization problems.

Keywords: Multiobjective Optimization, Evolutionary Algorithm, MOEA/D, Network Topology Planning.

1 Introduction

Multiobjective optimization problems (MOPs) present a greater challenge than single-objective optimization problems since objectives in a MOP contradict one another so that no single solution in the decision space can optimize all the objectives simultaneously. Therefore, a decision maker often wants to find an optimal tradeoff among these objectives. Pareto optimal solutions are best tradeoffs if there is no decision makers' preference information. A number of different Multiobjective evolutionary algorithms (MOEAs), such as NSGA-II [1] and MOEA/D [2][3], have been developed for approximating the Pareto optimal solution set.

* This work was supported in part by the National Natural Science Foundation of China under grant No.61070199, 61103188, 61103189 and 61100223, and Hunan Provincial Natural Science Foundation of China under grant No.11JJ7003.

C.A. Coello Coello et al. (Eds.): PPSN 2012, Part II, LNCS 7492, pp. 62–71, 2012.

It is widely believed that problem-specific knowledge should be utilized in designing of an evolutionary algorithm in order to improve the algorithm performance. In single objective evolutionary optimization, many successful applications of problem-specific knowledge have been reported. However, not much work have been done along this direction in MOEAs. One of the major reasons is that most existing problem-specific techniques are for single objective optimization. It is not very natural to use them in Pareto dominance based MOEAs, which are most popular methods now. By decomposing a MOP into many single objective optimization subproblems, the recent MOEA/D algorithm provides a good framework for using single objective optimization techniques for multiobjective optimization. Many different variants have been proposed and applied on different MOPs [4][5]. In this paper, we take topology planning problem as an example and study how to incorporate problem-specific knowledge into MOEA/D.

Planning a network topology involves multiple objectives including minimizing the total network cost, maximizing the transport efficiency and the network reliability, etc. In transparent optical networks (TONs), more objectives should be considered, such as the security under intentional attacks [6] and energy-efficiency for a green network planning [7][8].

Network topology planning is normally formulated as bi-objective optimization problems. Kumar et al. firstly used the Pareto Converging Genetic Algorithm (PCGA) to solve the problem [9], then applied a multi-island approach with Pareto ranking method [10]. The PCGA was also used in the problem with consideration of realistic traffic models [11][12].

In this paper, we study a network topology planning problem in TONs. It is formulated as a tri-objective optimization problem. The MOEA/D algorithm with generic graph operators is firstly designed to tackle the problem. Then, we propose *objective-guided operators* and use them in the MOEA/D algorithm for this problem. The main idea is to design operators using problem-specific knowledge for all objectives, and then use them with different probabilities for each scalar subproblem in MOEA/D. Due to the decomposition approach in MOEA/D, the probabilities can be easily determined with the weight vectors associated with subproblems. The proposed algorithms are evaluated on two networks with different sizes. The experimental results demonstrate the effectiveness of the MOEA/D algorithm using the *objective-guided operators*. Our algorithm provides a new approach for using problem-specific knowledge in MOEAs.

The rest of the paper is organized as follows. The problem is formulated in Section 2. Section 3 describes the MOEA/D algorithm using generic graph operators. The MOEA/D algorithm using objective-guided operators is proposed in Section 4. The experimental results are presented in Section 5 and the paper is concluded in Section 6.

2   Problem Formulation

Topological design of a transparent optical networks (TON) for meeting the requirements of consumers is a fundamental task before its deployment. Network topology design involves determining the layout of links between nodes to satisfy the requirements of average delay, cost and reliability. In packet-switching networks, the average delay between a source and a destination can be estimated from queuing theory. Normally, the transferring delay is affected by the number of intermediate hops and the traffic load on links made of a path. However, due to the circuit-switching nature of optical networks, the average delay is mainly determined by the hops a lightpath traverses.

Besides the traditional design objectives, energy and security-related issues have gained much attention in recent years. For the green sustainable communication purpose, energy consumption of telecom networks should be reduced as much as possible. Since both switching and transmission on fibers consume power, one should try to minimize the number of intermediate hops and reuse the network link which has been already "on" in routing of lightpaths. From the view of topological design, we should try to minimize both the hops and the number of links. Apparently, they are two contradictory objectives.

Random failures of nodes and links are main factors for reliability of a network. When security is demanded, intentional attacks should be emphasized in topology design. It has been shown that a scale-free complex network is robust from random failures, but fragile under intentional attacks, e.g., removing nodes by node degrees in descending order. So, in this paper, we consider the topology design problem with new objectives including energy-saving and security.

The network topology problem is formally defined below.

1) Design Parameters

- N: the total number of nodes in a network;
- *Cost*: a cost matrix in which $Cost(i, j)$ provides the cost of the link between node i and j. Normally, the cost of link between node i and j can be estimated by their physical distance and the cost factor per unit distance;

2) Objectives

- *network cost*: the sum of cost of all links;
- *average path length*: the average hops of each path. Since there may be unreachable node pairs in a network, we use the concept of network efficiency to calculate the objective value. The network efficiency is defined as the mean value of inverse values of shortest path lengths in a graph. Given a graph $G=(V, E)$, the average path length is measured by:

$$Lp(G) = 1 - \frac{1}{n(n-1)} \sum_{i \neq j \in V} \frac{1}{l_{i,j}} \tag{1}$$

where $n=|V|$ and $l_{i,j}$ is the shortest path length from node i to node j;
- *Vulnerability under intentional attacks*: we use the robustness measure R proposed by Schneider et al. [13], which is defined as:

$$R = \frac{1}{N+1} \sum_{Q=0}^{N} s(Q) \tag{2}$$

where N is the number of nodes in a network and $s(Q)$ is the fraction of nodes in the largest connected cluster after removing Q nodes. The range of R is $[0, 0.5]$, where $R = 0$ corresponds to a network with all nodes isolated, and $R = 0.5$ corresponds to a fully connected network. The vulnerability of a network under intentional attacks is calculated by:

$$Vu(G) = 1 - 2R \tag{3}$$

The average delay is measured by the average path length, while minimizing the average path length also contributes to the saving of energy. To calculate the shortest path lengths, the Dijkstra's shortest-path algorithm is used. To calculate the robustness measure R, a greedy attacking strategy is applied. That is, each time the node with the maximal nodal degree is selected and removed from the network and the size of the largest connected cluster is calculated.

3   MOEA/D for Network Topology Planning

The original MOEA/D algorithm [2] can be directly applied to solve the network topology planning problem. We call it as *MOEA/D-direct* algorithm. Each individual in the population encodes a possible network topology. A network topology is represented by its binary adjacency matrix A where $A_{i,j} = 1$ if there is a link between node i and j. Generic graph operators are used.

Let the adjacency matrices of two parents be A and B and the adjacency matrix of offspring be C. The crossover operator produces an offspring C as follows:

$$C_{i,j} = \begin{cases} A_{i,j} & , \quad r \leq p_c \\ B_{i,j} & , \quad otherwise \end{cases} \tag{4}$$

where r is a uniformly random value in $[0, 1]$. The parameter p_c is used to control the amount of information inherited from each parent. The offspring inherits from A with a probability of p_c and from B with a probability of $(1 - p_c)$.

In the mutation operator used in this paper, the bits in the adjacency matrix are flipped with a mutation probability p_m. Since undirected networks are considered in this paper, the adjacency matrices are symmetric. Thereafter, the crossover and mutation operators are applied only for elements when $i < j$, and we always let $C_{j,i} = C_{i,j}$.

The generated network topology may not be connected. In reality, a network may not be connected at its initial construction stage due to insufficient finance. Thus, disconnected networks are considered as valid solutions to the problem in this paper so that the algorithm can be simplified.

Parents for crossover are selected using a strategy slightly different from that in the original MOEA/D algorithm, whose details can be found in [2]. To generate the i-th offspring, we select an index k randomly from the neighborhood $B(i)$, and then use the i-th individual as the first parent and the k-th individual as the second parent. After the offspring is generated, its objective values are calculated. Then, the i-th and the k-th individual are updated if the new individual is better than them.

The population is initialized uniformly at random. When an individual is initialized, any two nodes are connected with a probability p_r. Assume N_{pop} individuals will be generated at the initial stage, then the probability p_r for the i-th individual is set to i/N_{pop}. After an individual is generated, its objective values are calculated and all individuals in the population are updated according to their weighted objective values.

4   MOEA/D with Objective-Guided Operators

It is commonly believed that algorithms using problem-specific knowledge can achieve much better performance than a generic algorithm. Therefore, heuristic local search algorithms should be used within a MOEA or genetic operators should be designed specially for a specific application problem. However, heuristic operations often optimize only one objective at a time. For MOPs, we need to optimize multiple objectives simultaneously. Using an operator to optimize one objective has been proposed in [14] for multiobjective 0/1 knapsack problems. Using different operators for different parts of the Pareto Front (PF) in MOEA/D has been investigated in [15].

However, There is still lack of general guidelines for designing problem-specific operators in a MOEA. In this paper, we propose *objective-guided operators* to utilize problem-specific knowledge in MOEAs. Since an operator that optimizes only one objective can often be designed easily, the main idea is to design one operator for each objective and then use them alternatively.

More specifically, we design the following operators for the studied problem.

- **Operator for objective 1:** The first objective is to minimize the total network cost. Thus, the operator selects a node i_0 randomly at first. The most expensive link connected with it is then removed;
- **Operator for objective 2:** The second objective is to minimize the average path length. For this objective, we randomly select a node from the network and compare its degree with its neighboring nodes. The node with the maximal degree is called a *local hub node*. We select two local hub nodes which are not connected, then add a link between them. By connecting 'hub' nodes, the average path length can be shortened with a few new links;
- **Operator for objective 3:** The third objective is to reduce the vulnerability of the network to the maximal extent. We firstly find the node with the minimal degree in the network, then connect it to an unconnected node with the minimal cost. Since the link cost is proportional to the distance between two nodes, the node with the minimal cost will be the nearest node from it.

MOEA/D decomposes a MOP into a number of scalar optimization subproblems. Each subproblem has a weight vector which sets weights on different objectives. The weight vector represents the preference of the subproblem on different objectives. Since different subproblems have different preference on objectives, we can not apply the objective-guided operators on different subproblems in the same way. Instead, we use them based on the preference of the objectives. We design an *objective-guided mutation* operator which is illustrated as follows.

In MOEA/D, the i-th individual x^i is to find the optima of the i-th subproblem with the weight vector $\lambda^i = (\lambda_1^i, \ldots, \lambda_m^i)$ where m is the number of objectives. Normally, $\lambda_j^i \in [0, 1]$ and $\sum_{j=1}^m \lambda_j^i = 1$. The j-th element in the weight vector represents the preference on the j-th objective. Thus, in the *objective-guided mutation* operator, we use different operators according to the value of the weight vector. In implementation, we generate a new offspring y from the i-th individual using the following steps:

For $k = 1, \ldots, r_{num}$, **do**
    Apply the j-th operator on x^i with
    probability= λ_j^i;
---

where the j-th operator is designed for the optimization of the j-th objective. r_{num} is a control parameter which determines the number of iteration in the objective-guided mutation operator.

5 Evaluation

5.1 Experiment Setting

To demonstrate the effectiveness of the MOEA/D algorithm using objective-guided operators (called as *MOEA/D-guided*), we conduct experiments on two networks with different sizes. The parameter setting is shown in Table 1.

Table 1. Parameters of Algorithms

Variable	Description	Value
N	total number of nodes in a network	33, 340
N_{pop}	population size (also number of subproblems)	66
N_{eval}	number of function evaluations	50000
T	size of neighborhood in MOEA/D	5
p_c	probability in crossover operation	0.5
p_m	probability in mutation operation	0.05
r_{num}	iteration number in objective-guided mutation	10

The performance metrics include [2]:

- Set Coverage (C-metric): Let A and B be two approximations to the PF of a MOP. $C(A, B)$ is defined as the percentage of the solutions in B that are dominated by at least one solution in A, i.e.

$$C(A, B) = \frac{|\{u \in B | \exists v \in A : v \ dominates \ u\}|}{|B|} \tag{5}$$

- Distance from Representatives in the PF (IGD-metric): Let P^* be a set of uniformly distributed points along the PF. Let A be an approximation to the PF, the average distance from A to P^* is defined as:

$$D(A, P^*) = \frac{\sum_{v \in P^*} d(v, A)}{|A|} \qquad (6)$$

where $d(v, A)$ is the minimum Euclidean distance between v and the points in A. Here the definition is slightly different than that in [2] in order to consider the effect of size of solution set.
- Size of Solution Set: number of non-dominated solutions found.

The quality of two solution sets with respect to Pareto dominance can be compared using the C-metric. The D-metric could measure both the diversity and convergence of a solution set in a sense. Since the actual Pareto fronts of the test networks are not known, we use an approximation of the PF as P^*. The approximation of PF is obtained from all non-dominated solutions found in all the runs of the two algorithms. The size of solution set reveals the ability of algorithm to find non-dominated solutions.

5.2   Experiment Results

Figure 1 shows the distribution of solution set found in one run on a network with 33 nodes. It can be seen that the MOEA/D-guided algorithm has obtained better distributed solutions and most solutions obtained by the MOEA/D-direct algorithm are dominated by those by the MOEA/D-guided algorithm.

On a network with 340 nodes, the results are quite different. As shown in Figure 2, the solutions of the two algorithms have very different distributions. The MOEA/D-direct algorithm has produced very few solutions with low values of objective 1 (network cost). The MOEA/D-guided algorithm has generated a lot of solutions distributed in the objective space with low values of objective 1 and relatively large values of objective 2 and 3. Thus, we can still conclude that the MOEA/D-guided algorithm has better exploration ability than the MOEA/D-direct algorithm on this test instance.

The performance metrics are averaged over 10 independent runs. The results are shown in Figure 3. With the increase of the function evaluation number, the C-metric of MOEA/D-guided vs MOEA/D-direct increases too. It implies that more solutions of MOEA/D-direct are dominated by solutions of MOEA/D-guided if more computational efforts are made. In the case of network with 33 nodes, the C-metric of MOEA/D-direct versus MOEA/D-guided is zero and not shown in Figure 3a.

The D-metric values of MOEA/D-guided are lower than those of MOEA/D-direct in the case of 33 nodes. It means that the solution set of MOEA/D-guided is more close to the approximated Pareto front. The D-metric values of MOEA/D-guided are higher than those of MOEA/D-direct in the case of 340 nodes. However, the values are getting closer with the increase of function evaluation number. Obviously, the complexity of the problem has increased when the

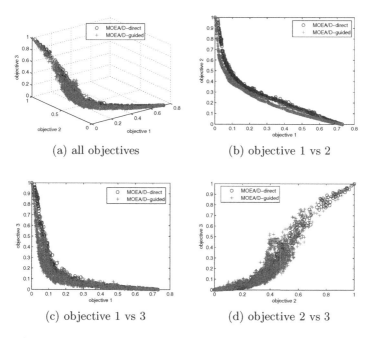

(a) all objectives

(b) objective 1 vs 2

(c) objective 1 vs 3

(d) objective 2 vs 3

Fig. 1. Results on network with 33 nodes

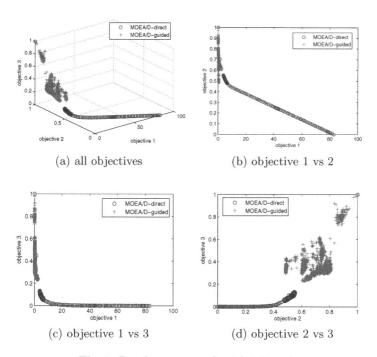

(a) all objectives

(b) objective 1 vs 2

(c) objective 1 vs 3

(d) objective 2 vs 3

Fig. 2. Results on network with 340 nodes

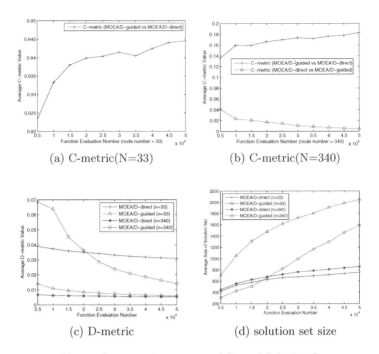

(a) C-metric(N=33)   (b) C-metric(N=340)

(c) D-metric   (d) solution set size

Fig. 3. C-metric, D-metric and Size of Solution Set

network scale increases. The higher D-metric values in MOEA/D-guided may be explained by the solution set size. AS shown in Figure 3d, the final solution sets of MOEA/D-guided in both cases are larger than those of MOEA/D-direct. With solutions distributed more widely, the distance from one solution to the approximated Pareto front is more likely large. The larger solution set of MOEA/D-guided also demonstrates its stronger exploration ability.

6   Conclusion

In this paper, we have investigated how to utilize problem-specific knowledge in multi-objective evolutionary algorithms. Specifically, we have studied the network topology planning problem using the MOEA/D algorithm. The algorithm was firstly applied on the problem using generic graph operators. Then the objective-guided operators were proposed and a way to use them in MOEA/D was proposed. The main idea is to design one operator for each objective and use the operators based on the weight vectors of subproblems in MOEA/D. Experimental results on networks of different scale have shown the superiority of the MOEA/D-guided algorithm which uses objective-guided operators.

Future work includes further improvement of the algorithm on large-scale problem instances. Other problem-specific operators may be incorporated and other approaches to exploit problem-specific knowledge can be investigated.

References

1. Deb, K., Pratap, A., Agarwal, S., Meyarivan, T.: A fast and elitist multi-objective genetic algorithm: NSGA-II. IEEE Trans. on Evol. Comp. 6(2), 182–197 (2002)
2. Zhang, Q., Li, H.: MOEA/D: A Multiobjective Evolutionary Algorithm Based on Decomposition. IEEE Tran. on Evol. Comp. 11(6), 712–731 (2007)
3. Li, H., Zhang, Q.: Multiobjective Optimization Problems with Complicated Pareto Sets, MOEA/D and NSGA-II. IEEE Trans. on Evol. Comp. 12(2), 284–302 (2009)
4. Zhang, Q., Liu, W., Tsang, E., Virginas, B.: Expensive Multiobjective Optimization by MOEA/D with Gaussian Process Model. IEEE Trans. on Evol. Comp. 14(3), 456–474 (2010)
5. Ishibuchi, H., Sakane, Y., Tsukamoto, N., Nojima, Y.: Evolutionary many-objective optimization by NSGA-II and MOEA/D with large populations. In: Proc. of 2009 IEEE Int. Conf. on Systems, Man, and Cybernetics, pp. 1820–1825 (2009)
6. Skorin-Kapov, N., Chen, J., Wosinska, L.: A New Approach to Optical Networks Security: Attack-Aware Routing and Wavelength Assignment. IEEE/ACM Trans. on Networking 18(3), 750–760 (2010)
7. Tomkos, I., Angelou, M.: New Challenges in Next-Generation Optical Network Planning. In: Proc. of 12th Int. Conf. on Transparent Optical Networks (ICTON 2010), Munich, Germany, pp. 1–4 (2010)
8. Bathula, B.G., Elmirghani, J.M.H.: Green Networks: Energy Efficient Design for Optical Networks. In: Proc. of Sixth IEEE/IFIP Int. Conf. on Wireless and Optical Communications Networks (WOCN 2009), Cairo, Egypt, pp. 1–5 (2009)
9. Kumar, R., Parida, P.P., Gupta, M.: Topological Design of Communication Networks using Multiobjective Genetic Optimization. In: Proc. of the 2002 Congress on Evolutionary Computation, Honolulu, HI, USA, pp. 425–430 (2002)
10. Kumar, R., Banerjee, N.: Multicriteria Network Design Using Evolutionary Algorithm. In: Cantú-Paz, E., Foster, J.A., Deb, K., Davis, L., Roy, R., O'Reilly, U.-M., Beyer, H.-G., Kendall, G., Wilson, S.W., Harman, M., Wegener, J., Dasgupta, D., Potter, M.A., Schultz, A., Dowsland, K.A., Jonoska, N., Miller, J., Standish, R.K. (eds.) GECCO 2003, Part II. LNCS, vol. 2724, pp. 2179–2190. Springer, Heidelberg (2003)
11. Banerjee, N., Kumar, R.: Multiobjective Network Design for Realistic Traffic Models. In: Proc. of GECCO 2007, London, United Kingdom, pp. 1904–1911 (2007)
12. Kumar, R., Banerjee, N.: Multiobjective network topology design. Applied Soft Computing 11(8), 5120–5128 (2011)
13. Schneider, C.M., Moreira, A.A., Andrade, J.S., Havlin, S., Herrmann, H.J.: Mitigation of Malicious Attacks on Networks. Proc. of the National Academy of Sciences of the United States of America 108(10), 3838–3841 (2011)
14. Ishibuchi, H., Hitotsuyanagi, Y., Tsukamoto, N., Nojima, Y.: Use of Heuristic Local Search for Single-Objective Optimization in Multiobjective Memetic Algorithms. In: Rudolph, G., Jansen, T., Lucas, S., Poloni, C., Beume, N. (eds.) PPSN 2008. LNCS, vol. 5199, pp. 743–752. Springer, Heidelberg (2008)
15. Konstantinidis, A., Zhang, Q., Yang, K.: A Subproblem-dependent Heuristic in MOEA/D for the Deployment and Power Assignment Problem in Wireless Sensor Networks. In: IEEE Congress on Evolutionary Computation, pp. 2740–2747 (2009)

Elitist Archiving for Multi-Objective Evolutionary Algorithms: To Adapt or Not to Adapt

Hoang N. Luong and Peter A.N. Bosman

Centrum Wiskunde & Informatica (CWI)
P.O. Box 94079, 1090 GB Amsterdam, The Netherlands
{Hoang.Luong,Peter.Bosman}@cwi.nl
http://www.cwi.nl

Abstract. Objective-space discretization is a popular method to control the elitist archive size for evolutionary multi-objective optimization and avoid problems with convergence. By setting the level of discretization, the proximity and diversity of the Pareto approximation set can be controlled. This paper proposes an adaptive archiving strategy which is developed from a rigid-grid discretization mechanism. The main advantage of this strategy is that the practitioner just decides the desirable target size for the elitist archive while all the maintenance details are automatically handled. We compare the adaptive and rigid archiving strategies on the basis of a performance indicator that measures front quality, success rate, and running time. Experimental results confirm the competitiveness of the adaptive method while showing its advantages in terms of transparency and ease of use.

Keywords: Multiobjective optimization, estimation of distribution algorithms, elitist archive.

1   Introduction

Optimization problems in practice may involve more than a single objective, and often conflicting ones. A *utopian* solution, that optimizes all objectives at the same time, is unachievable. A solution x can be better than another solution y in some objectives, but worse in others. The optimum for such multi-objective optimization (MO) problems is thus a set of equally preferable trade-off solutions rather than a single optimal point. We formalize the terminologies and notations for MO used in this paper as follows:

1. **Multi-objective optimization.** m objective functions $f_i(x), i \in \mathcal{M} = \{0, 1, \ldots, m-1\}$, without loss of generality, must all be minimized. A solution vector $x = (x_0, x_1, \ldots, x_{k-1})$ in the decision space has an corresponding image vector $f(x) = (f_0(x), f_1(x), \ldots, f_{m-1}(x))$ in the objective space.
2. **Pareto dominance.** A solution x^0 *dominates* a solution x^1 (denoted $x^0 \succ x^1$) if and only if $(\forall i \in \mathcal{M} : f_i(x^0) \leq f_i(x^1)) \wedge (f(x^0) \neq f(x^1))$.

C.A. Coello Coello et al. (Eds.): PPSN 2012, Part II, LNCS 7492, pp. 72–81, 2012.
© Springer-Verlag Berlin Heidelberg 2012

3. **Pareto set**. A Pareto set $\mathcal{P}$ of size n is called a Pareto set if and only if $\neg \exists \boldsymbol{x}^0, \boldsymbol{x}^1 \in \mathcal{P} : \boldsymbol{x}^0 \succ \boldsymbol{x}^1$.
4. **Pareto optimality**. A solution $\boldsymbol{x}^0$ is said to be *Pareto optimal* if and only if $\neg \exists \boldsymbol{x}^1 : \boldsymbol{x}^1 \succ \boldsymbol{x}^0$.
5. **Pareto-optimal set**. The set $\mathcal{P}_S$ of all Pareto-optimal solutions: $\mathcal{P}_S = \{\boldsymbol{x}^0 | \neg \exists \boldsymbol{x}^1 : \boldsymbol{x}^1 \succ \boldsymbol{x}^0\}$.
6. **Pareto-optimal front**. The set $\mathcal{P}_F$ in the objective space of all image vectors corresponding to the solutions in $\mathcal{P}_S$ in the decision space: $\mathcal{P}_F = \{\boldsymbol{f}(\boldsymbol{x}) = (f_0(\boldsymbol{x}), f_1(\boldsymbol{x}), \dots, f_{m-1}(\boldsymbol{x})) | \boldsymbol{x} \in \mathcal{P}_S\}$.

The optimal solution for a multi-objective optimization problem is the Pareto-optimal set $\mathcal{P}_S$ and its corresponding image $\mathcal{P}_F$. The actual size of $\mathcal{P}_S$ and $\mathcal{P}_F$ may be infinite or too numerous to be obtained by finite computational resources. In practice the desired result is often a representative solutions subset $\mathcal{S}$ of $\mathcal{P}_S$ having a reasonable size, from which decision makers are able to consider and make their final choice. This subset $\mathcal{S}$ should have its image $\boldsymbol{f}(\mathcal{S})$ well-spread along the Pareto-optimal front $\mathcal{P}_F$, which means diversity in the quality of trade-off solutions regarding all related objectives.

Different works have shown that elitism is crucial for the convergence of multi-objective optimization evolutionary algorithms (MOEAs) [1,2]. While elitist preservation for single objective optimization is a trivial task, in which the only best solution needs to be kept and updated along the run, multi-objective optimization requires more complicated elitism strategies. A separate data structure, called the elitist archive, is often used to keep track of the best Pareto set, in which every solution is not dominated by any other solution in the whole population nor by any other elitist solution in earlier generations. When the number of solutions on the Pareto front is large, the archive may grow to an extreme size. Large archives are furthermore computationally expensive to maintain. Because computational resources are always limited, an upper bound for the archive size is definitely compulsory. Problems occur when this upper bound is reached and new non-dominated solutions continue to be found. One way to differentiate MOEAs is how they treat this elitist archiving problem.

Laumanns et al. [3] propose ϵ-dominance and ϵ-Pareto set to address the problem of convergence and diversity of the approximate set. The ϵ-Pareto set is proved to have bounded size. However, Hernández-Díaz et al. [4] point out that the box-domination scheme for maintaining an ϵ-Pareto set prevents the archive from achieving its intended upper bound. The authors then present Pareto-adaptive ϵ-dominance ($pa\epsilon$-dominance) with a curve-fitting scheme to determine several parameters in order to generate a more suitable grid depending on the problem being solved at hand. $pa\epsilon$-dominance limits the types of Pareto fronts it can handle to the curves of the family $\{x_1^p + x_2^p + \dots + x_n^p = 1 : 0 \le x_1, x_2, \dots, x_n \le 1, 0 < p < \infty\}$, and the objective space should thus be continuous. A notable adaptive grid archiving (AGA) strategy is presented by Knowles and Corne [5]. AGA uses a grid, which can adapt its position and size, to estimate the density of the archived solutions in the objective space. When the archive is full, and a new non-dominated solution is generated in a less crowded region, a solution in a more crowded region will be removed. If the new solution lies in an already crowded region, it will be ignored. AGA thus can control the size of the elitist archive, but its convergence cannot be guaranteed [5]. Furthermore, without *a priori* knowledge of true Pareto front ranges, determining how many regions the objective space should be divided into (to generate the grid) before the run is started is an uninformed decision, and thus could easily make the grid become too coarse-grained or too fine-grained.

In this paper, we present a new adaptive elitist archiving strategy for MOEAs. The work is based on a straightforward rigid objective-space discretization approach that was already used in earlier research [6]. With the proposed adaptive elitist archiving strategy, an optimization practitioner can straightforwardly decide her desirable archive size, and let the algorithm automatically adapt its structure. Our paper is organized as follows. In Section 2 we describe two archiving strategies: the rigid-grid discretization, and the adaptive grid discretization. Section 3 shows experimental results comparing the performance of the two strategies under different parameter settings. Section 4 concludes our paper.

2   Elitists Archiving Strategies

2.1   Rigid Grid Discretization

MOEAs with competent operators (e.g., selection, modelling, and variation operators) can generate good solutions which are distributed along the true Pareto-optimal front. Because the number of non-dominated solutions may exceed the capacity of the elitist archive, archiving strategies are needed to decide which solutions should be stored and which solutions can be discarded. To limit the elitist archive to reasonable sizes while ensuring that non-dominated solutions are potentially well distributed across their ranges, the objective space is discretized into equal hypercubes, and each hypercube is allowed to contain only one solution at a time (see, e.g. [6]). The discretization is performed by dividing each objective dimension f_i into equal segments of unit length λ_i; for the sake of simplicity, here λ_i are set to the same λ for all objectives. Because the edge-lengths λ_i of hypercubes are determined before an MOEA run, and are fixed during the run, we refer to this method as rigid-grid discretization (RGD).

When the MOEA generates a non-dominated solution, it will go through an acceptance test to enter the elitist archive. If the new solution is (Pareto) dominated by any archive solutions, it is discarded. A new non-dominated solution can enter the elitist archive if and only if it occupies an empty hypercube or it dominates the solution that currently resides in the same hypercube. If the new non-dominated solution does not dominate the occupant, that new solution is considered as a dominated solution and is discarded as well. When a new solution is accepted into the archive, all solutions dominated by it are removed from the archive to ensure that the archive is always a Pareto set. The pseudo-code for adding a non-dominated solution into the archive is described in Fig. 1.

While keeping non-dominated solutions well-spread, RGD also prevents the elitist archive from degeneration. Degeneration happens if an MOEA prunes a

```
UPDATEELITISTARCHIVE($\mathcal{A}, \boldsymbol{x}^0, \boldsymbol{\lambda}$)          ISSAMEBOX($\boldsymbol{f}^0, \boldsymbol{f}^1, \boldsymbol{\lambda}$)
   $\boldsymbol{\lambda} = (\lambda_0, \lambda_1, \ldots, \lambda_{m-1})$              $\boldsymbol{f}^0 = (f_0^0, f_1^0, \ldots, f_{m-1}^0)$,
1  if $\exists \boldsymbol{x}^1 \in \mathcal{A} : \boldsymbol{x}^1 \succ \boldsymbol{x}^0$ then   $\boldsymbol{f}^1 = (f_0^1, f_1^1, \ldots, f_{m-1}^1)$,
   1.1  $\mathcal{A}' \leftarrow \mathcal{A}$                                         $\boldsymbol{\lambda} = (\lambda_0, \lambda_1, \ldots, \lambda_{m-1})$
2  else if $\exists \boldsymbol{x}^1 \in \mathcal{A} :$ ISSAMEBOX($\boldsymbol{f}(\boldsymbol{x}^0), \boldsymbol{f}(\boldsymbol{x}^1), \boldsymbol{\lambda}$) $\wedge \boldsymbol{x}^0 \not\succ \boldsymbol{x}^1$   1  for $i \leftarrow 0$ to $m - 1$ do
   2.1  $\mathcal{A}' \leftarrow \mathcal{A}$                                            1.1  if $\lfloor f_i^0/\lambda_i \rfloor \neq \lfloor f_i^1/\lambda_i \rfloor$ then
3  else                                                                                     1.1.1  return false
   3.1  $\mathcal{D} \leftarrow \{\boldsymbol{x}^1 \in \mathcal{A} \mid \boldsymbol{x}^0 \succ \boldsymbol{x}^1\}$   2  return true
   3.2  $\mathcal{A}' \leftarrow \mathcal{A} \cup \{\boldsymbol{x}^0\} \setminus \mathcal{D}$
4  return $\mathcal{A}'$
```

Fig. 1. Pseudo-code for adding a non-dominated solution $\boldsymbol{x}^0$ into the elitist archive $\mathcal{A}$

non-dominated solution $\boldsymbol{x}^g$ from its elitist archive at iteration g, and at a later generation g', the archive accepts a solution $\boldsymbol{x}^{g'}$ which would be dominated by $\boldsymbol{x}^g$ if $\boldsymbol{x}^g$ were still in the archive. This for instance is the case for the archiving strategies adopted in the well-known MOEAs NSGA-II and SPEA2 [7]. With RGD, there is no need to additionally prune the elitist archive; RGD just decides whether or not a solution is qualified to enter the archive. An occupant of a hypercube is removed if and only if it is dominated by a better solution which is newly accepted into the elitist archive. Because the ranges of the Pareto-optimal front are limited, the maximal number $m_{\boldsymbol{\lambda}}$ of non-dominated solutions which can be put into the grid corresponding to a discretization level $\boldsymbol{\lambda}$ is bounded. The elitist archive size is thus always less than or equal to $m_{\boldsymbol{\lambda}}$. If the MOEA does not generate any better solutions, the elitist archive will stay the same. The MOEA thus converges in this sense.

Experimental results showed the effectiveness of this rigid grid discretization technique on various benchmark problems [6]. However, it requires practitioners to set the discretization levels (i.e. the hypercube sizes) before the run. If information about the ranges of feasible solutions in the objective space is not prior knowledge, then setting fixed values is problematic and raises problems such as making the archive too coarse-grained or too fine-grained. A too coarse-grained discretized objective space misses many valuable solutions, and a too fine-grained archive requires a considerable amount of computational resources to maintain. Furthermore, manually setting the hypercube sizes is not a transparent manner to control the ultimate archive size from the perspective of decision makers.

2.2 Adaptive Grid Discretization

In real-life scenarios, a practitioner may not have prior knowledge about the ranges of the Pareto-optimal front, and she still needs to control the elitist archive size around an allowable size t due to limitations in computational resources. We resolve this problem by proposing an adaptive grid discretization mechanism (AGD). Regarding available resources, the practitioner can decide her *budget* for the elitist archive before an MOEA run, and the objective space will be adaptively discretized to maintain the archive around the target size t.

The archive functions much like in the rigid case: non-dominated solutions enter the archive, and dominated solutions are removed. When the archive size deviates

---

MULTI-OBJECTIVEEVOLUTIONARYALGORITHM()
1   $\mathbf{P} \leftarrow$ INITIALIZE()
2   $\mathbf{F} \leftarrow \{\boldsymbol{f}(\boldsymbol{x}) = (f_0(\boldsymbol{x}), f_1(\boldsymbol{x}), \ldots, f_{m-1}(\boldsymbol{x}))| \boldsymbol{x} \in \mathbf{P}\}$
3   $\boldsymbol{\mathcal{A}} \leftarrow \emptyset$
4   $\boldsymbol{\lambda} \leftarrow (0, 0, \ldots, 0)$
5   while ¬TERMINATIONCONDITIONSSATISFIED() do
        5.1 $(\mathbf{S}, \mathbf{F}^{\mathbf{S}}) \leftarrow$ MAKESELECTION($\mathbf{P}, \mathbf{F}$)
        5.2 $\mathbf{O} \leftarrow$ GENERATENEWSOLUTIONS($\mathbf{S}, \boldsymbol{\mathcal{A}}$)
        5.3 $\mathbf{F}^{\mathbf{O}} \leftarrow \{\boldsymbol{f}(\boldsymbol{x}) = (f_0(\boldsymbol{x}), f_1(\boldsymbol{x}), \ldots, f_{m-1}(\boldsymbol{x}))| \boldsymbol{x} \in \mathbf{O}\}$
        5.4   for $i \leftarrow 0$ to $|\mathbf{O}| - 1$ do
                5.4.1 $\boldsymbol{\mathcal{A}} \leftarrow$ UPDATEELITISTARCHIVE($\boldsymbol{\mathcal{A}}, \mathbf{O}[i], \boldsymbol{\lambda}$)
        5.5   if $|\boldsymbol{\mathcal{A}}| > t^{up}$ then
                5.5.1 $(\boldsymbol{\mathcal{A}}, \boldsymbol{\lambda}) \leftarrow$ ADAPTGRIDDISCRETIZATION($\boldsymbol{\mathcal{A}}$)
        5.6 $(\mathbf{P}, \mathbf{F}) \leftarrow$ MAKEREPLACEMENT$((\mathbf{S}, \mathbf{F}^{\mathbf{F}}), (\mathbf{O}, \mathbf{F}^{\mathbf{O}}))$

ADAPTGRIDDISCRETIZATION($\boldsymbol{\mathcal{A}}$)
1   $low \leftarrow 1$
2   $high \leftarrow MAX$
3   $\boldsymbol{\mathcal{A}}' \leftarrow \boldsymbol{\mathcal{A}}$
4   $max_i \leftarrow \max\{f_i^0, f_i^1, \ldots, f_i^{|\boldsymbol{\mathcal{A}}|-1}\}, i \in \{0, 1, \ldots, m-1\}$
5   $min_i \leftarrow \min\{f_i^0, f_i^1, \ldots, f_i^{|\boldsymbol{\mathcal{A}}|-1}\}, i \in \{0, 1, \ldots, m-1\}$
6   for $count \leftarrow 0$ to $N - 1$ do
        6.1 $mid = \frac{low+high}{2}$
        6.2 $\lambda_i = \frac{max_i - min_i}{mid}, i \in \{0, 1, \ldots, m-1\}$
        6.3 $\boldsymbol{\lambda} \leftarrow (\lambda_0, \lambda_1, \ldots, \lambda_{m-1})$
        6.4 $\boldsymbol{\mathcal{A}} \leftarrow \emptyset$
        6.5   for $j \leftarrow 0$ to $|\boldsymbol{\mathcal{A}}'| - 1$ do
                6.5.1 $\boldsymbol{\mathcal{A}} \leftarrow$ UPDATEELITISTARCHIVE($\boldsymbol{\mathcal{A}}, \boldsymbol{\mathcal{A}}'[j], \boldsymbol{\lambda}$)
        6.6   if $|\boldsymbol{\mathcal{A}}| < t^{low}$ then
                6.6.1 $low \leftarrow mid$
        6.7   else
                6.7.1 $high \leftarrow mid$
7   return $\boldsymbol{\mathcal{A}}, \boldsymbol{\lambda}$

Fig. 2. Pseudo-code for adaptive grid discretization. Initially, $\boldsymbol{\lambda}$ is assigned a zero vector $(0, 0, \ldots, 0)$, which means that no objective-space discretization is used. MAX is the maximal number of segments which a dimension can be divided into. N is the maximum number of steps in the binary search for objective-space discretization. In this paper, we set $MAX = 2^{25}$ and $N = 25$.

too much from the target size t, the edge-lengths need to be re-determined. Because we do not want to perform the objective-space discretization every time a single non-dominated solution is generated, we allow the elitist archive to grow to an upper bound t^{up} before pruning it. To prevent the adaptation process from deleting too many solutions, we set an lower bound of t^{low} for the elitist archive size. As soon as the archive size reaches the upper bound, the objective space adaptation process is triggered. AGD first determines the ranges of all current archived solutions in the objective space, and then performs a binary search, targeted at t^{low}, for how many segments each range should be divided into. The final discretization must satisfy the condition that the archive size is greater than the lower bound and less than the upper bound (i.e. $t^{low} < t < t^{up}$). The details of AGD are described in Fig. 2. In this paper, we set t^{low} and t^{up} as $0.75*t$ and $1.25*t$, respectively. We calibrated these values by hand taking into account that if t^{up} is too large or t^{low} is too small then the actual archive size may deviate too much from the target, while setting the bounds closer to t increases the computational overhead for re-discretization too much.

AGD can be seen as a sequence of RGDs with different discretization levels $\boldsymbol{\lambda}$. When changing to a new discretization, degeneration of the elitist Pareto front can happen because some non-dominated solutions are removed, but during an RGD, degeneration does not happen. Ultimately, there is an iteration g when solutions in the elitist archive already cover the ranges of the Pareto-optimal front, i.e. the MOEA is nearing the Pareto-optimal front, and the current discretization level $\boldsymbol{\lambda}^g$ ensures that the maximal number of non-dominated solutions which can be put into the grid is close to the target size t of the elitist archive (i.e., $m_{\boldsymbol{\lambda}^g} \approx t$, and $m_{\boldsymbol{\lambda}^g} \leq t^{up}$). From that iteration g, there is no need to re-discretize the objective space any more. If the MOEA does not generate any better solutions, the elitist archive will stay the same. The MOEA thus again converges in this sense.

3 Experiments

3.1 Benchmark Problems

In this paper, we select the MAMaLGaM (Multi-objective Adapted Maximum-Likelihood Gaussian Model [6]) as the MOEA to be combined with the two elitist archiving strategies above. It should be noted however that because these archiving mechanisms work independently from the generation of new solutions, they can be readily implemented in other MOEAs, including those aimed at discrete parameter spaces. We carry out a performance assessment of the two archiving strategies over 8 benchmark problems described in Table 1. ZDT$_i$, $i \in \{1, 2, 3, 6\}$ are well-known test problems proposed by Zitzler et al. [2]. GM$_i$, $i \in \{1, 2\}$ are generalizations of the MED (Multiple Euclidean Distances) problems [8]. Developed from the well-known Rosenbrock function, BD$_i$, $i \in \{1, 2\}$ were recently introduced in the MOEA literature [9]. Details about construction and difficulty of these benchmarks can be found in the referenced research.

We refer to an approximation set $\mathcal{S}$ as the combination of the elitist archive and all the non-dominated solutions in the current population. We consider the outcome of an MOEA to be the final approximation set upon termination. To compare the quality of approximation sets, we use a performance indicator, denoted $D_{\mathcal{P}_F \to \mathcal{S}}$.

$$D_{\mathcal{P}_F \to \mathcal{S}}(\mathcal{S}) = \frac{1}{|\mathcal{P}_F|} \sum_{\boldsymbol{f}^0 \in \mathcal{P}_F} \min_{\boldsymbol{x} \in \mathcal{S}} \{d(\boldsymbol{f}(\boldsymbol{x}), \boldsymbol{f}^0)\} \tag{1}$$

where $\mathcal{P}_F$ is the Pareto-optimal front, $\boldsymbol{f}(\boldsymbol{x})$ is a point in objective space, which is the objective value vector of a solution $\boldsymbol{x} \in \mathcal{S}$, and $d(\cdot, \cdot)$ computes Euclidean distance. $D_{\mathcal{P}_F \to \mathcal{S}}$ is also referred to in literature as inverse generational distance. It can be inferred from Equation 1 that the proximity and diversity of $\mathcal{S}$ with respect to the Pareto-optimal set $\mathcal{P}_S$ is measured in the objective space with regard to the Pareto-optimal front $\mathcal{P}_F$. Because the Pareto-optimal fronts of all test problems here are continuous and thus are infinitely large, for the sake of computability, we approximated the true $\mathcal{P}_F$ by uniformly sampling along it a subset of 5000 points.

The smaller $D_{\mathcal{P}_F \to \mathcal{S}}$ value an approximation set $\mathcal{S}$ has, the better its quality is. In practice the Pareto-optimal front may not be known, and thus the performance indicator $D_{\mathcal{P}_F \to \mathcal{S}}$ cannot be used. However, for benchmarking purposes, where $\mathcal{P}_F$ is available, this indicator has a two-fold advantage: it can measure both the proximity and diversity of $\mathcal{S}$ with respect to $\mathcal{P}_F$. In our experiments, an MOEA run with its final $\mathcal{S}$ having $D_{\mathcal{P}_F \to \mathcal{S}} \leq 0.01$ is considered as a successful run because such approximation set are quite close to the true Pareto-optimal fronts. Fig. 3 shows the default Pareto-optimal fronts and the adaptive elitist archives with different desirable target sizes t for all the problems.

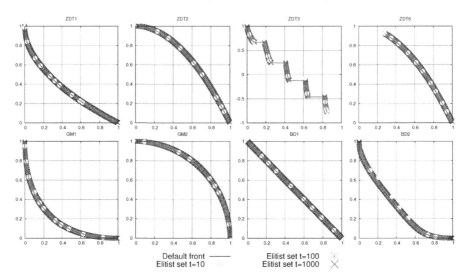

Fig. 3. For all problems: the default front and the elitist archive of 3 different archive size settings. Horizontal axis is f_0 objective value. Vertical axis is f_1 objective value.

Table 1. The MO problem test suite

Name	Objectives		IR
GM$_1$	$f_0 = \left\| \frac{1}{2}\left(\boldsymbol{x} - \boldsymbol{c}^0\right)\right\|^d$, $\quad f_1 = \left\| \frac{1}{2}\left(\boldsymbol{x} - \boldsymbol{c}^1\right)\right\|^d$		$[-1;1]^{10}$
	$\boldsymbol{c}^0 = (1,0,0,\ldots)$, $\quad \boldsymbol{c}^1 = (0,1,0,0,\ldots)$, $\quad d = 2$		$(l = 10)$
GM$_2$	$f_0 = \left\| \frac{1}{2}\left(\boldsymbol{x} - \boldsymbol{c}^0\right)\right\|^d$, $\quad f_1 = \left\| \frac{1}{2}\left(\boldsymbol{x} - \boldsymbol{c}^1\right)\right\|^d$		$[-1;1]^{10}$
	$\boldsymbol{c}^0 = (1,0,0,\ldots)$, $\quad \boldsymbol{c}^1 = (0,1,0,0,\ldots)$, $\quad d = \frac{1}{2}$		$(l = 10)$
ZDT$_1$	$f_0 = x_0$, $\quad f_1 = \gamma\left(1 - \sqrt{f_0/\gamma}\right)$		$[0;1]^{30}$
	$\gamma = 1 + 9\left(\sum_{i=1}^{l-1} x_i/(l-1)\right)$		$(l = 30)$
ZDT$_2$	$f_0 = x_0$, $\quad f_1 = \gamma\left(1 - (f_0/\gamma)^2\right)$		$[0;1]^{30}$
	$\gamma = 1 + 9\left(\sum_{i=1}^{l-1} x_i/(l-1)\right)$		$(l = 30)$
ZDT$_3$	$f_0 = x_0$, $\quad f_1 = \gamma\left(1 - \sqrt{f_0/\gamma} - (f_0/\gamma)\sin(10\pi f_0)\right)$		$[0;1]^{30}$
	$\gamma = 1 + 9\left(\sum_{i=1}^{l-1} x_i/(l-1)\right)$		$(l = 30)$
ZDT$_6$	$f_0 = 1 - e^{-4x_0}\sin^6(6\pi x_0)$, $\quad f_1 = \gamma\left(1 - (f_0/\gamma)^2\right)$		$[0;1]^{10}$
	$\gamma = 1 + 9\left(\sum_{i=1}^{l-1} x_i/(l-1)\right)^{0.25}$		$(l = 10)$
BD$_1$	$f_0 = x_0$, $\quad f_1 = 1 - x_0 + \gamma$		$[0;1] \times$ $[-5.12;5.12]^9$
	$\gamma = \sum_{i=1}^{l-2}\left(100(x_{i+1} - x_i^2)^2 + (1 - x_i)^2\right)$		$(l = 10)$
BD$_2^S$	$f_0 = \frac{1}{l}\sum_{i=0}^{l-1} x_i^2$		$[-5.12;5.12]^{10}$
	$f_1 = \frac{1}{l-1}\sum_{i=0}^{l-2}\left(100(x_{i+1} - x_i^2)^2 + (1 - x_i)^2\right)$		$(l = 10)$

3.2   Results

All the presented results here are averaged over 30 runs. Each run has a budget of 10^6 fitness evaluations. MAMaLGaM terminates when it uses all the allowable evaluations, or when all the distribution multipliers ≤ 0.5. Details about the operations of MAMaLGaM and its components can be found in [6].

Fig. 4 shows convergence graphs of the $D_{\mathcal{P}_F \to \mathcal{S}}$ indicator values from the beginning until termination for MAMaLGaM with the two elitist archiving strategies on all benchmark problems. When the elitist archive has limited volume (i.e., the target size is too small, $t = 10$, or the grid is too coarse-grained, $\lambda = 0.1$), it is less likely to achieve the desirable convergence ($D_{\mathcal{P}_F \to \mathcal{S}} \leq 0.01$). Otherwise, when having archives of adequate capacity, the MOEA achieves good convergence behavior for both variants of archiving mechanisms. Fig. 4 also shows that the greater the elitist archive is, the better $D_{\mathcal{P}_F \to \mathcal{S}}$ indicator values it can obtain. MAMaLGaM with rigid grid of $\lambda = 0.001$ shows its superiority in most of problems because it maintains the largest number of non-dominated solutions. Table 2 shows the average numbers of solutions in the archive for each benchmark problems. Because of allowing more solutions in the elitist archive, and thus in the approximation sets, the rigid grid MOEAs do perform slightly better than the their relatively corresponding adaptive versions (i.e., $\lambda = 0.1$ vs $t = 10$, $\lambda = 0.01$ vs $t = 100$, $\lambda = 0.001$ vs $t = 1000$). The $D_{\mathcal{P}_F \to \mathcal{S}}$ indicator values of RGD are thus slightly better than those of AGD. Note however, that this is a consequence of our choice for setting λ and t, and not because of inferiority of AGD. Doubling t would give similar $D_{\mathcal{P}_F \to \mathcal{S}}$ indicator values. Also, if we

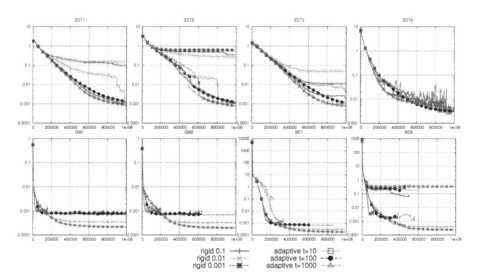

Fig. 4. Average performance of MAMaLGaM with two archiving strategies on all problems. Horizontal axis: number of evaluations (both objectives per evaluation). Vertical axis: $D_{\mathcal{P}_F \to \mathcal{S}}$. For each algorithm averages are shown both for successful runs and unsuccessful runs, giving double occurrences of lines if some runs were unsuccessful.

Table 2. Elitist archive sizes, success rates (i.e. the percentage of times MAMaLGaM obtained $D_{\mathcal{P}_F \to \mathcal{S}}$ indicator ≤ 0.01), and average running times (measured in seconds) of MAMaLGaM with 2 variants of elitist archiving strategies on all problems.

	BD_1	BD_2^S	GM_1	GM_2	ZDT_1	ZDT_2	ZDT_3	ZDT_6
ELITIST ARCHIVE SIZES								
$\lambda = 0.1$	22	16	21	21	21	16	24	17
$\lambda = 0.01$	200	166	201	197	200	194	210	165
$\lambda = 0.001$	2000	1751	1975	1954	1996	1852	2020	1576
$t = 10$	8	8	8	8	8	8	8	9
$t = 100$	98	97	100	108	90	102	103	105
$t = 1000$	1017	974	1035	1052	957	1003	1045	964
SUCCESS RATES								
$\lambda = 0.1$	100	3	100	100	93	80	90	100
$\lambda = 0.01$	100	43	100	100	100	96	100	100
$\lambda = 0.001$	100	66	100	100	100	93	100	100
$t = 10$	90	0	100	100	26	80	50	93
$t = 100$	100	66	100	100	100	96	100	100
$t = 1000$	100	66	100	100	100	93	100	100
AVERAGE RUNNING TIMES								
$\lambda = 0.1$	542	358	241	286	1101	1070	1073	1774
$\lambda = 0.01$	2015	1069	2589	2417	1161	1127	1126	2677
$\lambda = 0.001$	4892	4492	5211	5161	1903	1714	1778	4196
$t = 10$	536	552	160	160	1110	1105	1087	1224
$t = 100$	1180	719	1045	1139	1116	1106	1094	2530
$t = 1000$	4231	3834	4502	4498	1479	1424	1458	3925

terminate an MOEA run when it reaches the successful threshold ($D_{\mathcal{P}_F \to \mathcal{S}} \leq 0.01$), it can be seen that the adaptive and rigid archives have similar convergence behavior.

Table 2 shows the percentage of runs that an MOEA finds a final approximation set with performance indicator value $D_{\mathcal{P}_F \to \mathcal{S}} \leq 0.01$, which is considered as successful. It can be seen that MOEAs with tiny archives have lower success rate, which also means poorer reliability, in convergence. When the elitist archive has adequate capacity, regardless of being a rigid or adaptive, the optimization process will, in most of the times, converge successfully to fronts that are quite close to the true Pareto-optimal front. This is however more dependent on the capabilities of the other operators in MAMaLGaM rather than AGD.

Table 2 also demonstrates that the rigid and the adaptive archiving strategies have similar running times, which can partly reflect their computational costs. It is apparent that the greater elitist archive an MOEA has, the more *expensive* it is to maintain. While the match is tied for performance (indicator values), reliability (success rate), and efficiency (computational cost), the adaptive strategy wins over the rigid grid in terms of *transparency* with respect to desired archive size. For the rigid-grid discretization, the practitioner can indirectly and relatively influence the archive size by adjusting the λ value, but she hardly controls its actual growth without prior knowledge about the ranges of the objectives for the Pareto-optimal fronts. If our adaptive archive is employed, the practitioner simply decides the desirable target size of the elitist archive, thus its capacity, and then let all the details be handled behind the scenes.

4   Conclusions

In this paper, we have presented and compared two elitist archiving techniques for evolutionary multi-objective optimization: a rigid objective space discretization and an adaptive version. The two variants are showed to have similar convergence behavior, success rate, and running time on various benchmark problems. The advantage of the adaptive archiving strategy resides in its straightforwardness and transparency through which the practitioners can decide their desirable archive size and all the archiving processes are then automatically handled. According to the ranges of different dimensions in the objective space, our technique is able to select appropriate discretizations such that the final approximation set is well-spread with good proximity concerning the Pareto-optimal front provided that the MOEA is capable of generating such good solutions. Experimental results on benchmark problems support our above claims.

Although we only tested our adaptive archiving technique with the MAMaL-GaM, it can be implemented into other state-of-the-art MOEAs straightforwardly because it works independently from how new solutions are generated. Our technique is not limited to continuous search spaces as its design is not based on any assumptions about the continuity of functions. Our technique has potential to be applied successfully to a broad spectrum of optimization problems.

References

1. Bosman, P.A.N., Thierens, D.: The Balance between Proximity and Diversity in Multi–Objective Evolutionary Algorithms. IEEE Trans. Evol. Comput. 7(2), 174–188 (2003)
2. Zitzler, E., Deb, K., Thiele, L.: Comparison of multiobjective evolutionary algorithms: empirical results. Evolutionary Computation 8(2), 173–195 (2000)
3. Laumanns, M., Thiele, L., Deb, K., Zitzler, E.: Combining Convergence and Diversity in Evolutionary Multiobjective Optimization. Evolutionary Computation 10(3), 263–282 (2002)
4. Hernandez-Diaz, A.G., Santana-Quintero, L.V., Coello Coello, C.A.: Pareto-adaptive -dominance. Evolutionary Computation 15(4), 493–517 (2007)
5. Knowles, J., Corne, D.: Properties of an adaptive archiving algorithm for storing nondominated vectors. IEEE Trans. Evol. Comput. 7(2), 100–116 (2003)
6. Bosman, P.A.N.: The Anticipated Mean Shift and Cluster Registration in Mixture-based EDAs for Multi-Objective Optimization. In: Proceedings of the 12th Annual Conference on Genetic and Evolutionary Computation, GECCO 2010, pp. 351–358. ACM, New York (2010)
7. López-Ibáñez, M., Knowles, J., Laumanns, M.: On Sequential Online Archiving of Objective Vectors. In: Takahashi, R.H.C., Deb, K., Wanner, E.F., Greco, S. (eds.) EMO 2011. LNCS, vol. 6576, pp. 46–60. Springer, Heidelberg (2011)
8. Harada, K., Sakuma, J., Kobayashi, S.: Local Search for Multiobjective Function Optimization: Pareto Descent Method. In: Proceedings of the 8th Annual Conference on Genetic and Evolutionary Computation, GECCO 2006, pp. 659–666 (2006)
9. Bosman, P.A.N., de Jong, E.D.: Combining Gradient Techniques for Numerical Multi–Objective Evolutionary Optimization. Proceedings of the 8th Annual Conference on Genetic and Evolutionary Computation, GECCO 2006, 627–634 (2006)

An Improved Multiobjectivization Strategy for HP Model-Based Protein Structure Prediction*

Mario Garza-Fabre, Eduardo Rodriguez-Tello, and Gregorio Toscano-Pulido

Information Technology Laboratory, CINVESTAV-Tamaulipas. Km. 5.5 carretera
Cd. Victoria-Soto La Marina. Cd. Victoria, Tamaulipas, 87130, MÉXICO
{mgarza,ertello,gtoscano}@tamps.cinvestav.mx

Abstract. Through multiobjectivization, a single-objective problem is restated in multiobjective form with the aim of enabling a more efficient search process. Recently, this transformation was applied with success to the hydrophobic-polar (HP) lattice model, which is an abstract representation of the protein structure prediction problem. The use of alternative multiobjective formulations of the problem has led to significantly better results. In this paper, an improved multiobjectivization for the HP model is proposed. By decomposing the HP model's energy function, a two-objective formulation for the problem is defined. A comparative analysis reveals that the new proposed multiobjectivization evaluates favorably with respect to both the conventional single-objective and the previously reported multiobjective formulations. Statistical significance testing and the use of a large set of test cases support the findings of this study. Both two-dimensional and three-dimensional lattices are considered.

Keywords: Multiobjectivization, protein structure prediction, HP model.

1 Introduction

Protein structure prediction, PSP, is the problem of finding the native (energy-minimizing) conformation for a protein given only its amino acid sequence. The *hydrophobic-polar* (HP) *model* is an abstraction of this problem, where hydrophobicity is assumed to be the main stabilizing force in protein folding [6]. Even under this rather simplified model, PSP remains a challenging problem in combinatorial optimization [1, 3]. An extensive literature exists on the use of metaheuristics to address this problem, some of which is reviewed in [17, 22].

Multiobjectivization refers to the reformulation of single-objective problems in terms of two or more objective functions [15]. This transformation has been successfully used to deal with difficult optimization problems. Among them, there can be mentioned the traveling salesman problem [12, 13, 15], job-shop scheduling [13, 16], and problems in the fields of mobile communications [19], computational mechanics [9] and computer vision [21]. Multiobjectivization has also been proposed for the PSP [4, 5, 10, 20]. However, it was not until recently that this concept was applied to the particular HP model of this problem [7, 8].

* This research has been partially funded by CONACyT projects 105060 and 99276.

C.A. Coello Coello et al. (Eds.): PPSN 2012, Part II, LNCS 7492, pp. 82–92, 2012.
© Springer-Verlag Berlin Heidelberg 2012

In [7], the originally single-objective HP model was restated in multiobjective form by decomposing the conventional energy (objective) function into two separate objectives. Such a decomposition relies on the fact that topological interactions on the lattice are only possible between amino acids whose sequence positions are of opposite parity.[1] This alternative formulation, called the *parity decomposition* (PD), showed very promising results, leading to an increased search performance in most of the conducted experiments. More recently, an improved multiobjectivization strategy for the HP model was proposed, the *locality decomposition* (LD) [8]. In LD, the decomposition of the HP model's objective is carried out by segregating local from nonlocal amino acid interactions. This locality notion is based on the sequence distance between the interacting amino acids.

Motivated from previous findings [7, 8], this paper introduces a novel multiobjectivization for the HP model, the *H-subsets decomposition* (HD). HD organizes the hydrophobic amino acids into different groups, the H-subsets. Then, the HP model's energy function is decomposed based on the correspondence of amino acids to the H-subsets. The suitability of this proposal is investigated. Through a comparative analysis, HD is evaluated with respect to the conventional single-objective formulation and the preceding PD and LD multiobjectivizations.

This paper is organized as follows. Background concepts are covered in Sect. 2. In Sect. 3, the new proposed multiobjectivization is described. Section 4 details the implemented algorithms and the performance assessment methodology. Results are given in Sect. 5. Finally, Sect. 6 provides some concluding remarks.

2   Background and Notation

2.1   The Hydrophobic-Polar (HP) Model

Proteins are chain-like molecules composed from 20 different building blocks called amino acids. The hydrophobicity of amino acids is a dominant force determining the functional, three-dimensional conformation of proteins. In the HP model [6], amino acids are classified either as hydrophobic (H) or polar (P). Protein sequences are thus of the form $S \in \{H, P\}^L$, where L is the length of the sequence. Valid protein conformations are modeled as *Self-Avoiding Walks* of the HP chain on a lattice; *i.e.*, each lattice node can be assigned to at most one amino acid and consecutive amino acids in S are to be also adjacent in the lattice.

The HP model aims to maximize the interaction among H amino acids in the lattice. Formally, protein structure prediction under the HP model is defined as the problem of finding $c^* \in C$ such that $E(c^*) = \min\{E(c) \mid c \in C\}$, being C the set of all valid conformations. $E(c)$ denotes the energy of conformation c:

$$E(c) = \sum_{s_i, s_j \in S} e(s_i, s_j) \ , \tag{1}$$

where $e(s_i, s_j) = -1$ if s_i and s_j form a *hydrophobic topological contact*, denoted by $htc(s_i, s_j)$. Otherwise, $e(s_i, s_j) = 0$. In hydrophobic topological contacts, two H amino acids $s_i, s_j \in S$ are nonconsecutive in S but adjacent in the lattice.

---

[1] This is true for the two-dimensional square and the three-dimensional cubic lattices.

2.2   Single-Objective and Multiobjective Optimization

A *single-objective optimization problem* can be stated as the problem of minimizing an objective function $f : \mathcal{F} \to \mathbb{R}$, where $\mathcal{F}$ denotes the set of all feasible solutions. The aim is to find those $x^* \in \mathcal{F}$ such that $f(x^*) = \min\{f(x) \mid x \in \mathcal{F}\}$.

Similarly, a *multiobjective optimization problem* can be defined as the problem of minimizing an objective vector $\mathbf{f}(x) = [f_1(x), \ldots, f_k(x)]^T$, where $f_i : \mathcal{F} \to \mathbb{R}$ is the i-th objective function, $i \in \{1, \ldots, k\}$. The goal is to find a set of *Pareto-optimal solutions* $\mathcal{P}^* \subset \mathcal{F}$ such that $\mathcal{P}^* = \{x^* \in \mathcal{F} \mid \nexists x \in \mathcal{F} : x \prec x^*\}$. The symbol "$\prec$" denotes the *Pareto-dominance* relation, which is defined as follows: $x \prec y \;\Leftrightarrow\; \forall i : f_i(x) \leq f_i(y) \wedge \exists j : f_j(x) < f_j(y),\; i, j \in \{1, \ldots, k\}$. If $x \prec y$, then x is said to *dominate* y. Otherwise, y is said to be *nondominated* with respect to x, denoted by $x \nprec y$. The image of $\mathcal{P}^*$ in the objective space is the so-called *Pareto-optimal front*, also referred to as the *trade-off surface*.

2.3   Multiobjectivization

Multiobjectivization concerns the reformulation of single-objective problems as multiobjective ones [15]. This is done either by adding *supplementary objectives* [2, 13], or through the *decomposition* of the original objective function [11, 15]. In either case, multiobjectivization introduces fundamental changes in the search landscape, usually leading algorithms to perform a more efficient exploration. However, the goal remains to solve the original problem, so that the original optima are to be also Pareto-optimal in the multiobjective version of the problem.

The present study is based on the decomposition approach. A single-objective problem, with a given objective function $f : \mathcal{F} \to \mathbb{R}$, is restated in terms of $k \geq 2$ objectives $f_i : \mathcal{F} \to \mathbb{R}, i \in \{1, \ldots, k\}$ such that $f(x) = \sum_{i=1}^{k} f_i(x), \forall x \in \mathcal{F}$. As the only possible effect [11], plateaus may be defined in the search landscape. That is, originally comparable solutions may become incomparable (mutually nondominated) with regard to the decomposed formulation. Decomposition has been proven to be effective as a means of escaping from local optima [11, 15].

3   The H-Subsets Decomposition

In this section, an improved multiobjectivization by decomposition proposal for the HP model is presented. First, all H amino acids in the protein sequence are assigned to one of two groups, namely H_1 or H_2. The H_1 and H_2 groups are to be referred to as the H-subsets. From this, a two-objective problem formulation, $\mathbf{f}(c) = [f_1(c), f_2(c)]^T$, is defined over the set of valid protein conformations $c \in C$:

$$f_1(c) = \sum_{s_i, s_j \in H_1} e(s_i, s_j) \;+\; \sum_{s_i, s_j \in H_2} e(s_i, s_j) \;, \tag{2}$$

$$f_2(c) = \sum_{s_i \in H_1, s_j \in H_2} e(s_i, s_j) \;, \tag{3}$$

where $f_1(c)$ and $f_2(c)$ are to be minimized and $e(s_i, s_j)$ was defined in Sect. 2.1.

That is, the objective function f_1 accounts for hydrophobic topological contacts $htc(s_i, s_j)$ where both the s_i and s_j amino acids belong to the same H-subset. On the contrary, f_2 is defined for those cases where s_i and s_j belong to different H-subsets. Note that $E(c) = f_1(c) + f_2(c)$ for all $c \in C$, which is consistent with the decomposition approach for multiobjectivization, see Sect. 2.3.

The organization of H amino acids into the H-subsets can be accomplished following different strategies, several of which are evaluated in Sect. 5.1.

4 Experimental Setup

4.1 Algorithms

A basic evolutionary algorithm (EA), the so-called (1+1) EA, is used to investigate the suitability of the proposed multiobjectivization (see pseudo-code below). First, an initial parent individual c is generated at random. Iteratively, an offspring c' is created by randomly mutating c at each encoding position with probability $p_m = \frac{1}{L-1}$. The new individual c' is rejected only if it is strictly worse than the parent individual c, otherwise c' is accepted as the starting point for the next generation. Such a discrimination between c and c' can be based either on the conventional, single-objective energy evaluation, or it can be based on the Pareto-dominance relation if using a multiobjective problem formulation. Only solutions representing valid protein conformations are accepted during the search.

Basic (1+1) EA

> *choose $c \in C$ uniformly at random*
> **repeat**
> > $c' \leftarrow mutate(c)$
> > **if** *c' not worse than c* **then**
> > > $c \leftarrow c'$
> > **end if**
> **until** *< stop condition >*

Archiving (1+1) EA

> *choose $c \in C$ uniformly at random*
> $A \leftarrow \{c\}$
> **repeat**
> > $c' \leftarrow mutate(c)$
> > **if** $\nexists \hat{c} \in A : \hat{c} \prec c'$ **then**
> > > $A \leftarrow \{\hat{c} \in A : c' \nprec \hat{c} \wedge \mathbf{f}(\hat{c}) \neq \mathbf{f}(c')\} \cup \{c'\}$
> > > $c \leftarrow c'$
> > **end if**
> **until** *< stop condition >*

It was also considered an archiving variant of the above described (1+1) EA (see pseudo-code above). In this variant, an external archive stores the nondominated solutions found along the evolutionary process. At each generation, the offspring c' is only accepted if it is not dominated by any individual in the archive. If accepted, c' is included in the archive and all individuals dominated by c', and those mapping to the same objective vector $\mathbf{f}(c')$, are removed. Note that this archiving strategy makes only sense for the multiobjective problem formulations.

A representation of absolute moves was adopted. That is, conformations are encoded as sequences in $\{U, D, L, R, F, B\}^{L-1}$, denoting the up, down, left, right, forward and backward lattice positions for an amino acid with regard to the preceding one. Only directions $\{U, D, L, R\}$ are used in the two-dimensional case.

4.2 Test Cases and Performance Assessment

A total of 30 HP instances are used in this study (15 are for the two-dimensional square lattice and 15 are for the three-dimensional cubic one). Due to space limitations, details of these instances are not provided here, but they are available online at http://www.tamps.cinvestav.mx/~mgarza/HPmodel/. For all the experiments, 100 independent executions were performed and the algorithms were run until a maximum number of 10^5 solution evaluations was reached. The results are evaluated in terms of the best (lowest) obtained energy (β), the number of times this solution was found (f) and the average energy (μ). Additionally, the *overall average performance* (OAP) measure was adopted. OAP is defined as the average ratio of the obtained μ values to the optimum (E^*). Formally:

$$\mathrm{OAP} = \frac{100\%}{|T|} \left(\sum_{t \in T} \frac{\mu(t)}{E^*(t)} \right) \,, \tag{4}$$

where T is the set of all the test cases. Larger OAP values are preferred. A value of OAP $= 100\%$ suggests the ideal situation where the optimum solution for each benchmark sequence was reached during all the performed executions.

Statistical significance analysis was conducted as follows. First, *D'Agostino-Pearson's omnibus K^2* test was used to evaluate the normality of data distributions. For normally distributed data, either *ANOVA* or the *Welch's t* parametric tests were used depending on whether the variances across the samples were homogeneous or not (*Bartlett's* test). For non-normal data, the non-parametric *Kruskal-Wallis* test was adopted. A significance level of $\alpha = 0.05$ was considered.

5   Results

In this section, the (1+1) EA is used in order to evaluate and compare the four different formulations of the HP model: the conventional single-objective formulation (SO), the recently reported parity (PD) [7] and locality (LD) [8] decompositions, and the H-subsets decomposition (HD) proposed in this paper.[2]

Given the importance that the H-subsets formation process has for the HD, different strategies are first investigated in Sect. 5.1. Then, Sect. 5.2 analyzes the impact of using the archiving strategy within the (1+1) EA for all the studied formulations. Finally, a detailed comparative analysis is presented in Sect. 5.3.

5.1   H-Subsets Formation

An important issue for the proposed HD is how H amino acids are organized into the H-subsets (H_1 and H_2). Therefore, the following strategies are investigated:

- FIX: the first half of H amino acids in S are assigned to H_1, all others to H_2. For an odd number of Hs, the one in the middle is assigned randomly.
- RND: each H amino acid is assigned to H_1 or to H_2 with equal probability.
- DYN_k: based on RND, but the H-subsets are dynamically and independently recomputed after k iterations of the algorithm without achieving an improvement. Different values for k are explored, $k = \{0, 10, 20, 25, 30, 50\}$, where $k = 0$ refers to the recomputation of the H-subsets at each iteration.

---

[2] LD depends on parameter δ. This parameter was set to $\delta = 7$ as suggested in [8].

Figure 1 presents the OAP measure obtained by the HD when using the above described strategies. Results are provided for both the basic and the archiving (1+1) EA. Also, the performance of the SO formulation is shown as a baseline.

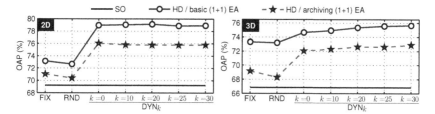

Fig. 1. Evaluating different strategies to form the H-subsets

It is evident from Fig. 1 that the proposed HD performed better in all cases compared to the conventional SO formulation. The highest OAP values were obtained when using the DYN_k strategy. That is, the ability of decomposition for allowing the algorithms to escape from local optima is further enhanced by changing the search landscape dynamically throughout the evolutionary process.

For the two-dimensional instances, no important differences in performance can be observed when varying k. Regarding the three-dimensional test cases, the algorithms responded positively to the increased value for k. The DYN_k strategy with $k = 30$ was adopted for the experiments presented in Sects. 5.2 and 5.3.

5.2   The Impact of Archiving

This section aims at investigating the impact of using the archiving (1+1) EA rather than the basic version of this algorithm. The results are presented in Fig. 2, which contrasts the performance of these algorithms (in terms of the OAP measure) when using the four studied HP model's formulations.[3]

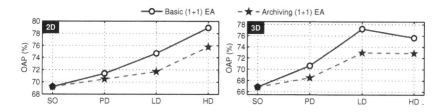

Fig. 2. Evaluating the impact of using the archiving strategy

From Fig. 2, it can be seen that an important increase in performance was obtained through multiobjectivization. The three multiobjective proposals (PD, LD and HD) improved the results for both the basic and the archiving (1+1) EA

---

[3] Although archiving is only useful in multiobjective scenarios, results of the archiving (1+1) EA applied to the SO formulation are shown only for illustrative purposes.

with respect to the conventional SO formulation. The proposed HD reached the highest OAP values at solving the two-dimensional instances. In contrast, the previously reported LD scored better results for the three-dimensional test cases.

Although competitive, the performance of PD, LD and HD was negatively affected by the use of the archiving strategy within the (1+1) EA. This is contrary to what is expected in multiobjective optimization, where archiving is essential for converging towards a set of trade-offs among the conflicting problem objectives [14, 18]. Nevertheless, in spite of being alternatively modeled and treated as a multiobjective problem, the HP model is actually a single-objective problem. Therefore, maintaining an approximation set of nondominated solutions becomes not as important. In addition, the archiving strategy influences the acceptance criterion of the algorithm in such a way that the introduction of plateaus, the only achievable effect of decomposition, may be partially reversed [11]. That is, some of the mutually incomparable solutions can be comparable to those in the archive. This could lead some parts of the plateaus to become inaccessible, thus restricting the exploration behavior of the algorithm.

5.3   Comparative Analysis

In all the cases, better results were obtained by using the basic (1+1) EA rather than the archiving (1+1) EA, see Sect. 5.2. Thus, the basic (1+1) EA is used for comparing in detail the four studied HP model's formulations. Tables 1 and 2 present, 2D and 3D respectively, the best (lowest) energy (β), its frequency (f) and the average energy (μ) obtained for each instance when using the different formulations. Also, the OAP measure is used to evaluate the overall performance of the approaches, see Sect. 4.2. The best (lowest) μ achieved for each instance, as well as the best (highest) OAP values, appear **shaded** in these tables.

Table 1. Results for the basic (1+1) EA on two-dimensional benchmarks

Seq.	L	E^*	SO β (f)	μ	PD β (f)	μ	LD β (f)	μ	HD β (f)	μ
2d1	18	-4	-4 (4)	-2.70	-4 (6)	**-2.71**	-4 (3)	-2.69	-4 (4)	-2.70
2d2	18	-8	-8 (18)	-6.81	-8 (24)	-7.04	-8 (31)	-7.16	-8 (66)	**-7.65**
2d3	18	-9	-8 (11)	-7.00	-8 (48)	-7.45	-9 (2)	-7.39	-9 (28)	**-8.27**
2d4	20	-9	-9 (8)	-6.84	-9 (4)	-6.95	-9 (11)	-7.23	-9 (48)	**-8.19**
2d5	20	-10	-9 (3)	-6.92	-10 (2)	-7.08	-9 (1)	-7.06	-9 (7)	**-7.51**
2d6	24	-9	-8 (14)	-6.81	-9 (1)	-6.87	-9 (2)	**-7.30**	-9 (6)	-7.26
2d7	25	-8	-7 (26)	-5.79	-8 (6)	-5.90	-8 (7)	-6.17	-8 (21)	**-6.51**
2d8	36	-14	-13 (1)	-9.97	-13 (1)	-10.23	-13 (4)	-10.61	-12 (30)	**-11.00**
2d9	48	-23	-18 (5)	-14.23	-19 (2)	-15.20	-20 (2)	-16.29	-20 (3)	**-17.46**
2d10	50	-21	-18 (2)	-13.79	-18 (1)	-14.06	-19 (1)	-15.07	-18 (14)	**-16.25**
2d11	60	-36	-30 (2)	-24.39	-30 (7)	-25.43	-32 (1)	-27.80	-32 (2)	**-29.11**
2d12	64	-42	-29 (1)	-23.82	-30 (1)	-25.12	-30 (4)	-26.61	-32 (2)	**-27.99**
2d13	85	-53	-41 (1)	-33.81	-41 (1)	-34.54	-44 (1)	-38.09	-45 (1)	**-39.35**
2d14	100	-48	-41 (1)	-30.80	-39 (3)	-32.18	-39 (2)	-34.41	-40 (2)	**-35.40**
2d15	100	-50	-40 (1)	-31.71	-40 (3)	-32.70	-39 (7)	-34.97	-40 (7)	**-36.68**
OAP				69.22%		71.39%		74.70%		**78.93%**

Table 2. Results for the basic (1+1) EA on three-dimensional benchmarks

Seq.	L	E^*	SO β (f)	μ	PD β (f)	μ	LD β (f)	μ	HD β (f)	μ
3d1	20	-11	-11 (57)	-10.48	-11 (69)	-10.64	-11 (94)	-10.94	-11 (100)	**-11.00**
3d2	24	-13	-13 (23)	-11.30	-13 (34)	-11.70	-13 (66)	-12.53	-13 (78)	**-12.75**
3d3	25	-9	-9 (57)	-8.48	-9 (70)	-8.65	-9 (95)	-8.95	-9 (99)	**-8.99**
3d4	36	-18	-18 (10)	-15.19	-18 (13)	-15.74	-18 (46)	**-16.97**	-18 (38)	-16.89
3d5	46	-35	-30 (2)	-23.87	-30 (1)	-25.38	-31 (1)	**-27.53**	-30 (1)	-27.26
3d6	48	-31	-29 (1)	-22.79	-29 (2)	-24.42	-31 (1)	**-26.66**	-29 (4)	-25.98
3d7	50	-34	-25 (6)	-20.64	-27 (1)	-22.07	-28 (1)	**-24.31**	-26 (4)	-23.58
3d8	58	-44	-35 (1)	-27.34	-36 (1)	-29.02	-36 (2)	**-31.98**	-35 (1)	-30.99
3d9	60	-55	-46 (1)	-37.20	-47 (1)	-40.03	-47 (3)	-42.88	-49 (1)	**-43.02**
3d10	64	-59	-45 (1)	-35.59	-46 (1)	-37.69	-50 (1)	**-43.29**	-46 (1)	-40.84
3d11	67	-56	-38 (2)	-30.17	-39 (2)	-32.65	-41 (1)	**-36.10**	-40 (1)	-35.16
3d12	88	-72	-47 (1)	-36.22	-49 (1)	-39.85	-53 (1)	**-46.13**	-48 (5)	-42.84
3d13	103	-58	-40 (1)	-29.97	-41 (1)	-31.31	-40 (1)	**-35.42**	-40 (1)	-34.25
3d14	124	-71	-43 (4)	-34.51	-48 (1)	-36.97	-50 (2)	**-43.98**	-46 (1)	-41.07
3d15	136	-83	-51 (1)	-37.26	-52 (1)	-42.11	-57 (1)	**-47.42**	-51 (4)	-45.47
OAP			66.84%		70.64%		**77.23%**		75.65%	

As shown in Table 1, the proposed HD reached the best average performance for 13 out of the 15 two-dimensional instances. This is reflected as an OAP increase of $(78.93 - 69.22) = 9.71\%$ with respect to the conventional SO formulation. HD allowed the OAP measure to be improved by 7.54% and by 4.23% with regard to the previously reported PD and LD multiobjectivizations, respectively.

The LD formulation achieved the lowest average energy for 11 out of the 15 three-dimensional benchmarks, see Table 2. The best results for the remaining four instances were obtained by the proposed HD. Although HD was inferior to LD in most of the three-dimensional test cases, with an OAP decrease of -1.58%, the results of this proposal are quite competitive; HD increased the OAP measure by 8.81% and by 5.01% over the SO and PD formulations, respectively.

Table 3 outlines how the formulations compare statistically with respect to each other in all the test cases. Each row in this table compares two formulations, say A and B, which is denoted as "A/B". If a significant performance difference exists between A and B, the corresponding cells are marked either as + or − depending on whether such a difference was in favor of or against A. Empty

Table 3. Statistical analysis for comparing the four studied HP model's formulations

	2D benchmarks 2d1	2d2	2d3	2d4	2d5	2d6	2d7	2d8	2d9	2d10	2d11	2d12	2d13	2d14	2d15	3D benchmarks 3d1	3d2	3d3	3d4	3d5	3d6	3d7	3d8	3d9	3d10	3d11	3d12	3d13	3d14	3d15	Overall
PD/SO	+							+			+	+		+	+	+	+	+	+	+	+	+	+	+	+	+	+	+	+		20+ 0−
LD/SO	+	+	+			+	+	+	+	+	+	+	+	+	+	+	+	+	+	+	+	+	+	+	+	+	+	+	+	+	28+ 0−
HD/SO	+	+	+	+	+	+	+	+	+	+	+	+	+	+	+	+	+	+	+	+	+	+	+	+	+	+	+	+	+		29+ 0−
LD/PD				+					+	+	+	+	+	+	+	+	+	+	+	+	+	+	+	+	+	+	+	+	+	+	23+ 0−
HD/PD	+	+	+	+	+	+	+	+	+	+	+	+	+	+	+	+	+	+	+	+	+	+	+	+	+	+	+	+	+		29+ 0−
HD/LD	+	+	+	+			+	+	+	+	+	+	+	+	+	+	+		−	−	−		−	−	−	−	−	−			15+ 9−

cells indicate that there was not a statistically important difference between the approaches. The rightmost column shows the overall results of this analysis.

As can be seen from Table 3, the three multiobjective approaches significantly outperformed the conventional SO formulation in most of the cases. The proposed HD performed significantly better than SO in 29 out of the 30 test instances. Among the previously reported decompositions, the results of LD for 23 of the benchmarks were statistically superior to those obtained by PD. Compared with respect to PD, the proposed HD formulation significantly increased the performance of the algorithm for all but one of the instances (2d1). Finally, the proposed HD was statistically better than LD in 15 of the instances, while there was a significant difference in favor of LD for 9 of the three-dimensional test cases.

6   Conclusions

Multiobjectivization has proven to be a promising approach for solving difficult optimization problems. When applied to the hydrophobic-polar (HP) model, a simplified version of the protein structure prediction problem (PSP), this transformation has significantly improved the performance of search algorithms.

In this paper, a novel multiobjectivization for the HP model was proposed, called the H-subsets decomposition (HD). To the best of authors' knowledge, the HD formulation, together with the multiobjectivizations reported in [7, 8], represent the first efforts on the use of multiobjective optimization methods to address the HP model for protein structure prediction. The aim of this study was to investigate the impact of using the proposed HD multiobjectivization on the resolution of this problem. Through a comparative analysis, it has been shown that the HD formulation evaluates favorably in most of the cases with respect to the previously proposed multiobjectivizations for the HP model [7, 8].

Only basic evolutionary algorithms were adopted for the experiments presented in this paper. Nevertheless, from the obtained results it is expected that multiobjectivization can improve also the performance of established state-of-the-art algorithms for solving the HP model of the PSP. This issue needs to be thoroughly investigated in order to derive more general conclusions.

References

1. Berger, B., Leighton, T.: Protein Folding in the Hydrophobic-Hydrophilic (HP) Model is NP-complete. In: International Conference on Research in Computational Molecular Biology, pp. 30–39. ACM, New York (1998)
2. Brockhoff, D., Friedrich, T., Hebbinghaus, N., Klein, C., Neumann, F., Zitzler, E.: Do Additional Objectives Make a Problem Harder? In: Genetic and Evolutionary Computation Conference, pp. 765–772. ACM, London (2007)
3. Crescenzi, P., Goldman, D., Papadimitriou, C., Piccolboni, A., Yannakakis, M.: On the Complexity of Protein Folding. In: ACM Symposium on Theory of Computing, pp. 597–603. ACM, Dallas (1998)

4. Cutello, V., Narzisi, G., Nicosia, G.: A Multi-Objective Evolutionary Approach to the Protein Structure Prediction Problem. Journal of The Royal Society Interface 3(6), 139–151 (2006)
5. Day, R., Zydallis, J., Lamont, G.: Solving the Protein Structure Prediction Problem Through a Multi-Objective Genetic Algorithm. In: IEEE/DARPA International Conference on Computational Nanoscience, pp. 32–35 (2002)
6. Dill, K.: Theory for the Folding and Stability of Globular Proteins. Biochemistry 24(6), 1501–1509 (1985)
7. Garza-Fabre, M., Rodriguez-Tello, E., Toscano-Pulido, G.: Multiobjectivizing the HP Model for Protein Structure Prediction. In: Hao, J.-K., Middendorf, M. (eds.) EvoCOP 2012. LNCS, vol. 7245, pp. 182–193. Springer, Heidelberg (2012)
8. Garza-Fabre, M., Toscano-Pulido, G., Rodriguez-Tello, E.: Locality-based Multiobjectivization for the HP Model of Protein Structure Prediction. In: Genetic and Evolutionary Computation Conference, ACM, Philadelphia (2012)
9. Greiner, D., Emperador, J.M., Winter, G., Galván, B.: Improving Computational Mechanics Optimum Design Using Helper Objectives: An Application in Frame Bar Structures. In: Obayashi, S., Deb, K., Poloni, C., Hiroyasu, T., Murata, T. (eds.) EMO 2007. LNCS, vol. 4403, pp. 575–589. Springer, Heidelberg (2007)
10. Handl, J., Lovell, S.C., Knowles, J.D.: Investigations into the Effect of Multiobjectivization in Protein Structure Prediction. In: Rudolph, G., Jansen, T., Lucas, S., Poloni, C., Beume, N. (eds.) PPSN X. LNCS, vol. 5199, pp. 702–711. Springer, Heidelberg (2008)
11. Handl, J., Lovell, S.C., Knowles, J.D.: Multiobjectivization by Decomposition of Scalar Cost Functions. In: Rudolph, G., Jansen, T., Lucas, S., Poloni, C., Beume, N. (eds.) PPSN X. LNCS, vol. 5199, pp. 31–40. Springer, Heidelberg (2008)
12. Jähne, M., Li, X., Branke, J.: Evolutionary Algorithms and Multi-Objectivization for the Travelling Salesman Problem. In: Genetic and Evolutionary Computation Conference, pp. 595–602. ACM, Montreal (2009)
13. Jensen, M.: Helper-Objectives: Using Multi-Objective Evolutionary Algorithms for Single-Objective Optimisation. Journal of Mathematical Modelling and Algorithms 3, 323–347 (2004)
14. Knowles, J., Corne, D.: Properties of an Adaptive Archiving Algorithm for Storing Nondominated Vectors. IEEE Transactions on Evolutionary Computation 7(2), 100–116 (2003)
15. Knowles, J.D., Watson, R.A., Corne, D.W.: Reducing Local Optima in Single-Objective Problems by Multi-objectivization. In: Zitzler, E., Deb, K., Thiele, L., Coello Coello, C.A., Corne, D.W. (eds.) EMO 2001. LNCS, vol. 1993, pp. 269–283. Springer, Heidelberg (2001)
16. Lochtefeld, D., Ciarallo, F.: Helper-Objective Optimization Strategies for the Job-Shop Scheduling Problem. Applied Soft Computing 11(6), 4161–4174 (2011)
17. Lopes, H.S.: Evolutionary Algorithms for the Protein Folding Problem: A Review and Current Trends. In: Smolinski, T.G., Milanova, M.G., Hassanien, A.-E. (eds.) Comp. Intel. in Biomed. & Bioinform. SCI, vol. 151, pp. 297–315. Springer, Heidelberg (2008)
18. López-Ibáñez, M., Knowles, J., Laumanns, M.: On Sequential Online Archiving of Objective Vectors. In: Takahashi, R.H.C., Deb, K., Wanner, E.F., Greco, S. (eds.) EMO 2011. LNCS, vol. 6576, pp. 46–60. Springer, Heidelberg (2011)

19. Segredo, E., Segura, C., Leon, C.: A Multiobjectivised Memetic Algorithm for the Frequency Assignment Problem. In: IEEE Congress on Evolutionary Computation, New Orleans, LA, USA, pp. 1132–1139 (2011)
20. Soares Brasil, C., Botazzo Delbem, A., Ferraz Bonetti, D.: Investigating Relevant Aspects of MOEAs for Protein Structures Prediction. In: Genetic and Evolutionary Computation Conference, pp. 705–712. ACM, Dublin (2011)
21. Vite-Silva, I., Cruz-Cortés, N., Toscano-Pulido, G., de la Fraga, L.G.: Optimal Triangulation in 3D Computer Vision Using a Multi-objective Evolutionary Algorithm. In: Giacobini, M. (ed.) EvoWorkshops 2007. LNCS, vol. 4448, pp. 330–339. Springer, Heidelberg (2007)
22. Zhao, X.: Advances on Protein Folding Simulations Based on the Lattice HP models with Natural Computing. Applied Soft Computing 8(2), 1029–1040 (2008)

MOEA/D with Iterative Thresholding Algorithm for Sparse Optimization Problems

Hui Li[1], Xiaolei Su[1], Zongben Xu[1], and Qingfu Zhang[2]

[1] Institute for Information and System Sciences & Ministry of Education
Key Lab for Intelligent Networks and Network Security,
Xi'an Jiaotong University, Xi'an, Shaanxi,710049, China
{lihui10,zbxu}@mail.xjtu.edu.cn, suxl062641@stu.xjtu.edu.cn
[2] School of Computer Science & Electronic Engineering,University of Essex,
Wivenhoe Park, Colchester, CO4 3SQ, UK
qzhang@essex.ac.uk

Abstract. Currently, a majority of existing algorithms for sparse optimization problems are based on regularization framework. The main goal of these algorithms is to recover a sparse solution with k non-zero components(called k-sparse). In fact, the sparse optimization problem can also be regarded as a multi-objective optimization problem, which considers the minimization of two objectives (i.e., loss term and penalty term). In this paper, we proposed a revised version of MOEA/D based on iterative thresholding algorithm for sparse optimization. It only aims at finding a local part of trade-off solutions, which should include the k-sparse solution. Some experiments were conducted to verify the effectiveness of MOEA/D for sparse signal recovery in compressive sensing. Our experimental results showed that MOEA/D is capable of identifying the sparsity degree without prior sparsity information.

Keywords: sparse optimization, multi-objective optimization, hard/ half thresholding algorithm, evolutionary algorithm.

1 Introduction

Compressive sensing (CS) is a novel sampling theory for reconstructing sparse signals or images from incomplete information. In recent years, it has found numerous applications, such as signal recovery, image processing as well as medical imaging [1]. A fundamental problem in CS is to find a sparse solution for underdetermined linear systems, which generally have infinite number of solutions. A sparse solution is often defined as the one with the minimal number of nonzero components among all solutions. Finding sparse solution involves the following NP-hard sparse optimization problem [2]:

$$\min \|x\|_0, \quad \text{s.t. } Ax = y \tag{1}$$

where $x \in R^N$ is a N-dimensional signal vector, A is a $M \times N$ measurement matrix with $M \ll N$, $y \in R^M$ is a measurement vector, and $\|x\|_0$ stands for the number of nonzero components of x.

C.A. Coello Coello et al. (Eds.): PPSN 2012, Part II, LNCS 7492, pp. 93–101, 2012.
© Springer-Verlag Berlin Heidelberg 2012

In the area of sparse optimization, greedy strategies and regularization methods are two commonly-used methods for finding sparse solutions [3–6]. The well-known greedy methods include matching pursuit (MP) [3] and orthogonal matching pursuit (OMP) [5]. In both algorithms, a k-sparse solution is iteratively constructed component by component in a greedy manner until k nonzero components are determined. Greedy strategies only provide approximate solutions for sparse optimization problems. In contrast, sparse optimization methods based on regularization frameworks, such as ℓ_0, ℓ_1 and $\ell_{0.5}$ [7], are more efficient to recover k-sparse solutions since they can recover the exact sparse solution. Iterative hard thresholding algorithm (iHardT) [8] and iterative half thresholding algorithm (iHalfT) [7] are two representative thresholding algorithms based on regularization frameworks.

The main difficulty in previous sparse optimization methods lies in the fact that the sparsity degree k is unknown in many real applications. To overcome this problem, an estimate of sparsity degree is often used both in greedy methods and in thresholding algorithms. In fact, a sparse optimization problem can also be modeled as a multi-objective optimization problem. Two conflicting objectives - the loss term ($\|Ax - y\|$) and the penalty term ($\|x\|_0$) should be minimized simultaneously. The solutions balancing both objectives are called trade-offs. In both greedy methods and regularization methods, the main goal is only to find one k-sparse solution, which belongs to the set of trade-off solutions for the multi-objective optimization problem. So far, not much work has been done for solving sparse optimization problems by multi-objective methods.

Since the early 1990s, evolutionary multi-objective algorithms (MOEAs) have received a lot of research interests [9]. The well-known representatives are NSGA-II [10] (Pareto-based), MOEA/D [11](decomposition-based), and IBEA [12](indicator-based). The main advantages of MOEAs lie in (i) the ability of finding multiple trade-off solutions with even spread in a single run, and (ii) the high possibility of finding global optima. In this work, we proposed a revised version of MOEA/D with thresholding algorithm for sparse optimization. In the proposed algorithm, the sparse multi-objective optimization problem is decomposed into multiple single objective subproblems. Each subproblem is associated with one sparsity level and a trade-off solution. It is optimized by existing thresholding algorithms in each generation. Moreover, the sparsity levels of subproblems are adaptively changed during the search. In our experiments, we tested the performance of the revised MOEA/D for sparse signal recovery in CS.

The remainder of this paper is organized as follows. In Section 2, the sparse optimization problem in CS is introduced. Section 3 gives an overview on two well-known iterative thresholding algorithms. MOEA/D with iterative thresholding algorithm for sparse optimization is presented in Section 4. The experimental results are reported and analyzed in Section 5. The final section concludes the paper.

2 Sparse Multi-objective Optimization

2.1 Sparse Optimization

A general sparse optimization problem in the CS can be formulated as the following bi-objective optimization problem

$$\min\{\|y - Ax\|^2, J(x)\} \tag{2}$$

where $\|y - Ax\|^2$ is the loss function. $J(x)$ is the penalty term for sparsity. The typical examples of $J(x)$ are $\|x\|_0$, $\|x\|_1 = \sum_{i=1}^{N} |x_i|$ and $\|x\|_{0.5}^{0.5} = (\sum_{i=1}^{N} |x_i|^{\frac{1}{2}})^2$, which correspond to three well-known regularization frameworks, denoted by ℓ_0, ℓ_1, and $\ell_{0.5}$, respectively.

Over the last a few years, thresholding algorithms based on regularization have been widely used in sparse optimization. In these algorithms, a regularization optimization problem is obtained by combining the loss function and the sparsity function of (2) in a linear manner:

$$\min \|y - Ax\|^2 + \lambda J(x) \tag{3}$$

where A and y are the same as above. $\lambda > 0$ is the regularization parameter, which is very sensitive to the performance of thresholding algorithms. The larger the value of λ, the solution of (3) is more sparse.

Among the aforementioned regularization frameworks, the solutions of ℓ_0 regularization problem are sparsest. But ℓ_0 regularization problem is difficult to solve since it is a NP-hard combinatorial optimization problem. To overcome this difficulty, ℓ_1 regularization, the relaxation of ℓ_0 regularization, was suggested [13]. Since ℓ_1 regularization problem is a convex quadratic optimization problem, there exist efficient algorithms for sparse solutions. $\ell_{0.5}$ regularization, a special case of $\ell_q (0 < q < 1)$, is the other popular framework for sparse optimization, which allows fast method for sparse solutions as they can be analytically expressed. Compared with ℓ_1 regularization, $\ell_{0.5}$ thresholding algorithms need less measurements to recover sparse signals, but it is more difficult to solve .

2.2 Pareto Optimality

As shown in (2), a sparse optimization problem is in nature a bi-objective optimization problem, which should have many trade-off solutions. In this work, we focus on the following bi-objective sparse optimization problem:

$$\min_{x \in R^N} \{(f_1(x), f_2(x))\} \tag{4}$$

where $f_1(x) = \|x\|_0$ and $f_2(x) = \|y - Ax\|^2$. A and y are the same as in (1).

In the context of multi-objective optimization, the optimality of solutions is defined in terms of *Pareto dominance*. For any two solutions $x^{(1)}$ and $x^{(2)}$ in R^N, $x^{(1)}$ is said to *dominate* $x^{(2)}$ if and only if $f_i(x^{(1)}) \le f_i(x^{(2)})$ for all $i \in \{1, 2\}$, and there exists at least one index $j \in \{1, 2\}$ such that $f_j(x^{(1)}) < f_j(x^{(2)})$. A solution

x^* is said to be *Pareto-optimal* if there doesn't exist such a solution in R^N which dominates x^*. The set of all Pareto-optimal solutions in the objective space is called Pareto-optimal front. A solution x^* is said to be *weakly Pareto-optimal* if no solution in R^N is strictly better than x^* regarding all objectives.

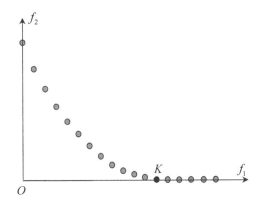

Fig. 1. Weakly Pareto-optimal solutions in sparse optimization

Fig. 1 illustrates the distribution of weakly Pareto-optimal solutions in the sparse optimization problem (4). Note that the number of these solutions is finite because the first objective $\|x\|_0$ takes integer numbers in $[0, N]$. Note that part of trade-off solutions are not Pareto-optimal but weakly Pareto-optimal. For example, all points on the right side of point K in Fig. 1 are only weakly Pareto-optimal. In many existing sparse optimization methods, the goal is to find the 'knee' Pareto-optimal solution K (k-sparse) with $y = Ax$. Unfortunately, the value of sparsity k is usually unknown.

3   Iterative Thresholding Algorithms

In this section, we briefly introduce two efficient iterative thresholding algorithms for sparse optimization - iterative hard thresholding algorithm (iHardT) [14] and iterative half thresholding algorithm [7] (iHalfT), which are based on ℓ_0 regularization and $\ell_{0.5}$ regularization respectively.

- In iHardT, an iterative procedure is performed as follows:

$$x^{(n+1)} = H_k(x^{(n)} + A^T(y - Ax^{(n)}))$$ (5)

  where $H_k(\cdot)$ is a nonlinear thresholding operator that retains the largest k components of a vector in magnitude and sets others as zeros. Note that $A^T(Ax^{(n)} - y)$ is actually the gradient vector of $\|Ax - y\|^2$.
- In iHalfT, the important components of a vector are retained by a more complex rule as follows:

$$x^{(n+1)} = H_{\lambda,\mu,0.5}(x^{(n)} - \mu A^T(y - Ax^{(n)}))$$ (6)

where $H_{\lambda,\mu,0.5}(x) = (\rho_{\lambda,\mu,0.5}(x_1)), \rho_{\lambda,\mu,0.5}(x_2), \cdots, \rho_{\lambda,\mu,0.5}(x_N)))$ is the thresholding operator. The details of $\rho_{\lambda,\mu,1/2}(\cdot)$ is referred to the literature [7].

4   MOEA/D for Sparse Optimization

4.1   Motivations

In most existing regularization methods, the bi-objective optimization problem (2) is often converted into a single objective regularization optimization problem, which could be NP-hard, or have many local optimal solutions. Due to these characteristics, these methods may suffer from being local optimizers and having sensitivity to regularization parameter, and replying on the estimation of sparsity degree. To determine the sparsity degree, regularization methods should be repeatedly applied to solve regularization problems with various sparsity levels. The set of all resultant solutions should be weakly Pareto-optimal and may contain the 'knee' Pareto-optimal solution shown in Fig. 1.

Since sparse optimization problem is a multi-objective optimization problem, the use of MOEAs to find multiple weakly Pareto-optimal solutions should be a straightforward choice. Among all MOEAs, a recent popular algorithm - MOEA/D is very suited for multi-objective sparse optimization problem due to the fitness assignment based on decomposition. The main idea in MOEA/D is to optimize multiple subproblems. Each subproblem is associated with one Pareto-optimal solution. For sparse optimization, the objective functions of subproblems can be defined by:

$$\min g^{(s)}(x) = \|y - Ax\|^2, \tag{7}$$
$$\text{s.t.}\quad \|x\|_0 = s,$$

where the sparsity level s ranges from 0 to N. Note that only the optimal solutions of $g^{(s)}$ with sparsity level s close to sparsity degree k are preferred.

In this work, we suggested a revised version of MOEA/D based on thresholding algorithms to find part of weakly Pareto-optimal solutions near the 'knee' solution. Unlike previous variants of MOEA/D, this version evolves all subproblems such that the corresponding solutions are in the neighborhood of the preferred 'knee' solution. The detail of MOEA/D is described in the following subsection.

4.2   MOEA/D with Thresholding Algorithm

In MOEA/D, a set of *pop* subproblems are defined by $g^{(s_i)}(x), i = 1, \ldots, pop$, where s_i is an integer number between an estimation interval $[s_{min}, s_{max}]$ including sparsity degree k. For each subproblem $g^{(s_i)}(x)$, a solution $x^{(i)}$ is associated and maintained. The general framework of MOEA/D for sparse optimization is outlined in Algorithm 1. The following are the detailed illustrations for the major steps in MOEA/D.

Algorithm 1. MOEA/D for Sparse Optimization

1: **Input** pop- population size, $\#ls$- maximal no. of steps in local search
2: **Output** P - sparse solutions, S- sparsity levels
3: **Step 1: Initialization**
4:   initialize $S = \{s_1, \ldots, s_{pop}\}$ and $P = \{x^{(1)}, \ldots, x^{(pop)}\}$ randomly.
5: **Step 2: Variation and Local Search**
6: **for all** $i \in \{1, \ldots, pop\}$ **do**
7:   perturb $x^{(i)}$ via mutation;
8:   apply iterative thresholding algorithm to improve $x^{(i)}$ w.r.t sparsity level s_i.
9: **end for**
10: **Step 3: Non-dominated Sorting**
11:   determine all non-dominated solutions in P and save them into Q.
12: **Step 4: Sparsity Update**
13:   **Step 4.1** sort S in an increasing order, i.e., $s_{i_1} < s_{i_2} < \ldots < s_{i_{pop}}$.
14:   **Step 4.2** determine the set of new potential sparsity levels $\tilde{S} = \{\lfloor s_{i_j} + 0.5(s_{i_{j+1}} -$
15:       $s_{i_j}) \rfloor | |s_{i_{j+1}} - s_{i_j}| \geq 2, j = 1, \ldots, pop - 1\}$
16:   **Step 4.3** insert sparsity level in $\tilde{S}$ if $|\tilde{S}| \neq 0$:
17:     **Step 4.3.1** $s_{i_1} \leftarrow$ a level in $\tilde{S}$ randomly if $|Q| > 0.5pop$.
18:     **Step 4.3.2** $s_{i_{pop}} \leftarrow$ a level in $\tilde{S}$ randomly if $|Q| \leq 0.5pop$.
19:   **Step 4.4** offset sparsity level if $|\tilde{S}| = 0$:
20:     **Step 4.4.1** $s_{i_1} \leftarrow \min\{s_{i_{pop}} + 0.5pop, s_{max}\}$ if $|Q| = pop$.
21:     **Step 4.4.2** $s_{i_{pop}} \leftarrow \max\{s_{i_1} - 0.5pop, s_{min}\}$ if $|Q| = 1$.
22:   **Step 4.5** replace $x^{(i_1)}$ or $x^{(i_{pop})}$ by one solution in the half best in P if updated.
23: **Step 5: Stopping Criteria**
24:   If stopping criteria is fulfilled, then output P and S; otherwise go to **Step 2**.

- In **Step 2**, each solution $x^{(i)}$ is first perturbed by a mutation operator, and then improved by a local search procedure, i.e., iterative thresholding algorithm. In this work, we use either iHardT in (5) or iHalfT in (6) for this purpose. The parameter $\#ls$ is used to control the maximal number of iterations allowed in each step of local search. $x^{(i)}$ is used as the starting solution and then updated by the improved solution obtained in the previous step. To perturb the starting solution, one of its non-zero component is set to zero randomly. Then, we use the greedy constructive strategy in OMP to complete the solution until s_i non-zero components are determined.

- In **Step 3**, the set of all non-dominated solutions in P are determined and saved in Q. Note that weakly Pareto-optimal solutions for sparse optimization problem are excluded in Q. The size of Q will tell us if the sparsity k is among S.

- **Step 4** is the core step of MOEA/D for sparse optimization, which adaptively updates the set S of pop sparsity levels. **Step 4.1** and **Step 4.2** determine the set $\tilde{S}$ of candidate sparsity levels. Each of them is located in the middle of adjacent two sparsity levels in S.

- In **Step 4.3** and **Step 4.4**, the minimum s_{i_1} or the maximum $s_{i_{pop}}$ among all sparsity levels in S are updated by the candidate sparsity levels randomly chosen from $\tilde{S}$. As the search progresses, subproblems with pop consecutive sparsity levels are expected to obtain.

– In **Step 4.4**, we also increase the maximal sparsity level $s_{i_{pop}}$ or decrease the minimal sparsity level s_{i_1} since all sparsity levels in S may locate on one side of the sparsity degree k. If all members of S are on the left side of k, then we need to increase the maximal sparsity level. In this case, all members of population are non-dominated. Otherwise, the minimal sparsity level should be decreased if only one is non-dominated.

– **Step 4.5** updates the solutions of the selected candidate sparsity level by the member of P with the better value of f_2.

5   Computational Experiments

5.1   Experimental Settings

In our experiments, we considered to recover noiseless real-valued signals. The elements in sensor matrix $A_{M \times N}$ and a k-sparse signal x^* are randomly generated. MOEA/D was tested on four small instances with the length of signal 512 and sparsity degree 130, and four large instances the length of signal 1024 and sparsity degree 130. All instances are named by N-M-k. The initial range of sparsity degree is assumed to be $[50, 250]$. For all instances, pop in MOEA/D is set to 10. In local search, $\#ls$ used in iHardT or iHalfT is set to 20. The total number of iterations is set to 10000 for the small instances and 20000 for the large instances. MOEA/D was implemented by C++ and tested on the operating system Windows XP with Intel Quad CPU 2.66 GHz.

5.2   Experimental Results

Table 1 summarizes the average values of mean square error (MSE) between the sparse signals and the recovered signals found by MOEA/D with two thresholding algorithms in 20 runs. The comparison of MOEA/D with iHardT and iHalfT was also provided. From these results, we can see that MOEA/D with both thresholding algorithms can find 130-sparse solutions for the first three

Table 1. Comparison of MOEA/D with iHardT and iHalfT for 8 instances with sparsity level 130 in terms of average mse to the true sparse solution

Instance	MOEA/D+iHardT	MOEA/D+iHalfT	iHardT	iHalfT
512-350-130	3.23255e-014	3.98776e-014	3.17919e-014	3.74779e-014
512-320-130	4.82847e-014	4.83276e-014	4.39184e-014	4.21504e-014
512-290-130	4.79143e-014	5.60058e-014	3.25566e-014	4.80787e-014
512-260-130	N/A	7.26465e-014	N/A	N/A
1024-500-130	5.81117e-014	6.25171e-014	6.53619e-014	5.54127e-014
1024-400-130	8.05214e-014	8.43726e-014	8.79100e-014	8.54657e-014
1024-350-130	7.95822e-014	8.87359e-014	N/A	6.20355e-014
1024-300-130	N/A	N/A	N/A	N/A

instances since the MSE values of the solutions found by MOEA/D are quite small. This indicates the obtained solutions are very close to the sparse solutions (less than 10^{-13}). However, for the instance 512-260-130, all four algorithms except iHalfT failed to find the 130-sparse solution. When M is small, the sparse optimization problem becomes too difficult to solve. This also happened for the instance 1024-300-130, where all four algorithms failed to find 130-sparse solution. Overall, MOEA/D with iHalfT works better than MOEA/D with iHardT.

Fig. 5.2 plots the weakly solutions found by MOEA/D with two thresholding algorithms for two instances 512-350-130 and 512-260-130 in one of 20 runs. It can be seen from Fig. 5.2 (a) that the 130-sparse solution is also the 'knee' solution along the weakly Pareto front for the instance 512-350-130. Fig. 5.2(b) shows that MOEA/D with iHalfT still found that 130-sparse solution for the instance 512-260-130 while MOEA/D with iHardT failed. This indicates that iHalfT is superior to iHardT in MOEA/D for the instances with less measurements. This observation is also consistent with the results in Table 1.

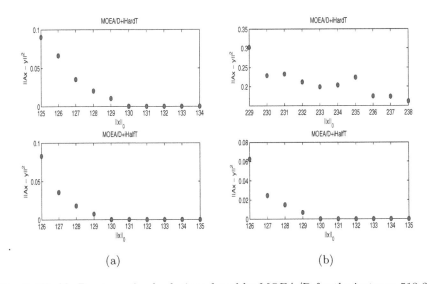

(a)   (b)

Fig. 2. Weakly Pareto-optimal solutions found by MOEA/D for the instance 512-350-130 (a) and the instance 512-260-130 (b).

6   Conclusions

In this work, we suggested a revised version of MOEA/D for sparse signal recovery in CS. It attempts to find a local part of Pareto front, which should include the k-sparse solution. Our experimental results showed that MOEA/D with both iHardT and iHalfT is effective for sparse optimization without prior sparsity information. In the future work, we plan to apply the proposed algorithm to deal with nonlinear sparse optimization problems.

Acknowledgment. The authors would like to thank the anonymous reviewers for their insightful comments. This work was supported by National Natural Science Foundation of China (NSFC) grants 61175063 and 11131006 .

References

1. Donoho, D.: Compressed sensing. IEEE Trans. on Information Theory 52, 1289–1306 (2006)
2. Natarajan, B.K.: Sparse Approximate Solutions to Linear Systems. SIAM Journal on Computing 24, 227–234 (1995)
3. Mallat, S., Zhang, Z.: Matching pursuits with time-frequency dictionaries. IEEE Trans. on Signal Processing 41, 3397–3415 (1993)
4. Efron, B., Hastie, T., Johnstone, I., Tibshirani, R.: Least angle regression. Annals of Statistics 23, 407–499 (2004)
5. Tropp, J., Gilbert, A.: Signal recovery from partial information via orthogonal matching pursuit. IEEE Trans. on Information Theory 53, 4655–4666 (2006)
6. Candés, E., Romberg, J., Tao, T.: Stable signal recovery from incomplete and inaccurate measurements. Communications on Pure Applied Mathematics 59, 1207–1223 (2006)
7. Xu, Z., Chang, X., Xu, F., Zhang, H.: L1/2 Regularization: A Thresholding Representation Theory and A Fast Solver. Technical report, Xi'an Jiaotong University (2010)
8. Bredies, K., Lorenz, D.: Iterated hard shrikage for minimization problems with sparsity constraints. SIAM Journal of Scientific Computing 30, 657–683 (2008)
9. Zhou, A., Qu, B.Y., Li, H., Zhao, S.Z., Suganthan, P.N., Zhang, Q.: Multiobjective evolutionary algorithms: A survey of the state of the art. Swarm and Evolutionary Computation 1(1), 32–49 (2011)
10. Deb, K., Agrawal, S., Pratap, A., Meyarivan, T.: A fast and elitist multiobjective genetic algorithm: NSGA-II. IEEE Trans. Evolutionary Computation 6(2), 182–197 (2002)
11. Zhang, Q., Li, H.: MOEA/D: A Multiobjective Evolutionary Algorithm Based on Decomposition. IEEE Trans. Evolutionary Computation 11(6), 712–731 (2007)
12. Zitzler, E., Künzli, S.: Indicator-Based Selection in Multiobjective Search. In: Yao, X., Burke, E.K., Lozano, J.A., Smith, J., Merelo-Guervós, J.J., Bullinaria, J.A., Rowe, J.E., Tiño, P., Kabán, A., Schwefel, H.-P. (eds.) PPSN VIII. LNCS, vol. 3242, pp. 832–842. Springer, Heidelberg (2004)
13. Tibshirani, R.: Regression shrinkage and selection via the lasso. Journal of the Royal Statistical Society 46, 431–439 (1996)
14. Blumensath, T., Davies, M.: Iterative hard thresholding for compressed sensing. Appl. Comput. Harmon. Anal. 27, 265–274 (2009)

A Study on Evolutionary Multi-Objective Optimization with Fuzzy Approximation for Computational Expensive Problems

Alessandro G. Di Nuovo[1,2], Giuseppe Ascia[2], and Vincenzo Catania[2]

[1] Centre of Robotics and Neural Systems, Plymouth University, United Kingdom
[2] Dipartimento di Ingegneria Elettrica, Elettronica e Informatica,
Universita´ degli Studi di Catania, Italy

Abstract. Recent progress in the development of Evolutionary Algorithms made them one of the most powerful and flexible optimization tools for dealing with Multi-Objective Optimization problems. Nowadays one challenge in applying MOEAs to real-world applications is that they usually need a large number of fitness evaluations before a satisfying result can be obtained. Several methods have been presented to tackle this problem and among these the use of approximate models within MOEA-based optimization methods proved to be beneficial whenever dealing with problems that need computationally expensive objective evaluations. In this paper we present a study on a general approach based on an inexpensive fuzzy function approximation strategy, that uses data collected during the evolution to build and refine an approximate model. When the model becomes reliable it is used to select only promising candidate solutions for real evaluation. Our approach is integrated with popular MOEAs and their performance are assessed by means of benchmark test problems. Numerical experiments, with a low budget of fitness evaluations, show improvement in efficiency while maintaining a good quality of solutions.

Keywords: Evolutionary Multi-objective Optimization, Expensive Optimization Problems, Fuzzy Function Approximation.

1   Introduction

Evolutionary algorithms (EAs) proved to be very powerful and flexible techniques for finding solutions to many real-world search and optimization problems. In fact they have been used in science and engineering as adaptive algorithms for solving practical problems and as computational models of natural evolutionary systems. In particular great effort was recently devoted to develop EAs to solve Multi-Objective Optimization (MOO) problems. Algorithms in this particular class of problems are named Multi-objective evolutionary algorithms (MOEAs) and they aim at finding a set of representative Pareto optimal solutions in a single run, see [16] for details and examples. Despite the great successes achieved, evolutionary algorithms have also encountered many challenges. For most evolutionary algorithms, a large number of fitness evaluations (performance calculations) are needed before a well acceptable solution can be found. In many real-world applications, fitness evaluation is not trivial. There are several situations in

C.A. Coello Coello et al. (Eds.): PPSN 2012, Part II, LNCS 7492, pp. 102–111, 2012.

which fitness evaluation becomes difficult and computationally efficient approximations of the fitness function have to be adopted. A detailed survey on proposals to speedup the evaluation of a single configuration can be found in [8]. The most popular models for fitness approximation are polynomials (often known as response surface methodology), the Kriging model, whereby Gaussian process models are parameterized by maximum likelihood estimation, most popular in the design and analysis of computer experiments, and artificial neural networks (ANNs). In particular in [8], it is stated that ANNs are recommended under the condition that a global model is targeted and that the dimension is high. The reason is that ANNs need a lower number of free parameters compared to polynomials or Gaussian models. As an example, in [6] an inverse neural network is used to map back from a desired point in the objective space (beyond the current Pareto surface) to an estimate of the decision parameters that would achieve it. The test function results presented look particularly promising, though fairly long runs (of 20,000 evaluations) are considered.

A multi-objective evolutionary algorithm, called ParEGO [9], was devised to obtain an efficient approximation of the Pareto-optimal set with a budget of a few hundred evaluations. The ParEGO algorithm is based on Kringing model, and it begins with solutions in a latin hypercube and updates a Gaussian process surrogate model of the search landscape after every function evaluation, which it uses to estimate the solution with the largest expected improvement. A recent work, [15], presents MOEA/D-EGO, that is based on Fuzzy clustering and Gaussian stochastic process modeling extends the ParEGO algorithm to generate many candidate solutions at the same time, in such a way it is possible to evaluate them all using parallel computing. In [12] an improved Archive-based Micro Genetic Algorithm (referred to as AMGA2) for constrained MOO is proposed. AMGA2 borrows and improves upon several concepts from existing MOEAs. Benchmarking and comparison demonstrate its improved search capability in the range between 5000 and 20000 function evaluations.

Recently, in [2] a MOEA with hierarchical fuzzy approximation was studied to speed-up the Design Space Exploration (DSE) of embedded computer architectures. The Evolutionary-Fuzzy methodology, named MOEA+FUZZY, exploits the knowledge of the embedded computer architecture with a hierarchical design of the fuzzy approximator system, this way, in comparisons with other MOEA for computational expensive optimization problems, like ParEGO, showed to save a great amount of time and also gives more accurate results.

In this work we present a study on a general implementation of the MOEA+FUZZY approach that can be applied to every optimization problem, using a general strategy to efficiently build a fuzzy function approximator. Details on MOEA+FUZZY approach are given in section 2. Our study aims to assess efficiency and performance of MOEAs combined with our +FUZZY approach when only a low budget of fitness evaluations is available. To this end we integrated proposed approach with popular MOEAs and tested four synthetic benchmarks. The setup of experiments is described in Section 3, while numerical results are presented in Section 4. Finally Section 5 gives our conclusion and directions for future work.

2   MOEA+FUZZY : Multi-Objective Evolutionary Optimization with Fuzzy Function Approximation

In this section we give a detailed presentation of our Evolutionary-Fuzzy strategy which has the ability to avoid the real evaluation of individuals that it foresees to be not good enough to belong to the Pareto-set and to give them fitness values according to a fast estimation of the objectives obtained by means of a Fuzzy System (FS). The main idea is that data collected for previously evaluated candidate solutions can be used during the evolution to build and refine an approximate model, and through it to avoid evaluating less promising candidate solutions. By doing so, expensive evaluations are only necessary for the most promising population members and the saving in computational costs is considerable.

The proposed approach could be informally described as follows: in a first phase the MOEA evolves normally; in the meanwhile the FS learns from real fitness function evaluations until it becomes expert and reliable. From this moment on the MOEA stops using the real function evaluation and uses the FS to estimate the objectives. From this moment on only if the estimated objective values are good enough to enter the Pareto-set will the associated individual be exactly evaluated. This avoids the insertion in the Pareto set of non-Pareto system individuals. It should be pointed out, however, that a "good" individual might be erroneously discharged due to the approximation and estimation error. At any rate, this could affect the overall quality of the solution found only after long runs as will be shown in Section 4. The reliability condition is essential in this flow. It assures that the approximator is reliable and that it can be used in place of the real function evaluation. To test reliability during the training phase the difference (approximation error) between the actual fitness function output and the predicted (approximated) fuzzy system output is evaluated. If this difference is below a user defined threshold and a minimum number of samples have been presented, the approximator is considered to be reliable. This strategy avoid to pre-set the number of samples needed by the approximator before the EA exploration starts, that is difficult when the objective function is not known.

The MOEA and the FS represent the main components of the proposed approach. Whereas the first one is used to select individuals to be explored, the second one is used to evaluate them. In our approach MOEAs could be chosen among ones presented in the literature, while the next subsection focus on fuzzy system generation strategy.

2.1   Strategy to Build a Fuzzy Approximation System during Evolutionary Process

The MOEA+Fuzzy approach uses a Fuzzy System (FS), which has been demonstrated to be a universal approximator [14]. In this work fuzzy systems are generated with a method that is based on the well-known Wang and Mendel method [13]. It consists of five steps:

- *Step 1* Divides the input and output space of the given numerical data into fuzzy regions;
- *Step 2* Generates fuzzy rules from the given data;

- *Step 3* Assigns a degree to each of the generated rules for the purpose of resolving conflicts among them (rule with higher degree wins);
- *Step 4* Creates a combined fuzzy rule base based on both the generated rules and, if there were any, linguistic rules previously provided by human experts;
- *Step 5* Determines a mapping from the input space to the output space based on the combined fuzzy rule base using a defuzzifying procedure.

The main advantages of this method are that allows to build rules step by step and that do not require a priori knowledge about function to be approximated. In addition from Step 1 to 5 it is evident that this method is simple and straightforward, in the sense that it is a one-pass build-up procedure that does not require time-consuming training. Our single-threaded implementation needs just about 10^{-2} seconds both to add a new rule and to perform an evaluation of an individual even a relatively big system, with thousands of fuzzy rules and tens of input variables, on an Intel Core i5 machine. In our implementation the output space could not be divided in Step 1, because we had no information about boundaries. For this reason we used Takagi-Sugeno fuzzy rules [11], in which each i-th rule has as consequent M real numbers s_{iz}, with $z \in [1, M]$, associated with all the M outputs. TS_j being the set of fuzzy sets associated with the variable x_j, the fuzzy rules R_i of the single fuzzy subsystem are defined as:

$$R_i : \text{if } x_1 \text{ is } S_{i1} \text{ and} \ldots \text{and } x_N \text{ is } S_{iN} \text{ then } y_{i1} = s_{i1}, \ldots, y_{im} = s_{iM}$$

where $S_{ij} \in TS_j$. Let α_{jk} be the degree of truth of the fuzzy set S_{jk} belonging to TS_j corresponding to the input value $\bar{x}_j$. If m_j is the index such that α_{jm_j} is the greatest of the α_{jk}, the rule R_i will contain the antecedent x_j is S_{jm_j}. After constructing the set of antecedents the consequent values y_{iz} equal to the values of the outputs are associated. The rule R_i is then assigned a degree equal to the product of the N highest degrees of truth associated with the fuzzy sets chosen S_{ij}. The rules generated in this way are "and" rules, i.e., rules in which the condition of the IF part must be met simultaneously in order for the result of the THEN part to occur. For the problem considered in this paper, i.e., generating fuzzy rules from numerical data, only "and" rules are required since the antecedents are different components of a single input vector. In this work fuzzy sets shape is Gaussian with normal distribution. Steps 2 to 4 are iterated with the MOEA: after every evaluation a fuzzy rule is created and inserted into the rule base, according to its degree in case of conflicts. More specifically, if the rule base already contains a rule with the same antecedents, the degrees associated with the existing rule are compared with that of the new rule and the one with the highest degree wins. In Step 5 the defuzzifying procedure to calculate the approximated output value $\hat{y}$ is the one suggested in [13]. According to this method the defuzzified output is determined as follows

$$\hat{y}_j = \frac{\sum\limits_{r=1}^{K} m_r \bar{y}_{rz}}{\sum\limits_{r=1}^{K} m_r} \tag{1}$$

where K is the number of rules in the fuzzy rule base, $\bar{y}_{rz}$ is the output estimated by the r-th rule for the z-th output and m_r is the degree of truth of the r-th rule. In our implementation the defuzzifying procedure and the shape of the fuzzy sets were chosen a priori. This choice proved to be effective as well as a more intelligent implementation, which could embed a selection procedure to choose the best defuzzifying function and shape to use online. The advantage of our implementation is a lesser computational requirement of the algorithm and a faster evaluation.

3   Experimental Setup

For implementation of the MOEA+Fuzzy strategy described above we used the PISA suite [3]. PISA stands for *Platform and programming language Independent interface for Search Algorithms* and allows to implement application-specific parts (representation, variation, objective function calculation) separately from the actual search strategy (fitness assignment, selection). Several multi-objective evolutionary algorithms as well as other well-known benchmark problems, such the widely used set of continuous test functions the ZTL [17] and DTLZ [5], are available for download at the PISA website [1]. Due to space restrictions, not all results can be presented here. Instead, we will focus on four test problems ZDT3, ZDT6, DTLZ3 and DTLZ6, that summarize main issues encountered in our tests. Table 1 lists the synthetic test problems chosen for this study. Test problem ZDT3 was selected because it is discontinuous, while DTLZ3 because it is multi-modal and difficult to solve. ZDT6 and DTLZ6 were selected because they involve a highly skewed distribution of points in the search space corresponding to a uniform distribution of points in the objective space, thus challenge an optimization algorithm's ability to find the global Pareto-optimal frontier. Detailed description of test problems can be found in their respective references. Using problems presented above we tested the proposed methodology integrating it with the 2 most popular MOEAs, SPEA2 [18] and NSGA-II [4], and a novel version of ϵ-constraint evolutionary algorithm ECEA [10]. In this work we tested two different set-ups of our approach in order to assess it after different ranges of real function evaluations:

1. +FUZZY$_1$. Fuzzy system has 9 sets for each input variable and reliability thresholds are distance of 1.0 and maximum of 5000 evaluations. The minimum number of evaluations is 200.
2. +FUZZY$_2$. Fuzzy system has 25 sets for each input variable and reliability thresholds are distance of 0.5 and maximum of 10000 evaluations. The minimum number of evaluations is 1000.

The first set-up is intended for a very low budget of real evaluations, from some hundreds to few thousand, while the second should perform better with longer runs.

Table 1. Test problems

Name	Variables	Objectives	Remarks
ZDT3	10	2	Discontinuous
ZDT6	10	2	Skewed
DTLZ3	12	3	Multi-modal
DTLZ6	12	3	Skewed

The performance measure we considered is the *Hypervolume* [19], that is the only one widely accepted and, thus, used in many recent similar works. This index measures the hypervolume of that portion of the objective space that is weakly dominated by the Pareto set to be evaluated. In order to measure this index the objective space must be bounded, then a bounding reference point that is (at least weakly) dominated by all points should be used. In this work we applied a standard linear normalization procedure, i.e. all values are normalized to the minimum and maximum value observed on the test problem. We took as bounding point vectors [1.1,1.1] and [1.1,1.1,1.1] for two and three objectives, respectively.

To present an uniform comparison between different problems, in Section 4 we show the percentage of a reference hypervolume covered by algorithms under investigation.

The reference hypervolume is calculated from a reference Pareto-set, that was obtained in following way: first, we combined all approximations sets generated by the algorithms under consideration after 50000 function evaluations (i.e. 500 generations with a population of 100 individuals), and then the dominated objective vectors are removed from this union. At last, the remaining points, which are not dominated by any of the approximations sets, form the reference set. The advantage of this approach is that the reference set weakly dominates all approximation sets under consideration.

Identical setting is used for all the algorithms. The parameter settings used for each algorithm are as follows: number of generations = 500; population size = 100; number of parents = 100; number of offsprings = 100; individual mutation probability = 1.0; individual recombination probability = 1.0; variable mutation probability = 1.0; variable swap probability = 0.5; variable recombination probability = 1.0; mutation distribution index = 20.0; recombination distribution index = 15.0.

4   Numerical Results

Using the experimental setup described in section 3, for each test problem, algorithms ran twenty times with different random seed. Median values for performance indicators are presented to represent the expected (mid-range) performance. For the analysis of multiple runs, we compute the quality measures of each individual run, and report the median and the standard deviation of these. Since the distribution of the algorithms we compare are not necessarily normal, we use the Kruskal-Wallis test [7] to indicate if there is a statistically significant difference between distributions. We recall that the significance level of a test is the maximum probability α, assuming the null hypothesis, which the statistic will be observed, i.e. the null hypothesis will be rejected in error when it is true. The lower the significance level the stronger the evidence. In this work we assume that the null hypothesis is rejected if $\alpha < 0.01$.

Table 2 presents median number of real function evaluations for MOEA+FUZZYs. As expected for ZDT problems MOEAs need less function evaluations to converge, for this reason we chose different pre-fixed amounts of real function evaluations for comparison reported in Table 3. To make an uniform comparison we calculated median values of +FUZZY algorithms taking into account all runs, even those with a number of function evaluations lower than maximum threshold selected. This means that absolute performance of MOEA+FUZZYs is sometimes slightly underestimated.

Table 2. Real function evaluations of MOEA+Fuzzy after 500 generations with a population of 100 individuals

Algorithm	Test Problem / Real function evaluations of MOEA+Fuzzy							
	ZDT3		ZDT6		DTLZ3		DTLZ6	
	median	stddev	median	stddev	median	stddev	median	stddev
SPEA2+FUZZY$_1$	3095	3710	374	72	5354	60	3480	378
SPEA2+FUZZY$_2$	2246	509	1184	119	10571	112	10001	2107
NSGA-II+FUZZY$_1$	2387	4305	338	126	5256	75	3661	339
NSGA-II+FUZZY$_2$	2251	418	1218	131	10594	110	8549	1871
ECEA+FUZZY$_1$	297	38	298	32	10001	0	684	77
ECEA+FUZZY$_2$	1268	155	1294	203	10001	2	1969	2645

Table 3. Comparison of median hypervolume percentage covered after a fixed amount of real function evaluations

Algorithm	Test problem / Maximum number of real function evaluations*											
	ZDT3			ZDT6			DTLZ3			DTLZ6		
	400	1200	2000	400	1200	2000	2000	3000	5000	2000	3000	5000
SPEA2	81.26	95.90	99.30	32.80	58.61	77.43	99.38	99.52	99.64	81.06	87.65	93.68
+FUZZY$_1$	**94.27**	95.42	95.59	**58.08**	59.04	59.04	99.35	99.48	99.65	81.14	**89.34**	91.63
+FUZZY$_2$	81.25	**97.75**	99.00	34.19	**72.40**	74.61	99.35	99.48	99.64	81.00	87.61	93.91
NSGA-II	81.57	95.85	99.34	33.38	56.28	75.58	99.47	99.58	99.70	83.62	90.34	95.55
+FUZZY$_1$	**94.01**	96.56	96.65	**59.91**	60.96	60.96	99.46	99.57	99.71	83.91	**92.11**	93.73
+FUZZY$_2$	81.56	**97.66**	99.04	33.32	**72.36**	75.23	99.46	99.57	99.70	83.91	90.43	95.99
ECEA	74.48	76.16	76.39	28.87	32.94	33.59	97.36	97.40	97.53	62.48	62.72	68.24
+FUZZY$_1$	76.25	76.25	76.25	31.17	31.17	31.17	97.35	97.39	97.51	63.09	63.09	63.09
+FUZZY$_2$	74.85	76.25	77.47	28.53	33.08	33.90	97.35	97.39	97.54	63.04	63.11	63.15

* Median values of +FUZZY algorithms are reported taking into account also runs with a number of function evaluations lower than the threshold used for the comparison.
Results **in bold** are better than others with statistical significance level $\alpha < 0.01$, according to the Kruskal-Wallis test.

In particular, looking at Table 2 we remark that none of +FUZZY algorithms reached 2000 real evaluations in ZDT6 problem. From the results of the benchmark study, we can see that the SPEA2+FUZZYs and NSGA-II+FUZZYs perform comparably well in ranges considered, while ECEA has a slower convergence that impacts performance of ECEA+FUZZYs. Results in Table 3 show that fuzzy system in scenario +FUZZY$_1$ is able to speedup the convergence of MOEAs after a very low number of real fitness evaluations for ZDT problems. On the other hand +FUZZY$_2$ improvement is smaller, but its performance on longer runs is more reliable, as shown also in Table 4. Figure 1 shows two examples of hypervolume improvement during the evoluationary process. Speedup of +FUZZYs approach is evident in Figure 1(a). In DTLZ3 there is no significant improvement thanks to fuzzy approximation strategy, while in DTLZ6 our fuzzy approach help to improve only SPEA2 and NSGA-II evolution in the range between 2000 and 3000 as also shown in Figure 1(b).

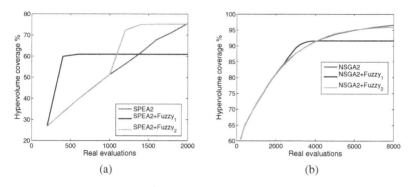

(a)   (b)

Fig. 1. Hypervolume coverage: (a) ZDT6 - SPEA2 ; (b) DTLZ6 - NSGA-II

Table 4. Comparison of hypervolume percentage covered after 500 generations

Algorithm	Test Problem			
	ZDT3	ZDT6	DTLZ3	DTLZ6
SPEA2	**100**	**100**	99.95	**99.85**
SPEA2+FUZZY$_1$	96.67	60.96	99.72	93.74
SPEA2+FUZZY$_2$	99.15	75.23	99.82	97.98
NSGA-II	**100**	**100**	99.96	**99.85**
NSGA-II+FUZZY$_1$	95.68	59.04	99.65	91.63
NSGA-II+FUZZY$_2$	99.11	74.61	99.81	96.39
ECEA	**87.63**	**39.61**	98.22	**94.59**
ECEA+FUZZY$_1$	76.25	31.17	97.98	63.09
ECEA+FUZZY$_2$	77.47	33.90	97.99	63.32

Results **in bold** are better than others with statistical significance level $\alpha < 0.01$, according to the Kruskal-Wallis test.

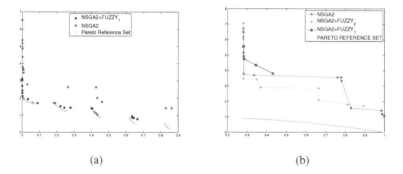

(a)   (b)

Fig. 2. Pareto set comparison: (a) ZDT3 after 400 real function evaluation ; (b) ZDT6 after 1200 real function evaluations

As expected +FUZZY$_1$ is more efficient that +FUZZY$_2$, but the second setup is able to achieve a better quality of solutions. Table 4 present results after 500 generations. MOEA+FUZZY solutions maintain a good quality even if they are not able to find the Pareto optimal set. ZDT6 is the only problem in which, despite the great efficiency improvement, +FUZZYs are not able to find a good approximation of the Pareto optimal set, this is because MOEAs converge slowly. However this problem could be solved using a higher minimum threshold for the fuzzy system learning strategy. Figure 2 presents two comparison: in (a) Pareto sets for ZDT3 problem are shown, demonstrating that NSGA-II+FUZZY$_1$ (as well as NSGA-II+FUZZY$_1$) is able to obtain a very good approximation of the Pareto optimal set after 400 real function evaluation; Pareto sets for ZDT6 problem are drawn in (b), where it is shown NSGA+FUZZY$_2$ outperform NSGA-II+FUZZY$_1$ and classic NSGA-II, but they are still quite far from the Pareto optimal set.

5   Conclusion and Future Work

In this paper we have presented an empirical study on use of fuzzy function approximation to speed up evolutionary multi-objective optimization. The methodology uses a MOEA for heuristic exploration of the search space and a fuzzy system to evaluate the candidate system individuals to be visited. Our methodology works in two phases: firstly all individuals are evaluated using computationally expensive evaluations and their results are used to train the fuzzy system until it becomes reliable; in the second phase the system is used to estimate fitness of all individuals and only promising individuals are actually evaluated to improve the accuracy of the fuzzy system.

Empirical results with low budgets of real evaluations (i.e. from hundreds to three thousand) encourage the use of a fuzzy system as approximate model to improve efficiency of MOEAs. This is because to the strategy used to build the fuzzy system, that allows generating an efficient fitness function approximator without any previous learning phase and knowledge of real function. Strengths of our approach are the inexpensive learning procedure, that could be easily integrated with every MOEA giving the opportunity to take advantage of novel algorithms, and the possibility to set-up the fuzzy system according to a maximum number of real function evaluations.

On the other hand fuzzy systems have the characteristic to allow to embed prior knowledge about the function to be approximated, this could be useful in problems where there is an expert that knows at least part of the behaviours of the objective function to be evaluated. This will be matter of our future work along with a study on improvement of fuzzy system learning and evaluation strategies in order to maximize approximation performance and, thus, tackle the problem of loss of solution quality in longer runs.

References

1. http://www.tik.ee.ethz.ch/pisa/
2. Ascia, G., Catania, V., Di Nuovo, A., Palesi, M., Patti, D.: Performance evaluation of efficient multi-objective evolutionary algorithms for design space exploration of embedded computer systems. Applied Soft Computing 11(1), 382–398 (2011)

3. Bleuler, S., Laumanns, M., Thiele, L., Zitzler, E.: PISA – A Platform and Programming Language Independent Interface for Search Algorithms. In: Fonseca, C.M., Fleming, P.J., Zitzler, E., Deb, K., Thiele, L. (eds.) EMO 2003. LNCS, vol. 2632, pp. 494–508. Springer, Heidelberg (2003)

4. Deb, K., Agrawal, S., Pratap, A., Meyarivan, T.: A Fast Elitist Non-Dominated Sorting Genetic Algorithm for Multi-objective Optimization: NSGA-II. In: Deb, K., Rudolph, G., Lutton, E., Merelo, J.J., Schoenauer, M., Schwefel, H.-P., Yao, X. (eds.) PPSN VI. LNCS, vol. 1917, pp. 849–858. Springer, Heidelberg (2000)

5. Deb, K., Thiele, L., Laumanns, M., Zitzler, E.: Scalable test problems for evolutionary multi-objective optimization. In: Abraham, A., Jain, R., Goldberg, R. (eds.) Evolutionary Multiobjective Optimization: Theoretical Advances and Applications, ch. 6, pp. 105–145. Springer (2005)

6. Gaspar-Cunha, A., Vieira, A.S.: A hybrid multi-objective evolutionary algorithm using an inverse neural network. In: Hybrid Metaheuristics, Workshop at ECAI 2004 (August 2004)

7. Gibbons, J.D.: Nonparametric Statistical Inference, 2nd edn. M. Dekker (1985)

8. Jin, Y.: A comprehensive survey of fitness approximation in evolutionary computation. Soft Computing Journal 9, 3–12 (2005)

9. Knowles, J.: ParEGO: A hybrid algorithm with on-line landscape approximation for expensive multiobjective optimization problems. IEEE Transactions on Evolutionary Computation 10(1), 50–66 (2006)

10. Laumanns, M., Thiele, L., Zitzler, E.: An efficient, adaptive parameter variation scheme for metaheuristics based on the epsilon-constraint method. European Journal of Operational Research 169(3), 932–942 (2006)

11. Takagi, T., Sugeno, M.: Fuzzy identification of systems and its application to modeling and control. IEEE Transactions on System, Man and Cybernetics 15, 116–132 (1985)

12. Tiwari, S., Fadel, G., Deb, K.: Amga2: improving the performance of the archive-based micro-genetic algorithm for multi-objective optimization. Engineering Optimization 43(4), 377–401 (2011)

13. Wang, L.X., Mendel, J.M.: Generating fuzzy rules by learning from examples. IEEE Transactions on System, Man and Cybernetics 22, 1414–1427 (1992)

14. Zeng, K., Zhang, N.Y., Xu, W.L.: A comparative study on sufficient conditions for takagi-sugeno fuzzy systems as universal approximators. IEEE Transactions on Fuzzy Systems 8(6), 773–778 (2000)

15. Zhang, Q., Liu, W., Tsang, E.P.K., Virginas, B.: Expensive multiobjective optimization by moea/d with gaussian process model. IEEE Trans. Evolutionary Computation 14(3), 456–474 (2010)

16. Zhou, A., Qu, B.Y., Li, H., Zhao, S.Z., Suganthan, P.N., Zhang, Q.: Multiobjective evolutionary algorithms: A survey of the state of the art. Swarm and Evolutionary Computation 1(1), 32–49 (2011)

17. Zitzler, E., Deb, K., Thiele, L.: Comparison of multiobjective evolutionary algorithms: Empirical results. Evolutionary Computation 8(2), 173–195 (2000)

18. Zitzler, E., Laumanns, M., Thiele, L.: SPEA2: Improving the strength pareto evolutionary algorithm for multiobjective optimization. In: Proceedings of the EUROGEN 2001 Conference, Athens, Greece, September 19-21, pp. 95–100 (2002)

19. Zitzler, E., Thiele, L., Laumanns, M., Fonseca, C.M., da Fonseca, V.G.: Performance assessment of multiobjective optimizers: an analysis and review. IEEE Transactions Evolutionary Computation 7(2), 117–132 (2003)

Multi-objective Optimization
for Selecting and Scheduling Observations
by Agile Earth Observing Satellites

Panwadee Tangpattanakul[1,2], Nicolas Jozefowiez[1,3], and Pierre Lopez[1,2]

[1] CNRS, LAAS, 7 avenue du colonel Roche, F-31400 Toulouse, France
[2] Univ de Toulouse, LAAS, F-31400 Toulouse, France
[3] Univ de Toulouse, INSA, LAAS, F-31400 Toulouse, France
{panwadee.tangpattanakul,nicolas.jozefowiez,pierre.lopez}@laas.fr

Abstract. This paper presents a biased random-key genetic algorithm for solving a multi-objective optimization problem concerning the management of agile Earth observing satellites. It addresses the selection and scheduling of a subset of photographs from a set of candidates in order to optimize two objectives: maximizing the total profit, and ensuring fairness among users by minimizing the maximum profit difference between users. Two methods, one based on dominance, the other based on indicator, are compared to select the preferred solutions. The methods are evaluated on realistic instances derived from the 2003 ROADEF challenge.

Keywords: Multi-objective optimization, Earth observing satellite, scheduling, genetic algorithm.

1 Introduction

This paper studies the use of multiobjective optimization applied to the scheduling of one Earth observing satellite in a context where multiple users request photographs from the satellite. Genetic algorithms are proposed to solve the problem and experiments are conducted on realistic instances.

The mission of Earth Observing Satellites (EOSs) is to obtain photographs of the Earth surface satisfying users' requirements. When the ground station center receives requests from several users, it has to consider all users' requirements and output an order consisting of a sequence of selected photographs to be transmitted to the satellites. The management problem of EOSs is to select and schedule a subset of photographs from a set of candidates. Among the various types of EOSs, only agile satellites are considered in our study.

An agile EOS has one on-board camera that can move in three axes: roll, pitch, and yaw. It has more efficient capabilities for taking photographs than for example, SPOT5, a non-agile satellite. The selection and scheduling of taking photographs with agile EOSs is more complicated because there are several possible schedules for the same set of selected photographs. The starting time of

C.A. Coello Coello et al. (Eds.): PPSN 2012, Part II, LNCS 7492, pp. 112–121, 2012.

each photograph is not fixed; nonetheless, it must be within a given time interval. This problem is a scheduling problem and it is the issue under consideration in this paper.

Several algorithms including greedy algorithm, dynamic programming, constraint programming, and local search have been applied for solving agile EOSs scheduling problems [1]. The ROADEF 2003 challenge (see http://challenge. roadef.org/2003/en/) requires the scheduling solutions that maximize total profit of the acquired photographs and also satisfy all physical constraints of agile EOSs. The winner used an algorithm based on simulated annealing [2] and the second prize winner proposed an algorithm based on tabu search [3].

Our work considers agile EOSs scheduling problem where the requests emanate from different users. Hence an objective function to maximize the total profit is not sufficient. The ground station center should also share fairly the resources among users. Therefore, a multi-objective model is considered. The idea to use two objective functions related to fairness and efficiency was proposed in [4], and three ways were discussed for solving this sharing problem. The first one gives priority to fairness, the second one to efficiency, and the third one computes a set of trade-offs between fairness and efficiency. For the multi-criteria method, instead of building a complete set of non-dominated solutions, the authors only searched for a decision close to the line with a specified slope on the objective function plane. In [5], a tabu search was used for the multi-satellite, multi-orbit, and multi-user management to select and schedule requests. The upper bounds on the profit were derived by means of a column generation technique. They tested these algorithms with the data instances provided by the French Center for Spatial Studies (CNES).

This paper proposes a biased random-key genetic algorithm (BRKGA) in order to solve the multi-objective optimization problem for selecting and scheduling the subset of required photographs from multiple users. The two objective functions for this scheduling problem are to maximize the total profit and minimize the maximum difference of profit values between users. The second objective function represents the fairness of resources sharing among the users. The solutions must also satisfy the physical constraints of the agile EOSs.

The article is organized as follows. The problem is explained in Section 2. Section 3 describes the biased random-key genetic algorithm for solving the multi-objective optimization problem. The computational results are reported in Section 4. Finally, conclusions and future work are discussed in Section 5.

2 Multi-objective Optimization for Photograph Scheduling Problem of Agile Earth Observing Satellites

According to the mission and physical constraints of agile EOSs, the requests which are required from users cannot be assigned to a satellite directly. The shape of the area of candidate photographs can be either a spot or polygonal. A spot is a small circular area with a radius of less than 10 km. A polygonal area is an area ranging from 20 to 100 km. All requests (both spot and polygonal area) must be managed from the ground station center by transforming the requests into rectangular shapes called strips for which the camera can take a photograph at once. Each spot is considered as one single strip. Each polygonal area is decomposed into several strips with fixed width but variable length. Each strip can be acquired following two possible opposite directions as shown in Figure 1, only one of them will be selected in the scheduling results. Requests can be mono or stereo photographs. A mono photograph is taken only once, whereas a stereo photograph must be acquired twice in the same direction but from different angles.

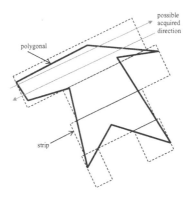

Fig. 1. A polygonal area is decomposed into several strips; each strip can be acquired according to two possible directions

In [6], a simplified version of the problem of managing the mission of agile Earth observing satellites was presented. An instance gives the set of candidate requests with shape type, mono or stereo characteristic, associated gain and surface areas. Let r be the set of requests. These requests are divided into the set of strips s. Each strip includes details, which consist of the identity of request $R[j]$ where that strip is split from, the useful surface area $Su[j]$, duration time $Du[j]$, and earliest and latest visible times from two ends $Te[j,0], Tl[j,0], Te[j,1]$, and $Tl[j,1]$. Each strip is possibly taken from two directions but only one can be selected. Thus, our scheduling problem is solved for selecting and scheduling the possible strip acquisition that is associated with the possible acquisition direction of each strip. If one possible strip acquisition is selected, the other one (possible acquisition in opposite direction) of the same strip is forbidden to be selected. For the profit calculation of each acquired request, its profit can be calculated by a piecewise linear function of gain depending on the fraction of taken useful area and the whole area of each request, as illustrated in Figure 2.

Hence, we extend the case to multiple users as in [5]. However, we solve the problem as a real bi-objective problem. The two objectives are to maximize total profit and ensure fairness between users. For the second objective, the defined function is to minimize the maximum difference in profit between the users. The imperative constraints for finding the feasible solutions are: take each strip within

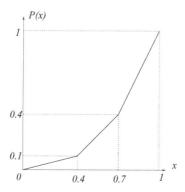

Fig. 2. Piecewise linear function of gain $P(x)$ depending on the effective ratio x of acquired area [6]

their associated time windows, no overlapping images, sufficient transition times, acquire only one direction for each strip, and satisfy the stereoscopic constraint for stereo requests.

3 Biased Random-key Genetic Algorithm for the Multi-user Photograph Scheduling

Genetic algorithm is a heuristic search method that mimics the process of natural evolution. The starting step of genetic algorithm is the initial population generation and the population consists of several chromosomes. Each chromosome, which is formed of several genes, represents one solution. The genetic algorithm involves three mechanisms (selection, crossover, and mutation) to generate the new chromosomes for the next generation and repeats to generate the new generation until the stopping criterion is satisfied.

We propose a genetic algorithm for selecting and scheduling the required photographs for the agile EOSs from multi-user requests. The biased random-key genetic algorithm (BRKGA) [7] is used to solve this scheduling problem with two important steps (encoding and decoding). Two methods are used to select the preferred solutions in each genetic algorithm iteration: i) fast nondominated sorting with crowding distance assignment [8]; ii) indicator based on the hypervolume concept [9]. Let p, p_e, and p_m be the sizes of the population, of the elite set, and of the mutation set, respectively.

3.1 Chromosome Generation in the Encoding Process

The initial population consisting of p chromosomes is generated. Each chromosome consists of genes which are encoded by real values randomly generated in the interval $(0, 1]$. For our problem, each strip can be taken following two opposite directions (but only one direction will be selected). Each gene is associated with one direction of a strip, which is a possible strip acquisition that we will just call *acquisition* in the sequel, for the sake of simplification. It is the reason why we define the number of genes to be equal to twice the number of strips.

3.2   Schedule Generation in the Decoding Process

Each chromosome is decoded in order to obtain one solution, which is a sequence of selected acquisitions of the scheduling problem. In this decoding step, the considered priority of each acquisition depends on the gene values: from high to low. The imperative constraints, except the stereo constraint, are verified during each considered acquisition. The stereo constraint is checked once all acquisitions have been treated. All constraints must be satisfied in order to obtain a feasible solution. The flowchart of these decoding steps is depicted in Figure 3.

The example of one solution from the modified instance, which needs to schedule two strips, is shown in Figure 4. Two strips are considered in this instance; therefore the number of genes equals 4. The random-keys are generated for all genes and each gene represents one acquisition. The decoding steps are used to obtain the sequence of selected acquisitions and the values of the two objective functions.

3.3   Biased Random-Key Genetic Algorithm

In BRKGA, the new population is combined from three parts (selection, crossover, and mutation) [7]. The first part is the selection part in which we can choose a selection method from several efficient algorithms, e.g., NSGA-II [8], IBEA [9], SMS-EMOA [10], etc. We propose two selection methods to choose p_e preferred chromosomes (elite set) from the current population. We copy these p_e chromosomes to the top part of the next population. The two methods are:

1. **Fast nondominated sorting and crowding distance assignment**
   Fast nondominated sorting and crowding distance assignment methods were proposed in the Nondominated Sorting Genetic Algorithm II (NSGA-II) [8]. In our work, the fast non-dominated sorting method is used to find the solutions in rank zero (nondominated solutions). If the number of nondominated solutions is more than the parameter setting value of maximum size of the elite set, the crowding distance assignment method is applied to select some solutions from the nondominated set to become the elite set. Otherwise all nondominated solutions will become the elite set. The concept of the crowding distance assignment method is to get an estimate of the density of solutions surrounding a particular solution in the population.

2. **Indicator based on the hypervolume concept**
   The use of an indicator based on the hypervolume concept was proposed in the Indicator-Based Evolutionary Algorithm (IBEA) [9]. The indicator based method is used to assign fitness values based on the hypervolume concept to the population members and some solutions in the current population are selected to become the elite set for the next population. The

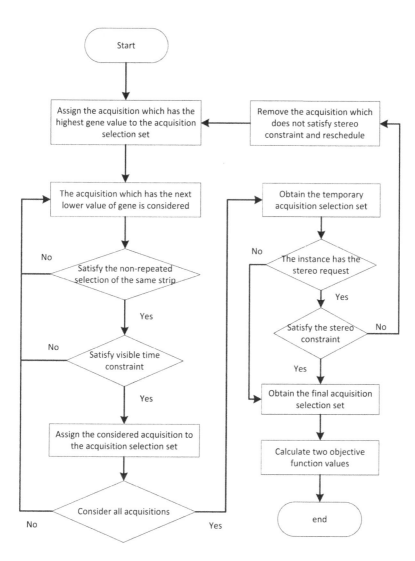

Fig. 3. Decoding steps flowchart of one chromosome into one solution

Random-key chromosome	Acquisition0	Acquisition1	Acquisition2	Acquisition3
	0.6984	0.9939	0.6885	0.2509
Sequence of selected acquisitions	1 2			
Total profit	1.04234e+007			
Maximum difference profit	5.21172e+006			

Fig. 4. Solution example from the modified instance, which needs to schedule two strips

indicator-based method performs binary tournaments for all solutions in the current population and implements environmental selection by removing the worst solution from the population and by updating the fitness values of the remaining solutions. The worst solution is removed repeatedly until the number of remaining solutions satisfies the recommended size of the elite set for BRKGA.

The second part is the bottom part which is the mutant set. It is the set of p_m chromosomes generated to avoid entrapment in a local optimum. These chromosomes are randomly generated by the same method used to generate the initial population. The last part is the crossover part for which each crossover offspring is built from one elite chromosome and one chromosome in the previous population. Each element in the crossover offspring is obtained from the element in elite chromosome with the probability ρ_e. The crossover offspring is stored in the middle part of the new population. Hence, the size of crossover offspring set is $p - p_e - p_m$ to fulfill the remaining space of chromosomes in the next population. The process for generating the next populations is applied repeatedly until the stopping criterion is satisfied.

4   Computational Results

The ROADEF 2003 challenge instances (subsetA) from ROADEF Challenge website (http://challenge.roadef.org/2003/en/sujet.php) are modified for 4-user requirements and the format of instance names are changed to a_b_c, where a is the number of requests, b is the number of stereo requests, and c is the number of strips. For the proposed biased random-key genetic algorithm, the recommended parameter value settings is displayed in Table 1 [7]. Two population sizes of n and $2n$, where n is the length of a chromosome, are tested. The best solutions are stored in the archive set. If there is at least one solution from the current population that can dominate some solutions in the archive set, the archive set will be updated. Thus, we use the number of iterations of the last archive set improvement to be the stopping criterion. The algorithms were experimentally tuned and the stopping value is set to 50. The size of the elite set is equal to the number of non-repeating photograph scheduling results from the nondominated solutions, but it is not over $0.15p$. The size of the mutant set is $0.3p$. The probability of elite element inheritance for crossover operation

Table 1. Recommended parameter values of BRKGA [7]

Parameter	Recommended value
p	$p = a.n$, where $1 \leq a \in \mathbb{R}$ is a constant and n is the length of the chromosome
p_e	$0.10p \leq p_e \leq 0.25p$
p_m	$0.10p \leq p_m \leq 0.30p$
ρ_e	$0.5 \leq \rho_e \leq 0.8$

Table 2. Results of the modified ROADEF 2003 challenge instances subsetA

	Dominance-based			
Instance	Hypervolume		#	CPU
	Average	σ	solutions	time (s)
Population size n 2_0_2	-	-	-	-
4_0_7	4.6734×10^{15}	8.19837×10^{14}	4.8	0
12_2_25	5.00258×10^{16}	2.62563×10^{14}	11.1	1611.7
12_9_28	1.23618×10^{16}	1.06097×10^{15}	7.1	0.3
68_12_106	2.50509×10^{17}	2.22774×10^{16}	20.8	24.2
77_40_147	2.69889×10^{16}	3.93096×10^{15}	25.3	77.7
218_39_295	4.62395×10^{17}	1.11213×10^{17}	25.4	483.9
150_87_342	4.80749×10^{17}	3.59237×10^{16}	27.7	938.9
336_55_483	1.24056×10^{18}	1.58973×10^{17}	23.8	2068.2
375_63_534	1.09387×10^{18}	1.14147×10^{17}	28	1878.9
Population size $2n$ 2_0_2	5.43238×10^{13}	0	1	0
4_0_7	5.78803×10^{15}	6.11588×10^{13}	7.2	0.2
12_2_25	5.08453×10^{16}	6.69058×10^{13}	23.5	3489.9
12_9_28	1.24857×10^{16}	1.43998×10^{15}	7.2	0.6
68_12_106	2.51852×10^{17}	2.20262×10^{16}	37.5	68.8
77_40_147	2.96205×10^{16}	3.19952×10^{15}	36.4	172.2
218_39_295	4.50592×10^{17}	5.74569×10^{16}	26.7	937
150_87_342	5.05842×10^{17}	4.40302×10^{16}	30.2	2013.7
336_55_483	1.15971×10^{18}	1.58218×10^{17}	27.7	2381.2
375_63_534	1.24831×10^{18}	1.93988×10^{17}	26.5	4473.1

	Indicator-based			
Instance	Hypervolume		#	CPU
	Average	σ	solutions	time (s)
Population size n 2_0_2	-	-	-	-
4_0_7	5.36459×10^{15}	2.82301×10^{14}	5.5	0
12_2_25	4.49038×10^{16}	1.5282×10^{15}	10.2	0.2
12_9_28	1.16085×10^{16}	1.56771×10^{15}	5.4	0.3
68_12_106	2.38139×10^{17}	3.06587×10^{16}	9.4	24.3
77_40_147	2.56001×10^{16}	2.87972×10^{15}	12.5	58.1
218_39_295	4.97074×10^{17}	6.973×10^{16}	13.3	450.5
150_87_342	4.4792×10^{17}	2.9806×10^{16}	12.2	852.3
336_55_483	1.26768×10^{18}	1.07123×10^{17}	10.3	2666.4
375_63_534	1.34292×10^{18}	1.45157×10^{17}	10.3	4623.4
Population size $2n$ 2_0_2	5.43238×10^{13}	0	1	0
4_0_7	5.81005×10^{15}	5.93069×10^{13}	8.8	0
12_2_25	4.9165×10^{16}	1.27084×10^{15}	23.1	1.1
12_9_28	1.24013×10^{16}	9.66474×10^{14}	8.1	1.4
68_12_106	2.55655×10^{17}	2.32915×10^{16}	13.2	88.4
77_40_147	2.73546×10^{16}	2.72353×10^{15}	14.7	223.4
218_39_295	5.30441×10^{17}	3.9777×10^{16}	13.2	3017.3
150_87_342	4.8244×10^{17}	3.30219×10^{16}	14	5276.6
336_55_483	1.41375×10^{18}	8.84685×10^{16}	10.3	12904.2
375_63_534	1.39739×10^{18}	1.07724×10^{17}	13.4	28760

is 0.6. Two methods (dominance-based and indicator-based) for selecting some solutions to become the elite set are tested. They are implemented in C++. We test ten runs per instance. Hypervolumes of the approximate Pareto front are calculated by using a reference point of 0 for the first objective and the maximum of the profit summations of each user for the second one. The average and standard deviations values of hypervolumes, the average number of solutions, and average CPU times are reported in Table 2. Each method obtains a set of solutions, which considers both objective functions (maximize total profit and ensure fairness among users) and all constraints of agile EOSs are satisfied. For both methods, when comparing the results from two different population sizes, most of them show that the methods with the population size $2n$ obtain the better average and standard deviation values of hypervolumes and acquire more solutions, but CPU times are higher. In the other way, when we compare results between dominance-based and indicator-based, the average values of hypervolumes cannot show exactly which one obtains the better solutions. However, the standard deviation for the population size $2n$ of indicator-based is better than dominance-based. On the number of solutions and CPU time, dominance-based obtains more solutions and spends less CPU times, especially for large instances. Except for instance 12_2_25, dominance-based takes very high CPU time and this is strange. Hence, more tests are done to check the number of iterations until the stopping criterion is satisfied for instance 12_2_25 and instance 12_9_28. The average number of iterations for instance 12_2_25 and instance 12_9_28 are 2657535.7 and 177.7, respectively. Therefore, instance 12_2_25 spends very high CPU time, because it uses a huge number of iterations until the stopping criterion is satisfied. For instance 2_0_2 when using the population size n, both methods cannot reach any result, because the population size is too small for generating the new generation from 3 parts in BRKGA. Nevertheless, the computation times for large instances are quite high, that means that the efficiency of the decoding methods certainly deserves to be improved.

5   Conclusions and Future Work

Multi-objective optimization is applied to solve the problem of selecting and scheduling the observations of agile Earth observing satellites. The instances of ROADEF 2003 challenge are modified in order to take account explicitly of 4-user requirements. Two objective functions are considered to maximize the total profit and to minimize the maximum difference profit between users for the fairness of resource sharing. Moreover, all constraints have to be satisfied. A biased random-key genetic algorithm (BRKGA) is applied to solve this problem. Random-key encoding generates each chromosome in the population and all of them are decoded to be the solutions. Thus, two methods, fast nondominated sorting with crowding distance assignment on the one hand and indicator based on the hypervolume concept on the other hand, are used for selecting the elite set of solutions from the population. An elite set, a crossover offsprings set, and a mutant set are combined to become the next population. The results of the dominance-based and indicator-based methods with two population sizes are compared. The approximate solutions are obtained but the computation times for large instances are quite high.

This work is still in progress. As a future work, we plan to use other random-key decoding methods in order to reduce the computation times. Moreover, we will apply indicator-based multi-objective local search (IBMOLS) to solve this problem and compare the IBMOLS results with the BRKGA results which are proposed in this paper.

Acknowledgment. This work was supported by THEOS Operational Training Programme (TOTP). The authors gratefully acknowledge this support. Thanks are also due to the referees for their valuable comments.

References

1. Lemaître, M., Verfaillie, G., Jouhaud, F., Lachiver, J.M., Bataille, N.: Selecting and Scheduling Observations of Agile Satellites. Aerospace Science and Technology 6, 367–381 (2002)
2. Kuipers, E.J.: An Algorithm for Selecting and Timetabling Requests for an Earth Observation Satellite. Bulletin de la Société Française de Recherche Opérationnelle et d'Aide à la Décision, 7–10 (2003)
3. Cordeau, J.F., Laporte, G.: Maximizing the Value of an Earth Observation Satellite Orbit. Journal of the Operational Research Society 56, 962–968 (2005)
4. Bataille, N., Lemaître, M., Verfaillie, G.: Efficiency and Fairness when Sharing the Use of a Satellite. In: Proc. Fifth International Symposium on Artificial Intelligence, Robotics and Automation in Space, pp. 465–470 (1999)
5. Bianchessi, N., Cordeau, J.F., Desrosiers, J., Laporte, G., Raymond, V.: A Heuristic for the Multi-Satellite, Multi-Orbit and Multi-User Management of Earth Observation Satellites. European Journal of Operational Research 177, 750–762 (2007)
6. Verfaillie, G., Lemaître, M., Bataille, N., Lachiver, J.M.: Management of the Mission of Earth Observation Satellites Challenge Description. Technical report, Centre National d'Etudes Spatiales, France (2002)
7. Gonçalves, J.F., Resende, M.G.C.: Biased Random-Key Genetic Algorithms for Combinatorial Optimization. Journal of Heuristics 17, 487–525 (2011)
8. Deb, K., Pratep, A., Agarwal, S., Meyarivan, T.: A Fast and Elite Multiobjective Genetic Algorithm: NSGA-II. IEEE Transactions on Evolutionary Computation 6, 182–197 (2002)
9. Zitzler, E., Künzli, S.: Indicator-Based Selection in Multiobjective Search. In: Yao, X., Burke, E.K., Lozano, J.A., Smith, J., Merelo-Guervós, J.J., Bullinaria, J.A., Rowe, J.E., Tiño, P., Kabán, A., Schwefel, H.-P. (eds.) PPSN VIII. LNCS, vol. 3242, pp. 832–842. Springer, Heidelberg (2004)
10. Beume, N., Naujoks, B., Emmerich, M.: SMS-EMOA: Multiobjective selection based on dominated hypervolume. European Journal of Operational Research 181, 1653–1669 (2007)

Tailoring ε-MOEA to Concept-Based Problems

Amiram Moshaiov and Yafit Snir

Faculty of Engineering, Tel-Aviv University, Israel
moshaiov@eng.tau.ac.il, yafit_co@eng.tau.ac.il

Abstract. Concept-based MOEAs are tailored MOEAs that aim at solving problems with a-priori defined subsets of solutions that represent conceptual solutions. In general, the concepts' subsets may be associated with different search spaces and the related mapping into a mutual objective space could have different characteristics from one concept to the other. Of a particular interest are characteristics that may cause premature convergence due to local Pareto-optimal sets within at least one of the concept subsets. First, the known ε-MOEA is tailored to cope with the aforementioned problem. Next, the performance of the new algorithm is compared with C_1-NSGA-II. Concept-based test cases are devised and studied. In addition to demonstrating the significance of premature convergence in concept-based problems, the presented comparison suggests that the proposed tailored MOEA should be preferred over C_1-NSGA-II. Suggestions for future work are also included.

1 Introduction

In the concept-based approach a design concept (in short – concept) is represented by a set of potential solution alternatives [1]. Such a representation has been termed Set-Based Concept (SBC). In contrast to the traditional way of evaluating concepts, the SBC approach allows concept selection to be based not only on optimality considerations, but also on performance variability, which is inherent to the SBC representation [2].

The SBC approach unfolds various ways to compare concepts by their associated sets of performances in objective space [3]. The most studied approach is known as the s-Pareto approach [4]. It involves finding which particular solutions, of which concepts, are associated with the Pareto-front that is obtained by domination comparisons among all individual solutions from all concepts. The interested reader is referred to [5] for some concrete engineering examples of the s-Pareto approach. The current study focuses on such an approach, yet it is restricted to algorithmic aspects rather than to engineering examples.

Concept-based Multi-Objective Evolutionary Algorithms (C-MOEAs) have been originated as a part of the development of a concept-based approach to support conceptual design [1]. C-MOEAs can be obtained by modifying existing MOEAs. This, however, should be done with care. Classical MOEAs are tested for problems where the decision space is common to all solutions. C-MOEAs have to deal with situations where some or all concepts may have, each their own search space. A concept-related premature convergence problem is highly expected when SBCs are

C.A. Coello Coello et al. (Eds.): PPSN 2012, Part II, LNCS 7492, pp. 122–131, 2012.
© Springer-Verlag Berlin Heidelberg 2012

evolved. This is due to situations where at least one SBC may exhibit a local Pareto whereas the other has none. In such a case the algorithm might, at the extreme case, abandon a good concept. Current C-MOEAs have neither been designed to specially cope with this problem, nor have they been tested to examine their performance under such conditions (e.g., [3], [6]). The current work attempts to fill this gap, by tailoring ε-MOEA, [7], to finding the global s-Pareto front for SBCs. It also includes a comparison of the proposed algorithm with a previously reported C-MOEA (of [6]). The comparison is executed with a special focus on the aforementioned computational problem. To simulate situations each concept may have a different decision space, we adapt the common testing approach of MOEAs by running each concept with a different test function. As presented here, the proposed algorithm is proven to be promising for dealing with the local Pareto problem in the context of concept-based problems.

The rest of this paper is organized as follows. Section 2 provides the background for this paper. Section 3 describes the fundamental issues concerning our methodology. Section 4 presents the suggested algorithm, and section 5 provides the details of the executed tests. Finally, section 6 concludes this paper.

2   Background

2.1   MOEAs' Coping with Local Pareto

In complex problems, and in particular those with a large number of local optima, many existing algorithms are likely to return a sub-optimal solution. This phenomenon is termed premature convergence. In multi-objective problems, MOEAs might get stuck at a local Pareto, and hence, could fail to find the global one. There have been several MOEAs developed in recent years, which show promising results concerning the problem of premature convergence. Nevertheless, none promises convergence to the global Pareto-front. One way for tackling this issue is to use epsilon dominance (e.g. [7]). According to [7], the ε-dominance does not allow two solutions within any of predefined hyper-cubes (using ε_i in the i-th objective) to be non-dominated to each other, thereby allowing a good diversity to be maintained in a population. Furthermore, as pointed out in [7], the method is quite pragmatic because it allows the user to choose a suitable ε_i depending on the desired resolution in the i-th objective.

As explained in the introduction concept-related premature convergence problem is highly expected when SBCs are evolved. Due to their promising characteristics, epsilon-based MOEAs are potential candidates to be transformed into C-MOEAs. As demonstrated here, such tailored algorithms can cope with the peculiarities of the concept-based premature convergence problem.

2.2   Overview of Relevant Algorithms

C_1-NSGA-II and C_2-NSGA-II, which are presented in [6], are C-MOEAs that involve tailoring of the original NSGA-II, of [8], to deal with SBCs. Both are based on a simultaneous approach to the search of optimal concepts. Instead of sequentially evolving a single concept in each run using a classical MOEA such as NSGA-II, in C_1-NSGA-II, and also in C_2-NSGA-II, the population contains solutions from several concepts, and they evolve simultaneously. In C_1-NSGA-II solutions of a more fitted concepts spread on the expense of a less fitted concepts, whereas C_2-NSGA-II involves a reduction of the population size on the expense of the less fitted concepts. These algorithms have been investigated for several interesting computational aspects [6]; however, the issue of concept-based local Pareto has neither been examined in testing C_1-NSGA-II, nor in testing C_2-NSGA-II. In the current study, we use C_1-NSGA-II to compare the proposed algorithm with. No comparisons are made with C_2-NSGA-II since that its search mechanism is in principle the same as that of C_1-NSGA-II.

The ε-MOEA, presented in [7], is a classical MOEA, which is computationally fast and capable of finding a well-converged and well-distributed set of solutions. It uses two co-evolving populations: an EA population $P(t)$ and an archive population $A(t)$ (where t is the iteration counter). The run begins with an initial population $P(0)$. The initial archive population $E(0)$ is assigned with the ε-non-dominated solutions of $P(0)$. Thereafter, two solutions, one from $P(t)$ and one from $A(t)$, are chosen for mating and an offspring solution c is created. Thereafter, the offspring solution c can enter either one of the two populations with different strategies. In section 4 ε-MOEA is modified into the proposed C-ε-MOEA.

3   Fundamentals

Section 3.1, which is provided here for the sake of clarity and completeness, briefly describes the concept-based problem that is dealt with in this paper (based on [4], and [6]). Next, section 3.2 provides a discussion on the need to tailor existing MOEAs into C-MOEAs. This discussion is required since that, in general, existing MOEAs can also be used, as-is, to find the s-Pareto.

3.1   Problem Description

In the following, we consider a finite set C of SBCs, namely of candidate-sets of particular solutions, where $|C| = cs$ is the number of the examined concepts (SBCs). Each $S_m \in C$, m= 1,...., cs, represents the solutions belonging to the m-th SBC. Also considered, for each $S_m \in C$, is the feasible-set $X_m \subseteq S_m$ resulting from possible constraints on using members of S_m. Next, let any i-th member of any X_m be denoted as $x_i^m \in X_m$, and let the set X be the union of the feasible members from all candidate-sets. In general, $X_m \cap X_n$ is an empty set for any $m \neq n$. It is noted that for each S_m there is an associated decision-variable space. In general, for any $m \neq n$, it should be assumed that S_m and S_n do not have a mutual decision-variable-space. For a given mapping $F: X \to Y$, the members of the union X are mapped into a mutual multi-objective space $Y \subseteq R^k$, such that for any $x_i^m \in X$ there is one and only one associated vector $y_i^m \in Y$, where $y_i^m = (y_1^{m,i}, ... y_j^{m,i}, ..., y_k^{m,i})$.

A concept-based problem involves determining, for each $S_m \in C$, which represents the m-th concept, an evaluation-set E_m that consists of "passed-members" among the members of X_m. For the s-Pareto approach, [4], [6], the "passed-members" are members of the Pareto-optimal set of X based on domination comparisons in Y among all members of X.

In other words, without loss of generality, the problem amounts to $min \quad y_i^m \in Y$,

over all m and i, to find the s-Pareto front and the associated optimal set. Implicit to the above is that the evaluation sets are meant to be used for concept selection. Another implicit aspect is that the mapping, F, may involve numerical characteristics, which may vary from one concept to the other.

It can be argued that the s-Pareto optimality is essentially no different from the Pareto-optimality [6]. Hence it is valid to ask why C-MOEAs are needed or why traditional MOEAs cannot be used as are.

3.2   Why Tailoring Is Needed?

The intention in using the s-Pareto approach is to find all the Pareto-optimal concepts, where each such concept has at least one member of its set being a non-dominated solution with respect to the entire feasible set of solutions. A tailored MOEA for finding the s-Pareto should ensure adequate representation of the concepts along the s-Pareto-front [6]. This means that the resulting set should contain individuals from all the Pareto-optimal concepts. Furthermore, an adequate representation means that the resulting subsets are well distributed on the front.

As seen in the above section, a concept-based problem is almost equivalent to a classical MOP. It is therefore legitimate to ask why we cannot use traditional MOEAs, as are, to solve a concept-based problem. A sequential search approach is certainly possible, where the front of each SBC is separately found. Yet, as discussed in [6], the use of such an approach could mean the waste of resources on finding the fronts of inferior concepts. In contrast, while carrying efficiency promise, the simultaneous SBC search approach, involves the numerical risk of a concept-related premature convergence problem (see introduction).

A simultaneous search technique could be conceived, in which the entire set of solutions from all concepts is treated by a traditional MOEA without any special tailoring to the problem. This assumes that, posterior to the evolutionary run, the obtained Pareto-optimal set and front can be analyzed to identify the parts associated with each concept. Under the assumption that no crossover can take place among individuals from different concepts, such a search approach is restrictive. Namely, a large part of existing MOEAs use a genetic algorithm approach rather than an evolutionary strategy one and therefore cannot be used as-is. Furthermore, as discussed in [6], the use of any traditional MOEA, without some tailoring, may fail to provide adequate representation of the concepts along the s-Pareto front even under the case of a mutual decision space. This is further explained below.

Even in the case of a mutual search space and assuming that individuals from different concepts can mate, the search is inherently divided into different regions to explore the behavior of concepts rather than just specific solutions. The end result, which is the s-Pareto optimal set and its associated front, should provide an understanding of the distribution of the concepts' representatives on the front, rather than just the distribution of particular solutions without their associated concepts' labels. A simple tailoring of a MOEA, for a simultaneous search of the s-Pareto, would amount to making sure that each concept has sufficient representatives in the population, such that even if its proportional part in the s-Pareto-front is relatively small, it will be adequately found. The use of a classical MOEA could fail to ensure that such an optimal concept will be found [6]. Even under the simple case where convergence characteristics of all concepts are the same, the use of a classical evolutionary strategy-based MOEA, with no distinction among solutions of different concepts may occasionally fail to produce the s-Pareto. This is especially because of the possible existence of "overlapping" regions in the s-Pareto-front where solutions from several concepts are mapped into the same or similar performances in the front. Such a phenomenon may become profound under a situation with a local Pareto.

In summary, different concepts are associated with different decision spaces, or with different regions within a mutual decision space. This may lead to the possibility of a local Pareto within a concept. In a sequential search approach any MOEA that can overcome local Pareto would be sufficient, since that the sequential approach does not involve a simultaneous search within several concepts. In a simultaneous search approach, the existence of a local Pareto-front, within any of the concepts, could be detrimental, as it can cause an improper balance among the search resources given to each concept.

4   The Proposed Algorithm

4.1   Tailoring Requirements

Generally, any state-of-the-art MOEA can be adapted to suit a simultaneous search for the s-Pareto. The main features of the required modifications are: 1. The division of the population to subsets according to the concepts; 2. The restrictions imposed namely no crossover among individuals of different concepts; and 3. The mechanism for resource distribution among the concepts.

A tailored algorithm, termed C-ε-MOEA is introduced below. C-ε-MOEA is a variant of the ε-MOEA algorithm of [7] with some modifications to handle concepts. The following refers to meeting the tailoring requirements by the proposed algorithm. The first two issues are explicitly dealt with as follows. In the proposed C-ε-MOEA the population is divided into sub-populations; each of them represents a different concept. The recombination operator allows recombination only among members of the same sub-population. In contrast, the third requirement concerning resources is only implicitly involved such that a concept that has better performance compared to another concept will be allocated more resources than the second according to the proposed selection process.

4.2   C-ε-MOEA

4.2.1   Main Steps of the Algorithm

The suggested procedure is described as follows:

1 Randomly initialize a population $P(0)$ with equally sized subpopulations for each concept. The concept-based $\mathcal{E}$-non-dominated solutions of $P(0)$, over the entire population, are copied to an archive population $A(0)$ (as detailed in section 4.2.2). Set the iteration counter $t = 0$.

2 One solution p is chosen from the population $P(t)$ using the "pop selection method" (detailed in section 4.2.3).

3 One solution a is chosen from the archive population $A(t)$ using the "concept archive selection method" (detailed in section 4.2.4).

4 One offspring solution c is created using p and a.

5 Solution c is included in $P(t)$ using the "concept pop acceptance method" (detailed in section 4.2.5).

6 Solution c is included in $A(t)$ using the "concept archive acceptance method" (detailed in section 4.2.6).

7 If termination criterion is not satisfied, set $t = t + 1$ and go to Step 2, else report $A(t)$.

4.2.2   Concept-Based Population and Archiving

Similar to the original algorithm of [7], C-ε-MOEA uses two co-evolving populations including a population $P(t)$ and an archive population $A(t)$ (where t is the iteration counter). The proposed MOEA begins with an initial population $P(0)$, which is composed of cs subsets of p solutions each. To meet the first tailoring requirement, as detailed in section 4.2.1, the $A(t)$ and $P(t)$ are maintained such that:

$$A(t) = \cup_{i=1}^{cs} A_i(t) \tag{1}$$

$$P(t) = \cup_{i=1}^{cs} P_i(t) \tag{2}$$

Where the sub-archive $A_i(t)$ and sub-population $P_i(t)$ contains individuals associated only with the i-th concept, and t is the iteration counter.

The archive population $A(0)$ is assigned with the concept-based $\mathcal{E}$-non-dominated solutions of $P(0)$. The concept-based $\mathcal{E}$-non-dominated solutions are obtained ("defined") as follows: for each hyper-box, which has at least one non-dominated solution from the entire P associated with it, we keep neither one such solution nor all. Rather, for each such hyper-box and for each concept i, which has one or more solutions with performances in that hyper-box, we save one solution which is selected randomly from the non-dominated solutions of the concept within that hyper-box.

4.2.3   Pop Selection Method ([7])

This procedure repeats the procedure in [7]. To choose a solution p from $P(t)$, two population members from $P(t)$ are picked up at random, regardless of their concept association, and a domination check is made. If one solution dominates the other, the former is chosen. Otherwise, the event indicates that these two solutions are non-dominated to each other and in such a case we simply choose one of them at random.

4.2.4   Concept Archive Selection Method

In this method we randomly pick a solution a from $A(t)$, from the subset $A_i(t)$ which corresponds to the subset $P_i(t)$ that p was chosen from, namely

$$\{p \in P_i(t) \ and \ a \in A_j(t) \mid i = j\} \tag{3}$$

If $A_i(t)$ is empty, we select another solution a at random from $P_i(t)$. This ensures that in step 4 (see section 4.2.1) the mating is done meeting the second tailoring requirement as detailed in section 4.1

4.2.5   Concept Pop Acceptance Method

This method defines the decision criteria for an offspring c to replace any population member. We compare the offspring with all population members, regardless of their concept association. If any population member dominates the offspring, the offspring is not accepted. Otherwise, if the offspring dominates one or more population members, then the offspring replaces one of the dominated ones (chosen at random). This means that in such a case a change in the allocated resources occurs; no longer the concepts have equal resources (see section 4.1). When both the above tests fail (that is, the offspring is non-dominated by the population members), the offspring replaces a randomly chosen population member from its' own concept sub-population.

4.2.6   Concept Archive Acceptance Method

For the offspring c to be included in the archive population, the offspring is compared with each member of the archive, in the ε-dominance sense, as follows:

1. If the offspring is ε-dominated by a member of the archive – it is not accepted.
2. If the offspring ε-dominates a member of the archive – it replaces that member.
3. If none of the following exists then the offspring is ε-non-dominated with all archive members.
    a. If the offspring shares a hyper-box with an archive member, who is from the same concept as the offspring, then they are compared in the usual dominance sense – and the member which dominates is chosen. Otherwise they are non-dominated and the member which is closer to the B vector, as defined in [7], (in the Euclidean sense) is chosen. If they have the same distance – one is chosen at random.
    b. If none of the archive members, which are associated with the same concept, share the same hyper-box as the offspring, then the offspring is accepted.

It is interesting to note that the suggested procedure ensures that only one solution per concept may exist in each hyper-box.

4.2.7   Algorithm Properties

The following properties of the C-ε-MOEA procedure are derived from the basic ε-MOEA algorithm ([7]):

1. It is a steady-state MOEA.
2. It emphasizes concept-based non-dominated solutions, and by so emphasizes concepts with better performing solutions.

3. It maintains the diversity in the archive by allowing only one solution per concept to be present in each pre-assigned hyper-box on the Pareto-optimal front.
4.   It is an elitist approach.
5. It solves for the s-Pareto front within the pre-defined resolution.

5   Results

Two tests are reported. In each of the tests two concepts are simultaneously evolved. Different test functions are used for the different concepts, where one concept involves a multi-modal behavior, and the other exhibits a single-modality behavior. The used functions include: the ZDT4 multi-modal function, the discrete function, and the SCH function. The definitions of the above can be found in [8]. In the first test, ZDTt4 and SCH are used, respectively, for each of the two concepts tested. In the second test, SCH is replaced by FON. The decision spaces are kept for each of these functions (per concept) as in [8]. All tests, which are described below, are done with a population size of 100 and for 250 generations. We use the real-parameter SBX recombination operator with probability of 0.9 and $\eta c=15$ and a polynomial mutation operator with probability of $1/n$ (n is the number of decision variables) and $\eta m=20$ [8]. The results of C_1-NSGA-II are taken after elite preserving operator is applied. Epsilon values were chosen after several trials to be $\varepsilon_1 = \varepsilon_2 = 0.05$. A too large epsilon will result in a low granulation front – small set of solutions found. A too small epsilon will not make the desired effect on diversity and convergence.

Figures 1 and 2 show typical results of the s-Pareto fronts for the two tests. Clearly in both cases, C-ε-MOEA overcame the numerical difficulty whereas C_1-NSGA-II failed to cope with it. Both tests were run 30 times each with random initial population. The statistics are included in Table 1. We use convergence and sparsity metrics [6] to compare between the two algorithms. It can be observed that while the sparsity metric is similar, there is a significant improvement in convergence when C-ε MOEA is used. Moreover, this is done with better efficiency as the time (measured in seconds) is also significantly decreased.

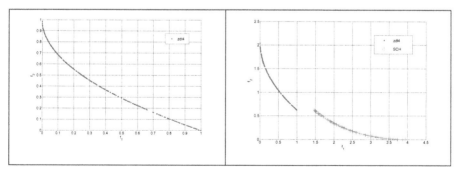

Fig. 1. ZDT4 & SCH
Left and Right: Front by C-ε-MOEA and by C_1-NSGA-II respectively
(ZDT4 designated by dots and SCH by pluses)

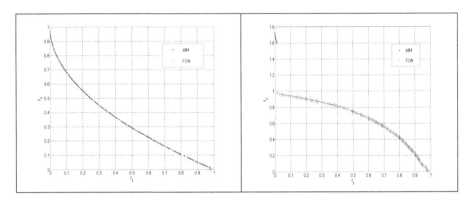

Fig. 2. ZDT4 & FON
Left and right– Resulted front using C-ε-MOEA and C_1-NSGA-II respectively
(ZDT4 designated by dots and FON by pluses)

Table 1. Statistics of the runs

MOEA	Convergence		Sparsity		Time	
	Avg.	SD	Avg.	SD	Avg.	SD
ZDT4 & SCH						
C1-	2.8113	1.8815	0.945	0.0321	1190	115
C-ε-	0.01016	2.43e-03	0.914	0.0710	32	0.8
ZDT4 & FON						
C1-	0.07395	0.00692	0.931	0.0560	1719	123
C-ε-	0.00361	1.92e-04	0.901	0.0462	42	1.7

6   Conclusions and Future Work

Although C1-NSGA-II has been shown to produce good results for many test cases involving SBCs, [6], it is shown here that it often fails to converge in the case of multi-modal concept-based problems. Solving such problems can result in sub-optimal front and may lead to undesired results. This can restrict the application of C1-NSGA-II to real-world problems. In this paper a tailored algorithm, C-ε-MOEA, is proposed, based on [7], in order to deal with the premature convergence difficulty, which is expected in concept-based problems. The experimental results show that C-ε-MOEA is able to obtain the s-Pareto front on hard multi-modal test cases, where C_1-NSGA-II fails to do so.

It should be noted that the current study, which focuses on the s-Pareto approach, is likely to also be most relevant for the future extension of this work to support concept selection by other SBCs methods (e.g. [3]). This is expected since that, in the context of SBCs, the problem of concept-based premature convergence is a generic one.

Future work should include an expansion of the tests done here, and sensitivity analysis to different epsilons. It may also be beneficial to compare the proposed algorithm with others that could be developed based on newer algorithms such as in [9].

References

1. Moshaiov, A., Avigad, G.: Concept-based Multi-objective Problems and Their Solution by EC. In: Proc. of the GECCO 2007 Workshop on User Centric Evolutionary Computation, GECCO, London (2007)
2. Avigad, G., Moshaiov, A.: Set-based Concept Selection in Multi-objective Problems: Optimality versus Variability Approach. J. of Eng. Design 20(3), 217–242 (2009)
3. Denenberg, E., Moshaiov, A.: Evolutionary Search of Optimal Concepts using a Relaxed-Pareto-optimality Approach. In: Proc. of the IEEE Congress on Evolutionary Computations, CEC 2009, Trondheim, Norway, pp. 1093–1100 (2009)
4. Mattson, C.A., Messac, A.: Pareto Frontier based Concept Selection under Uncertainty with Visualization. Optimization and Engineering 6, 85–115 (2005)
5. Mattson, C.A., Mullur, A.A., Messac, A.: Case Studies in Concept Exploration and Selection with s-Pareto Frontiers. Int. J. of Product Development, Special Issue on Space Exploration and Design Optimization 9(1/2/3), 32–59 (2009)
6. Avigad, G., Moshaiov, A.: Simultaneous Concept based Evolutionary Multiobjective Optimization. Applied Soft Computing 11(1), 193–207 (2011)
7. Deb, K., Mohan, M., Mishra, S.: Evaluating the ε-domination based Multi-objective Evolutionary Algorithm for a Quick Computation of Pareto-optimal Solutions. Evolutionary Computation 16(3), 355–384 (2005)
8. Deb, K., Pratap, A., Agarwal, S., Meyarivan, T.: A Fast and Elitist Multiobjective Genetic Algorithm: NSGA-II. IEEE Tran. on Evolutionary Computation 6, 182–197 (2002)
9. Deb, K., Tiwari, S.: Omni-optimizer: A Generic Evolutionary Algorithm for Single and Multi-objective Optimization. European J. of Operational Research 185(3), 1062–1087 (2008)

Recombination of Similar Parents in SMS-EMOA on Many-Objective 0/1 Knapsack Problems

Hisao Ishibuchi, Naoya Akedo, and Yusuke Nojima

Department of Computer Science and Intelligent Systems, Graduate School of Engineering,
Osaka Prefecture University, 1-1 Gakuen-cho, Naka-ku, Sakai, Osaka 599-8531, Japan
{hisaoi@,naoya.akedo@ci.,nojima@}cs.osakafu-u.ac.jp

Abstract. In the evolutionary multiobjective optimization (EMO) community, indicator-based evolutionary algorithms (IBEAs) have rapidly increased their popularity in the last few years thanks to their theoretical background and high search ability. Hypervolume has often been used as an indicator to measure the quality of solution sets in IBEAs. It has been reported in the literature that IBEAs work well on a wide range of multiobjective problems including many-objective problems on which traditional Pareto dominance-based EMO algorithms such as NSGA-II and SPEA2 do not always work well. In this paper, we examine the behavior of SMS-EMOA, which is a frequently-used representative IBEA with a hypervolume indicator function, through computational experiments on many-objective 0/1 knapsack problems. We focus on the effect of two mating strategies on the performance of SMS-EMOA: One is to select extreme parents far from other solutions in the objective space, and the other is to recombine similar parents. Experimental results show that the recombination of similar parents improves the performance of SMS-EMOA on many-objective problems whereas the selection of extreme parents is effective only for a two-objective problem. For comparison, we also examine the effect of these mating strategies on the performance of NSGA-II.

Keywords: Evolutionary multiobjective optimization, evolutionary many-objective optimization, SMS-EMOA, mating schemes, knapsack problems.

1 Introduction

Evolutionary multiobjective optimization (EMO) has been one of the most active research areas in the field of evolutionary computation in the last two decades. Since Goldberg's suggestion in 1989 [9], Pareto dominance-based fitness evaluation has been the mainstream in the EMO community. Almost all of well-known traditional EMO algorithms such as NSGA-II [7], PAES [21], SPEA [35] and SPEA2 [34] are categorized as Pareto dominance-based EMO algorithms. Whereas Pareto dominance-based EMO algorithms have been successfully applied to multiobjective problems in various application fields [5], [6], [28], they do not always work well on many-objective problems with four or more objectives as repeatedly pointed out in the literature [10], [20], [23], [37]. This is because almost all solutions in the current population become non-dominated with each other in early generations in the

C.A. Coello Coello et al. (Eds.): PPSN 2012, Part II, LNCS 7492, pp. 132–142, 2012.

application of EMO algorithms to many-objective problems [16], [17], [24]. When all solutions in the current population are non-dominated, Pareto dominance-based fitness evaluation cannot generate any selection pressure towards the Pareto front. As a result, the convergence property of Pareto dominance-based EMO algorithms is deteriorated in their application to many-objective problems. Motivated by strong intentions to overcome such an undesirable behavior, evolutionary many-objective optimization has become a hot issue in the EMO community in the last few years [1], [26], [27].

Recently, two classes of nontraditional EMO algorithms have attracted a lot of attention as promising approaches to many-objective optimization. One is indicator-based EMO algorithms where an indicator function is used to measure the quality of solution sets [3], [4], [29], [33], [36]. EMO algorithms in this class are referred to as indicator-based evolutionary algorithms (IBEAs). Hypervolume has been frequently used as an indicator function in EMO algorithms in this class because it has a good theoretical background such as Pareto compliance [2], [32]. By using a fast calculation method of the exact hypervolume [30] or an efficient approximation method [3], the applicability of indicator-based EMO algorithms to many-objective problems has been improved. SMS-EMOA [4] in this class has often been used in the literature. Its high search ability on many-objective problems has been clearly demonstrated [29].

The other class is scalarizing function-based EMO algorithms where a number of scalarizing functions with different weight vectors are used to search for a wide variety of Pareto optimal solutions [11], [13], [19], [31]. One advantage of this class is the computational efficiency of scalarizing function calculation. MOEA/D [31] in this class has been frequently used as a high-performance EMO algorithm [12], [22].

Through the use of an indicator or scalarizing functions, these two classes of EMO algorithms overcome the main difficulty in the handling of many-objective problems by traditional EMO algorithms (i.e., the deterioration in their convergence property).

Another difficulty in the handling of many-objective problems, which has not been stressed in the literature, is negative effects of a large solution diversity on the effectiveness of recombination operators. In general, the increase in the number of objectives in a multiobjective problem leads to the increase in the number of its Pareto-optimal solutions and their diversity. As a result, the diversity of solutions in the current population becomes very large in the application of EMO algorithms to many-objective problems. That is, solutions in the current population are totally different from each other. Since good solutions are not likely to be generated from the recombination of totally different solutions, large solution diversity seems to have negative effects on the performance of EMO algorithms on many-objective problems. Actually, it was shown by Sato et al. [25] that the performance of NSGA-II on many-objective 0/1 knapsack problems was improved by local recombination. It was also shown that the performance of MOEA/D on many-objective 0/1 knapsack problems was deteriorated by increasing the size of a neighborhood structure for parent selection [15]. These reported results in the literature suggest the importance of the recombination of similar parents in EMO algorithms on many-objective problems.

The use of mating schemes has been proposed to improve the performance of traditional Pareto dominance-based EMO algorithms in the literature (e.g., [14], [25]). However, their use has not been discussed for MOEA/D or SMS-EMOA. This is because (i) these two algorithms usually show high search ability on a wide range of multiobjective problems, (ii) MOEA/D inherently has a local recombination mechanism based on a neighborhood structure of solutions, and (iii) efficient hypervolume calculation has been the main issue in hypervolume-based IBEAs. The aim of this paper is to clearly demonstrate the usefulness of mating schemes in SMS-EMOA on many-objective 0/1 knapsack problems.

This paper is organized as follows. First we briefly explain a mating scheme in our former study [14] in Section 2, which is used to implement two mating strategies: extreme parent selection and similar parent recombination. Next we explain our many-objective 0/1 knapsack problems in Section 3. Then we show the setting of our computational experiments in Section 4. In Section 5, it is demonstrated that only the similar parent recombination improves the performance of SMS-EMOA and NSGA-II on many-objective problems with four or more objectives while the extreme parent selection as well as the similar parent recombination improves their performance on two-objective problems. Finally Section 6 summarizes this paper.

2   Mating Scheme with Two Mating Strategies

In our former study [14], we proposed a mating scheme in Fig. 1 to examine the effect of the following two mating strategies on the performance of NSGA-II:

(1) Selection of extreme solutions far from other solutions in the objective space.
(2) Recombination of similar parents in the objective space.

In the left part of Fig. 1, α candidates are selected by iterating binary tournament selection with replacement α times. Their average vector is calculated in the objective space. The farthest candidate with the largest distance from the average vector is chosen as Parent A. In the right part, β candidates are selected in the same manner. The closest candidate to Parent A in the objective space is chosen as Parent B.

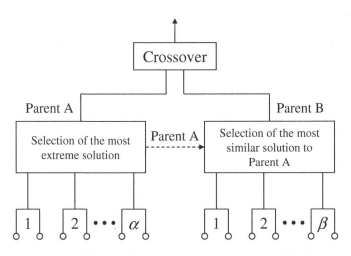

Fig. 1. Mating scheme for NSGA-II with binary tournament selection [14]. Binary tournament selection is replaced with random selection when the mating scheme is used in SMS-EMOA.

When our mating scheme is used in SMS-EMOA, α and β candidates are randomly selected from the population because SMS-EMOA randomly chooses parents from the population. The values of α and β can be viewed as showing the strength of the tendency to choose extreme candidates and to recombine similar candidates, respectively. When $\alpha = \beta = 1$, our mating scheme has no effects on EMO algorithms. In our computational experiments, we examine four values of α and β: 1, 5, 10 and 20.

In our former study [15], we obtained good results by dynamically changing the values of α and β during the execution of NSGA-II. However, we handle α and β as pre-specified constants in this paper to clearly examine the effect of each mating strategy. Experimental results in this paper can be improved by dynamically changing the values of α and β during the execution of NSGA-II and SMS-EMOA.

3   Multiobjective 0/1 Knapsack Problems

As test problems, we use multiobjective 0/1 knapsack problems with two, four, six and eight objectives. Our two-objective test problem is the same as the two-objective 500-item 0/1 knapsack problem with two constraint conditions in Zitzler and Thiele [35]. This two-objective test problem can be written as follows:

$$\text{Maximize} \ \ \mathbf{f}(\mathbf{x}) = (f_1(\mathbf{x}), \ f_2(\mathbf{x})), \tag{1}$$

$$\text{subject to} \ \ \sum_{j=1}^{n} w_{ij} x_j \le c_i, \ \ \ i = 1, 2, \tag{2}$$

$$x_j = 0 \text{ or } 1, \ \ \ j = 1, 2, ..., n, \tag{3}$$

$$\text{where} \ \ f_i(\mathbf{x}) = \sum_{j=1}^{n} p_{ij} x_j, \ \ \ i = 1, 2. \tag{4}$$

In (1)-(4), $n = 500$ (i.e., 500 items), $\mathbf{x}$ is a 500-bit binary string, p_{ij} is the profit of item j according to knapsack i, w_{ij} is the weight of item j according to knapsack i, and c_i is the capacity of knapsack i. This problem is referred to as the 2-500 problem.

As in Zitzler and Thiele [35], we can easily generate other objective functions $f_i(\mathbf{x})$ in the form of (4) for $i = 3, 4, ..., 8$ by randomly specifying each value of p_{ij} as an integer in [10, 100]. In this manner, we have generated 500-item 0/1 knapsack problems with four or more objectives (i.e., 4-500, 6-500 and 8-500 problems). The constraint conditions in (2) of the 2-500 problem are always used in our test problems independent of the number of objectives. This means that all of our test problems with a different number of objectives have exactly the same set of feasible solutions.

For the 2-500 problem, we use the same greedy repair method as in Zitzler and Thiele [35] for handing infeasible solutions. Infeasible solutions are repaired by removing items one by one until all the constraint conditions are satisfied. The order of the items to be removed is specified based on the maximum profit/weight ratio (see [35] for details). The same greedy repair method is used for all of our test problems because they have the same constraint conditions.

4   Setting of Computational Experiments

SMS-EMOA [4] and NSGA-II [7] are used under the following setting:

Population size: 100,
Termination condition: Evaluation of 400,000 solutions,
Crossover probability: 0.8 (Uniform crossover),
Mutation probability: 1/500 (Bit-flip mutation),
Reference point for hypervolume calculation: Origin of the objective space.

SMS-EMOA and NSGA-II are based on a $(\mu+\lambda)$-ES generation update mechanism. Random selection and binary tournament selection are used for parent selection in SMS-EMOA and NSGA-II, respectively. In our computational experiments, μ and λ are specified as $\mu=100$ and $\lambda=1$ in SMS-EMOA and $\mu=\lambda=100$ in NSGA-II. An initial population is randomly generated in each algorithm. Since only a single solution is newly generated for generation update in SMS-EMOA with $\lambda=1$, 100 generations of SMS-EMOA are counted as one generation of NSGA-II when experimental results of SMS-EMOA and NSGA-II at some generations are shown. This is to show their experimental results after the same computation load. Average results are calculated over 100 runs of each algorithm on each test problem except for SMS-EMOA on the 6-500 problems (20 runs) and the 8-500 problems (10 runs).

5   Experimental Results

First we report experimental results on the 2-500 problem. In Fig. 2, we show experimental results of a single run of SMS-EMOA for each of the four settings of our mating scheme: $(\alpha, \beta) = (1, 1), (10, 1), (1, 10), (10, 10)$. Since our mating scheme with $\alpha=1$ and $\beta=1$ does not change SMS-EMOA, Fig. 2 (a) can be viewed as experimental results by SMS-EMOA without the mating scheme. The diversity of solutions is improved by the extreme parent selection with $\alpha=10$ in Fig. 2 (b) and the similar parent recombination with $\beta=10$ in Fig. 2 (c). In Fig. 2 (d), the diversity is further improved by the simultaneous use of these two mating strategies. Experimental results in Fig. 2 are consistent with reported results on the 2-500 problem in the literature [14], [18], [23] where the importance of diversity maintenance was demonstrated.

For comparing the average experimental results over 100 runs between SMS-EMOA and NSGA-II, we show the 50% attainment surface [8] in Fig. 3 for each algorithm with $(\alpha, \beta) = (1, 1), (10, 10)$. In Fig. 3, our mating scheme with $\alpha=10$ and $\beta=10$ has similar effects on SMS-EMOA and NSGA-II. That is, the diversity of solutions is clearly improved while the convergence is slightly degraded.

We further examine the effect of our mating scheme with various settings of α and β on the two algorithms. Experimental results are summarized in Fig. 4 where the average hypervolume over 100 runs of each algorithm is shown for the 4×4 combinations of the four values of α and β: $\alpha=1, 5, 10, 20$ and $\beta=1, 5, 10, 20$. As in Fig. 2, we can see that the performance of SMS-EMOA and NSGA-II is improved by the extreme solution selection $(\alpha>1)$ and the similar parent recombination $(\beta>1)$.

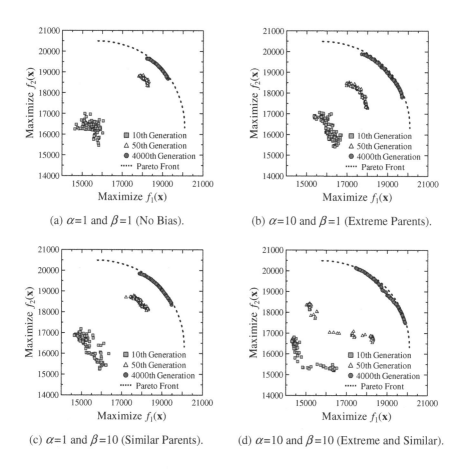

(a) $\alpha=1$ and $\beta=1$ (No Bias).

(b) $\alpha=10$ and $\beta=1$ (Extreme Parents).

(c) $\alpha=1$ and $\beta=10$ (Similar Parents).

(d) $\alpha=10$ and $\beta=10$ (Extreme and Similar).

Fig. 2. Experimental results of a single run of SMS-EMOA on the 2-500 problem

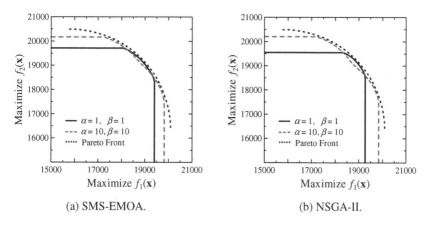

(a) SMS-EMOA.

(b) NSGA-II.

Fig. 3. 50% attainment surface over 100 runs of each EMO algorithm on the 2-500 problem

Experimental results on other test problems are shown in Figs. 5-7. From these figures, we can see that the increase in the number of objectives leads to (i) better results of SMS-EMOA over NSGA-II, (ii) positive effects of the similar parent recombination with $\beta>1$, and the negative effects of the extreme parent selection with $\alpha>1$. Fig. 8 illustrates the effects of the two mating strategies on the behavior of SMS-EMOA on the 4-500 problem by projecting the final population in the four-dimensional objective space onto the f_1-f_2 plane. Fig. 8 (c) suggests the improvement in the diversity and the convergence by the similar parent recombination.

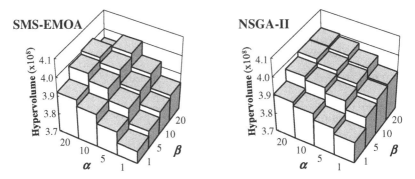

Fig. 4. Average results of SMS-EMOA (left) and NSGA-II (right) on the 2-500 problem

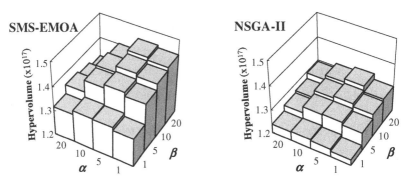

Fig. 5. Average results of SMS-EMOA (left) and NSGA-II (right) on the 4-500 problem

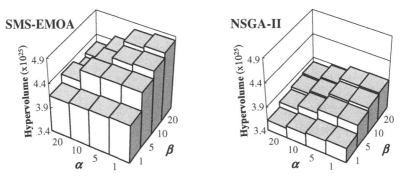

Fig. 6. Average results of SMS-EMOA (left) and NSGA-II (right) on the 6-500 problem

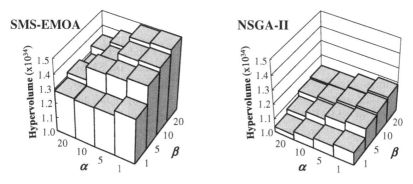

Fig. 7. Average results of SMS-EMOA (left) and NSGA-II (right) on the 8-500 problem

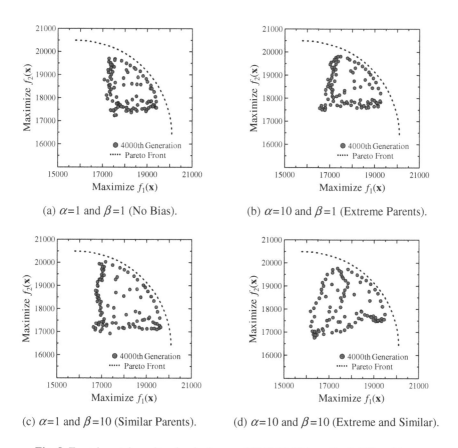

(a) $\alpha=1$ and $\beta=1$ (No Bias).

(b) $\alpha=10$ and $\beta=1$ (Extreme Parents).

(c) $\alpha=1$ and $\beta=10$ (Similar Parents).

(d) $\alpha=10$ and $\beta=10$ (Extreme and Similar).

Fig. 8. Experimental results of a single run of SMS-EMOA on the 4-500 problem

6   Conclusions

We examined the effect of the two mating strategies (i.e., extreme parent selection and similar parent recombination) on the performance of SMS-EMOA through computational experiments on multiobjective 500-item 0/1 knapsack problems with two, four, six and eight objectives. These two mating strategies improved the performance of SMS-EMOA on the 2-500 problem with two objectives. The best results on the 2-500 problem were obtained when the two mating strategies were simultaneously used. The similar parent recombination improved the performance of SMS-EMOA on our test problems independent of the number of objectives. However, the extreme parent selection improved the performance of SMS-EMOA only on the 2-500 problem. Its negative effects were observed on the performance of SMS-EMOA on our many-objective test problems. The performance of SMS-EMOA on the 2-500 problem was similar to that of NSGA-II. By increasing the number of objectives, the advantage of SMS-EMOA over NSGA-II became clear. Moreover, much larger improvements in the average hypervolume measure by the similar parent recombination were obtained in Figs. 5-8 by SMS-EMOA than NSGA-II. In Fig. 2 (c) and Fig. 8 (c), the similar parent recombination increased the diversity without deteriorating the convergence.

References

1. Adra, S.F., Fleming, P.J.: Diversity Management in Evolutionary Many-Objective Optimization. IEEE Trans. on Evolutionary Computation 15, 183–195 (2011)
2. Auger, A., Bader, J., Brockhoff, D., Zitzler, E.: Theory of the Hypervolume Indicator: Optimal μ-Distributions and the Choice of the Reference Point. In: Proc. of FOGA 2009, pp. 87–102 (2009)
3. Bader, J., Zitzler, E.: HypE: An Algorithm for Fast Hypervolume-Based Many-Objective Optimization. Evolutionary Computation 19, 45–76 (2011)
4. Beume, N., Naujoks, B., Emmerich, M.: SMS-EMOA: Multiobjective Selection based on Dominated Hypervolume. European J. of Operational Research 181, 1653–1669 (2007)
5. Coello Coello, C.A., Lamont, G.B.: Applications of Multi-Objective Evolutionary Algorithms. World Scientific, Singapore (2004)
6. Deb, K.: Multi-Objective Optimization using Evolutionary Algorithms. John Wiley & Sons, Chichester (2001)
7. Deb, K., Pratap, A., Agarwal, S., Meyarivan, T.: A Fast and Elitist Multiobjective Genetic Algorithm: NSGA-II. IEEE Trans. on Evolutionary Computation 6, 182–197 (2002)
8. Fonseca, C.M., Fleming, P.J.: On the Performance Assessment and Comparison of Stochastic Multiobjective Optimizers. In: Ebeling, W., Rechenberg, I., Voigt, H.-M., Schwefel, H.-P. (eds.) PPSN IV. LNCS, vol. 1141, pp. 584–593. Springer, Heidelberg (1996)
9. Goldberg, D.E.: Genetic Algorithms in Search, Optimization, and Machine Learning. Addison-Wesley, Reading (1989)
10. Hughes, E.J.: Evolutionary Many-Objective Optimization: Many Once or One Many? In: Proc. of CEC 2005, pp. 222–227 (2005)

11. Hughes, E.J.: MSOPS-II: A General-Purpose Many-Objective Optimizer. In: Proc. of CEC 2007, pp. 3944–3951 (2007)
12. Ishibuchi, H., Akedo, N., Ohyanagi, H., Nojima, Y.: Behavior of EMO Algorithms on Many-Objective Optimization Problems with Correlated Objectives. In: Proc. of CEC 2011, pp. 1465–1472 (2011)
13. Ishibuchi, H., Murata, T.: A Multi-Objective Genetic Local Search Algorithm and Its Application to Flowshop Scheduling. IEEE Trans. on SMC - Part C 28, 392–403 (1998)
14. Ishibuchi, H., Narukawa, K., Tsukamoto, N., Nojima, Y.: An Empirical Study on Similarity-Based Mating for Evolutionary Multiobjective Combinatorial Optimization. European J. of Operational Research 188, 57–75 (2008)
15. Ishibuchi, H., Sakane, Y., Tsukamoto, N., Nojima, Y.: Simultaneous Use of Different Scalarizing Functions in MOEA/D. In: Proc. of GECCO 2010, pp. 519–526 (2010)
16. Ishibuchi, H., Tsukamoto, N., Hitotsuyanagi, Y., Nojima, Y.: Effectiveness of Scalability Improvement Attempts on the Performance of NSGA-II for Many-Objective Problems. In: Proc. of GECCO 2008, pp. 649–656 (2008)
17. Ishibuchi, H., Tsukamoto, N., Nojima, Y.: Evolutionary Many-Objective Optimization: A Short Review. In: Proc. of CEC 2008, pp. 2424–2431 (2008)
18. Ishibuchi, H., Tsukamoto, N., Nojima, Y.: Diversity Improvement by Non-Geometric Binary Crossover in Evolutionary Multiobjective Optimization. IEEE Trans. on Evolutionary Computation 14, 985–998 (2010)
19. Jaszkiewicz, A.: On the Computational Efficiency of Multiple Objective Metaheuristics: The Knapsack Problem Case Study. European J. of Operational Research 158, 418–433 (2004)
20. Khare, V.R., Yao, X., Deb, K.: Performance Scaling of Multi-objective Evolutionary Algorithms. In: Fonseca, C.M., Fleming, P.J., Zitzler, E., Deb, K., Thiele, L. (eds.) EMO 2003. LNCS, vol. 2632, pp. 376–390. Springer, Heidelberg (2003)
21. Knowles, J.D., Corne, D.W.: Approximating the Nondominated Front using the Pareto Archived Evolution Strategy. Evolutionary Computation 8, 149–172 (1999)
22. Li, H., Zhang, Q.: Multiobjective Optimization Problems with Complicated Pareto Sets, MOEA/D and NSGA-II. IEEE Trans. on Evolutionary Computation 13, 284–302 (2009)
23. Purshouse, R.C., Fleming, P.J.: On the Evolutionary Optimization of Many Conflicting Objectives. IEEE Trans. on Evolutionary Computation 11, 770–784 (2007)
24. Sato, H., Aguirre, H.E., Tanaka, K.: Controlling Dominance Area of Solutions and Its Impact on the Performance of MOEAs. In: Obayashi, S., Deb, K., Poloni, C., Hiroyasu, T., Murata, T. (eds.) EMO 2007. LNCS, vol. 4403, pp. 5–20. Springer, Heidelberg (2007)
25. Sato, H., Aguirre, H.E., Tanaka, K.: Local Dominance and Local Recombination in MOEAs on 0/1 Multiobjective Knapsack Problems. European J. of Operational Research 181, 1708–1723 (2007)
26. Schütze, O., Lara, A., Coello Coello, C.A.: On the Influence of the Number of Objectives on the Hardness of a Multiobjective Optimization Problem. IEEE Trans. on Evolutionary Computation 15, 444–455 (2011)
27. Singh, H.K., Isaacs, A., Ray, T.: A Pareto Corner Search Evolutionary Algorithm and Dimensionality Reduction in Many-Objective Optimization Problems. IEEE Trans. on Evolutionary Computation 15, 539–556 (2011)
28. Tan, K.C., Khor, E.F., Lee, T.H.: Multiobjective Evolutionary Algorithms and Applications. Springer, Berlin (2005)
29. Wagner, T., Beume, N., Naujoks, B.: Pareto-, Aggregation-, and Indicator-Based Methods in Many-Objective Optimization. In: Obayashi, S., Deb, K., Poloni, C., Hiroyasu, T., Murata, T. (eds.) EMO 2007. LNCS, vol. 4403, pp. 742–756. Springer, Heidelberg (2007)

30. While, L., Bradstreet, L., Barone, L.: A Fast Way of Calculating Exact Hypervolumes. IEEE Trans. on Evolutionary Computation 16, 86–95 (2012)
31. Zhang, Q., Li, H.: MOEA/D: A Multiobjective Evolutionary Algorithm Based on Decomposition. IEEE Trans. on Evolutionary Computation 11, 712–731 (2007)
32. Zitzler, E., Brockhoff, D., Thiele, L.: The Hypervolume Indicator Revisited: On the Design of Pareto-compliant Indicators Via Weighted Integration. In: Obayashi, S., Deb, K., Poloni, C., Hiroyasu, T., Murata, T. (eds.) EMO 2007. LNCS, vol. 4403, pp. 862–876. Springer, Heidelberg (2007)
33. Zitzler, E., Künzli, S.: Indicator-Based Selection in Multiobjective Search. In: Yao, X., Burke, E.K., Lozano, J.A., Smith, J., Merelo-Guervós, J.J., Bullinaria, J.A., Rowe, J.E., Tiňo, P., Kabán, A., Schwefel, H.-P. (eds.) PPSN VIII. LNCS, vol. 3242, pp. 832–842. Springer, Heidelberg (2004)
34. Zitzler, E., Laumanns, M., Thiele, L.: SPEA2: Improving the Strength Pareto Evolutionary Algorithm. TIK-Report 103, Department of Electrical Engineering, ETH, Zurich (2001)
35. Zitzler, E., Thiele, L.: Multiobjective Evolutionary Algorithms: A Comparative Case Study and the Strength Pareto Approach. IEEE Trans. on Evolutionary Computation 3, 257–271 (1999)
36. Zitzler, E., Thiele, L., Bader, J.: On Set-Based Multiobjective Optimization. IEEE Trans. on Evolutionary Computation 14, 58–79 (2010)
37. Zou, X., Chen, Y., Liu, M., Kang, L.: A New Evolutionary Algorithm for Solving Many-Objective Optimization Problems. IEEE Trans. on SMC - Part B 38, 1402–1412 (2008)

An Artificial Bee Colony Algorithm
for the Unrelated Parallel Machines Scheduling
Problem

Francisco J. Rodriguez[1], Carlos García-Martínez[3], Christian Blum[2],
and Manuel Lozano[1]

[1] Department of Computer Science and Artificial Intelligence,
University of Granada, Granada, Spain
[2] ALBCOM Research Group, Technical University of Catalonia, Barcelona, Spain
[3] Department of Computing and Numerical Analysis,
University of Córdoba, Córdoba, Spain
fjrodriguez@decsai.ugr.es, cgarcia@uco.es,
cblum@lsi.upc.edu, lozano@decsai.ugr.es

Abstract. In this work, we tackle the problem of scheduling a set of jobs
on a set of non-identical parallel machines with the goal of minimising
the total weighted completion times. Artificial bee colony (ABC) algorithm is a new optimization technique inspired by the intelligent foraging
behaviour of honey-bee swarm. These algorithms have shown a better
or similar performance to those of other population-based algorithms,
with the advantage of employing fewer control parameters. This paper
proposes an ABC algorithm that combines the basic scheme with two
significant elements: (1) a local search method to enhance the exploitation capability of basic ABC and (2) a neighbourhood operator based on
iterated greedy constructive-destructive procedure. The benefits of the
proposal in comparison to three different metaheuristic proposed in the
literature are experimentally shown.

Keywords: discrete optimisation, metaheuristics, artificial bee colony,
unrelated parallel machines schedulling problem.

1 Introduction

The *unrelated parallel machine scheduling with minimising total weighted completion times* (UPMSP) considers a set J of n independent jobs that have to be
processed on a set M of m parallel non-identical machines. Each job $j \in J$ has
to be processed by exactly one of the m parallel machines and no machine can
process more than one job at the same time. A job j is processed on a given
machine until completion, i.e., without pre-emption. If a job j is processed on
a machine i, it will take a positive integral processing time p_{ij} whose value is
determined arbitrarily. The objective is to schedule the jobs in such a way that
the sum of the weighted completion times of the jobs is minimised:

$$\text{Minimise } \sum_{i=1}^{n} w_j * C_j,$$

where C_j represents the completion time of job j for a given schedule.

C.A. Coello Coello et al. (Eds.): PPSN 2012, Part II, LNCS 7492, pp. 143–152, 2012.

It is important to note that the jobs assigned to a specific machine are processed in non-decreasing order with respect to the ratio between processing time (p_{ij}) and weight (w_j). This order is known as the *weighted shortest processing time* order. According to [1], sequencing the jobs in each machine following this ordering produces an optimal scheduling for this machine.

According to the standard notation proposed by Azizoglu et al. [2] and Allahverdi et al. [3], the family of problems considered in this work is notated in the literature in the following manner: $R_m || \sum w_j * C_j$. Different real-world applications of scheduling on parallel machines can be found in the literature, covering a wide variety of fields. Some of these fields are human resources [4], production management [5,6,7], mail facilities [8], robotized systems [9], sport tournaments [10], and chemical processes [11].

A mixed integer linear programming formulation for the UPMSP problem is provided for the sake of completeness. Let $x_{ji}^t = 1$ if the job j is processed in the tth position (unit time) on machine i and 0, otherwise. And a variable C_{ji}^t denotes the completion time of the job j scheduled in the tth position on machine i. The model is stated by Azizoglu and Kirca [12] as:

$$\min \sum_j \sum_i \sum_t w_j \cdot C_{ji}^t \cdot x_{ji}^t \tag{1}$$

$$\text{subject to: } \sum_i \sum_t x_{ji}^t = 1 \ \forall j, \tag{2}$$

$$\sum_j x_{ji}^t \leq 1 \ \forall t, i, \tag{3}$$

$$C_{ji}^t = \sum_{r=1}^{n} \sum_{s=1}^{t-1} p_{ir} \cdot x_{ri}^s + p_{ij} \ \forall j, t, i, \tag{4}$$

$$x_{ji}^t \in \{0, 1\} \ \forall j, t, i. \tag{5}$$

In this paper, an approach using the Artificial Bee Colony (ABC) [13,14] method is explored for solving the UPMSP. The ABC algorithm is a new swarm optimization approach that is inspired by the intelligent foraging behaviour of honey-bee swarm. It consists of three essential components: food source positions, nectar-amount and three honey-bee classes (employed bees, onlookers and scouts). Each food source position represents a feasible solution for the problem under consideration. The nectar-amount for a food source represents the quality of such solution (represented by an objective function value). Each bee-class symbolizes one particular operation for generating new candidate food source positions. Specifically, employed bees search the food around the food source in their memory; meanwhile they pass their food information to onlooker bees. Onlooker bees tend to select good food sources from those founded by the employed bees, and then further search the foods around the selected food source. If the employed bee and onlookers associated with a food source cannot find a better neighboring food source, the latter is abandoned and the employed bee associated with this food source becomes a scout bee that performs a search for discovering new food sources. After the scout finds a new food source, it becomes an employed bee again. Due to its simplicity and ease of implementation, the ABC algorithm has captured much attention. Besides, ABC has exhibited state-of-the-art performances for a considerable number of problems [15,16,17].

The proposed ABC extends the basic scheme by considering two significant elements in order to improve its performance when dealing with the UPMSP. In the first place, after producing neighbouring food sources (in the employed and onlooker bees phases), a local search is applied with a predefined probability to further improve the quality of some of the solutions. Secondly, we propose a new neighbourhood operator based on the solution constructive-destructive process performed by iterated greedy (IG) algorithms [18,19].

The remainder of this paper is organized as follows. In Section 2, we outline different metaheuristics proposed in the literature for the UPMSP. In Section 3, we present in detail the proposed ABC for the UPMSP. In Section 4, we present an empirical study that compares the behaviour of the ABC algorithm with regards to those of other metaheuristics from the literature. Finally, in Section 5, we discuss conclusions and further work.

2   Metaheuristics for the UPMSP

Since the introduction of this problem by McNaughton in [20], it has received much attention and many papers have been published in this area. The research efforts to deal with the SNIM-WCT problem have focused on three main research lines: exact procedures, approximation algorithms through solving relaxations of the problem, and metaheuristics procedures. Concerning the latter, Vredeveld et al. [21] presented two types of neighbourhood functions. The first function is called the *jump neighbourhood*. It consists of selecting a job j and a machine i so that job j is not scheduled on machine i. Then job j is moved to machine i. The second one is called *swap neighbourhood*. For this neighbourhood, two jobs j and k must be selected and assigned to different machines. The corresponding neighbouring solution is obtained by interchanging the machine allocations of the two selected jobs. These two neighbourhood functions are applied in two metaheuristic, a multistart local search and a tabu search.

Recently, Li et al. [22] presented a *genetic algorithm* approach to deal with unrelated parallel machines scheduling using three different performance criteria. In particular, the proposed approach initialises the population adding some solutions generated by heuristics methods. The remaining ones are generated randomly to provide enough diversity. Roulette wheel selection is used to choose a new population with respect to a fitness-proportional probability distribution. The crossover and mutation schemes are those proposed by Cheng et al. [23]. Elitism is considered by removing two chromosomes and adding the best two previous chromosomes to the new generation if they are not selected through the roulette-wheel-selection process. The experimental study performed compares the proposed genetic algorithm with a set of heuristics. Results show that the proposed algorithm outperforms the competing heuristics.

3   Proposed ABC for the UPMSP

In this section, we describe the proposed ABC algorithm for the UPMSP. The general scheme of the proposed approach is outlined in Figure 1. It starts by initialising a population P with $|P| - 1$ random solutions (Initialise()). The remaining solution is initialised by means of a greedy procedure (GreedyProcedure). Then, the following steps are repeated until time limit t_{max} is reached:

- *Employed bees phase.* In this step, employed bees produce new solutions by means of a neighbourhood operator (GenerateNeighbour()). In order to enhance the exploitation capability of ABC, a local search method (LocalSearch()) is applied to the solution obtained by the neighbourhood operator with a probability $probLS$.
- *Onlooker bees phase.* Onlookers bees look for new solutions from solutions of the population selected by means of a binary tournament selection (BinaryTournament()). Specifically, an onlooker bee selects the best food source among two food sources that were randomly picked up from the population. Later, each onlooker bee performs, as in employed bees phase, the neighbourhood operator on the selected solution and a local search procedure.
- *Scout bees phase.* In this phase, the scout bees determine the solutions that has not been improved for *limit* iterations and replace them by new random solutions (Initialise()).

At the end of execution, the best solution found (S_b) is returned by the algorithm (BestSolutionFound()).

3.1   The Greedy Procedure

The greedy constructive procedure used for the initialisation of a solution of the population and the re-construction of partial solutions in the neighbourhood operator works as follows. At each step, it considers placing a unassigned job in any of the machines. For each of these options, it calculates a contribution value according to a predefined heuristic (H). The option which causes the best heuristic value is selected. The procedure stops once all the unassigned jobs are allocated. The heuristic H that guided the greedy procedure is that proposed for this problem in [22]. This heuristic distinguishes two steps. In the first one, a job $\bar{j}_* \in \bar{J}$ (set of unassigned jobs) is selected according to:

$$\bar{j}_* = \mathrm{argmin}\{t_i + p_{ij}/w_j : i = 1, \ldots, m; j = 1, \ldots, n\},$$

once the job $\bar{j}_*$ is selected, machine i_* in which it will be processed is the one that holds:

$$i_* = \mathrm{argmin}\{t_i + p_{i\bar{j}_*}/w_{\bar{j}_*} : i = 1, \ldots, m\},$$

being t_i the completion time of the last job scheduled on the machine i.

```
      Input: t_max, P, n_d, probLS, limit,H
      Output: S_b
      //Number of employed and onlooker bees
  1   NE ← |P|;
  2   NO ← |P|;
      //Initialisation phase
  3   for i ← 2 to |P| do
  4   |   S_i ← Initialise() ;
  5   end
  6   S_1 ← GreedyProcedure(H);
  7   while computation time limit t_max not reached do
          //Employed bees phase
  8   |   for i ← 1 to NE do
  9   |   |   E ← GenerateNeighbour(S_i, n_d, H);
 10   |   |   E' ← LocalSearch(E);
 11   |   |   if E' is better than S_i then
 12   |   |   |   S_i ← E' ;
 13   |   |   end
 14   |   end
          //Onlooker bees phase
 15   |   for i ← 1 to NO do
 16   |   |   j ← BinaryTournament(S_1, ..., S_{|P|});
 17   |   |   O ← GenerateNeighbour(S_j, n_d, H);
 18   |   |   O' ← LocalSearch(O);
 19   |   |   if O' is better than S_j then
 20   |   |   |   S_i ← O' ;
 21   |   |   end
 22   |   end
          //Scout bees phase
 23   |   for i ← 1 to |P| do
 24   |   |   if S_i does not change for limit iterations then
 25   |   |   |   S_i ← Initialise() ;
 26   |   |   end
 27   |   end
 28   |   S_b ← BestSolutionFound();
 29   end
```

Fig. 1. ABC scheme

3.2 The Neigbourhood Operator

The proposed neighbourhood operator is based on the constructive-destructive procedure used in IG algorithms [18,19]. IG iteratively tries to refine a solution by removing elements from this solution, by means of a destructive procedure, and reconstructing the resulting partial solution using a greedy constructive procedure. In this case, the proposed neighbourhood operator consists of a unique

Table 1. 12 instance types considered concerning the unrelated parallel machines scheduling problem. The last table column provides the maximum CPU time limit for each instance type (in seconds).

Number of jobs (n)	Number of machines (m)	Time limit (s)
20	5	40
	10	40
50	5	100
	10	100
	20	100
100	5	200
	10	200
	20	200
200	5	400
	10	400
	20	400
	50	400

iteration of the constructive-destructive procedure. In the first place, n_d elements of the current solution are removed (destruction). Then, the partial solution obtained before is reconstructed using a greedy procedure (construction, see Section 3.1). The greedy procedure is guided by the heuristic commented in section 3.1.

4   Computational Experiments

This section describes the computational experiments performed to assess the performance of the ABC model presented in the previous section. Our own algorithms (ABC) as well as all competitor algorithms have been implemented in C++ and the source code has been compiled with gcc 4.5. All experiments were conducted on a computer with a 2.8 GHz Intel i7 processor with 12 GB of RAM running Fedora Linux V15. In this work we considered problem instances from 12 different combinations of the number of jobs (n) and the number of machines (m). These 12 instance types are shown in the first two columns of Table 1. Moreover, the same table shows—in the 3rd column—the maximum CPU time allotted for each instance type ($2 \cdot n$ seconds). For each of the 12 instance types ten problem instances were randomly generated, which is a common choice in recent works dealing with this and related problems [24,25]. The weights of the n jobs were selected uniformly at random from $\{1, \ldots, 10\}$ and the processing time of job j on machine i (p_{ij}, $i = 1, \ldots, n$ and $j = 1, \ldots, m$) was chosen uniformly at random from $\{1, \ldots, 100\}$.

Non-parametric tests [26] have been used to compare the results of the different optimization algorithms under consideration. The only condition to be fulfilled for the use of non-parametric tests is that the algorithms to be compared should have been tested under the same conditions (that is, the same set of problem instances, the same stopping conditions, the same number of runs, etc). Specifically, Wilcoxon's matched-pairs signed-ranks test is used to compare the results of two algorithms.

4.1   Tuning Experiments

In the first place, we have performed an experimental study in order to perform a fine-tuning of the ABC presented above. The goal of the first preliminary experiment is to identify the best combination of the values for the following algorithm parameters:

1. **Population size** ($|P|$). Experiments with $|P|$ in $\{10, 15, 20, 25, 30\}$ were develped.
2. **Destruction size** (n_d). For the percentage of elements dropped from a solution during the neighbourhood operator, values from $\{5\%, 10\%, 15\%, 20\% \ 25\%\}$ were considered.
3. **Probability of performing local search** ($probLS$). This parameter takes values from $\{0.05, 0.1, 0.2, 0.5, 1\}$).
4. **Local search procedure** ($typeLS$). Three different local search procedures [21] were tested: first-improvement local search with jump moves ($FI - JM$), first-improvement local search with swap moves ($FI - SW$), and first improvement local search with both kinds of moves ($FI - JSM$).
5. **Iterations to determine an exhausted food source** ($limit$). Values from $\{0.25 \cdot n, 0.5 \cdot n, n, 2 \cdot n\}$) were considered, where n is the number of jobs of the instance.

For each combination of values for the different parameters (full factorial design), we applied ABC to each of the 120 problem instances. Through a rank-based analysis on results obtained, we identified the parameter combination with the best average rank over all testing instances. This combination is specified in Table 2.

Table 2. Parameters values

Parameter	Value
Population size ($|P|$)	15
Elements dropped (n_d)	25%
Local search procedure ($typeLS$)	FI-JSM
Local search probability ($probLS$)	0.2
Iterations to abandon a food source ($limit$)	$0.25 \cdot n$

4.2   Comparison with Other Metaheuristics

In this section, we compare ABC to different approaches found in the literature for tackling the UPMSP. More specifically, we considered the following approaches (see also Section 2):

- Iterative multistart method (**MultiS**) [21].
- Tabu search (**Tabu**) [21].
- Genetic algorithm (**GA**) [22].

The parameter values used for each considered algorithm are those recommended in the original works. In order to assure a fair comparison, each algorithm was applied under the same conditions as ABC that is, each algorithm was applied exactly once to each of the 120 problem instances. Moreover, the same CPU time limits were used as with ABC (see Table 1).

Table 3. ABC versus competitors using Wilcoxon's test (level of significance $\alpha = 0.05$, critical value $= 13$)

Competitor	$R+$	$R-$	Diff.?
GA	78	0	yes
Tabu	76.5	1.5	yes
MultiS	73	5	yes

The results of the considered algorithms in Tables 3 and 4 allow us to make the following observations:

- The proposed ABC statistically outperforms all competing algorithms (see Table 3).
- Concerning the results shown in Table 4, it is important to highlight that ABC obtains the best average results in all the instances. Moreover, the most significant differences with respect to their competitors are obtained on large instances.
- Only MultiS and Tabu are able to match the results of ABC on the smallest problem instances.

Table 4. Results of the studied algorithms averaged over the 10 instances of each of the 12 instance types

n	m	ABC	GA	Tabu	MultiS
200	50	**5004**	5138	5062	5063
200	20	**16988**	17345	17066	17045
200	10	**52919**	53677	53039	52979
200	5	**182102**	183400	182166	182130
100	20	**5601**	5858	5657	5608
100	10	**14921**	15138	14993	14926
100	5	**45961**	46412	46014	45062
50	20	**1833**	1908	1858	**1833**
50	10	**4611**	4737	4655	**4611**
50	5	**12625**	12787	12643	**12625**
20	10	**1299**	1330	**1299**	**1299**
20	5	**2512**	2570	**2512**	**2512**

5   Conclusions and Future Work

In this paper, we presented an ABC algorithm for the UPMSP. The proposed algorithm add a local search procedure and a novel IG-based neighbourhood operator to the basic ABC scheme. This neighbourhood operator is based on the constructive-destructive procedure of IG algorithms. The resulting ABC algorithm has proved to be superior, especially in the case of larger instances, to three different metaheuristic existing in the literature for this problem. We can conclude from the experiments performed that this algorithm represents a very competitive alternative to the existing methods for the UPMSP.

We believe that the ABC algorithm presented in this paper is a significant contribution, worthy of future study. We will mainly focus on the following avenues of possible research: (1) to adapt the ABC approach for its application to other variants of scheduling problems on parallel machines and (2) to employ the IG-based neighbourhood operator in ABC approaches dealing with other challenging optimisation problems.

Acknowledgements. This work was supported by grant TIN2011-24124 of the Spanish government and by grant P08-TIC-4173 of the Andalusian regional government.

References

1. Elmaghraby, S., Park, S.: Scheduling jobs on a number of identical machines. AIIE Transactions 6(1), 1–13 (1974)
2. Azizoglu, M., Kirca, O.: On the minimization of total weighted flow time with identical and uniform parallel machines. European Journal of Operational Research 113(1), 91–100 (1999)
3. Allahverdi, A., Gupta, J., Aldowaisan, T.: A review of scheduling research involving setup considerations. Omega 27(2), 219–239 (1999)
4. Rosenbloom, E., Goertzen, N.: Cyclic nurse scheduling. European Journal of Operational Research 31, 19–23 (1987)
5. Buxey, G.: Production scheduling: Practice and theory. European Journal of Operational Research 39, 17–31 (1989)
6. Dodin, B., Chan, K.H.: Application of production scheduling methods to external and internal audit scheduling. European Journal of Operational Research 52(3), 267–279 (1991)
7. Pendharkar, P., Rodger, J.: Nonlinear programming and genetic search application for production scheduling in coal mines. Annals of Operations Research 95(1), 251–267 (2000)
8. Jarrah, A.I.Z., Bard, J.F., de Silva, A.H.: A heuristic for machine scheduling at general mail facilities. European Journal of Operational Research 63(2), 192–206 (1992)
9. Rochat, Y.: A genetic approach for solving a scheduling problem in a robotized analytical system. Journal of Heuristics 4, 245–261 (1998)
10. Croce, F.D., Tadei, R., Asioli, P.: Scheduling a round robin tennis tournamentunder courts and players availability constraints. Annals of Operations Research 92, 349–361 (1999)
11. Brucker, P., Hurink, J.: Solving a chemical batch scheduling problem by local search. Annals of Operations Research 96(1), 17–38 (2000)
12. Azizoglu, M., Kirca, O.: Scheduling jobs on unrelated parallel machines to minimize regular total cost functions. IIE Transactions 31(2), 153–159 (1999)
13. Karaboga, D., Basturk, B.: A powerful and efficient algorithm for numerical function optimization: artificial bee colony (abc) algorithm. J. of Global Optimization 39(3), 459–471 (2007)

14. Karaboga, D., Basturk, B.: On the performance of artificial bee colony (ABC) algorithm. Applied Soft Computing 8(1), 687–697 (2008)
15. Sundar, S., Singh, A.: A swarm intelligence approach to the quadratic minimum spanning tree problem. Information Sciences 180(17), 3182–3191 (2010)
16. Kashan, M.H., Nahavandi, N., Kashan, A.H.: DisABC: A new artificial bee colony algorithm for binary optimization. Applied Soft Computing 12(1), 342–352 (2012)
17. Akbari, R., Hedayatzadeh, R., Ziarati, K., Hassanizadeh, B.: A multi-objective artificial bee colony algorithm. Swarm and Evolutionary Computation 2, 39–52 (2012)
18. Jacobs, L., Brusco, M.: A local-search heuristic for large set-covering problems. Naval Research Logistics 42, 1129–1140 (1995)
19. Ruiz, R., Stützle, T.: A simple and effective iterated greedy algorithm for the permutation flowshop scheduling problem. European Journal of Operational Research 177(3), 2033–2049 (2007)
20. McNaughton, R.: Scheduling with deadlines and loss functions. Management Science 6(1), 1–12 (1959)
21. Vredeveld, T., Hurkens, C.: Experimental comparison of approximation algorithms for scheduling unrelated parallel machines. Informs Journal on Computing 14(2), 175–189 (2002)
22. Lin, Y., Pfund, M., Fowler, J.: Heuristics for minimizing regular performance measures in unrelated parallel machine scheduling problems. Computers & Operations Research 38(6), 901–916 (2011)
23. Cheng, R., Gen, M., Tozawa, T.: Minmax earliness/tardiness scheduling in identical parallel machine system using genetic algorithms. Computers & Industrial Engineering 29(1-4), 513–517 (1995)
24. Fanjul-Peyro, L., Ruiz, R.: Iterated greedy local search methods for unrelated parallel machine scheduling. European Journal of Operational Research 207(1), 55–69 (2010)
25. Zaidi, M., Jarboui, B., Loukil, T., Kacem, I.: Hybrid meta-heuristics for uniform parallel machine to minimize total weighted completion time. In: Proc. of 8th International Conference of Modeling and Simulation, MOSIM 2010 (2010)
26. Garcia, S., Molina, D., Lozano, M., Herrera, F.: A study on the use of non-parametric tests for analyzing the evolutionary algorithms' behaviour: A case study on the CEC'2005 special session on real parameter optimization. Journal of Heuristics 15, 617–644 (2008)

Controlling the Parameters of the Particle Swarm Optimization with a Self-Organized Criticality Model

Carlos M. Fernandes[1,2], Juan J. Merelo[2], and Agostinho C. Rosa[1]

[1] Technical University of Lisbon
c.m.fernandes.photo@gmail.com, acrosa@laseeb.org
[2] University of Granada
jjmerelo@gmail.com

Abstract. This paper investigates a Particle Swarm Optimization (PSO) with a Self-Organized Criticality (SOC) strategy that controls the parameter values and perturbs the position of the particles. The algorithm uses a SOC system known as *Bak-Sneppen* for establishing the inertia weight and acceleration coefficients for each particle in each time-step. Besides adjusting the parameters, the SOC model may be also used to perturb the particles' positions, thus increasing exploration and preventing premature convergence. The implementation of both schemes is straightforward and does not require hand-tuning. An empirical study compares the *Bak-Sneppen PSO* (BS-PSO) with other PSOs, including a state-of-the-art algorithm with dynamic variation of the weight and perturbation of the particles. The results demonstrate the validity of the algorithm.

1 Introduction

Inspired by the swarm and social behavior of bird flocks and fish schools, Kennedy and Eberhart proposed in [6] the Particle Swarm Optimization (PSO) algorithm for binary and real-valued function optimization. Since its inception, PSO has been applied with success to a number of problems and motivated several lines of research that investigate its main working mechanisms. One of these research trends deals with PSO's parameters and aims at devising methods for controlling those parameters and improve the algorithms' performance and robustness. Self-Organized Criticality (SOC), proposed in [2], provides interesting schemes for controlling PSO's working mechanisms. In fact, SOC has been used in the past in population-based metaheuristics, like Evolutionary Algorithms ([5] and [7]) and even PSO [8].

This paper proposes a versatile method, inspired by the SOC theory [2], for controlling the parameters of PSO, and demonstrates that it is a viable and effective method. The algorithm is based on a SOC system known as the *Bak-Sneppen model of co-evolution between interacting species* (or simply *Bak-Sneppen*), proposed by Bak and Sneppen in [3]. The *Bak-Sneppen PSO* (BS-PSO) uses the fitness values of the population of co-evolving species for regulating the parameters of the algorithm. Furthermore, the exact same values are used for perturbing the particle's position, thus introducing a kind of mutation in the PSO equations. The potentiality of the

C.A. Coello Coello et al. (Eds.): PPSN 2012, Part II, LNCS 7492, pp. 153–163, 2012.
© Springer-Verlag Berlin Heidelberg 2012

proposed method as a stochastic (although with predictable global behavior) seed for varying the parameters is investigated here, postponing a study of a stronger hybridization of the SOC model and the PSO for a future work.

A simple experimental setup was designed as a proof-of-concept. BS-PSO is compared with methods for controlling the inertia weight, as well as with a state-of-the-art PSO that also combines dynamic control of the parameters with perturbations of the particles' positions. The tests are conducted in a way such that each new component of BS-PSO is examined separately in order to investigate its effects on the performance of the algorithm. The results demonstrate the validity of the approach and show that BS-PSO, without requiring the hand-tuning of the traditional parameters or any additional one, is competitive with other PSOs. Furthermore, the base-model is simple and well-studied by the SOC theory, and may be treated as a black-box system that outputs batches of values for the parameters.

2   Particle Swarm Optimization

PSO is a population-based algorithm in which a group of solutions travels through the search space according to a set of rules that favor their movement towards optimal regions of the space. PSO is described by a simple set of equations that define the velocity and position of each particle. The position vector of the i-th particle is given by $\vec{X}_i = (x_{i,1}, x_{i,2}, \dots x_{i,D})$, where D is the dimension of the search space. The velocity is given by $\vec{V}_i = (v_{i,1}, v_{i,2}, \dots v_{i,D})$. The particles are evaluated with a fitness function $f(\vec{X}_i)$ in each time step and then their positions and velocities are updated by:

$$v_{i,d}(t) = v_{i,d}(t-1) + c_1 r_1 \big(p_{i,d} - x_{i,d}(t-1)\big) + c_2 r_2 \big(p_{g,d} - x_{i,d}(t-1)\big) \qquad (1)$$

$$x_{i,d}(t) = x_{i,d}(t-1) + v_{i,d}(t) \qquad (2)$$

were p_i is the best solution found so far by particle i and p_g is the best solution found so far by the neighborhood. Parameters r_1 and r_2 are random numbers uniformly distributed in the range $[0,1.0]$ and c_1 and c_2 are acceleration coefficients that tune the relative influence of each term of the formula (usually set within the range $[1.0,2.0]$). The first, influenced by the particles best solution, is known as the *cognitive part*, since it relies on the particle's own experience. The last term is the *social part*, since it describes the influence of the community in the velocity of the particle.

The neighborhood of the particle may be defined by a great number of schemes but most PSOs use one of two simple sociometric principles. The first connects all the members of the swarm to one another, and it is called *gbest*, were *g* stands for *global*. The second, called *lbest* (*l* stands for local), creates a neighborhood that comprises the particle itself and its *k* nearest neighbors. In order to prevent particles from stepping out of the limits of the search space, the positions $x_{i,d}(t)$ of the particles are limited by constants that, in general, correspond to the domain of the problem: $x_{i,d}(t) \in [-Xmax, Xmax]$. Velocity may also be limited within a range in order to prevent the *explosion* of the velocity vector: $v_{i,d}(t) \in [-Vmax, Vmax]$.

Although the basic PSO may be very efficient on numerical optimization, it requires a proper balance between local and global search. If we look at equation 1, we see that the first term on the right-hand side of the formula provides the particle with global search abilities. On the other hand, the second and third terms act as a local search mechanism and it is trivial to demonstrate that without the first term the swarm shrinks around the best position found so far. Therefore, by weighting these two parts of the formula it is possible to balance local and global search. In order to achieve a balancing mechanism, Shi an Eberhart [10] introduced the inertia weight ω, which is adjusted — usually within the range [0, 1.0] — together with the constants c_1 and c_2, in order to achieve the desired balance. The modified velocity equation is:

$$v_{i,d}(t) = \omega . v_{i,d}(t-1) + c_1 r_1 \big(p_{i,d} - x_{i,d}(t-1)\big) + c_2 r_2 \big(p_{g,d} - x_{i,d}(t-1)\big) \tag{3}$$

The parameter may be used as a constant that is defined after an empirical investigation of the algorithm's behavior. Another possible strategy is to use *time-varying inertia weights* (TVIW-PSO) [11]: starting with an initial and pre-defined value, the parameter value decreases linearly with time, until it reaches the minimum value. Later, Eberhart and Shi [4] found that the TVIW-PSO is not very effective on dynamic environments and proposed a random inertia weight for tracking dynamic systems. In the remainder of this paper, this method is referred to as RANDIW-PSO.

An adaptive approach is proposed in [1]. The authors describe the *global local best inertia weight PSO* (GLbestIW-PSO), an on-line variation strategy that depends on the p_i and p_g values. The strategy is defined in a way that better solutions use lower inertia weight values, thus increasing their local search abilities. The worst particles are modified with higher ω values and therefore they tend to explore the search space.

In [9], Ratnaweera *et al.* describe new parameter automation strategies that act upon several working mechanisms of the algorithm. The authors propose the concept of time-varying acceleration coefficients. They also introduce the concept of mutation, by adding perturbations to randomly selected modulus of the velocity vector. Finally, the authors describe a *self-organizing hierarchical particle swarm optimizer with time-varying acceleration coefficients* (HPSO-TVAC), which restricts the velocity update policy to the influence of the cognitive and social part, reinitializing the particles whenever they are stagnated in the search space.

Another method for controlling ω is given by Suresh *et al.* in [12]. The authors use the Euclidean distance between the particle and *gbest* for computing ω in each time-step for each particle. Particles closer to the best global solution tend to have higher ω values, while particles far from *gbest* are modified with lower inertia. The algorithm introduces a parameter ω_0 that restricts the inertia weight to working values. In addition, Suresh *et al.* also uses a perturbation mechanism of the particles' positions that introduces a random value in the range $[1, \rho]$, where ρ is a new parameter for the algorithm (see equation 4, which replaces equation 2). The authors report that the *Inertia-Adaptive PSO* (IA-PSO) outperforms several other methods in a 12-function benchmark, including the abovereferred state-of-the-art HPSO-TVAC.

$$x_{i,d}(t) = (1 + \rho) . x_{i,d}(t-1) + v_{i,d}(t) \tag{4}$$

Like HPSO-.TVAC and IA-PSO, the method proposed in this paper also aims at controlling the balance between local and global search by dynamically varying the inertia weight and/or the acceleration coefficients, while introducing perturbations in the particles' positions (like IA-PSO, but with ρ controlled by the SOC model). The main objective is to construct a simple scheme that does not require complex parameter tuning or pre-established strategies.

3   SOC and the Bak-Sneppen Particle Swarm

In complex adaptive systems, complexity and self-organization usually arise in the transition region between order and chaos. SOC systems are dynamical system with a critical point in that region as an attractor. However, and unlike many physical systems which have a parameter that needs to be tuned for criticality, a SOC model is able to self-tune to that critical state.

One of the properties of SOC systems is that small disturbances can lead to *avalanches*, i.e., events that are spatially or temporally spread through the system. Moreover, the same perturbation may lead to small or large avalanches, which in the end show a power-law proportion between the size of the events and its abundance.

The Bak-Sneppen is a model of co-evolution that displays SOC properties. Different species in the same ecosystem are related trough several features; they co-evolve, and the extinction of one species affects the species that are related to it, in a chain reaction that can affect large segments of the population. In the model, each species has a fitness value assigned to it and it is connected to other species in a ring topology (i.e., each one has two neighbors). Every time step, the species with the worst fitness and its neighbors are replaced by individuals with random fitness. When plotting the size of extinctions over their frequency in a local segment of the population and below a certain threshold close to a critical value, a power-law relationship is observed.

This description may be translated to a mathematical model. The system is defined by n^d fitness values f_i arranged on a d-dimensional lattice with n cells. At each time step, the smallest f value and its neighbors are replaced by uncorrelated random values drawn from a uniform distribution. The system is thus driven to a critical state where most species reach a fitness value above a certain threshold, with avalanches producing non-equilibrium fluctuations in the configuration of the fitness values.

The behavior of the numerical values of the Bak-Sneppen model — power-law relationships, increasing average fitness of the population, periods of stasis in segments of the population punctuated by intense activity — are the motivation behind this study. By linking a Bak-Sneppen model to the particles and then using the species' fitness values as input for adjusting the algorithm's parameters, it is expected that the resulting strategy is able to control PSO. To the extent of our knowledge, this is the first proposal of a scheme linking the Bak-Sneppen model and PSO in such a way. However, SOC has been applied to this field of research in the past.

In [7], Krink *et al.* proposed SOC-based mass extinction and mutation operator schemes for Evolutionary Algorithms. A *sandpile model* [2] is used here and its equations are previously computed in order to obtain a record of values with a power-law relationship. Those values are then used during the run to control the number of individuals that will be replaced by randomly generated solutions or the mutation probability of the Evolutionary Algorithm. Tinós and Yang [13] were also inspired by the Bak-Sneppen model to create the *Self-Organized Random Immigrants Genetic Algorithm* (SORIGA). The authors apply the algorithm to time-varying fitness landscapes and claim that SORIGA is able to outperform other algorithms in the proposed test set. In [5], Fernandes *et al.* describe an Evolutionary Algorithm attached to a sandpile model. The avalanches dynamically control the algorithm's mutation operator. The authors use the proposed scheme in time-varying fitness functions and claim that the algorithm is able to outperform other state-of-the-art methods in a wide range of dynamic problems. Finally, Løvbjerg and Krink [8] apply SOC to PSO in order to control the diversity to the population. The authors introduce a *critical value* associated to each particle and define a rule that increments that value when two particles are closer than a *threshold distance*. When the critical value of a particle exceeds a globally set *criticality limit*, the algorithm disperses the criticality of the particle within a certain neighborhood and mutates it. The algorithm also uses the particle's critical value to control the inertia. The authors claim that the method attains better solutions than the basic PSO. However, it introduces working mechanisms that can complicate its design. Overall, there are five parameters that must be tuned or set to *ad hoc* values.

The proposed BS-PSO uses the Bak-Sneppen model without introducing complicated control mechanisms. The only exception is an upper limit for the number of mutations in each time-step, a practical limitation due to the nature of the model and the requirements of numerical optimization. Besides that, the model is run in its original form, feeding the PSO with values in the range [0,1.0] (species' fitness values) that are used by the algorithm to control the parameters. Please note that if the PSO does not interact directly with the model (which is the case in this paper), the model can be executed prior to the optimization process and its fitness values stored in order to be used later in any kind of problem (meaning also that the running times are exactly the same of a basic PSO). However, in order to generalize the system and describe a framework that can easily be adapted to another level of hybridization of the SOC model and PSO, it is assumed that the model evolves on-line with the swarm.

In the Bak-Sneppen model, a population of species is placed in a ring topology and a random value between 0 and 1.0 is assigned to each individual. In BS-PSO, the number of species is equal to the size of the swarm. Therefore, the algorithm may be implemented just by assigning a secondary fitness, called *bak-sneppen fitness* (*bs_fitness*), to each individual in the swarm. This way, each individual is both the particle of the PSO and the species of the co-evolutionary model, with two independent fitness values: a quality measure fitness $f(\vec{X})$, computed by the objective function, and the *bs_fitness* $f_{bs}(\vec{X})$, modified according to Algorithm 1.

The main body of the BS-PSO is very similar to the basic PSO. The differences are: Algorithm 1 is called in each time-step, and modifies three or more *bs_fitness* values; the inertia weigh of each particle is defined in each time-step and for each particle i with equation 5, where $\overrightarrow{X_i}$ is the position of particle i; the acceleration coefficients c_1 and c_2 are defined in each time-step by equation 6; the position's update is done using equation 7, where $\rho_i(t) = random[0,1 - bs\_fitness(\overrightarrow{X_i})]$.

$$\omega_i(t) = 1 - bs\_fitness(\overrightarrow{X_i}) \tag{5}$$

$$c_1(t) = c_2(t) = 1 + bs\_fitness(\overrightarrow{X_i}) \tag{6}$$

$$x_{i,d}(t) = (1 + \rho_i(t)). x_{i,d}(t-1) + v_{i,d}(t) \tag{7}$$

Algorithm 1 is executed in each time-step. At $t = 0$, the *bs_fitness* values are randomly drawn from a uniform distribution in the range $[0, 1.0]$. Then, the algorithm searches for the worst species in the population (lowest *bs_fitness*), stores its fitness value (*minFit*) and mutates it by replacing the fitness with a random value in $[0, 1.0]$. In addition, the neighbors of the worst are also mutated (remember that a ring topology connects each species with index j to its two neighbors with indexes $j + 1$ and $j - 1$). Then, the algorithm searches again for the worst. If the fitness of that species is lower than *minFit*, the process repeats: species and its neighbors are mutated. This cycle proceeds while the worst fitness in the population is bellow *minFit* (and the number of mutations is below the limit). When the worst is found to be above *minFit*, the algorithm proceeds to PSO's standard procedures (see Algorithms 1 and 2).

Algorithm 1 .(Bak-Sneppen Model)

---

1. Set $mutations = 0$; set $max\_mutation = 2 \times swarm\_size$
2. Find the index j of the species with lowest bak-sneppen *fitness*
3. Set $minFit = bs\_fitness(\overrightarrow{X_j})$
4. Replace the fitness of individuals with indices $j, j - 1$, and $j + 1$ by random values
5. Increment mutations: $++ mutations$
6. Find the index j of the species with lowest fitness
7. If $bs\_fitness(\overrightarrow{X_j}) < minFit$ or $mutations = max\_mutation$, return to 4; else, end

---

Algorithm 2. (BS-PSO)

---

1. Initialize velocity and position of each particle.
2. Evaluate each particle i: $fitness\left(\overrightarrow{X_i}\right) = f(\overrightarrow{X_i})$
3. Initialize bak-sneppen fitness values: $bs\_fitness(\overrightarrow{X_i}) = random[0, 1.0]$
4. Update Bak-Sneppen Model (Algorithm 1).
5. For each particle i:
     6. Set $\omega = \rho = 1 - bs\_fitness(\overrightarrow{X_i})$
     7. Set $c_1 = c_2 = 1 + bs\_fitness(\overrightarrow{X_i})$
     8. Update velocity (equation 3) and position (equation 7); evaluate $fitness\left(\overrightarrow{X_i}\right) = f(\overrightarrow{X_i})$
9. If (stop criteria not met) return to 4; else, end.

As stated above, a stop criterion is introduced in Algorithm 1 in order to avoid long mutation cycles that would slow down BS-PSO after a certain number of iterations. If the number of mutations reaches a maximum pre-defined value, Algorithm 1 ends. In this paper, the critical value is set to twice the swarm's size. This value was intuitively fixed, not tuned for optimization of the performance. It is treated as a constant and its effects on the algorithm are beyond the scope of this paper. It is even possible that other strategies for avoiding long intra-time-steps mutation cycles that do not require a constant can be devised. However, such a study is left for future work. The main objective here is to demonstrate that controlling the inertia, acceleration coefficients and particles' positions with values given by a SOC model is viable and effective.

4   Testbed Set and Results

The experiments were designed with five benchmark functions (see Table 1). The minimum of all functions is in the origin with fitness 0. The dimension of the search space is set to $D = 30$ (except f_6). TVIW-, RANDIW-, GLbestIW- and IA-PSO were included in the tests in order to evaluate the efficiency of the method. (It is not our intention to prove that BS-PSO is better than the best PSOs in a wide range of functions. This simple experiment is mainly a proof-of-concept, and the peer-algorithms were chosen so that the different mechanism of BS-PSO can be properly evaluated.)

The population size n is set to 20 for all algorithms; *lbest* topology is used. The acceleration coefficients were set to 1.494, a value suggested in [4] for RANDIW-PSO. However, and since we are using algorithms with varying parameters, it is expected that other PSOs require different c values. Therefore, the coefficients c were also set to 1.2 and 2.0 (as in the studies that introduce GLbestIW-PSO and IA-PSO). $Xmax$ is defined as usual by the domain's upper limit and $Vmax = Xmax$. TVIW-PSO uses linearly decreasing inertia weight, from 0.9 to 0.4. The maximum number of generations is 3000 (except f_6, for which the limit is 1000); 50 runs for each experiment are conducted. Since PSO takes advantage of the fact that the optima are located in the centre of search space, asymmetrical initialization is often used for testing PSO. The initialization range for each function is given in Table 1.

Table 1. Benchmarks for the experiments. Dynamic and initialization range.

function	mathematical representation	Range of search	Range of initialization
Sphere f_1	$f_1(\vec{x}) = \sum_{i=}^{D} x_i^2$	$(-100, 100)^D$	$(50, 100)^D$
Rosenbrock f_2	$f_2(\vec{x}) = \sum_{i=1}^{D-1}(100(x_{i+1} - x_i^2)^2 + (x_i - 1)^2$	$(-100, 100)^D$	$(15, 30)^D$
Rastrigin f_3	$f_3(\vec{x}) = \sum_{i=1}^{D}(x_i^2 - 10\cos(2\pi x_i) + 10)$	$(-10, 10)^D$	$(2.56, 5.12)^D$
Griewank f_4	$f_4(\vec{x}) = 1 + \frac{1}{4000}\sum_{i=1}^{D} x_i^2 - \prod_{i=1}^{D}\cos\left(\frac{x_i}{\sqrt{i}}\right)$	$(-600, 600)^D$	$(300, 600)^D$
Schaffer f_6	$f_6(\vec{x}) = 0.5 + \frac{\left(\sin\sqrt{x^2 + y^2}\right)^2 - 0.5}{\left(1.0 + 0.001(x^2 + y^2)\right)^2}$	$(-100, 100)^2$	$(15, 30)^2$

The first test compares BS-PSO with different degrees of parameter control (i.e., the control of the acceleration coefficients and the perturbation of position were disabled in order to evaluate the effects of introducing the schemes). Table 2 summarizes the results. In the table's header, bs means that ω, c or ρ are controlled by $bs\_fitness$ values; otherwise, the control is disabled and the parameter is set to the corresponding value. As seen in Table 2, BS-PSO was tested with inertia control enabled and different c values (with $\rho = 0$); in general, higher c lead to a better performance. When the dynamic control of c is enabled (i.e, $(\omega, c, \rho) = (bs, bs, 0)$) the performance on f_1 and f_2 is improved, while for the other functions the fitness value decreases when compared to the best configuration with fixed c. However, the results are better than those attained by suboptimal configurations, which means that it may be an alternative to fine-tuning the parameter. Introducing a perturbation of the positions with the ρ parameter clearly improves the results, especially when ρ is controlled by the model. (Please note that ρ is set to 0.25, as in [12], in order to compare not only fixed and SOC-based perturbation, but also BS-PSO and IA-PSO later in in this section).

Table 2. BS-PSO: average and standard deviation of the optimal value for 50 trials

$(\omega, c, \rho) \rightarrow$	$(bs, 1.2, 0)$	$(bs, 1.49, 0)$	$(bs, 2.0, 0)$	$(bs, bs, 0)$	$(bs, bs, 0.25)$	(bs, bs, bs)
f_1	2.21e+04	3.35e+01	1.38e-15	8.30e-32	**0.00e+00**	**0.00e+00**
	(7.72e+03)	(1.90e+02)	(3.21e-15)	(3.47e-31)	**(0.00e+00)**	**(0.00e+00)**
f_2	9.76e+07	1.67e+05	1.88e+02	8.56e+01	2.61e+01	**2.60e+01**
	(2.83e+07)	(1.17e+06)	(2.53e+02)	(7.98e+01)	(2.66e-01)	**(1.58e-01)**
f_3	3.57e+02	2.82e+02	1.11e+02	2.02e+02	4.88e+00	**3.32e+00**
	(4.91e+01)	(4.44e+01)	(2.75e+01)	(4.16e+01)	(7.73e+00)	**(7.09e+00)**
f_4	1.53e+02	1.63e+00	1.25e-02	1.65e-02	**3.79e-03**	4.51e-03
	(7.30e+01)	(5.93e+00)	(1.26e-02)	(2.24e-02)	**(2.29e-03)**	(4.00e-03)
f_6	9.63e-02	3.31e-02	4.05e-03	5.55e-03	1.55e-03	**3.89e-04**
	(9.43e-02)	(4.08e-02)	(4.72 e-03)	(4.80e-03)	(3.60e-03)	**(1.92e-04)**

Table 3. TVIW-PSO, RANDIW-PSO and GLbestIW-PSO

	TVIW $c = 1.2$	TVIW $c = 1.49$	TVIW $c = 2.0$	RANDIW $c = 1.2$	RANDIW $c = 1.49$	RANDIW $c = 2.0$	GLbestIW $c = 1.2$	GLbestIW $c = 1.49$	GLbestIW $c = 2.0$
f_1	1.22e-23	**8.64e-29**	2.81e-06	**1.12e-33**	1.22e-18	6.68e+02	1.14e+05	3.67e+04	**2.83e+03**
	(5.81e-23)	**(1.75e-28)**	(2.77e-06)	**(1.90e-33)**	(1.26E-18)	(2.60e+02)	(6.47e+03)	(8.25e+03)	**(1.92e+03)**
f_2	1.24e+02	**1.03e+02**	5.96e+02	7.53e+01	**7.28e+01**	2.07e+07	2.95e+08	**9.10e+07**	3.46e+08
	(1.66e+02)	**(9.31e+01)**	(1.72e+03)	(7.24e+01)	**(6.69e+01)**	(1.26e+07)	(4.80e+07)	**(3.41e+07)**	(9.03e+07)
f_3	9.82e+01	7.85e+01	**5.84e+01**	1.80e+02	**1.11e+02**	1.94e+02	4.37e+02	3.56e+02	**1.68e+02**
	(2.44e+01)	(2.01e+01)	**(1.39e+01)**	(3.01e+01)	**(2.51e+01)**	(2.77e+01)	(3.24e+01)	(3.56e+01)	**(2.79e+01)**
f_4	8.71e-03	**8.66e-03**	1.22e-02	1.25e-02	**1.04e-02**	5.96e+00	9.77e+02	3.08e+02	**2.34e+01**
	(1.06Ee-02)	**(1.14e-02)**	(1.26e-02)	(1.64e-02)	**(1.50e-02)**	(1.62e+00)	(5.30e+01)	(6.63e+01)	**(1.53e+01)**
f_6	2.34e-03	**2.18e-03**	2.34e-03	**4.14e-03**	4.48e-03	2.60e-03	6.99e-02	8.12e-03	**0.00e+00**
	(4.19e-03)	**(3.94e-03)**	(4.07e-03)	**(4.80e-03)**	(4.88e-03)	(4.18e-03)	(1.10e-01)	(4.05e-02)	**(0.00e+00)**

Table 4. Kolmogorov-Smirnov statistical tests comparing the best configurations of each algorithm. '+' sign means that PSO 1 is statistically better than PSO 2, '~' means that the PSOs are equivalent, and '−' means that PSO 1 is worse.

PSO 1 *vs.* PSO 2	f_1	f_2	f_3	f_4	f_6
BS-PSO $(\mathbf{bs, bs, bs})$ *vs* TVIW-PSO	+	+	+	+	+
BS-PSO $(\mathbf{bs, bs, bs})$ *vs* RANDIW-PSO	+	+	+	+	+
BS-PSO $(\mathbf{bs, bs, bs})$*vs* GLbestIW-PSO	+	+	+	+	−
BS-PSO $(\mathbf{bs, bs}, 0)$ *vs* TVIW-PSO	+	+	−	−	−
BS-PSO $(\mathbf{bs, bs}, 0)$ *vs* RANDIW-PSO	−	~	−	~	~
BS-PSO $(\mathbf{bs, bs}, 0)$ *vs* GLbestIW-PSO	+	+	~	+	−

Table 5. IA-PSO: average and standard deviation of the optimal value for 50 trials

	$c = 1.2$ $\rho = 0$	$c = 1.2$ $\rho = 0.25$	$c = 1.2$ $\rho = bs$	$c = 1.49$ $\rho = 0$	$c = 1.49$ $\rho = 0.25$	$c = 1.49$ $\rho = bs$	$c = 2.0$ $\rho = 0$	$c = 2.0$ $\rho = 0.25$	$c = 2.0$ $\rho = bs$
f_1	1.16e+04 (8.32e+03)	**0.00e+00** (0.00e+00)	**0.00e+00** (0.00e+00)	2.42e+02 (1.43e+03)	**0.00e+00** (0.00e+00)	**0.00e+00** (0.00e+00)	5.19e-02 (2.61e-02)	6.56e-03 (5.34e-03)	2.60e-02 (1.70e-02)
f_2	2.05e+07 (1.43e+07)	2.71e+01 (4.44e+00)	2.63e+01 (1.29e+00)	7.45e+04 (5.26e+05)	2.62e+01 (3.71e-01)	**2.60e+01** (**1.84e-01**)	4.26e+02 (8.30e+02)	3.97e+01 (2.14e+01)	7.21e+01 (8.25e+01)
f_3	3.74e+02 (3.27e+01)	1.16e+02 (2.10e+01)	7,79e+01 (2.17e+01)	2.82e+02 (3.47e+01)	5.26e+01 (3.08e+01)	3.96e+01 (2.02e+01)	8.87e+01 (2.66e+01)	1.81e+00 (3.12e+00)	**1.12e+01** (**1.42e+01**)
f_4	1.21e+02 (7.57e+01)	4.02e-03 (2.80e-03)	3,95e-03 (2.22e-03)	2,62e+00 (1.30e+01)	**3.72e-03** (**2.23e-03**)	4.71e-03 (3.12e-03)	1.84e+00 (1.27e+01)	1.11e-02 (7.74e-03)	1.30e-02 (7.08e-03)
f_6	2.45e-01 (7.60e-02)	9.52e-03 (1.37e-03)	5.26e-03 (4.87e-03)	8.90e-02 (8.39e-02)	5.44e-03 (4.87e-03)	7.77e-04 (2.66e-03)	3.75e-03 (4.73e-03)	**3.89e-04** (**1.92e-03**)	4.89e-04 (2.03e-03)

In order to assure fair comparisons, Table 3 shows the complete set of results attained by TVIW-, RANDIW- and GLbestIW-PSO. Apparently, BS-PSO outperforms the other algorithms in most of the scenarios. However, PSOs in Table 3 do not include perturbation of the particle's position and therefore they should be also compared to a BS-PSO with that scheme disabled ($(bs, bs, 0)$ in Table 2): Table 4 compares BS-PSO (with and without perturbation of the particles) to the other PSOs using statistical non-parametric tests (best configurations in Table 3 were chosen). It is confirmed that the fully enabled BS-PSO outperforms the other algorithms. As for the version restricted to parameter control, it is in general better than GLbestIW, while being competitive with the other methods. These are interesting results, since the performance of BS-PSO is attained without fine-tuning the parameters.

Table 6. Kolmogorov-Smirnov statistical tests comparing IA-PSO and BS-PSO

PSO 1 *vs.* PSO 2	f_1	f_2	f_3	f_4	f_6
BS-PSO $(\mathbf{bs}, 2.0, 0)$ *vs* IA-PSO $(\rho = 0)$	+	+	~	+	~
BS-PSO $(\mathbf{bs, bs, bs})$*vs* IA-PSO $(bs$ controlled $\rho)$	~	~	+	~	+

A final test compares BS-PSO with IA-PSO. The later was tested with different acceleration coefficients and three perturbation strategies: disabled ($\rho = 0$), set to 0.25 (as in [12]) and controlled by the Bak-Sneppen model (using a Bak-Sneppen controlled IA-PSO permits to compare only the parameter control scheme of both algorithms). The results are in Table 5 and the statistical tests in Table 6. BS-PSO is, in general, more efficient than IA-PSO, whether the schemes are fully enabled or not. At this point, a question arises: what are the mechanisms behind the control scheme that make BS-PSO efficient in adjusting the parameters? Figure 1 gives some hints. The plot in the figure represents the distribution of ω_i during a typical run of BS-PSO, and, although it is not the definitive answer, helps to clarify this issue. The values seem to keep within a range that is not only suited for ω but also appropriate to model a mutation scheme. If the system had higher values with more frequency, the effect would be destructive, since it would increase exploration beyond a reasonable point.

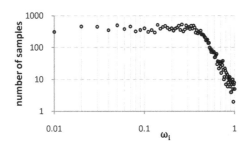

Fig. 1. Distribution of the ω_i values of all particles in a typical run

5   Conclusions and Future Work

The Bak-Sneppen Particle Swarm Optimization (BS-PSO) is a variation of the basic PSO that uses a Self-Organized Criticality (SOC) model to control the inertia weight and the acceleration coefficients, as well as the perturbation factor of the particles' positions. A single scheme controls three parameters making hand-tuning of the basis PSO unnecessary. An experimental setup demonstrates the validity of the algorithm and shows that the incorporation of each control mechanism may improve the performance or at least reduce the tuning effort. The BS-PSO is compared with other methods. In particular, the algorithm is able to attain better results than a recently proposed inertia weight PSO (IA-PSO) in most of the experimental scenarios. In a future work, a scalability analysis will be conducted, as well as study on the effects of the limit imposed to mutation events by the current algorithm, and possible alternatives to this *ad hoc* solution. The test set will also be extended and BS-PSO compared with the algorithms proposed in [8] and [9]. Finally, different levels of hybridization between the Bak-Sneppen model and PSO will be tested, in order to introduce information from the search into the variation scheme of the parameter values.

Acknowledgement. The first author wishes to thank FCT, *Ministério da Ciência e Tecnologia*, his Research Fellowship SFRH/BPD/66876/2009. This work is supported by project TIN2011-28627-C04-02 awarded by the Spanish Ministry of Science and Innovation and P08-TIC-03903 awarded by the Andalusian Regional Government.

References

1. Arumugam, M.S., Rao, M.V.C.: On the Performance of the Particle Swarm Optimization Algorithm with Various Inertia Weight Variants for Computing Optimal Control of a Class of Hybrid Systems. Discrete Dynamics in Nature and Society (2006), Article ID 79295, 17 pages (2006)
2. Bak, P., Tang, C., Wiesenfeld, K.: Self-organized Criticality: an Explanation of 1/f Noise. Physical Review Letters 59(4), 381–384 (1987)
3. Bak, P., Sneppen, K.: Punctuated Equilibrium and Criticality in a Simple Model of Evolution. Physical Review Letters 71(24), 4083–4086 (1993)
4. Eberhart, R.C., Shi, Y.: Tracking and optimizing dynamic systems with particle swarms. In: Proc. IEEE of the Congress on Evolutionary Computation 2001, pp. 94–97. IEEE Press (2001)
5. Fernandes, C.M., Merelo, J.J., Ramos, V., Rosa, A.C.: A Self-Organized Criticality Mutation Operator for Dynamic Optimization Problems. In: Proc. of the 2008 Genetic and Evolutionary Computation Conference, pp. 937–944. ACM (2008)
6. Kennedy, J., Eberhart, R.: Particle Swarm Optimization. In: Proceedings of IEEE International Conference on Neural Networks, vol. 4, pp. 1942–1948 (1995)
7. Krink, T., Rickers, P., René, T.: Applying Self-organized Criticality to Evolutionary Algorithms. In: Deb, K., Rudolph, G., Lutton, E., Merelo, J.J., Schoenauer, M., Schwefel, H.-P., Yao, X. (eds.) PPSN VI. LNCS, vol. 1917, pp. 375–384. Springer, Heidelberg (2000)
8. Løvbjerg, M., Krink, T.: Extending particle swarm optimizers with self-organized criticality. In: Proceedings of the 2002 IEEE Congress on Evolutionary Computation, vol. 2, pp. 1588–1593. IEEE Computer Society (2002)
9. Ratnaweera, A., Halgamuge, K.S., Watson, H.C.: Self-organizing Hierarchical Particle Swarm Optimizer with Time-varying Acceleration Coefficients. IEEE Transactions on Evolutionary Computation 8(3), 240–254 (2004)
10. Shi, Y., Eberhart, R.C.: A Modified Particle Swarm Optimizer. In: Proc. of IEEE 1998 Congress on Evolutionary Computation, pp. 69–73. IEEE Press (1998)
11. Shi, Y., Eberhart, R.C.: Empirical Study of Particle Swarm Optimization. In: Proc. of the IEEE. Congress on Evolutionary Computation, pp. 101–106 (1999)
12. Suresh, K., Ghosh, S., Kundu, D., Sen, A., Das, S., Abraham, A.: Inertia-Adaptive Particle Swarm Optimizer for Improved Global Search. In: Proceedings of the 8th International Conference on Intelligent Systems Design and Applications, vol. 2, pp. 253–258. IEEE, Washington, DC (2008)
13. Tinós, R., Yang, S.: A self-organizing Random Immigrants Genetic Algorithm for Dynamic Optimization Problems. In: Genetic Programming and Evolvable Machines, vol. 8(3), pp. 255–286 (2007)

The Apiary Topology: Emergent Behavior in Communities of Particle Swarms

Andrew McNabb and Kevin Seppi

Computer Science Department, Brigham Young University
{a,k}@cs.byu.edu

Abstract. In the natural world there are many swarms in any geographical region. In contrast, Particle Swarm Optimization (PSO) is usually used with a single swarm of particles. We define a simple new topology called Apiary and show that parallel communities of swarms give rise to emergent behavior that is fundamentally different from the behavior of a single swarm of identical total size. Furthermore, we show that subswarms are essential for scaling parallel PSO to more processors with computationally inexpensive objective functions. Surprisingly, subswarms are also beneficial for scaling PSO to high dimensional problems, even in single processor environments.

Keywords: Particle Swarm Optimization, parallel PSO, swarm topology, subswarms, multiple swarms, parallel computation.

1 Introduction

Particle Swarm Optimization (PSO) is a continuous function optimization algorithm inspired by the flocking behaviors of birds and insects. It is typically used with small swarms of 20 to 50 particles organized in simple topologies that do not fully reflect the complex social interactions of insects. In agriculture, for example, bees are managed in sets of hives called apiaries. The number of hives in an apiary usually ranges from 10 to 150.

Using conventional topologies, a single swarm of particles often fails to scale both to large numbers of processors and to high-dimensional problems. First, with a large number of processors and an inexpensive objective function, communication costs make parallel PSO with a single swarm impractical. Parallel PSO naturally works well for problems with computationally expensive function evaluations, but for inexpensive objective functions, the time to communicate a single position can exceed the time to perform a function evaluation. Second, for high-dimensional problems, particles are prone to premature convergence. Even for Sphere, the simplest of benchmark functions, standard PSO struggles to find the global optimum when the number of dimensions is 400 or greater.

Multiple swarms have been used to scale parallel PSO for inexpensive objective functions but have not been considered for scaling to high-dimensional problems. Semi-independent swarms of particles provide a natural way to parallelize the computation of PSO across a set of processors without requiring

C.A. Coello Coello et al. (Eds.): PPSN 2012, Part II, LNCS 7492, pp. 164–173, 2012.

instantaneous communication [1]. However, the behavior of subswarms has not been explored, particularly with respect to high-dimensional problems.

The Apiary topology, proposed in Section 3, spreads the population of particles among a set of small subswarms. In this topology, a subswarm is a social entity which lies between the individual particle and the full population and which serves as another source of emergent behavior. Each subswarm consists of a fixed set of particles and is mostly independent of other subswarms. Periodically, a single particle in each subswarm communicates with a few particles in other subswarms. This communication between subswarms is rare and limited, so computation is particularly well suited to parallel computation.

The Apiary topology helps PSO scale, both to large numbers of processors and to high-dimensional objective functions. Unlike some other proposed PSO techniques using subswarms, this topology is simple, clearly defined, and appropriate for parallel PSO. Experiments, described in Section 4, show significant improvements over standard PSO. Even in single processor environments, apiaries produce better results in the same time and are less prone to premature convergence for every benchmark function we tested. These results are presented and discussed in Section 4.1. The standard parameters are justified in Section 4.2, along with indications of when these parameters might be changed. Parallel PSO with Apiary is compared in Section 4.3. Despite inexpensive functions being particularly challenging for parallelization, the run time is reduced from 256 minutes with a single processor to 17 minutes with 40 processors.

2   Background Material: Particle Swarm Optimization

Particle Swarm Optimization, proposed by Kennedy and Eberhart [2], simulates the motion of particles in the domain of an objective function. These particles search for the global optimum by evaluating the function as they move. During each iteration, each particle is pulled toward the best position it has sampled, known as the *personal best*, and the best position of any particle in its neighborhood, known as the *neighborhood best*.

Constricted PSO is generally considered the standard variant [3]. Each particle's position x_0 and velocity v_0 are initialized to random values based on a function-specific feasible region. During iteration t, the following equations update the ith component of a particle's position x_t and velocity v_t with respect to the personal best p_{t-1} and neighborhood best n_{t-1} from the preceding iteration:

$$v_{t,i} = \chi \left[v_{t-1,i} + \phi^P u_{t-1,i}^P (x_{t-1,i}^P - x_{t-1,i}) + \phi^N u_{t-1,i}^N (x_{t-1,i}^N - x_{t-1,i}) \right] \quad (1)$$

$$x_{t,i} = x_{t-1,i} + v_{t,i} \quad (2)$$

where x^P is the personal best, x^N is the neighborhood best, ϕ^P and ϕ^N are usually set to 2.05, $u_{t,i}^P$ and $u_{t,i}^N$ are samples drawn from a standard uniform distribution, and $\chi = 2/\left|2 - \phi - \sqrt{\phi^2 - 4\phi}\right|$ where $\phi = \phi^P + \phi^N$ [4].

The neighborhoods within a swarm are defined by the *topology* graph. The choice of topology can have a significant effect on performance [5]. Additionally, the topology determines task dependencies and overhead in parallel PSO [6]. The $Ring_{50}$ topology, a swarm of 50 particles where each particle has a single neighbor on either side, is a standard starting point [3].

3   The Apiary Topology

The Apiary topology is a dynamic topology of independent subswarms which occasionally communicate with each other. Each subswarm has an *inner topology*, and the subswarms are connected in an *outer topology*. In most iterations, the neighbors of each particle are defined purely by the inner topology of its subswarm. After a fixed number of independent *subiterations*, each subswarm communicates with its neighboring subswarms, as defined by the outer topology. Each subswarm sends its neighbors the best value from any of its particles. It updates the neighborhood best of a fixed set of particles (the neighborhood of the first particle in the swarm) with the values from neighboring subswarms.

In the Apiary topology, subswarms share important characteristics with communities in nature. Just as each bee colony has its own social structure, each subswarm has its own particles and its own topology. Like bee colonies, the subswarms are independent and rarely interact. Curiously, bees occasionally allow foreign forage bees to enter a hive if they are fully loaded [7], and the native bees will be able to learn from those foreign bees if they are from another colony or even another species [8]. Likewise, a single particle in each subswarm occasionally engages in light communication with neighboring swarms. In this simple structure, subswarms are simple entities with a balance of independence and interaction that favors emergent behavior.

This approach contrasts with previous attempts to define subpopulations in PSO. Dynamic Multi-Swarm PSO [9] periodically shuffles by reassigning all particles to random subswarms. This global reshuffling increases the amount of communication required in parallel PSO and is incompatible with asynchronous parallel PSO [10]. In contrast, neighborhoods in the Apiary topology are deterministic and require very little communication. Section 4.1 compares the performance of Dynamic Multi-Swarm PSO with that of the Apiary topology. Most subswarm approaches have introduced strategies—some of them quite complex—to manage the migration of particles between subswarms [11,12,13,14]. Other works have used subswarm-style topologies within a limited context [6], including completely independent subswarms [1]. Romero and Cotta's island-structured swarms [11], is limited to small numbers of large subswarms and low-dimensional problems, and its conclusions do not seem to apply to high-dimensional problems. In contrast to other approaches, the Apiary topology is static and thus well suited to any implementation of parallel PSO, and it requires very little communication between subswarms.

The inner and outer topologies, as well as the number of subiterations, are changeable parameters. We recommend *Ring* for both the outer and inner topologies, with a starting point of 5 particles per subswarm, 40 total subswarms, and 100 subiterations. These recommendations are justified in Section 4.2.

4   Experimental Results

The Apiary topology provides significant improvements for both serial and parallel PSO with respect to a variety of benchmark functions. Benchmark functions are computationally inexpensive enough for large-scale experimentation but share interesting properties with challenging real-life problems. We use the Ackley, Rastrigin, Rosenbrock, Schwefel 1.2, and Sphere benchmark functions [15] with both 250 and 500 dimensions. Experiements were run on a Linux cluster consisting of 320 nodes (Dell PowerEdge M610). Each node is equipped with two quad-core Intel Nehalem processors (2.8 GHz) and 24 GB of memory.

Each experiment was repeated at least 40 times. We report the median instead of the mean because these distributions are skewed. The 10th and 90th percentiles illuminate both the variability and skewness. We determine statistical significance using a one-sided Monte Carlo permutation test [16]. A t-test would be inappropriate because it uses the mean statistic and assumes a normal distribution, which we can not assume in part because of skew. Each table cell is bolded if it is better than every other entry in its row with a p-value of 0.05.

Each table and plot presents either the median number of evaluations required to reach a threshold or the median best value at a fixed number of evaluations or iterations. The notation $Ring_n$ denotes a ring topology where each particle has one neighbor on each side, and $Ring_m-Ring_n$ denotes an Apiary topology with a $Ring_m$ outer topology and a $Ring_n$ inner topology. Each benchmark function is accompanied by its dimensionality, for example, "Sphere-500."

The balance of this section seeks to identify some of the most interesting observation and give greater clarity and meaning to these results. Section 4.1 compares the $Ring_{40}-Ring_5$ apiary with the standard recommendation of $Ring$. Section 4.2 justifies the particular choice of $Ring_{40}-Ring_5$ as a standard starting point. Finally, Section 4.3 demonstrates the suitability of the Apiary topology to parallel PSO by demonstrating its efficiency in a typical parallel environment.

4.1   Apiaries in Serial PSO

Limiting the interaction between subswarms to once every 100 iterations might be expected to compromise the performance of serial PSO in exchange for improved parallel efficiency, but this social organization in fact improves performance even in serial PSO. Figures 1 and 2 show the progress toward convergence for 500 dimensional Rastrigin and Sphere respectively. The $Ring_{40}-Ring_5$ apiaries require the same number of evaluations per iteration as the $Ring_{200}$ swarms, but they perform far better than the individual $Ring$ swarms. Note that the $Ring_{200}$ swarm in Figure 2 converges more slowly than the $Ring_{50}$ swarm because it requires more evaluations per iteration.

One might wonder whether the performance of the Apiary topology are dependent on the social interactions or whether they are merely due to the repetition of a high-variance experiment. After all, running 40 independent swarms of 5 particles would be expected to perform better than a single swarm of 5 particles. Figure 2 includes the abysmal results of such an $Independent_{40}-Ring_5$ topology,

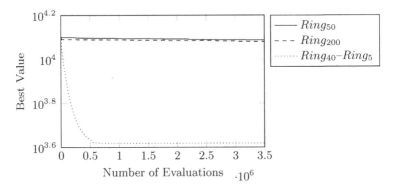

Fig. 1. Convergence plot for Rastrigin in serial PSO, comparing an apiary (using 100 subiterations) with a swarm of the same total number of total particles (200) and a swarm of 50 particles

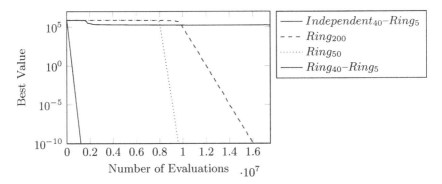

Fig. 2. Convergence plot for Sphere in serial PSO, comparing an apiary (using 100 subiterations) with a swarm of the same number of total particles (200) and a swarm of 50 particles

thus dispelling this possibility. The difference between the independent swarms and the apiary demonstrates emergent behavior.

We include results for the full range of benchmark functions in tabular form. In the case of Sphere, all runs of all PSO variants eventually converge to the global minimum. Table 1 reports the number of function evaluations to convergence. Table 2 reports the best value obtained at a fixed number of function evaluations for the other benchmark functions. The fixed number of evaluations for each function are equivalent to about 6 hours of computation, specifically: 6×10^6 for Ackley-250 and Ackley-500, 1×10^7 for Rastrigin-250, 3.5×10^6 for Rastrigin-500, 1×10^7 for Rosenbrock-250, 5×10^6 for Rosenbrock-500, 6×10^6 for Schwefel1.2-250, and 2×10^6 for Schwefel1.2-500. In all cases the apiary is best with statistical significance. Though the results for the Ackley function are statistically significant, the difference is small.

Table 1. Median number of function evaluations to reach a value of 10^{-10}. The best cell in each row is bolded if statistically significant

Function	$Ring_{50}$	$Ring_{200}$	$Ring_{40}$–$Ring_5$
Sphere-250	7.8×10^5	3×10^6	$\mathbf{5.9 \times 10^5}$
$(10^{th}, 90^{th})$	$(7.6 \times 10^5, 8.1 \times 10^5)$	$(3 \times 10^6, 3.1 \times 10^6)$	$\mathbf{(5.9 \times 10^5, 6 \times 10^5)}$
Sphere-500	9.5×10^6	1.6×10^7	$\mathbf{1.2 \times 10^6}$
$(10^{th}, 90^{th})$	$(3.5 \times 10^6, 2.6 \times 10^7)$	$(1 \times 10^7, 4 \times 10^7)$	$\mathbf{(1.2 \times 10^6, 1.2 \times 10^6)}$

Table 2. Median best value at a fixed number of function evaluations

Function	$Ring_{50}$	$Ring_{200}$	$Ring_{40}$–$Ring_5$
Ackley-250	20	20	**20**
$(10^{th}, 90^{th})$	$(20, 20)$	$(20, 20)$	$\mathbf{(20, 20)}$
Ackley-500	20	20	**20**
$(10^{th}, 90^{th})$	$(20, 21)$	$(20, 21)$	$\mathbf{(20, 20)}$
Rastrigin-250	2.6×10^3	2.3×10^3	$\mathbf{1.9 \times 10^3}$
$(10^{th}, 90^{th})$	$(2.1 \times 10^3, 2.9 \times 10^3)$	$(2 \times 10^3, 2.5 \times 10^3)$	$\mathbf{(1.7 \times 10^3, 2.1 \times 10^3)}$
Rastrigin-500	1.2×10^4	1.2×10^4	$\mathbf{4.1 \times 10^3}$
$(10^{th}, 90^{th})$	$(1.1 \times 10^4, 1.3 \times 10^4)$	$(8.5 \times 10^3, 1.2 \times 10^4)$	$\mathbf{(3.9 \times 10^3, 4.5 \times 10^3)}$
Rosenbrock-250	70	3.7×10^2	**0.0012**
$(10^{th}, 90^{th})$	$(0.029, 3.4 \times 10^2)$	$(2.8 \times 10^2, 4.5 \times 10^2)$	$\mathbf{(6.1 \times 10^{-9}, 4)}$
Rosenbrock-500	4.3×10^{12}	4.1×10^{12}	$\mathbf{8.8 \times 10^2}$
$(10^{th}, 90^{th})$	$(4 \times 10^{12}, 4.6 \times 10^{12})$	$(3.8 \times 10^{12}, 4.3 \times 10^{12})$	$\mathbf{(7 \times 10^2, 1.1 \times 10^3)}$
Schwefel1.2-250	7.4×10^4	2.3×10^5	$\mathbf{1.6 \times 10^4}$
$(10^{th}, 90^{th})$	$(4.6 \times 10^4, 1.5 \times 10^5)$	$(2.1 \times 10^5, 2.7 \times 10^5)$	$\mathbf{(1.2 \times 10^4, 2.1 \times 10^4)}$
Schwefel1.2-500	1.6×10^6	2.2×10^6	$\mathbf{8.1 \times 10^5}$
$(10^{th}, 90^{th})$	$(1.1 \times 10^6, 2.3 \times 10^6)$	$(1.8 \times 10^6, 2.8 \times 10^6)$	$\mathbf{(7.1 \times 10^5, 9.4 \times 10^5)}$

For some functions, the Apiary topology outperforms Dynamic Multi-Swarm PSO [9] (DMS-PSO) in serial, while for other functions, DMS-PSO outperforms the Apiary topology. For Sphere-500, the Apiary topology finds the minimum faster with a small but statistically significant advantage, while for Sphere-250, the situation is reversed (the full table is omitted due to space constraints). Table 3 shows similarly mixed results for the other benchmark functions.

4.2 Apiary Parameters

We now justify the basic apiary parameters of a $Ring_{40}$ outer topology, a $Ring_5$ inner topology, and 100 subiterations. General recommendations set $Ring_{50}$ as a standard swarm topology [3] or even higher for difficult problems [6]. Previous subswarm topologies have suggested that 50–100 particles per subswarm [11] or 32 particles per subswarm [1] give ideal performance. In contrast, we recommend starting with small subswarms of about 5 particles.

Increasing the number of particles per subswarm or the total number of subswarms provide improvements only in some circumstances. Table 4 compares

Table 3. Median best value at a fixed number of function evaluations. The topology is $Ring_{40}-Ring_{25}$ for Rastrigin and $Ring_{40}-Ring_5$ for Rosenbrock and Schwefel 1.2.

Function	Apiary	DMS-PSO
Rastrigin-250	1.5×10^3	$\mathbf{4.8 \times 10^2}$
$(10^{th}, 90^{th})$	$(1.4 \times 10^3, 1.6 \times 10^3)$	$(\mathbf{4 \times 10^2}, \mathbf{5.9 \times 10^2})$
Rastrigin-500	3.3×10^3	$\mathbf{1.3 \times 10^3}$
$(10^{th}, 90^{th})$	$(3 \times 10^3, 3.6 \times 10^3)$	$(\mathbf{1.2 \times 10^3}, \mathbf{1.6 \times 10^3})$
Rosenbrock-250	$\mathbf{0.0012}$	1.3×10^2
$(10^{th}, 90^{th})$	$(\mathbf{6.1 \times 10^{-9}}, \mathbf{4})$	$(1.4, 2.3 \times 10^2)$
Rosenbrock-500	8.8×10^2	8.3×10^2
$(10^{th}, 90^{th})$	$(7 \times 10^2, 1.1 \times 10^3)$	$(6.4 \times 10^2, 9.8 \times 10^2)$
Schwefel1.2-250	$\mathbf{1.6 \times 10^4}$	7×10^4
$(10^{th}, 90^{th})$	$(\mathbf{1.2 \times 10^4}, \mathbf{2.1 \times 10^4})$	$(5.1 \times 10^4, 8.9 \times 10^4)$
Schwefel1.2-500	$\mathbf{8.1 \times 10^5}$	1.4×10^6
$(10^{th}, 90^{th})$	$(\mathbf{7.1 \times 10^5}, \mathbf{9.4 \times 10^5})$	$(1.2 \times 10^6, 1.7 \times 10^6)$

Table 4. Median best value at n function evaluations

Function	$Ring_{40}-Ring_5$	$Ring_{200}-Ring_5$	$Ring_{40}-Ring_{25}$
Rastrigin-250	1.9×10^3	1.7×10^3	$\mathbf{1.5 \times 10^3}$
$(10^{th}, 90^{th})$	$(1.7 \times 10^3, 2.1 \times 10^3)$	$(1.6 \times 10^3, 1.9 \times 10^3)$	$(\mathbf{1.4 \times 10^3}, \mathbf{1.6 \times 10^3})$
Rastrigin-500	4.1×10^3	3.8×10^3	$\mathbf{3.3 \times 10^3}$
$(10^{th}, 90^{th})$	$(3.9 \times 10^3, 4.5 \times 10^3)$	$(3.6 \times 10^3, 4 \times 10^3)$	$(\mathbf{3 \times 10^3}, \mathbf{3.6 \times 10^3})$
Rosenbrock-250	$\mathbf{0.0012}$	2.6×10^2	0.27
$(10^{th}, 90^{th})$	$(\mathbf{6.1 \times 10^{-9}}, \mathbf{4})$	$(2 \times 10^2, 3.2 \times 10^2)$	$(0.0016, 76)$
Rosenbrock-500	$\mathbf{8.8 \times 10^2}$	5×10^3	9.4×10^2
$(10^{th}, 90^{th})$	$(\mathbf{7 \times 10^2}, \mathbf{1.1 \times 10^3})$	$(3.4 \times 10^3, 1.4 \times 10^4)$	$(8.2 \times 10^2, 1.2 \times 10^3)$
Schwefel1.2-250	$\mathbf{1.6 \times 10^4}$	1.5×10^5	3.7×10^4
$(10^{th}, 90^{th})$	$(\mathbf{1.2 \times 10^4}, \mathbf{2.1 \times 10^4})$	$(1.3 \times 10^5, 1.6 \times 10^5)$	$(2.8 \times 10^4, 4.8 \times 10^4)$
Schwefel1.2-500	$\mathbf{8.1 \times 10^5}$	1.6×10^6	1×10^6
$(10^{th}, 90^{th})$	$(\mathbf{7.1 \times 10^5}, \mathbf{9.4 \times 10^5})$	$(1.5 \times 10^6, 1.8 \times 10^6)$	$(8.8 \times 10^5, 1.2 \times 10^6)$

$Ring_{40}-Ring_5$ to $Ring_{200}-Ring_5$, an apiary with 5 times as many subswarms, and to $Ring_{40}-Ring_{25}$, an apiary with 5 times as many particles per subswarm. For most of the benchmark functions, the $Ring_{40}-Ring_5$ apiary performs significantly better than either of the larger topologies. Likewise, the $Ring_{40}-Ring_5$ topology significantly outperforms the others for Sphere-250 and Sphere-500 (the table is omitted due to space). For such functions, the increased number of evaluations per iteration offsets any increased exploration provided by the larger swarms. On the other hand, both of the larger topologies are better for Rastrigin-250, Rastrigin-500, and Rosenbrock-500. As the number of local minima in Rastrigin increases exponentially with the number of dimensions, we conclude that larger swarms are preferable for highly multimodal objective functions.

Changing the communication between swarms can affect performance dramatically. Using a more connected outer topology, such as *Complete*, gives poor

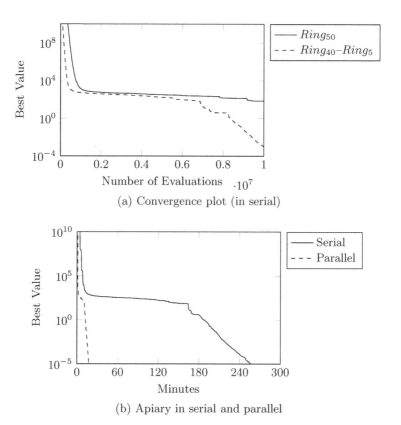

(a) Convergence plot (in serial)

(b) Apiary in serial and parallel

Fig. 3. Convergence plots of the Apiary topology for the Rosenbrock function with respect to function evaluations and time

performance in serial PSO in addition to requiring more communication in parallel PSO. Sharing the best value of an arbitrary member of each subswarm instead of the best particle of each subswarm also reduces performance. Setting the number of subiterations to 100 is high enough to provide reasonable task granularity even for the least expensive benchmark functions in parallel PSO.

4.3   Parallel Performance of Apiaries

Benchmark functions are extremely inexpensive, yet despite the high relative cost of communication, the Apiary topology performs extremely well in parallel. Figure 3 shows the results for the Rosenbrock function with both serial and parallel computation. Performing 100 iterations on 5 particles requires only 0.2 seconds, and parallel PSO took about 0.5 seconds per iteration. With any realistically expensive function, the overhead of 0.3 seconds would be negligible.

In a parallel context with a large number of spare processors, there may be limited additional overhead in increasing the number of subswarms. In this light, we revisit the conclusions from Section 4.2. In this context, the number of

Table 5. Median best value at n iterations

Function	$Ring_{40}$–$Ring_5$	$Ring_{200}$–$Ring_5$	$Ring_{40}$–$Ring_{25}$
Rastrigin-250	$1.9{\times}10^3$	$1.7{\times}10^3$	$\mathbf{1.5{\times}10^3}$
$(10^{th}, 90^{th})$	$(1.7{\times}10^3, 2.1{\times}10^3)$	$(1.6{\times}10^3, 1.9{\times}10^3)$	$\mathbf{(1.4{\times}10^3, 1.6{\times}10^3)}$
Rastrigin-500	$4.2{\times}10^3$	$3.8{\times}10^3$	$\mathbf{3.4{\times}10^3}$
$(10^{th}, 90^{th})$	$(3.9{\times}10^3, 4.5{\times}10^3)$	$(3.6{\times}10^3, 4{\times}10^3)$	$\mathbf{(3{\times}10^3, 3.6{\times}10^3)}$
Rosenbrock-250	$3.6{\times}10^2$	$\mathbf{2.5{\times}10^2}$	$2.9{\times}10^2$
$(10^{th}, 90^{th})$	$(2.6{\times}10^2, 4.4{\times}10^2)$	$\mathbf{(1.9{\times}10^2, 3.2{\times}10^2)}$	$(2.1{\times}10^2, 3.5{\times}10^2)$
Rosenbrock-500	$2{\times}10^4$	$\mathbf{4.2{\times}10^3}$	$1.5{\times}10^4$
$(10^{th}, 90^{th})$	$(4.5{\times}10^3, 3.1{\times}10^7)$	$\mathbf{(3{\times}10^3, 9.4{\times}10^3)}$	$(3.5{\times}10^3, 7.5{\times}10^6)$
Schwefel1.2-250	$1.7{\times}10^5$	$\mathbf{1.4{\times}10^5}$	$1.7{\times}10^5$
$(10^{th}, 90^{th})$	$(1.4{\times}10^5, 2{\times}10^5)$	$\mathbf{(1.3{\times}10^5, 1.6{\times}10^5)}$	$(1.3{\times}10^5, 2{\times}10^5)$
Schwefel1.2-500	$1.8{\times}10^6$	$\mathbf{1.6{\times}10^6}$	$1.8{\times}10^6$
$(10^{th}, 90^{th})$	$(1.6{\times}10^6, 2.2{\times}10^6)$	$\mathbf{(1.5{\times}10^6, 1.8{\times}10^6)}$	$(1.5{\times}10^6, 2.1{\times}10^6)$

iterations of PSO is a more appropriate measure than the number of function evaluations [6]. With respect to iterations, Table 4 compares $Ring_{40}$–$Ring_5$ with $Ring_{200}$–$Ring_5$ and $Ring_{40}$–$Ring_{25}$, which loosely represent the situations where additional processors or time are available, respectively. If extra resources are available, they clearly provide improvements in the pursuit of better answers.

5   Conclusions and Future Work

Organizing particle swarms into communities of subswarms significantly improves the performance of PSO. We attribute the improvement to emergent behavior from the social interaction of particles. We speculate that small groups of particles might make progress on implicit subproblems. Likewise, subswarms might help other subswarms get unstuck if they have prematurely converged in individual dimensions. In any case, the behavior of particle swarm apiaries is not explained by amount of communication, but rather the structure of the swarms.

Furthermore, we have shown that apiaries are particularly well-suited to parallel computation. With low communication and adjustable task granularity, the topology is easily adapted to varying computational architectures. With an inexpensive benchmark function, parallel PSO was able to perform about 2 outer iterations per second and provide a speedup of 15 on 40 processors. For any non-trivial function, the performance would be even more pronounced. Unlike other multi-swarm topologies like DMS-PSO [9], which requires frequent global communication, the Apiary topology requires very little communication.

We believe there are several interesting areas that are open to future work. In particular, organizing subswarms into hierarchies is a promising possibility. Apiaries are effective with extremely small subswarms, so a hierarchical structure can be built with a low branching factor. For example, a three-layer apiary would only have $5^3 = 125$ particles, and a four-layer apiary would have $5^4 = 625$ particles, well within the range that can be computed on a medium-size cluster.

Acknowledgments. The Fulton Supercomputing Lab at Brigham Young University generously provided 10 processor-years of resources.

References

1. Schutte, J.F., Reinbolt, J.A., Fregly, B.J., Haftka, R.T., George, A.D.: Parallel global optimization with the particle swarm algorithm. International Journal for Numerical Methods in Engineering 61(13), 2296–2315 (2004)
2. Kennedy, J., Eberhart, R.C.: Particle swarm optimization. In: Proc. International Conference on Neural Networks IV (1995)
3. Bratton, D., Kennedy, J.: Defining a standard for particle swarm optimization. In: Proc. IEEE Swarm Intelligence Symposium, pp. 120–127 (2007)
4. Clerc, M., Kennedy, J.: The particle swarm—explosion, stability, and convergence in a multidimensional complex space. IEEE Transactions on Evolutionary Computation 6(1), 58–73 (2002)
5. Mendes, R.: Population Topologies and Their Influence in Particle Swarm Performance. PhD thesis, Universidade do Minho, Guimaraes, Portugal (2004)
6. McNabb, A., Gardner, M., Seppi, K.: An exploration of topologies and communication in large particle swarms. In: Proc. IEEE Congress on Evolutionary Computation, pp. 712–719 (2009)
7. Sammataro, D., Avitabile, A.: Beekeeper's Handbook, 3rd edn. Cornell University Press (1998)
8. Sample, I.: Bees translate dances of foreign species. The Guardian (June 3, 2008)
9. Liang, J., Suganthan, P.: Dynamic multi-swarm particle swarm optimizer. In: Proc. IEEE Swarm Intelligence Symposium, pp. 124–129 (2005)
10. Venter, G., Sobieszczanski-Sobieski, J.: A parallel particle swarm optimization algorithm accelerated by asynchronous evaluations. In: Proc. 6th World Congresses of Structural and Multidisciplinary Optimization (2005)
11. Romero, J., Cotta, C.: Optimization by Island-Structured Decentralized Particle Swarms. In: Reusch, B. (ed.) Computational Intelligence, Theory and Applications. AISC, vol. 33, pp. 25–33. Springer, Heidelberg (2005)
12. Chu, S.C., Pan, J.S.: Intelligent parallel particle swarm optimization algorithms. Parallel Evolutionary Computations, 159–175 (2006)
13. Lorion, Y., Bogon, T., Timm, I., Drobnik, O.: An agent based parallel particle swarm optimization—APPSO. In: Swarm Intelligence Symposium (2009)
14. Wang, H., Qian, F.: An improved particle swarm optimizer with shuffled subswarms and its application in soft-sensor of gasoline endpoint. In: Proc. International Conference on Intelligent Systems and Knowledge Engineering (2007)
15. Tang, K., Li, X., Suganthan, P., Yang, Z., Weise, T.: Benchmark functions for the CEC 2010 special session and competition on large scale global optimization. Technical report, IEEE Congress on Evolutionary Computation (November 2009)
16. Dwass, M.: Modified randomization tests for nonparametric hypotheses. The Annals of Mathematical Statistics 28(1), 181–187 (1957)

ACO on Multiple Gpus
with CUDA for Faster Solution of Qaps

Shigeyoshi Tsutsui

Hannan University, Matsubara Osaka 580-8502, Japan
tsutsui@hannan-u.ac.jp

Abstract. In this paper, we implement ACO algorithms on a PC which has 4 GTX 480 Gpus. We implement two types of ACO models; the island model, and the master/slave model. When we compare the island model and the master/slave model, the island model shows promising speedup values on class (iv) QAP instances. On the other hand, the master/slave model showed promising speedup values on both classes (i) and (iv) with large-size QAP instances.

1 Introduction

Recently, GPU (Graphics Processing Unit) computation has become popular with great success, especially in scientific fields such as fluid dynamics, image processing, and visualization using particle methods [1]. As for parallel ACO on GPU, Bai et al. [2], Fu et al. [3], and Delévacqa et al. [4] implemented MMAS on GPU with CUDA and applied it to solve TSP. In [5], Diego et al. proposed a parallelization strategy to solve the VRP with ACO on a GPU.

In a previous paper [6], we proposed an ACO to solve large scale quadratic assignment problems (Qaps) on a GPU (GTX480) with CUDA. We used tabu search (TS) as a local search of solutions obtained by the ACO. In the implementation, we proposed a novel threads assignment method in CUDA, which we call *MATA* (Move-Cost Adjusted Thread Assignment), to reduce the idling time of threads caused by thread divergence in a warp (see Section 2.1). We tested the ACO using several large-size benchmark instances in QAPLIB [7]. The ACO was able to solve the QAP instances successfully with about 20x speedup compared with CPU computation (i7 965, 3.2GHz). As for the ACO algorithm, we use the Cunning Ant System (*c*AS) [8].

In this paper, we implement the previous ACO algorithm on a PC which has 4 GTX 480 Gpus. We implement two types of ACO models using multiple Gpus. One is the island model, and the other is the master/slave model. In the island model, we implement one colony on each GPU, agents (solutions) are exchanged among colonies at defined ACO iteration intervals using several types of topologies. In the master/slave model, we have only one colony in the CPU, and only local search (TS) processes are distributed to each GPU.

In the remainder of this paper, Section 2 reviews of the previous study of ACO on a GPU with MATA and shows revised results using newly tuned parameter

C.A. Coello Coello et al. (Eds.): PPSN 2012, Part II, LNCS 7492, pp. 174–184, 2012.
© Springer-Verlag Berlin Heidelberg 2012

settings. Then, Section 3 describes how the ACO is implemented on a PC with multiple GPUs in detail. In Section 4, experimental results and their analysis are given. Finally, Section 5 concludes this paper.

2   A Review of an ACO on a GPU with MATA and Revised Results

2.1   GPU Computation with CUDA

Processors in a CUDA GPU are grouped into multiprocessors (MPs). Each MP consists of thread processors (TPs). TPs in an MP exchange data via fast-shared memory (SM). On the other hand, data exchange among MPs is performed via VRAM. In a CUDA program, threads form two hierarchies: the *grid* and *thread blocks*. A block is a set of threads. A grid is a set of blocks with the same size. Each thread executes the same code specified by the *kernel function*.

Threads in a block are executed through a mode called *single instruction, multiple threads* (SIMT) [9]. In SIMT, each MP executes threads in groups of 32 parallel threads called *warps*. A warp executes one common instruction at a time, so full efficiency is realized when all 32 threads of a warp agree on their execution path.

2.2   ACO with TS on a GPU for Solving QAP

The QAP is the problem which assigns a set of facilities to a set of locations and can be stated as a problem to find a permutation ϕ which minimizes

$$cost(\phi) = \sum_{i=0}^{n-1} \sum_{j=0}^{n-1} a_{ij} b_{\phi(i)\phi(j)} \qquad (1)$$

where $A = (a_{ij})$ and $B = (b_{ij})$ are two $n \times n$ matrices and ϕ is a permutation of $\{0, 1, \cdots, n-1\}$. Matrix A is a flow matrix between facilities i and j, and B is the distance between locations i and j. Thus, the goal of the QAP is to place the facilities on locations in such a way that the sum of the products between flows and distances is minimized.

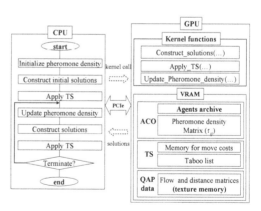

Fig. 1. ACO with TS on a GPU

Fig. 1 shows the configuration of the ACO with TS to solve GAPs on a GPU in [6]. As shown in the figure, each step of the algorithm is coded as a kernel function of CUDA. All of the data of the algorithm are located in VRAM of the GPU. As for the local search in Fig. 1, we implement TS based on Ro-TS [10].

Construction of a solution is performed by the kernel function "Construct-solutions($\cdots$)" in a single block. Then each m solutions are stored in VRAM. In the kernel function "Apply_tabu_search($\cdots$)", m solutions are distributed in m thread blocks. This function performs the computation of move-cost in parallel using a large number of threads in each block. Kernel function "Pheromone-update($\cdots$)" consists of 4 separate kernel functions to ease implementation. Thus, in this configuration, the CPU performs only loop control of the algorithm.

2.3 Move-cost Adjusted Thread Assignment (MATA)

More than 99% of computation time was used for execution of TS when we ran the algorithm using CPU with a single thread (see Table 3 in Section 3.2). MATA was proposed for efficient implementation of TS on a GPU.

As is well known, in TS we need to check all solutions neighboring the current solution to obtain the best move. This move-cost calculation is costly. Let $N(\phi)$ be the set of neighbors of the current solution ϕ. A neighbor, $\phi' \in N(\phi)$, is obtained by exchanging a pair of elements (i, j) of ϕ. Then, we need to compute move-costs $\Delta(\phi, i, j) = cost(\phi') - cost(\phi)$ for all the neighboring solutions. The neighborhood size of $N(\phi)$ ($|N(\phi)|$) is $n(n-1)/2$ where n is the problem size. When we exchange r-th and s-th elements of ϕ (i.e., $\phi(r)$ and $\phi(s)$), $\Delta(\phi, r, s)$ can be calculated in computing cost $\mathcal{O}(n)$ [6].

Let ϕ' be obtained from ϕ by exchanging r-th and s-th elements of ϕ, then fast computation of $\Delta(\phi', u, v)$ is obtained in computing cost $\mathcal{O}(1)$ if u and v satisfy the condition $u, v \cap r, s = \emptyset$ [11]. To use this fast update, additional memorization of the $\Delta(\phi, i, j)$ values for all pairs (i, j) in a table are required. For each move, we assign an index number as shown in Fig. 2. In this example, we assume a problem size of $n = 8$. Thus, the neighborhood size $|N(\phi)|$ is $8 \times 7/2 = 28$. As described in Section 2.2, each set of move-cost calculations of an solution is being done in one block. The simplest approach to computing the move-costs in parallel in a block is to assign each move indexed i to the corresponding sequential thread indexed i in a block.

Here, consider a case in which a solution ϕ' is obtained by exchanging positions 2 and 4 of a current solution ϕ in a previous TS iteration. Then the computation of $\Delta(\phi', u, v)$, the numbers shown in white font in black squares in Fig. 2, must be performed in $\mathcal{O}(n)$. The computation of the remaining moves are performed in $\mathcal{O}(1)$ fast. Thus, if we simply assign each move to the block thread, threads of a warp diverge via the conditional branch ($u, v \cap 2, 4 = \emptyset$) into two calculations; threads in one group run in $\mathcal{O}(n)$ and threads in the other

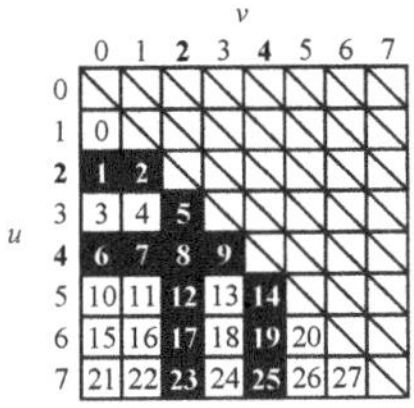

Fig. 2. Indexing of moves ($n = 8$)

group run in $\mathcal{O}(1)$. In threads of CUDA, all instructions are executed in SIMT (see Section 2.1). As a result, the computation time of each thread in a warp becomes longer, and we cannot receive the benefit of the fast calculation of $\mathcal{O}(1)$ in [11].

Thus, we should remove a situation where threads which run in $\mathcal{O}(1)$ and threads which run in $\mathcal{O}(n)$ co-exist in the same warp. In MATA, we assign move-cost computations of a solution ϕ which are in $\mathcal{O}(1)$ and in $\mathcal{O}(n)$ to threads which belong to different warps in a block, as shown in Fig. 3. Since the computation of a move-cost which is $\mathcal{O}(1)$ is smaller than the com-

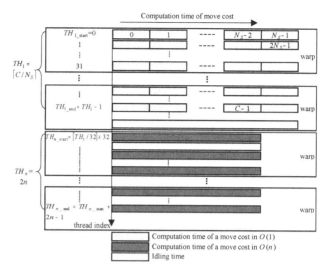

Fig. 3. Move-Cost Adjusted Thread Assignment

putation which is $\mathcal{O}(n)$, we assign a multiple number of N_S computations which are $\mathcal{O}(1)$ to a single thread in the block. Also, it is necessary to assign multiple calculations of the move-costs to a thread, because the maximum number of threads in a block is limited (1024 for GTX 480). Let C be $|N(\phi)|$ ($C = n(n-1)/2$). Here, each neighbor is numbered by $0, 1, 2, \cdots, C-1$ (see Fig. 2). Then, the thread indexed as $t = \lfloor k/N_S \rfloor$ computes moves for $k \in tN_S, tN_S + 1, \cdots, tN_S + N_S - 1$. In this computation, if k is a move in $\mathcal{O}(n)$, then the thread indexed as t skips the computation. The total number of threads assigned for computations in $\mathcal{O}(1)$ is $TH_1 = \lceil C/N_S \rceil$.

For the computation in $\mathcal{O}(n)$, we assign only one computation of move-cost to one thread in the block. Although the total number of moves in $\mathcal{O}(n)$ is $2n-3$, we used $TH_n = 2n$ threads for these computations for implementation convenience. Since the threads for these computations must not share the same warp with threads used for computations in $\mathcal{O}(1)$, the starting thread index should be a multiple of warp size (32), which follows the index of the last thread used for computation in $\mathcal{O}(1)$. Thus, the total number of threads in a block TH_{total} is $\lceil TH_1/32 \rceil \times 32 + TH_n$.

2.4 Revised Results

In this section, we present revised results from a previous study [6]. We tuned TS parameters so that we can get better performance as shown in Table 1. The machine is the same as before; i.e., a PC which has one Intel Core i7 965 (3.2 GHz) processor and a single NVIDIA GeForce GTX480 GPU. The OS was Windows XP Professional. We updated CUDA 4.0 SDK from previous 3.1 SDK.

The instances on which we tested our algorithm were taken from the QAPLIB benchmark library [7]. QAP instances in the QAPLIB can be classified into 4 classes; (i) randomly generated instances, (ii) grid-based distance matrix,

(iii) real-life in-stances, and (iv) real-life like instances [11]. In this revised experiment, we used 10 instances which were classified as either (i) or (iv) (please see Table 2). Here, note that instances classified into class (i) are much harder to solve than those in class (iv). 25 runs were performed for each instance. In Table 1, values in parentheses are values used in [6].

Let IT_{TS} be the length of TS applied to one solution which is constructed by ACO, and IT_{ACO} be the iterations of ACO, respectively. Then, $IT_{TOTAL} = m \times IT_{ACO} \times IT_{TS}$ represents a total length of TS in the algorithm. We define a value $IT_{TOTAL-MAX} = m \times n \times 3200$. In this revised experiment, if IT_{TOTAL} reaches $IT_{TOTAL-MAX}$ or a known optimal solution is found, the algorithm terminates. This $IT_{TOTAL-MAX}$ is larger

Table 1. Revised parameter values (γ is a control parameter of cAS [8])

Parameters		Values	
		class (i) QAPs	class (iv) QAPs
ACO	Number of ants: m	n	n $(4n)$
	Evaporation factor: ρ	0.2 (0.5)	0.2 (0.5)
	γ	0.4	0.5
TS	Length of TS: IT_{TS}	$64n$ $(16n)$	n $(4n)$
	Taboo list size: $T_{list\ s}$	$4n$ $(16n)$	$n/2$ (n)
	N_s	$n/4$	$n/4$

than the $IT_{TOTAL-MAX}$ in [6]. T_{avg} and $Error(\%)$ are mean run time and mean error over 25 runs, respectively. The revised results are summarized in Table 2. The effectiveness of using MATA is clearly observed as was shown in [6]. Values of $Error$ are smaller and values of $Speedup$ in T_{avg} are larger than observed in [6] due to revised parameter settings and longer runs.

3   Implementation of ACO on Multiple GPUs

Since there are four PCIe x16 slots in our system. we added an additional 3 GTX 480 GPUs and constructed a multi-GPU environment with a total of 4 GTX 480 GPUs. In this section, we propose two types of multi-GPU models for ACO with MATA, to attain a fast computation speed in solving QAPs. They are the island model and the master/slave model, the most popular parallel EAs [12,13].

3.1   Island Model

Island models for EAs in a massively parallel platform are intensively studied in [14]. The cAS, which is used as our ACO model in this study, has an archive which maintains m solutions (see Fig. 1). This archive is similar to a population in EAs. In our implementation, we exchange (immigrate) the solutions among GPUs. In our imple-

Table 2. Revised results with MATA

QAP instances		GPU Computation (GTX 480)				CPU Computation		Speedup in T_{avg}
		T_{avg} (sec)			Error (%)	T_{avg} (sec)	Error (%)	CPU
		MATA	non-MATA	non-MATA MATA				MATA
class (i)	tai40a	9.5	70.6	7.4	0.14	191.0	0.14	20.0
	tai50a	18.3	136.0	7.4	0.34	463.7	0.35	25.3
	tai60a	32.6	269.1	8.3	0.32	962.8	0.36	29.6
	tai80a	154.9	728.2	4.7	0.41	3108.7	0.35	20.1
	tai100a	431.8	2352.9	5.4	0.36	7894.3	0.33	18.3
class (iv)	tai50b	0.3	1.7	6.4	0	6.8	0	24.9
	tai60b	0.5	4.0	7.9	0	15.4	0	30.5
	tai80b	5.5	30.6	5.6	0	145.0	0	26.5
	tai100b	14.6	80.8	5.5	0	374.4	0	25.7
	tai150b	2893.2	16348.8	5.7	0.07	48948.6	0.05	16.9
average		-	-	6.4	-	-	-	23.8

mentation, one ACO model in a GPU in Section 2 composes one island. In the configuration of ACO on a GPU in Fig. 1, all m solutions are maintained in VRAM of the GPU. In an island model, we need to exchange solutions among islands (GPUs) depending on its topology.

It is possible to exchange solutions among GPUs using "cudaMemcpyPeer(···)" function with CUDA 4.x. without via CPU. However, to perform exchange solutions depending on a defined topology, the CPU needs to know which data should be exchanged. This means that the CPU can't execute the cudaMemcpyPeer(···) function without having solutions from each GPU. Since the data needs to be sent to the CPU anyway, it is most efficient to exchange this data through the CPU rather than doing a direct exchanged between GPUs. Thus, in our implementation of island models, solutions in VRAM are transferred between GPU and CPU using usual "cudaMemcpy(···)" function when immigrations are required as shown in Fig. 4. As for control multiple GPUs in the CPU, we use OpenMP API.

Although there are many topologies for island models [13], in this study we implement the following 4 models:

(1) Island model with independent runs (IM_INDP): Four GPUs are executed independently. When at least one ACO in a GPU finds an acceptable solution, then the algorithm terminates.

(2) Island model with elitist (IM_ELIT): In this model, at defined ACO iteration interval $I_{interval}$ the CPU collects the global best solution from the 4 GPUs, and then distributes it to all GPUs except the GPU that produced that best solution. In each GPU, the worst solution in each archive is replaced with the received solution.

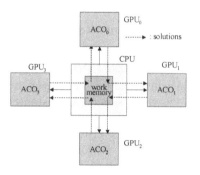

Fig. 4. Island model with 4 GPUs

(3) Island model with ring connected (IM_RING): The best solution in each GPU g ($g = 0, 1, 2, 3$) is distributed to its neighbor GPU $(g + 1)$ Mod 4 at $I_{interval}$. In each GPU, the worst solution in each archive is replaced with the received solution if the received one is better than the worst one.

(4) Island model with elitist and massive ring connected (IM_ELMR): In this model, first the global best solution is distributed, as performed in IM_ELIT. Then, in addition to this immigration operation, randomly selected $m \times d_{rate}$ of solutions in the archive in each GPU are distributed to its neighbor. Received solutions in each GPU are compared with randomly selected, non-duplicate solutions. We use d_{rate} of 0.5 in this study.

3.2   Master/Slave Model

As mentioned in Section 2.3, more than 99% of computation time was used for execution of TS when we ran the algorithm using CPU with a single thread (see Table 3). In the master/slave model in this study, the ACO algorithm is executed in the CPU as shown in Fig. 5. Let m be number of agents in the archive of cAS, then we assign $m/4$ number of solutions to each GPU. When new solutions are generated in the CPU, first, $m/4$ number of solutions are transferred to each

GPU, then "Apply_tabu_search(···)" kernel function is lunched to apply the TS with MATA to these solutions. The improved solutions are send back to the CPU from each GPU.

Note here that in practical implementation of the master/slave model, the value of m must be divisible by the number 4. So, we assigned $\lfloor m/4 \rfloor$ number of agents to each slave GPU and we used an agents number of $m' = \lfloor m/4 \rfloor \times 4$ instead of m.

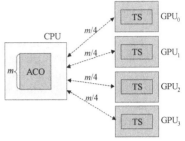

Fig. 5. Master/slave model with 4 GPUs

4   Experiment of Multiple GPUs in Solving QAPs

4.1   Experimental Setup

The machine is the same as in Section 2.4. We used 4 GTX 480 GPUs in 4 PCIe slots. We use the control parameter values shown in Table 1. We used the same QAP instances described in Section 2.4.

Termination criteria are slightly different from those in Section 2.4. When we perform a fair comparison of different algorithms, sometimes it is difficult to determine their termi-

Table 3. Computation time with sequential CPU run

QAP Instances	Construction of solusions	TS	Updating Trail
tai60a	0.004%	99.996%	0.000%
tai100a	0.002%	99.998%	0.000%
tai100b	0.008%	99.991%	0.000%
tai150b	0.005%	99.995%	0.000%

nation criteria. In this experiment, we run the algorithms until predetermined acceptable solutions are obtained and effectiveness of using 4 GPUs is measured by average time $(T_{avg,4})$ to obtained the solutions. We obtain the speedup by $T_{avg,1}/T_{avg,4}$, where $T_{avg,1}$ is average time to obtain acceptable solutions by ACO using a single GPU configured as described in Section 3. We performed 25 runs for each experiment.

We determined acceptable solutions as follows. For the class (i) instances, since it is difficult to obtain known optimal solutions 25 times in 25 runs with reasonable run time, we set the their acceptable solutions to be within 0.5% of the known optimal solutions. For the class (iv) instances, except tai150b, we set them to known optimal solutions. We set tai150b to be within 0.2% of the known optimal solution. We used $I_{interval}$ value of 1.

4.2   Results of the Island Models

Results of the island models are summarized in Table 4. The IM_INDP is the simplest of the island models. Thus, we use results of IM_INDP as bench marks for other island models. Except for the results from tai40a, all other island models had improved speedup values compared to IM_INDP. In the table, we showed the average number of iterations of the ACO to obtain the acceptable solutions (IT_{ACO}). On tai40a, this value is only 1.7. Thus, on this instance, there was no benefit from immigration operations.

The speedup values of IM_RING and IM_ELIT showed very similar results with each other. On tai80b and tai150b, we can see super-linear speedup values. We performed t-test between IM_ELIT and IM_INDP, showing a clear effect of using this topology, especially for class (iv) instances. Since we used long-tabu search length for class (i) instances (see Table 1), values of IT_{ACO} are smaller than those of class (iv) instances. This could have caused the reduced effect of immigration on class (i) instances, compared with (iv) instances.

Among the four island models, IM_ELMR showed the best speedup, except for tai40a. However, the t-test between IM_ELIT and IM_ELMR shows that the advantage of using IM_ELMR over IM_RING and IM_ELIT on class (i) instances again becomes smaller than on class (iv) instances. The speedup values are different among instances. Consider why

Table 4. Results of the island models with 4 GPUs

QAP instances	Acceptable Error (%)	$T_{avg, 1}$ (sec)	Average IT_{ACO}	Speedup ($T_{avg, 1}/T_{avg, 4}$)				p-values	
				IM_INDP	IM_RING	IM_ELIT	IM_ELMR	IM_INDP v.s. IM_ELIT	IM_ELIT v.s. IM_ELMR
tai40a	0.5	0.4	1.7	1.7	**1.8**	1.7	1.7	0.90	0.85
tai50a	0.5	4.2	11.4	2.1	2.6	2.9	**3.3**	0.07	0.44
tai60a	0.5	10.6	14.8	1.9	2.2	2.3	**2.5**	0.03	0.61
tai80a	0.5	58.3	18.9	2.4	2.5	2.7	**2.9**	0.45	0.75
tai100a	0.5	125.9	14.6	1.7	2.1	2.2	**2.5**	0.03	0.13
tai50b	0	0.2	31.8	1.5	2.3	2.5	**3.0**	6.50E-05	0.08
tai60b	0	0.4	25.7	1.2	1.4	1.9	**2.6**	4.68E-05	5.83E-05
tai80b	0	6.6	121.3	1.5	4.7	4.3	**6.5**	9.16E-11	4.00E-09
tai100b	0	10.1	67.0	1.4	2.3	2.3	**3.2**	6.00E-08	1.24E-05
tai150b	0.2	105.9	116.6	1.8	4.2	4.3	**4.9**	2.10E-05	0.22

(Rows tai40a–tai100a: class (i); rows tai50b–tai150b: class (iv))

these difference occur using IM_INDP as a parallel model. Let probability density function of the run time on a single GPU be represented by $f(t)$ and probability distribution function of $f(t)$ be $F(t)$. Here, consider an IM_INDP with p GPUs. Let the probability distribution function of run time of the IM_INDP with p GPUs be represented by $G(t,p)$. Since there is no interaction among GPUs in IM_INDP, the $G(t,p)$ can be obtained as

$$G(p,t) = 1 - (1 - F(t))^p, \qquad (2)$$

and the average run time $T_{avg,p}$ is obtained as

$$T_{avg,p} = \int_0^\infty t \cdot G'(p,t)dt \qquad (3)$$

Thus, the speedup with p GPUs is obtained as $Speedup(p) = T_{avg,1}/T_{avg,p}$. Table 5 shows the values of $Speedup(p)$ and $Speedup(4)$ for assuming various functions of $f(t)$. This analysis gives us a good understanding of the results of IM_INDP in Table 4. But for more detail analysis, we need to identify $f(t)$ by sampling the data of run times.

Table 5. Estimation of *Speedup*

$f(t)$	Speedup (p)	Speedup (4)
$f(t) = 1, \ 0 \le t \le 1$	$(p+1)/2$	2.5
$f(t) = 2(1-t), \ 0 \le t \le 1$	$(2p+1)/3$	3
$f(t) = 3(1-t)^2 \ 0 \le t \le 1$	$(3p+1)/4$	3.25
$f(t) = \lambda e^{-\lambda t} \ 0 \le t$	p	4

4.3   Results of the Master/Slave Model

Since computation times of TS occupy more than 99% of the algorithm (Table 3), we expected the master/slave model to show good results in the speedup. However, as shown in Fig. 6, results on the small-size instances in this study (tai40a, tai50a, tai60a, tai50b, tai60b) showed relatively small speedup values against the ideal speedup value of 4. In the figure, the *Speedup* values are defined in Section 4.1. Results on large-size instances (tai80a, tai100a, tai80b, tai100b tai150b), the speedup values were nearer to ideal speedup values.

Now we will analyze why these results were obtained on the master/slave model. Fig. 7 shows the average computation times of tai60a and tai150b over 10 runs for 10 ACO iterations by changing the number of agents m from 1 to 150 with step 1. Here, the ACO algorithm is the master/slave model in Section 3.2

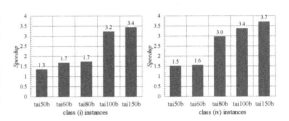

Fig. 6. Results of the master/slave model with 4 GPUs

with a GPU number setting of 1, and the computation time is normalized by the time of $m = 1$. Since the number of MPs of GTX 480 is 15, we can see the computation times increases nearly 15 interval of m. However, the increasing times are different between these two instances.

On tai60a ($n = 60$) instance, the difference of computation times among $1 \leq m \leq 15$, $16 \leq m \leq 30$, and $31 \leq m \leq 45$, and $46 \leq m \leq 60$ is very small. In our implementation of TS on a GPU, we assigned one thread block to each agent (solution), and thus number of agents is identical to number of thread blocks. In CUDA, multiple blocks are allocated to one MP if computation resources, such as registers, are available. In the execution of tai60a ($n = 60$), this situation occurs and multiple blocks are executed

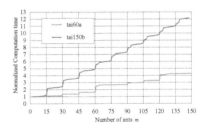

Fig. 7. Computation times for various number of agents

in parallel in one MP. Since in this experiment, we set $m = 60$ ($:m = n$, see Section 4.1 and Table 1), the solutions assigned to one slave GPU is only 15. This means the speedup using the master/slave model becomes very small as was seen in Fig. 6.

On the other hand, on tai150b ($n = 150$) instances, computation times proportionally increase according to m with every 15 intervals. This means that on tai150b, a single thread block is allocated to one MP at the same time, with the resulting speedup shown in Fig. 6. Note here that on this instances, the number of agents assigned to one GPU is $\lfloor 150/4 \rfloor = 37$ and the total agent number of $37 \times 4 = 148$ was used in this experiment.

5   Conclusion

In a previous paper, we proposed an ACO for solving QAPs on a GPU by combining TS local search in CUDA. There, we implemented an efficient thread assignment method, MATA. Based on this implementation, in this paper we implemented the algorithm on multiple GPUs to solve QAPs fast. We implemented two types of models on multiple GPUs; the island model and the the master/slave model. For these models, we experimented using QAP benchmark instances, and gave analysis on the results.

As for the island model, we used 4 types of topologies. Although the results of speedup much depend on the instances we used, we showed that the island model IM_ELMR has a good speedup feature. As for the master/slave model, we observed reasonable speedups for large-size of instances, where we used large number of agents.

When we compared the island model and the master/slave model, the island model showed promising speedup values on class (iv) instances of QAP. On the other hand, the master/slave model consistently showed promising speedup values both on classes (i) and (iv) with large-size QAP instances with large number of agents. As regards to this comparison, a more intensive analytical study is an interesting future research direction. Implementation using an existing massively parallel platform such as EASEA [14] is also an interesting future research topic.

References

1. Ryoo, S., Rodrigues, C.I., Stone, S.S., Stratton, J.A., Ueng, S.Z., Baghsorkhi, S.S., Mei, W., Hwu, W.: Program optimization carving for GPU computing. J. Parallel Distrib. Comput. 68(10), 1389–1401 (2008)
2. Bai, H., OuYang, D., Li, X., He, L., Yu, H.: MAX-MIN ant system on GPU with CUDA. In: Innovative Computing, Information and Control, pp. 801–804 (2009)
3. Fu, J., Zhou, G., Lei, L.: A parallel ant colony optimization algorithm with GPU-acceleration based on all-in-roulette selection. In: Workshop on Advanced Computational Intelligence, pp. 260–264 (2010)
4. Delévacqa, A., Delislea, P., Gravelb, M., Krajeckia, M.: Parallel ant colony optimization on graphics processing units. Journal of Parallel and Distributed Computing (in press, 2012)
5. Diego, F., Gómez, E., Ortega-Mier, M., García-Sánchez, Á.: Parallel CUDA architecture for solving de VRP with ACO. In: Industrial Engineering and Industrial Management, pp. 385–393 (2012)
6. Tsutsui, S., Fujimoto, N.: ACO with tabu search on a GPU for solving QAPs using move-cost adjusted thread assignment. In: GECCO 2011, pp. 1547–1554. ACM (2011)
7. Burkard, R.E., Çela, E., Karisch, S.E., Rendl, F.: QAPLIB - a quadratic assignment problem library (2009), www.seas.upenn.edu/qaplib
8. Tsutsui, S.: cAS: Ant Colony Optimization with Cunning Ants. In: Runarsson, T.P., Beyer, H.-G., Burke, E.K., Merelo-Guervós, J.J., Whitley, L.D., Yao, X. (eds.) PPSN IX. LNCS, vol. 4193, pp. 162–171. Springer, Heidelberg (2006)
9. NVIDIA: (2010), developer.download.nvidia.com/compute/cuda/3_2_prod/toolkit/docs/CUDA_C_Programming_Guide.pdf

10. Taillard, É.: Robust taboo search for quadratic assinment problem. Parallel Computing 17, 443–455 (1991)
11. Taillard, É.: Comparison of iterative searches for the quadratic assignment problem. Location Science 3(2), 87–105 (1995)
12. Alba, E.: Parallel Metaheuristics: A New Class of Algorithms. John Wiley and Sons (2005)
13. Cantú-Paz, E.: Efficient and Accurate Parallel Genetic Algorithms. Kuwer Academic Publishers (2000)
14. Maitre, O., Krüger, F., Querry, S., Lachiche, N., Collet, P.: EASEA: specification and execution of evolutionary algorithms on GPGPU. Soft Comput. 16(2), 261–279 (2012)

It's Fate: A Self-organising Evolutionary Algorithm

Jan Bim, Giorgos Karafotias, S.K. Smit, A.E. Eiben, and Evert Haasdijk

Vrije Universiteit Amsterdam

Abstract. We introduce a novel evolutionary algorithm where the centralized oracle –the selection-reproduction loop– is replaced by a distributed system of *Fate
Agents* that autonomously perform the evolutionary operations. This results in a
distributed, situated, and self-organizing EA, where candidate solutions and Fate
Agents co-exist and co-evolve. Our motivation comes from evolutionary swarm
robotics where candidate solutions evolve in real time and space. As a first proof-
of-concept, however, here we test the algorithm with abstract function optimization problems. The results show that the Fate Agents EA is capable of evolving
good solutions and it can cope with noise and changing fitness landscapes. Furthermore, an analysis of algorithm behavior also shows that this EA successfully
regulates population sizes and adapts its parameters.

1   Introduction

Evolutionary algorithms (EAs) offer a natural approach to provide adaptive capabilities
to systems that are by nature distributed in a real or virtual space. Examples of such systems are robot swarms that have to adapt to some dynamically changing environment,
or a collection of adaptive software agents that provide services at different locations in
a vast computer network. Such systems are becoming more and more important, and so
is the need to make them evolvable on-the-fly. The problem is that traditional EAs with
central control are not suited for these kinds of applications.

In traditional EAs we can distinguish two entities: the population of candidate solutions that undergo evolution and an omniscient oracle (the main EA loop) that decides
about all individuals and performs the evolutionary operators. In situated evolution in
general, and in evolutionary swarm robotics in particular, a single oracle has a number of drawbacks [11]. Firstly, it forms a single point of failure, secondly, it may limit
scalability as it may not be able to process all the information about the individuals
in a timely manner, and thirdly, it may not be reachable for certain individuals if the
distance exceeds the feasible (or cost effective) range of communication. The natural
solution would be a system with multiple, spatially distributed EA-oracles that provide sufficient coverage of the whole population. Furthermore, the inherently dynamic
circumstances in such applications require that the EA-oracles can adjust their own
settings on-the-fly [8].

The first objective of this paper is to introduce a system that combines two fundamental properties by design: 1) evolutionary operators are distributed and 2) the algorithmic settings are self-regulating. The key idea is to decompose the EA loop into three
separate functional components, parent selection, reproduction/variation, and survivor
selection, and create autonomous entities that implement these components. We name

C.A. Coello Coello et al. (Eds.): PPSN 2012, Part II, LNCS 7492, pp. 185–194, 2012.

these entities *Fate Agents* after the 'Moirai' of Greek mythology. For a maximally modular system we define three types of Fate Agents, one for each EA component, and add several Fate Agents of each type to the regular population of candidate solutions. These Fate Agents control the evolution of both regular candidate solutions and other Fate Agents in their direct surroundings. Because the Fate Agents also act on Fate Agents, the population of Fate Agents evolves itself and the EA configuration becomes adaptive.

Obviously, an elegant design does not by itself justify a new system. The second objective of this paper is an experimental assessment to answer three main questions:

1. Can this evolutionary system solve problems at all?
2. Can this evolutionary system cope with noise?
3. Can this evolutionary system cope with changing fitness landscapes?

Because the Fate Agents EA is new and has, to our knowledge, never been implemented before, we are also interested in system behavior. Therefore, we also inspect various run-time descriptors (e.g., the numbers of agents) that can help us understand what happens during a run.

2   Related Work

Existing work can be related to our research from the angles of the two main properties mentioned above: distributed, agent-based evolutionary operators and self-regulating algorithmic settings. The latter is a classic theme in EC, often labelled parameter control or on-line parameter setting [2,8]. In this context, our work is distinguished by the novel Fate Agent technique to modify EA parameters and the fact that it can handle *all* parameters regarding selection, reproduction and population size. This contrasts with the majority of related work where typically one or two EA parameters are handled. Furthermore, our system naturally handles population sizes, which is one of the toughest problems in EAs with autonomous selection [14].

The distributed perspective is traditionally treated in the context of spatially structured EAs [12], where the cellular EA variants are the closest to our system [1]. Nevertheless, there are important differences: spatially structured EAs are based on "oracles" outside the population that do not change over time, while our Fate Agents operate "from within" and –most importantly– undergo evolution themselves. The combination of spatial structure and parameter control has been studied in [5] and [4], where each location in the grid space has a different combination of parameter values. These, however, are all set by the user at initialization and do not change over time.

Finally, our system can be related to meta-evolution [3], in particular the so-called local meta-evolutionary approaches [10] (not the meta-GA lookalikes). Work in this sub-domain is scarce, we only know about a handful of papers. For instance, [10] provides a theoretical analysis, [6] demonstrates it in GP, while [9] eloquently discusses computational vs. biological perspectives and elaborates on algorithms with an artificial chemistry flavor.

3   The Fate Agents Evolutionary Algorithm

Our Fate Agents EA is situated in a (virtual) space where agents move and interact. The evolving population consists of two main types of agents: passive agents that represent candidate solutions to the problem being solved and active Fate Agents that embody EA operators and parameters. Fate Agents form evolving populations themselves because they act not only upon candidate solutions but also upon each other. This makes the Fate Agents EA entirely self-regulated. By design, Fate Agents have a limited range of perception and action: they can only influence other agents within this range. Consequently, the evolutionary process is fully distributed as there is no central authority that orchestrates evolution but different parts of the environment are regulated by different agents. Below we describe the agent types and functionalities and subsequently the algorithm's main cycle.

Candidate Solution Agents are the simplest type of agent: they only carry a genome which represents a solution to the given problem. The fitness of a candidate solution agent is the fitness of its genome according to this problem. In a swarm robotic application, for example, we could have working robots and Fate robots; the principal problem would then be to evolve controllers for the working robots. The candidate solutions would be the controllers in the working robots encoded by some appropriate data structure and the corresponding fitness would be based on the task the working robots have to solve. To solve an abstract function optimization problem, the candidate solutions' genome would be no different from that in a regular EA, but the candidate solution would be situated in and move about a virtual space, along with the Fate Agents. In general, we assume that candidate solution agents are able to move. However, they are passive in the algorithmic sense, being manipulated by Fate Agents.

Fate Agents personify and embody evolutionary operators: parent selection, variation/reproduction and survivor selection. Fate Agents have a limited range of operation so that each one can act only within its local neighborhood. Fate Agents themselves form an evolving population, hence they require a measure of fitness. We experimented with various approaches, such as (combinations of) measures like diversity, average and median fitness; we found that the use of the best candidate solution fitness in the area yields the best results. Thus, the fitness of a Fate Agent is set to the fitness of the fittest candidate solution in its neighborhood. There are three types of Fate Agents, each responsible for different evolutionary operators: *cupids* select and pair up parents, *breeders* create offspring while *reapers* remove agents to make room for new ones. Note that they perform these operations not only on candidate solutions but on each other as well, e.g. cupids make matches between cupids, reapers kill breeders, etc.

Cupids realize parent selection by determining who mates with whom. The selection procedure is the same for all kinds of agents. A cupid creates lists of potential parents by running a series of tournaments in its neighborhood. The number of tournaments held for each type of agent depends on two values: the number of agents of that type in the cupid's neighborhood and a probability that this type of agent is selected. The latter probability is different for each distinct cupid and subject to evolution in the cupid strain. The tournament sizes also evolve. Thus, a cupid's genome consists of four real values representing the selection probabilities for each agent type and one integer for tournament size.

Reapers realize survivor selection indirectly, by selecting who dies. The selection mechanism of reapers is identical to that of cupids (with the difference that reapers' tournaments select the worst of the candidates). Reapers' genomes also consist of four selection probabilities for the different agent types and a tournament size. In earlier versions of the algorithm we tried different mechanisms for cupids and reapers. One approach was allowing a cupid/reaper to examine each and every agent in its neighborhood and make a separate decision whether to select it or not. That selection decision was facilitated by a simple perceptron using various measures of the agent and its surroundings as input. The weights and the threshold of the perceptron evolved. We found this representation to be overly complicated and results suggested mostly random selection. A variation of the selection scheme we currently use was to also evolve the probability that the winner of a tournament would actually be selected. Results suggested that this probability had no effect, possibly because the size of the tournament already provides sufficient control of selection pressure.

Breeders realize reproduction and variation by producing a child for a given couple of parent agents. For all kinds of agents the breeder performs both recombination and mutation. Breeders, as opposed to cupids and reapers, have different mechanisms for acting upon themselves and upon other agent types. In general, a breeder is given two parents by a cupid in the neighborhood and applies averaging crossover to produce one offspring and then Gaussian/creep mutation (in our experiments) on that offspring. A breeder's genome consists of three values: the mutation step sizes for candidate solutions, cupids and reapers. Thus, mutation of these agents evolves in the breeder population. Mutation step sizes for breeders are mutated according to the following rule taken from Evolution Strategies' self-adaptation:

$$\sigma_{t+1} = \sigma_t e^{\tau N(0,1)}$$

The reason for this distinction is that if breeders' mutation step sizes were also to evolve then these values would be used to mutate themselves. Trial experiments showed that this approach leads to a positive feedback loop that results in exploding values. Note, that the implementation of the breeder depends on the application: the crossover and mutation operators must suit the genomic representation in the candidate solutions. An earlier version of the breeder was designed with the intention to control as much of the reproduction process as possible. The breeders' genome consisted of mutation rates, mutation sizes and different parameters of crossover if applicable. It also included meta-mutation values that were used to mutate the previous values. There were three layers in a breeder's genome: the lower level consisted of values involved in the variation of candidate solutions and other Fate Agents while the upper levels were used to variate the lower layers (and thus the breeders themselves). Results showed that this approach was too complex and inappropriate for evolution, especially since upper level mutation step sizes had a rather minor short-term effect on the fitness of candidate solution agents.

The Main Cycle. In the experiments for this paper, we used the Fate Agent EA to solve abstract function optimization problems, so we had to devise a virtual space and operations for movement. Obviously, applications in a swarm robotics or ALife setting would come with predefined space and movement, and parts of the cycle presented here would be superfluous.

All the agents, both candidate solutions and Fate Agents, are situated in the same spatially structured environment, a torus shaped grid. Each cell can either be empty or occupied by exactly one agent. The algorithm makes discrete steps in two phases.

The first phase takes care of movement. For these experiments, we have simply implemented random movement by randomly swapping contents between two neighboring cells with a certain probability.

In the second phase evolutionary operators are executed. First, both cupids and reapers make their selections. Subsequently, offspring is produced in iterations as follows: in each iteration, a cupid with available parents and a free cell in its neighborhood is randomly chosen. A breeder is randomly selected from the neighborhood of the selected cupid and it is provided with the parents that are then removed from the cupid's list. The breeder produces a single child which is then placed in the empty cell. This procedure is repeated until there are no cupids with remaining selected agents and unoccupied cells in their neighborhood.

When offspring production has completed, reaping is performed: reapers are activated in random sequence until there are no reapers left with non-empty selection lists. Notice that during each reaping iteration, a reaper kills only one agent (of any type). Hence, a reaper can kill and be killed in the same reaping iteration. When reaping is complete, the evolutionary phase is concluded and the algorithm starts a new cycle. The overall algorithm cycle is presented in Algorithm 1.

The random sequence and individual actions of cupids and reapers during offspring production and reaping approximate a distributed system with agents acting autonomously and concurrently. It might seem unorthodox that selection by the reapers is performed before offspring are produced, meaning that unfit offspring are allowed to survive and possibly reproduce. Our motivation for this order of operators is to give Fate Agents a 'free pass': before a Fate Agent is considered for removal it should have the chance to act upon its neighborhood at least once, so that its evaluation will closer reflect its true fitness. The designed order does give this free pass to Fate Agents (at least to cupids and breeders).

4   Experimental Setup

We conducted several experiments to validate our algorithm from a problem solving perspective and to observe its runtime behavior. To this end we used the test functions from the BBOB2012 test suite from the GECCO 2012 black box optimization contest, because they are well designed and proven by several other research groups [1]. Furthermore, we experimented on the Fletcher & Powell function, because it is very hard to solve but its landscape can be easily redefined for the tests on changing fitness functions. We performed three sets of experiments:

A As a proof-of-concept that the algorithm generally works and is capable of problem solving and self-regulation we used 6 functions: BBOB2012 f_3, f_{20}, f_{22}, f_{23}, f_{24}, and the Fletcher & Powell. We allowed the algorithm to run for 500 generations.

B To see if the algorithm can cope with noise we used 6 other BBOB2012 functions: f_{122}, f_{123}, f_{125}, f_{126}, f_{128}, f_{129} and let the algorithm run for 1000 generations.

C To examine how well the system can recover from and adapt to sudden cataclysmic changes we ran tests on the Fletcher & Powell function randomizing its matrices every 250 generations. Here we allowed the algorithm to run for 2000 generations.

---

[1] http://coco.gforge.inria.fr/doku.php?id=bbob-2012

Algorithm 1. The Fate Agent EA algorithm

```
generation ← 0;
while generation ≤ maxGeneration do
    doMovement;
    for all Cupid c do
        c.SelectParents;
        cupids.Add(c);
    end for
    for all Reaper r do
        r.SelectDeaths;
        reapers.Add(r);
    end for
    while cupids.NotEmpty do
        c ← cupids.GetRandomCupid;
        if c.HasNoFreeCell ∨ c.SelectedParents.Empty then
            cupids.Remove(c);
        else
            b ← c.GetRandomNeighborBreeder;
            cell ← c.GetRandomFreeCell;
            a ← b.Breed(c.GetParents);
            cell.placeAgent(a);
        end if
    end while
    while reapers.NotEmpty do
        r ← reapers.GetRandomReaper;
        if r.agentsToKill.Empty then
            reapers.remove(r);
        else
            r.killAgent;
        end if
    end while
    generation ← generation + 1;
end while
```

For the spatial embedding we used a 100×100 grid where each cell can either be empty or occupied by only one agent (thus there is a maximum of 10000 agents). The grid is initialized by filling all cells with random agents of random type with the following probabilities: 0.0625 for each Fate Agent type and 0.8125 for candidate solutions. Random movement is implemented by swapping the contents of two neighboring cells with probability 0.5 for each edge. Fate Agents have a neighborhood of radius 5 cells. All our experiments are repeatable, since we offer the code of the Fate Agent EA and the experiments through the webpage of the second author. [2]

We emphasize that the purpose of these experiments is not to advocate the Fate Agents EA as a competitive numeric optimizer but only to demonstrate its ability to

---

[2] See http://www.few.vu.nl/~gks290/downloads/PPSN2012Fate.tar.gz for the whole source code.

solve problems and investigate its dynamics and self-regulating behavior without time consuming robotics or ALife experiments. The numeric test suite was used merely as a convenient testbed for evolution, thus a comparison with benchmarks or BBOB champions would be out of context.

5 Does It Work?

The results in section A in Table 1 show that the Fate Agents EA is indeed able to solve problems, achieving good fitness on almost all BBOB test functions and a reasonably high fitness for the very difficult FP problem. Good success ratios are also achieved for two noiseless and three noisy functions.

Section B of Table 1 demonstrates that our system is able to cope with noise very well: it achieves high fitness for all problems and a good success ratio for three out of six noisy functions. As was explained in Section 4 the purpose of the experiments is not to propose the Fate Agents EA as a numeric optimizer, thus, we will not examine its performance on BBOB any further or determine how competitive it is.

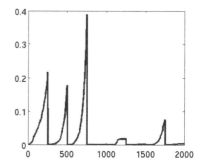

Fig. 1. Example of system recovery, fitness vs. time, experiment C, run 4

Finally, based on the results of experiment set C, we can give a positive answer to the third question we posed in Sec. 1: Fig. 1 presents the best fitness over time for an example run. The sudden drops in fitness mark the points in time when the matrix of the FP problem is randomly changed, drastically changing the fitness landscape. As can be seen, the system recovers from this catastrophic change, although it does not always succeed. In general, 24 out of 30 runs exhibited at least one successful recovery while, in total, we observed an equal number of successful and unsuccessful recoveries. It should be noted that the FP problem is very hard and the time provided between changes is quite short (250 generations). Nevertheless, results show that the Fate Agent EA does possess the ability to cope with radical change even though the design has no specific provisions for that purpose.

Table 1. Performance results for experiment sets A and B in terms of Average Best Fitness normalized between 0 (worst) and 1 (optimal), Average number of Evaluations to Best fitness achieved, Success Rate (success at 0.999 of normalized fitness) and Average number of Evaluations to Success (only successful runs taken into account)

	Set A - Static Noiseless					Set B - Static Noisy			
	ABF	AEB	SR	AES		ABF	AEB	SR	AES
F&P	0.53	289682	0.0	-	f_{122}	0.99	207857	0.03	240684
f_3	0.87	223341	0.76	228440	f_{123}	0.99	130996	0.96	131425
f_{20}	0.73	210527	0.03	266631	f_{125}	0.99	131383	0.0	-
f_{22}	0.90	309865	0.8	328891	f_{126}	0.99	126342	1.0	126342
f_{23}	0.90	204382	0.0	-	f_{128}	0.97	172836	0.40	195561
f_{24}	0.05	247797	0.0	-	f_{129}	0.99	137674	0.76	135225

6   System Behavior

One of the most important aspects of the system's behavior is that the population sizes for the different types of agents are successfully regulated. They reach a balance quite different to the initialization ratios (see Section 4) very quickly while no agent type becomes extinct or dominates the population even though there are no external limitations imposed. This is a very interesting result on its own right, since regulating population sizes in EAs with autonomous selection is an open issue [14]. An example run is shown in Fig. 2. Agent numbers are initialized to default values but populations soon converge and maintain a balance throughout the run. All runs across experiment sets demonstrate similar population dynamics.

Considering self-adaptation in EAs, one of the basic expectations is the adaptation of the size of search steps (the mutation step size σ in terms of evolution strategies). In our system, the mutation of agents is controlled by the breeder agents. The breeders' genome includes mutation step sizes for every other agent type. Fig. 3 presents examples of three different behaviors observed in breeder populations. Each graph illustrates the best and mean fitness of the candidate solution population and the mutation step size applied to

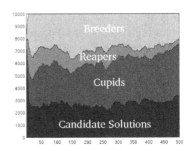

Fig. 2. Example of population breakdown over time, experiment A f_{22}, run 12

candidate solutions averaged over the whole breeder population. Case (a) is an example of typical evolution with mutation size slowly converging to zero as the search converges to the optimum. Case (b) demonstrates a successful response of the breeders to premature convergence to a local optimum: after around 100 generations the search gets stuck and the breeder population reacts with a steep increase of the mutation size which helps escape and progress. Case (c) shows a failed attempt to escape a local optimum, even though breeders evolve high mutation sizes after the search is stuck.

Note that mutation sizes are correlated to the average fitness, not to the best fitness. This is reasonable considering that Fate Agents have a limited range and are unaware of global values. In all cases, the mutation size converges to zero as soon as the whole population converges (mean fitness becomes equal to the best fitness). This implies that the system has the ability to respond and escape local optima as long as there is still diversity available but is unable to create new diversity after global convergence, as is the case in Fig. 3(c).

Finally, we made an interesting observation related to spatial dynamics: results show that, on average, cupids consistently have better fitness than reapers. Both agent types are evaluated according to the fittest candidate solution in their neighborhood and both agent types have the same range for these neighborhoods. We conclude that reapers are usually found in areas with less fit individuals while cupids frequent areas with fitter individuals. Since movement is random, this effect can only be the result of cupids' selection probabilities: cupids in 'bad' areas consistently evolve a preference for selecting reapers while cupids in 'good' areas develop a preference for selecting even more

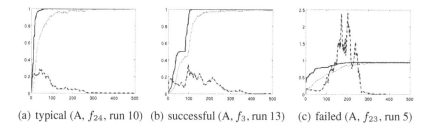

(a) typical (A, f_{24}, run 10)   (b) successful (A, f_3, run 13)   (c) failed (A, f_{23}, run 5)

Fig. 3. Three examples of breeders' evolution and response to premature convergence. Lines represent best fitness (solid), mean fitness (dotted) and mutation step size for candidate solutions (dashed) over time.

cupids. Furthermore, reapers almost always have very high preference for killing other reapers and, consequently, low ages (they mostly survive only one generation).[3] This implies that cupids in bad areas create 'reaper explosions' that eradicate low-fitness candidate solutions and also clean up after themselves as reapers select each other.

7 Conclusions and Future Work

Motivated by specific challenges in evolutionary swarm robotics, we introduced the Fate Agents Evolutionary Algorithm. It forms a new type of distributed and self-regulating EA where algorithmic operators are implemented through autonomous agents placed in the same space where the candidate solutions live and die. This provides a natural solution to the problems of the limited range and scalability a single oracle based EA would suffer from. Compared to alternative solutions where evolutionary operators are embodied in the robots [13,7] Fate Agents offer increased controllability for experimenters and users. Furthermore, our Fate Agents are not only operating on candidate solutions, but also on themselves. Hence, a Fate Agents EA has an inherent capability to regulate its own configuration.

Because proof-of-concept experiments with (simulated) robots would have taken very much time, we performed the first assessment of this new EA with synthetic fitness landscapes. To this end, we conducted experiments to explore our system's problem solving ability and self-regulating behavior on challenging numerical optimization problems. Results showed that the Fate Agents EA is capable of solving these problems and of coping with noise and disruptive changes. Furthermore, it successfully regulates population sizes and adapts its parameters.

In conclusion, the Fate Agents EA is a new kind of evolutionary algorithm that deserves further research from a number of angles. These include 1) applications in collective adaptive systems, such as swarm and evolutionary robotics (comparisons with other on-line on-board evolutionary mechanisms); 2) as a new paradigm for self-adapting EAs . Furthermore, though it may seem contradictory to our initial motivation, our results indicate that the Fate Agents EA may also deserve further investigation as a numeric function optimizer.

---

[3] These observations are true for almost every run we conducted. Due to lack of space we cannot present relevant graphs.

Acknowledgments. The authors would like to thank Istvan Haller and Mariya Dzhor-dzhanova for their valuable help in developing the Fate Agents EA prototype.

References

1. Alba, E., Dorronsoro, B.: Cellular Genetic Algorithms. Springer (2008)
2. Eiben, A., Michalewicz, Z., Schoenauer, M., Smith, J.: Parameter control in evolutionary algorithms. In: Lobo, et al. [8], pp. 19–46
3. Freisleben, B.: Meta-evolutionary approaches. In: Bäck, T., Fogel, D., Michalewicz, Z. (eds.) Handbook of Evolutionary Computation, pp. 214–223. Institute of Physics Publishing, Oxford University Press, Bristol, New York (1997)
4. Gong, Y., Fukunaga, A.: Distributed island-model genetic algorithms using heterogeneous parameter settings. In: IEEE Congress on Evolutionary Computation, pp. 820–827 (2011)
5. Gordon, V.S., Pirie, R., Wachter, A., Sharp, S.: Terrain-based genetic algorithm (TBGA): Modeling parameter space as terrain. In: Banzhaf, W., Daida, J., Eiben, A., Garzon, M., Honavar, V., Jakiela, M., Smith, R. (eds.) Proceedings of the Genetic and Evolutionary Computation Conference, GECCO 1999, pp. 229–235. Morgan Kaufmann, San Francisco (1999)
6. Kantschik, W., Dittrich, P., Brameier, M., Banzhaf, W.: Meta-Evolution in Graph GP. In: Langdon, W.B., Fogarty, T.C., Nordin, P., Poli, R. (eds.) EuroGP 1999. LNCS, vol. 1598, pp. 15–28. Springer, Heidelberg (1999)
7. Karafotias, G., Haasdijk, E., Eiben, A.: An algorithm for distributed on-line, on-board evolutionary robotics. In: Krasnogor, N., Lanzi, P.L., Engelbrecht, A., Pelta, D., Gershenson, C., Squillero, G., Freitas, A., Ritchie, M., Preuss, M., Gagne, C., Ong, Y.S., Raidl, G., Gallager, M., Lozano, J., Coello-Coello, C., Silva, D.L., Hansen, N., Meyer-Nieberg, S., Smith, J., Eiben, G., Bernado-Mansilla, E., Browne, W., Spector, L., Yu, T., Clune, J., Hornby, G., Wong, M.-L., Collet, P., Gustafson, S., Watson, J.-P., Sipper, M., Poulding, S., Ochoa, G., Schoenauer, M., Witt, C., Auger, A. (eds.) Proceedings of the 13th Annual Conference on Genetic and Evolutionary Computation, GECCO 2011, Dublin, Ireland, July 12-16, pp. 171–178. ACM (2011)
8. Lobo, F., Lima, C., Michalewicz, Z. (eds.): Parameter Setting in Evolutionary Algorithms. Springer (2007)
9. Nellis, A.: Meta evolution. Qualifying Dissertation (2009)
10. Samsonovich, A.V., De Jong, K.A.: Pricing the 'free lunch' of meta-evolution. In: Beyer, H.-G., O'Reilly, U.-M. (eds.) Proceedings of the Genetic and Evolutionary Computation Conference, GECCO 2005, pp. 1355–1362. ACM (2005)
11. Schut, M., Haasdijk, E., Eiben, A.E.: What is situated evolution? In: Proceedings of the 2009 IEEE Congress on Evolutionary Computation, Trondheim, May 18-21, pp. 3277–3284. IEEE Press (2009)
12. Tomassini, M.: Spatially Structured Evolutionary Algorithms: Artificial Evolution in Space and Time. Natural Computing Series. Springer-Verlag New York, Inc., Secaucus (2005)
13. Watson, R.A., Ficici, S.G., Pollack, J.B.: Embodied evolution: Distributing an evolutionary algorithm in a population of robots. Robotics and Autonomous Systems 39(1), 1–18 (2002)
14. Wickramasinghe, W., van Steen, M., Eiben, A.E.: Peer-to-peer evolutionary algorithms with adaptive autonomous selection. In: D.T., et al. (eds.) Proc of the 9th conference on Genetic and Evolutionary Computation, GECCO 2007, pp. 1460–1467. ACM Press (2007)

Guide Objective Assisted Particle Swarm Optimization and Its Application to History Matching

Alan P. Reynolds[1], Asaad Abdollahzadeh[2], David W. Corne[1], Mike Christie[2], Brian Davies[3], and Glyn Williams[3]

[1] School of Mathematical and Computer Sciences, Heriot-Watt University, Edinburgh, Scotland EH14 4AS
[2] Institute of Petroleum Engineering, Heriot-Watt University
[3] BP

Abstract. As is typical of metaheuristic optimization algorithms, particle swarm optimization is guided solely by the objective function. However, experience with separable and roughly separable problems suggests that, for subsets of the decision variables, the use of alternative 'guide objectives' may result in improved performance. This paper describes how, through the use of such guide objectives, simple problem domain knowledge may be incorporated into particle swarm optimization and illustrates how such an approach can be applied to both academic optimization problems and a real-world optimization problem from the domain of petroleum engineering.

1 Introduction

This paper describes a version of particle swarm optimization (PSO) that uses 'guide objectives' in addition to the overall objective in order to improve performance, in particular when the problem is 'roughly' separable. This introduction briefly describes the real-world problem that motivated this work, in order to give the reader an idea of what is meant by guide objectives and rough separability.

To be able to make effective decisions regarding the exploitation of an oil reservoir, it is necessary to create and update reservoir models. Initial models created using geological knowledge of the reservoir are improved using observations collected over time, in a process called *history matching*. This involves the adjustment of a reservoir model so that, when simulation software is applied, the simulated behaviour is similar to that observed in the real world. This can be posed as an optimization problem, minimizing a measure of misfit.

While we would like to automate the history matching process, incorporating reservoir experts' extensive domain knowledge into metaheuristic optimization algorithms in a generally applicable way has proved difficult. The avenue of research explored in this paper starts with the realization that, given a suitable model parameterization, certain model parameters will affect certain components of the misfit function to a greater degree than others. Indeed, if the reservoir

C.A. Coello Coello et al. (Eds.): PPSN 2012, Part II, LNCS 7492, pp. 195–204, 2012.

consists of distinct regions with little inter-region communication and if the model parameters describe regional features then the history matching problem may be (roughly) separated into a number of smaller, regional subproblems. This suggests that, given a suitable subset of model parameters, it may be possible to select a subset of the misfit components to create a *guide objective* for these parameters. We will show that, when using PSO, these guide objectives may be used in combination with the overall objective in a single optimization run.

In Sect. 2 we describe how metaheuristics perform when applied naively to separable problems and define more precisely what we mean by 'roughly separable' and 'guide objective'. Section 3 describes basic PSO, while Sect. 4 describes how PSO may be modified to exploit guide objectives, producing the guide objective assisted PSO algorithm (GuPSO). Results on simple function optimization problems are provided in Sect. 5.

In Sect. 6, reservoir history matching is described in more detail, including a description of the PUNQ-S3 case study used in this paper. Application of GuPSO to this problem and results obtained are described in Sect. 7. Finally, Sect. 8 presents conclusions and a discussion of potential areas of further research.

2   Optimization and Separable Problems

Consider the minimization of $f(x, y)$, where x and y can take any of a thousand values and nothing is known a priori about f. To guarantee finding the optimal solution one must evaluate all one million solutions. However, if it is known that $f(x, y) = g(x) + h(y)$ then the optimal solution can be found in two thousand evaluations by first optimizing the choice of x and then optimizing the choice of y. Now suppose a metaheuristic is naively applied to the minimization of f. A solution with the optimal value of x may be evaluated early in the search, but if it is coupled with a poor choice for y its significance will be missed. Clearly, knowledge of the problem's separability should be exploited to improve performance, typically by optimizing $g(x)$ and $h(y)$ separately.

Note that x and y may be vectors representing subsets of the decision variables. Also, it may not be obvious when this approach may be used. Suppose we wish to minimize $f(x, y) = x^4 + 2x^2y^2 + y^4$. We may not separate the problem directly, but we note that $f(x, y) = \left(x^2 + y^2\right)^2$ and that, since $x^2 + y^2 \geq 0$, this is equivalent to minimizing $x^2 + y^2$ — a clearly separable problem.

Now suppose we wish to minimize $f(x, y) = x^4 + 2x^2y^2 + y^4 + \epsilon x^3 y$, where ϵ is small. The problem may no longer be separated as above. However, the additional term may have limited impact on the quality of solutions: the problem may be thought of as being *roughly* separable. Minimizing $g(x) = x^2$ and $h(y) = y^2$ separately still leads to good values for $f(x, y)$. Therefore it makes sense to start the search by minimizing $g(x)$ and $h(y)$, rapidly finding a near optimal solution, before improving the result by optimizing f directly if desired.

In what follows, we will describe $g(x)$ and $h(y)$ as the *guide objectives* for x and y. These objectives are used to guide our search for good values for x and y and aid in the optimization of $f(x, y)$. So when minimizing $f(x, y) = x^4 + 2x^2y^2 + y^4(+\epsilon x^3 y)$, we use $g(x) = x^2$ and $h(y) = y^2$ as guide objectives.

3   Particle Swarm Optimization

To describe GuPSO we must first describe the basic PSO algorithm. PSO [7] is motivated by the collective behaviour of animals, such as the flocking of birds or swarming of bees. However, instead of a swarm of bees searching for a good source of nectar, PSO uses a swarm of particles moving through a multidimensional search space towards better quality solutions.

Each particle in the swarm has a position and a velocity, initialized at random. In each iteration of PSO, the velocity of each particle is adjusted by applying an acceleration towards the best solution visited by the particle in question and an acceleration towards the best solution visited by the swarm. The position is then adjusted according to the particle's velocity. In detail, if x_{ij} is the jth component of the position of particle i and v_{ij} is the jth component of its velocity, then these are updated as follows:

$$v_{ij} \leftarrow w v_{ij} + \alpha r_1 \left(p_{ij} - x_{ij} \right) + \beta r_2 \left(g_j - x_{ij} \right) \ , \tag{1}$$

$$v_{ij} \leftarrow \min \left(v_{ij}, V_{\max,j} \right) \ ,$$

$$v_{ij} \leftarrow \max \left(v_{ij}, -V_{\max,j} \right) \ ,$$

$$x_{ij} \leftarrow x_{ij} + v_{ij} \ .$$

Here w is the inertia weight, α and β control the amount of acceleration towards the particle's personal best and the global best solutions, r_1 and r_2 are randomly generated numbers between 0 and 1, p_{ij} is the jth component of the best solution visited by particle i and g_j is the jth component of the best solution visited by the swarm. $V_{\max,j}$ is the maximum velocity permitted in dimension j. Values for w, α and β are supplied by the user.

PSO may also use methods for ensuring that decision variables remain within their permitted bounds. In this paper we apply reflection with random damping. However, since boundary handling is unaffected by the use of guide objectives, we do not provide details but refer the reader to the PSO literature.

4   PSO and Guide Objectives

Now suppose we wish to apply PSO to a separable problem, for example the minimization of $f(x_1, x_2, x_3, x_4) = f_1(x_1) + f_2(x_2, x_3, x_4)$. We have suggested above that two separate optimizations should take place — the minimization of f_1 and the minimization of f_2. However, both these optimizations can be performed concurrently by adjusting the velocity update formula as follows:

$$v_{ij} \leftarrow w v_{ij} + \alpha r_1 \left(p_{ij}^{(j)} - x_{ij} \right) + \beta r_2 \left(g_j^{(j)} - x_{ij} \right) \ . \tag{2}$$

Here $g^{(j)}$ represents the best solution visited by the swarm according to the guide objective for the jth variable, while $p_i^{(j)}$ is the best solution visited by particle i according to the guide objective for variable j. f_1 acts as the guide objective for x_1 while f_2 acts as the guide objective for x_2, x_3 and x_4. Notice how this separates the optimization so that values taken by x_2, x_3 and x_4 have no affect on the choices for variable x_1 - the evolution of x_1 depends only on the values taken by its guide objective f_1, which is unaffected by the other parameters.

Merging two independent optimizations into a single run in this manner does not produce any immediate benefits in the case of separable problems. The advantage of the approach is that it allows for both guide objectives and the true objective to be used when the problem is only roughly separable, via the combination of update formulae (1) and (2) as follows:

$$v_{ij} \leftarrow w v_{ij} + \alpha r_1 \left(p_{ij} - x_{ij} \right) + \beta r_2 \left(g_j - x_{ij} \right) + \gamma r_3 \left(p_{ij}^{(j)} - x_{ij} \right) + \delta r_4 \left(g_j^{(j)} - x_{ij} \right) .$$
$$(3)$$

The selection of appropriate values for α, β, γ and δ allows the influence of the guide objectives on the search to be controlled. By changing the values of these parameters during the search, the algorithm may start by using only the guide objectives, but become increasingly influenced by the true objective until finally it behaves like standard PSO. This approach may be effective for roughly separable problems, where guide objectives are used to rapidly finding good solutions but where the final refinements can only be made with reference to the true objective. The resultant algorithm is Guided PSO or GuPSO.

5   Illustrative Results on Academic Problems

The operation of GuPSO on separable or roughly separable problems is best illustrated on academic problems. In this section the objective is always minimized and variables are constrained to lie between -10 and 10. We focus on variations of two functions: the multimodal function of Kvasnika et al. [8] and Rosenbrock's function [13]. The first of these is totally separable and is given by

$$f_1 (x_1, x_2, \ldots, x_n) = \sum_{i=1}^{n} g(x_i) \ ,$$

$$g(x) = 0.993851231 + e^{-0.01x^2} \sin(10x) \cos(8x) \ .$$

f_1 is used as our first test function, with $n = 20$ and an evaluation limit of 50,000. The guide objective for each variable, x_i, is simply $g(x_i)$.

Rosenbrock's function, given by

$$r(x_1, x_2, \ldots, x_n) = \sum_{i=1}^{n-1} \left[(1 - x_i)^2 + 100 \left(x_{i+1} - x_i^2 \right)^2 \right]$$

is inseparable. Our second test function is created by splitting fifty decision variables into ten equal sized blocks, summing 10 five variable Rosenbrock functions:

$$f_2 (x_1, x_2, \ldots, x_{50}) = \sum_{j=1}^{10} r(x_{5j-4}, \ldots, x_{5j}) \ .$$

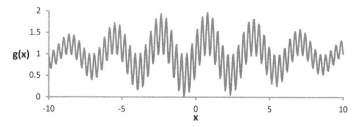

Fig. 1. The highly multimodal function $g\,(x)$

A limit of 500,000 evaluations is imposed. The guide objective for each variable is simply the Rosenbrock function to which the variable makes a contribution. A third test function combines f_1 with a twenty variable Rosenbrock function:

$$f_3\,(x_1, x_2, \ldots, x_{20}) = f_1\,(x_1, x_2, \ldots, x_{20}) + 0.001h\,(x_1, x_2, \ldots, x_{20})\,.$$

The result can be thought of as being roughly separable. An evaluation limit of 50,000 is imposed. The guide objective for x_i is simply $g\,(x_i)$, i.e. the contribution of the Rosenbrock function is ignored.

Experiments using the parameter values in table 1 were performed for each problem. Parameters α, β, γ and δ in (3) were set as follows:

$$\alpha = \beta = (1 - \lambda)A, \quad \gamma = \delta = \lambda A \,. \tag{4}$$

Here λ indicates the degree to which the guide objectives were used in preference to the overall objective. For the two separable functions, λ took the values 0 or 1. For f_3, values of 0, 0.2, 0.5, 0.8 and 1 were tried for λ. Experiments were also performed with λ decreasing linearly from 1.0 to 0.0 over the course of each run.

For each value of λ, thirty runs were performed for every combination of the remaining parameters, in order to find the best values. Thirty runs were then repeated using the best parameter set, allowing for a fair comparison between PSO ($\lambda = 0$) and GuPSO. Results are summarized in Table 2.

It is clear that, on the two separable problems, use of the guide objectives produces significantly better results. Indeed for f_1, using guide objectives resulted in the global optimum being found in all 30 runs, while it was never found using the true objective. However, as has been noted, identical results could be achieved by separating the problem into 20 sub-problems and optimizing each individually using standard PSO.

Table 1. Parameter values

Parameter	Values
Swarm size	10, 20, 50, 100
Inertia weight	0.75, 0.8, 0.85, 0.9, 0.95, 1.0
Acceleration (A)	0.5, 0.8, 1.0, 2.0

Table 2. Comparison of performance using just the true objective against using just the guide objectives. Figures in brackets indicate 95% confidence intervals.

Problem	True objective ($\lambda = 0$)	Guide objectives ($\lambda = 1$)
f_1	3.535 (3.190 – 3.880)	1.138×10^{-9} (1.138×10^{-9} – 1.138×10^{-9})
f_2	14.05 (12.06 – 16.03)	1.703 (0.785 – 2.622)
f_3	5.467 (4.939 – 5.996)	3.696 (3.687 – 3.707)

Results for f_3 also show significant improvements when using the guide objectives. However, the best results were only obtained when both guide objectives and the true objective were used to guide the search, as shown in Fig. 2.

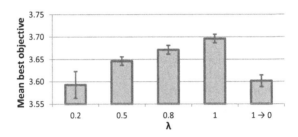

Fig. 2. Results for the third test function. Results obtained using just the true objective are of considerably poorer quality than those shown and are omitted to improve clarity. Error bars show the 95% confidence intervals.

6   The History Matching Problem

History matching, or petroleum reservoir model calibration, is the process of modifying a reservoir model so as to produce simulated outputs that closely match pressure, production and saturation data collected from a real world reservoir. Model parameters that may be modified include rock porosity, vertical and horizontal permeability, pore volume and aquifer volume. The reservoir is divided into regions or layers within which these factors can be assumed to be approximately constant. The location of the boundary between such regions may also be considered a modifiable model parameter. Furthermore, it may be appropriate to adjust various multipliers, rather than the physical characteristics directly.

The objective function is a measure of misfit. Although this paper focuses primarily on simply minimizing misfit, it is useful to obtain a range of different, low misfit models. The resulting ensemble of reservoir models can then be used not only to predict future output, but also to estimate the uncertainty of the prediction — a process known as *uncertainty quantification*.

A number of metaheuristics have been applied to the history matching problem, including simulated annealing [14], tabu search [15], genetic algorithms [12,5], estimation of distribution algorithms [11,3,2] and differential evolution [6]. Recent work has also suggested that PSO may be effectively applied to this

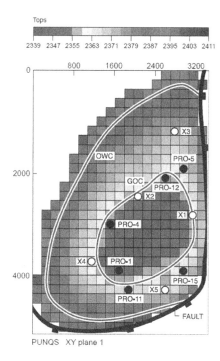

Fig. 3. The depth of reservoir rock and well locations in PUNQ-S3

problem [9,10]. This and the relative ease with which guide objectives may be incorporated provides the motivation for our use of PSO.

6.1 PUNQ-S3

We apply GuPSO to the history matching of the PUNQ-S3 reservoir [1] — a small industrial reservoir engineering model, adapted from a real field example and widely used for performance studies of history matching algorithms.

Problem: The simulation model contains 2660 (19x28x5) grid blocks, of which 1761 are active, and 6 production wells, numbered 1, 4, 5, 11, 12, and 15. The field is structurally bounded to the east and south by a fault, as shown in Fig. 3, while the link to a strong aquifer to the north and west means that no injection wells are required. The field initially has a small gas cap at the center of the structure, and production wells are located around this gas cap.

Porosity and permeability fields for the 'truth case' were generated using a Gaussian Random Fields model in such a way as to be, as much as possible, consistent with the geological model. The reservoir model was completed by using pressure, volume, temperature and aquifer data from the original model. Reservoir simulation was then used to generate production data (bottom-hole pressure

BHP, water cut WCT and gas oil ratio GOR), after which Gaussian noise was added to the well porosities/permeabilities and the synthetic production data to account for measurement error. (For further details see [1].)

Model Parameters and Objective: In this paper we use the parameterization of Hajizadeh et al. [6], with distinct porosity values in each of 9 homogeneous regions (labelled A to I) per layer for 5 layers, resulting in 45 model parameters. Parameter ranges and other details can be found in [6].

The objective function, to be minimized [4] is

$$M = \frac{1}{N_v} \sum_{j=1}^{N_v} \frac{1}{N_p} \sum_{i=1}^{N_p} W_{ij} \left(\frac{O_{ij} - S_{ij}}{\sigma_{ij}} \right)^2$$

where i runs over the time points at which observations are made, j indicates which of the 18 observations (BHP, WCT and GOR at each of the 6 wells) is being referred to, O_{ij} is the truth case value of observation j at time i, S_{ij} is the simulated value, σ_{ij} reflects the measurement error and W_{ij} is a weight factor.

7 Application to PUNQ-S3

Given a set of porosity parameters for a region, the wells that are primarily affected by these parameters and only marginally affected by the others may be selected. The misfit components for these wells then form the guide objective for these model parameters, as indicated in table 3.

Table 3. Guide objectives for PUNQ-S3 were taken to be the misfit over the set of wells most affected by model parameter in question

Region	A	B	C	D	E	F	G	H	I
Guide wells	5	5, 12	5, 12	5, 12	4, 5, 12	1, 4, 15	1, 4, 11, 15	1, 11, 15	1, 11

Despite the cost of solution evaluation, the basic PSO was tuned by experimenting with a range of swarm sizes (10, 20 and 50) and inertia weights (0.8, 0.85, 0.9, 0.95 and 1). The acceleration parameter A (and hence α and β in basic PSO) was set to one. For each combination of parameters, 30 runs were performed, of 3000 solution evaluations each. The PSO results that are compared with GuPSO in this paper were then obtained by performing additional runs of 3000 evaluations and 1000 evaluations with the best parameter combination.

Results for GuPSO were obtained using the best parameter set found for PSO, with the exception that α, β, γ and δ were set according to (4) using a range of values for λ. It can be seen from the results in Fig. 4 that GuPSO outperforms standard PSO, particularly in shorter runs.

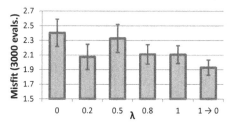 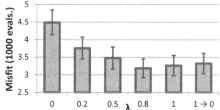

Fig. 4. GuPSO performance for different values of λ after 3000 and 1000 solution evaluations, averaged over 30 runs. Standard PSO is provided by setting λ to 0, while the last bar gives the results obtained by allowing λ to vary from 1 (guide objectives only) to 0 (true objective only) linearly over the course of the search.

8 Conclusions and Further Research

We have presented a modification to PSO whereby guide objectives are utilized in order to improve algorithm performance. The resulting algorithm has been shown to produce improved performance on some simple separable and roughly separable problems. More importantly, GuPSO outperforms standard PSO on a real-world reservoir history matching problem.

There are a number of areas of possible future research.

Other sources of guide objectives: Much of this paper assumes that guide objectives are found via the rough separability of the problem. However, *any* alternative objective that provides a better guide for the improvement of decision variables than the true objective could be used in this approach.

Multiple guide objectives: The approach need not be limited to a single guide objective for each decision variable. Multiple guide objectives may be used, either at different points in the search or through further modification of the velocity update formula.

Forgetfulness: History matching problems may be roughly separable, with the exception of one or two model parameters that affect the entire reservoir. GuPSO may remember a best solution for one guide objective that depends upon old, long discarded values for the 'global' parameters. It may be useful to allow GuPSO to 'forget' such solutions.

Other applications: In particular, GuPSO may be an appropriate approach to reservoir development optimization. When locating new wells, predicted oil recovery from the well should make a suitable guide objective for the well location. However, since placement of each well affects the output of both old wells and the other new wells, the problem is only roughly separable.

References

1. PUNQ-S3 test case. Department of Earth Science and Engineering. Imperial College, London (2010), http://www3.imperial.ac.uk/earthscienceandengineering/research/perm/punq-s3model (retrieved January 7, 2010)

2. Abdollahzadeh, A., Reynolds, A., Christie, M., Corne, D., Davies, B., Williams, G.: Bayesian optimization algorithm applied to uncertainty quantification. In: SPE EUROPEC/EAGE Annual Conf. and Exhibition, Vienna, Austria (2011)
3. Abdollahzadeh, A., Reynolds, A., Christie, M., Corne, D., Williams, G., Davies, B.: Estimation of distribution algorithms applied to history matching. In: SPE Reservoir Simulation Symposium, Houston, Texas, U.S.A. (2011)
4. Barker, J.W., Cuypers, M., Holden, L.: Quantifying uncertainty in production forecasts: Another look at the PUNQ-S3 problem. SPE Journal 6(4), 433–441 (2001)
5. Erbaş, D., Christie, M.A.: Effect of sampling strategies on prediction uncertainty estimation. In: SPE Reservoir Simulation Symposium, Houston, U.S.A. (2007)
6. Hajizadeh, Y., Christie, M., Demyanov, V.: Comparative study of novel population-based optimization algorithms for history matching and uncertainty quantification: PUNQ-S3 revisited. In: Abu Dhabi Int. Petroleum Exhibition and Conf. (2010)
7. Kennedy, J., Eberhart, R.: Particle swarm optimization. In: Proc. of the IEEE Int. Conf. on Neural Networks, vol. 4, pp. 1942–1948 (1995)
8. Kvasnicka, V., Pelikan, M., Pospichal, J.: Hill climbing with learning (an abstraction of genetic algorithm). Neural Network World 6(5), 773–796 (1995)
9. Mohamed, L., Christie, M.A., Demyanov, V.: Comparison of stochastic sampling algorithms for uncertainty quantification. SPE Journal 15(1), 31–38 (2010)
10. Mohammed, L., Christie, M., Demyanov, V.: Reservoir model history matching with particle swarms. In: SPE Oil and Gas India Conf. and Exhibition, Mumbai, India (2010)
11. Petrovska, I., Carter, J.N.: Estimation of distribution algorithms for history-matching. In: Proc. of the 10th European Conf. on the Mathematics of Oil Recovery. EAGE, Amsterdam (2006)
12. Romero, C.E., Carter, J.N., Gringarten, A.C., Zimmerman, R.W.: A modified genetic algorithm for reservoir characterisation. In: Int. Oil and Gas Conf. and Exhibition, Beijing, China (2000)
13. Rosenbrock, H.H.: An automatic method for finding the greatest or least value of a function. The Computer Journal 3(3), 175–184 (1960)
14. Sultan, A.J., Ouenes, A., Weiss, W.W.: Automatic history matching for an integrated reservoir description and improving oil recovery. In: Proc. of the Permian Basin Oil and Gas Recovery Conf., Midland, Texas, pp. 741–748 (1994)
15. Yang, C., Nghiem, L., Card, C., Bremeier, M.: Reservoir model uncertainty quantification through computer-assisted history matching. In: SPE Annual Technical Conf. and Exhibition, Anaheim, California, U.S.A. (2007)

Animal Spirits in Population Spatial Dynamics

Matylda Jabłońska and Tuomo Kauranne

Lappeenranta University of Technology,
P.O. Box 20, FI-53850, Lappeenranta, Finland

Abstract. Collective behavior of herding animals displays a balance
between conservative cohesive forces and "'innovation'", or moving towards a goal, as described by Couzin in 2005 (2005). We have given these
forces a quantitative mathematical form that is amenable to numerical
simulation. The simulations described herein reproduce the phenomena
that Couzin observed and indicate that a nearly 5 per cent critical mass
is sufficient to pull the whole population along, but smaller innovative
teams fail to attract a substantial following. The resulting non-linear dynamic equations have been applied also to modeling of financial market
dynamics, where they are seen to produce financial catastrophes by internal population dynamics alone, without any need for external forcing.
The equations can thereby also be interpreted as a model of John May-
nard Keynes' Animal Spirits (1936) that are often evoked to describe
market psychology.

1 Introduction

Animal behaviour in schools, swarms, herds, etc. have always been puzzling even
for biologists, who study their habits on a daily basis. For a long time scientists
were convinced that direction of, for example, a flock of birds cruising the sky
is driven by a single leader in front. But Couzin et al. (2005) showed through
numerical simulations that it has to be at least 5% of the total population heading
in a specific direction to pull the whole group behind them. That same fact was
verified two years later empirically during a big experiment in Cologne by Krause
and Dyer (see Mob mentality, 2009). There, a group of 200 people was told to
move freely around a large space, 400 by 230 feet, though without communicating
with each other and just staying close to their neighbors. After some time the
groups tended to move in two concentric circles rotating in opposite direction.
Then a small subgroup of all was told to head in a specific direction and it
appeared that still 2.5% did not influence the whole population movement, but
5% did interfere the circular motion, causing the main group follow in the same
direction.

This very idea has inspired a recent study in financial market modeling by
Jabłońska (2011) and Jabłońska and Kauranne (2011) where the connections of
the models of interest with the Keynes' animal spirits were widely discussed.
In those works, a one-dimensional model of population dynamics was used to
simulate the behavior of an ensemble of traders in electricity spot markets. Their

C.A. Coello Coello et al. (Eds.): PPSN 2012, Part II, LNCS 7492, pp. 205–214, 2012.
© Springer-Verlag Berlin Heidelberg 2012

results have shown that if a sufficiently big subgroup of the whole population was bidding far enough from the mean level, the others would follow that direction and skyrocket the price to the level a dozen times higher than the mean.

As the aforementioned model shows interesting properties, the aim of this work is to implement and analyze the same model in two dimensions for simulating dynamics of a group of individuals. The model is formed as a system of coupled stochastic differential equations, each representing one individual in the population. As such, most of the individuals have no intelligence of where they are heading. They only follow interactions within the whole group. However, a small subgroup is given a deterministic movement path, and the simulations verify what has to be the size of that escaping group to pull the other population members with them.

This article is structured as follows. Section 2 presents the family of population dynamics models being the background of the final model. Section 3 describes in detail the components of the final model and their physical and psychological interpretation. In Section 4, numerical simulations are presented for different model settings. Finally, Section 5 concludes and gives ideas for future work.

2 Population Spatial Dynamics

The aim of this study is to verify whether a specific population dynamics model can reproduce the natural fact that 5% of a population can pull the whole group towards a specific direction. This work presents an extended implementation of a model proposed by Jabłońska (2011) and Jabłońska and Kauranne (2011), which was then successfully used for simulation of electricity spot price behaviour. The model is based on the Capasso-Bianchi system of stochastic differential equations in a general form (1), used for modelling animal population dynamics by Morale et al. (2005) or price herding by Bianchi et al. (2003) and Capasso et al. (2005). In this approach, X^k are continuous stochastic processes representing movement of each particle k in the total population of N individuals, based on the location of each individual with respect to the whole population $f(X_t^k)$, as well as on its local interaction with the closest neighbors $h(k, \mathbf{X}_t)$. Also, dW^k represents randomness in the model through Wiener process increments, with volatility parameter σ.

$$dX_N^k(t) = [f(X_t^k) + h(k, \mathbf{X}_t)]dt + \sigma dW^k(t), \quad \text{for } k = 1, \ldots, N. \quad (1)$$

Jabłońska (2011) and Jabłońska and Kauranne (2011) extended this model with a Burgers'-type momentum component which catered for momentum in financial markets, and implemented it for one-dimensional price dynamics of an ensemble of traders. The referred model is presented in detail in Section 3 in Equations (2)-(4).

The following study presents a two-dimensional version of this model which, with suitable parameter values, can be used for simulating behaviours of large populations of individuals.

3   Model Formulation

This section presents in detail the model proposed in this study, as well as numerical simulations with different model parameter values. This work is based on the Capasso-Morale approach mentioned in Section 2. Each individual in the population is followed separately; together they form a system of coupled stochastic differential equations. The structure of each equation is the same and it is related to the main idea of an Ornstein-Uhlenbeck mean reverting process. However, the single constant mean reversion level is replaced by three individual components, each standing for a specific type of force acting on the population as a whole, and on its individuals separately.

The main components of the proposed model are:

Global mean: the whole population is expected to oscillate around its (moving) center of mass $\mathbf{X}_t^*$; this is related to the aggregation forces proposed by Morale et al. (2005). This component stands for the herding phenomenon, that is the willingness of the individuals to stay within a bigger group.

Momentum: in particular, the momentum effect $h(k, \mathbf{X}_t)$ should occur when a sufficiently big subgroup of the whole population has significantly different behavior (external information) that deviates from the total population mean. This has been noticed in studies by Couzin et al. (2005).

Local interaction: each individual in the population can perceive its neighbors up to a limited extent, which seems natural especially for big populations. Therefore, each population member will follow $g(k, \mathbf{X}_t)$, that is the furthest neighbor within a range that caters for the closest $p\%$ of the whole population. This will allow the emergence of a proper repulsion force and avoid overcrowding in any point in space. Also, individuals are deemed to follow the farthest units of their neighborhood, thinking that those have some distinct information in the 'big picture' and, therefore, are far for a good reason.

Randomness: each individual's move includes a Wiener increment to allow randomness in the system.

Hence, the model is defined as Equation (2)

$$d\mathbf{X}_t^k = [\gamma(\mathbf{X}_t^* - \mathbf{X}_t^k) + \theta(h(k, \mathbf{X}_t) - \mathbf{X}_t^k) + \xi(g(k, \mathbf{X}_t) - \mathbf{X}_t^k)]dt + \sigma_t d\mathbf{W}_t^k \quad (2)$$

where

$$h(k, \mathbf{X}_t) = \mathrm{M}(\mathbf{X}_t) \cdot [\mathrm{E}(\mathbf{X}_t) - \mathrm{M}(\mathbf{X}_t)] \quad (3)$$

having $\mathrm{M}(X)$ stand for the mode of a random variable X and $\mathrm{E}(X)$ being a classical expected value. Also,

$$g(k, \mathbf{X}_t) = \max_{k \in I}\{\mathbf{X}_t^k - \mathbf{X}_t\}, \quad \text{where } I = \{k|_{X^k \in N_{p\%}^k}\} \quad (4)$$

where $\mathbf{X}^k$ are continuous stochastic processes representing movement of each particle, $N_{p\%}^k$ means the neighborhood of the k-th individual formed by the closest $p\%$ of the population. $\mathbf{X}_t^*$ stands for the mean of the whole population at time

t, and parameters γ, θ and ξ are the forces with which each of the interactions takes place. In the original model by Jabłońska (2011) and Jabłońska and Kauranne (2011) these forces are allowed to vary in time, but in this work they are kept constant for simplicity. Also, this work extends previous implementation to the two-dimensional case.

4   Numerical Simulations

4.1   Model Parameters

As mentioned before, some of the model parameters are fixed for simplicity and, as there is no real life data available for calibration, they are chosen deterministically. The values are $\gamma = 0.05$ for the global mass center rate, $\theta = 10^{-5}$ for the momentum component and $\xi = 0.9$ for the strength of the local interaction. The percent for the range of the local interaction is chosen to be equal to the size of the escape group.

With the aforementioned parameter values, the model simulation will follow the movement of a total population of $N = 200$ individuals. Most of them will move only through model dynamics, with initial locations generated from a two-dimensional uniform distribution $U^2(-4h, 4h)$, where $h = 0.05$ is also used as a grid step factor for finding the mode of the population at every time step. That means that the grid for which we find the population mode (the location of the spatially most concentrated individuals) is recomputed at each time step with respect to total population spread. This two-dimensional world is representing the usual XY plane with Euclidean distance measure. In the following experiments there are no boundaries given for the plane. That means that the individuals are allowed to "escape" far away from the initial point if the system dynamics cause so.

4.2   Simulation Results

First, the simulation demonstrates the aggregative power of the model, which is apparent even when $\gamma = 0$. The individuals, originally uniformly distributed, quickly cluster into a few compact groups which then follow their own dynamics as presented in Figure 1. As a small subgroup (5%) of individuals moves in a deterministic way, one of the aggregated groups join them, and follow the deterministic path. Whenever any two groups get close enough to one another, they may merge. In the plots, D denotes the Euclidean distance between the centers of mass of the subgroup and the remaining population.

In the next simulation the size of the subpopulation, also called the *escape group*, is kept at the same 5% level. However, the global center of mass rate is set to $\gamma = 0.05$, and the escape group is now moving along a circle. As presented in Figure 2, the main stream of the population can easily follow the escape group, keeping the distance between their mass centers at a low level, that is, it remains smaller than the population radius. That is confirmed on the distance plot in Figure 3.

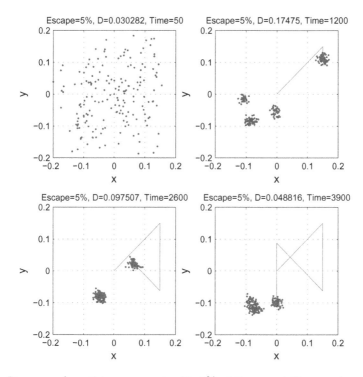

Fig. 1. Progress of particle movement with 5% of the population moving in a deterministic way and non-active center of mass attraction

Simulations have shown that any escape size bigger than 5% produces similar results, always keeping both groups close and well aggregated. However, if this value is decreased, the main stream of the population starts having trouble to keep close to the escape group, not even holding the distance constant. This is evident for the escape size of 4%, as depicted in Figures 4 and 5 for the path shapes and the distance plot, respectively. Apparently, the main population gets off the escape group course as the individuals try to cut the way short and catch the subgroup, but even that barely slows down the increase of the distance between the mass centers. Also, the 4% fails to attract even a single member of the main stream to their neighborhood, as opposite to the case from Figure 1.

To verify the threshold at which the main stream of the population has the probability to remain merged with the escape group equal to 1 we run the simulations multiple times for different population percent size with parameter `popperc` = {0.04,0.0405,...,0.05}, and verify how often the subgroup detaches from the whole population. With the total population size set now to $N = 2000$, for each subpopulation size the simulation is repeated 30 times (independent simulations). Then the distance between the groups is measured. Figure 6 presents the results, showing that the value of `popperc` = 0.0485 is the threshold from which onwards the main stream will always remain in contact with the escape group.

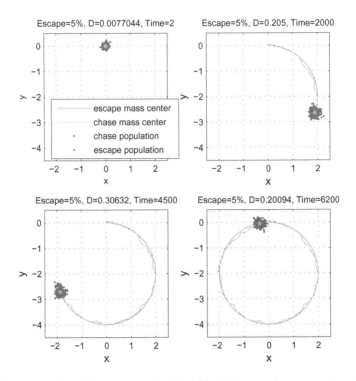

Fig. 2. Progress of particle movement with 5% of the population moving in a deterministic way and active center of mass attraction

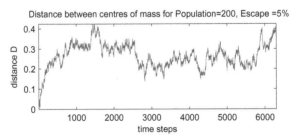

Fig. 3. Distance between the centers of mass of the escape group and the main stream with 5% of the population moving in a deterministic way and active center of mass attraction

The multiple runs allow us to make a rough approximation of probability of the main stream to remain merged with the escape group with respect to different subgroup sizes. As depicted in Figure 7, the probability is dramatically low for 4%, then stays moderate for most of the range, and reaches almost 1 at 4.8%, and 1 at 4.85%.

Finally, once again the global mass center attraction parameter is set to zero, to verify what percent size would suffice to hold the whole population compact

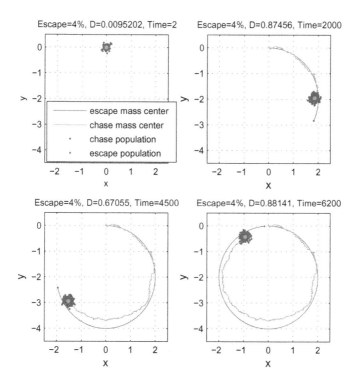

Fig. 4. Progress of particle movement with 4% of the population moving in a deterministic way and active center of mass attraction

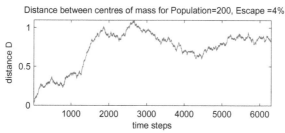

Fig. 5. Distance between the centers of mass of the escape group and the main stream with 4% of the population moving in a deterministic way and active center of mass attraction

when following the escape group. It appears, that there are two clear thresholds. At the level of 9% the probability of the whole population remaining as one group increases significantly from an average below 0.2 to an average of over 0.7, as presented in Figure 8. However, it is the subgroup size of 9.75% that guarantees the population remain merged with the escape group with probability 1. Otherwise, as presented in Figure 9 for the case of 8%, most of the the main stream of the population may loose interest in the subgroup after some time, and

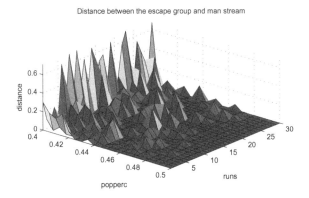

Fig. 6. Progress of particle movement with 4% of the population moving in a deterministic way

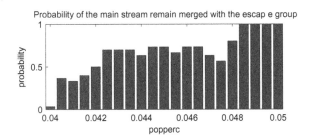

Fig. 7. Probability of the main stream to remain merged with the escape group with respect to different subgroup sizes, with active center of mass attraction

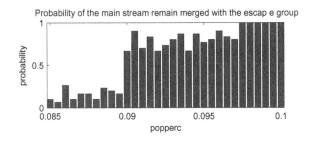

Fig. 8. Probability of the main stream to remain merged with the escape group with respect to different subgroup sizes, with active center of mass attraction

continue its movement only with respect to its own internal dynamics. However, some of the main stream particles stay attracted to the escape group, which is not the case in Figure 4 for the 4% subpopulation, even with the center of mass attraction active in the latter case.

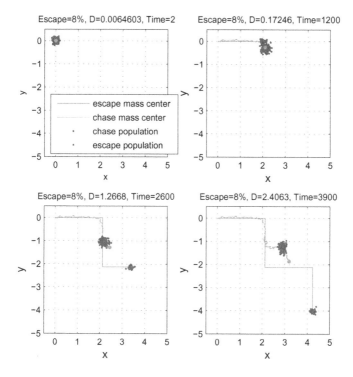

Fig. 9. Progress of particle movement with 8% of the population moving in a deterministic way and non-active center of mass attraction

The above results are also valid with other values of model parameters. The changes in strengths of particular interaction have mainly influence on the inner dynamics of the main stream, that is, on its spatial density and distribution.

The model seems to be a good interpretation of the Keynes' animal spirits. Though not set up in financial reality, it shows well how the forces of confidence in ones own knowledge and trust in somebody's else information can form population dynamics which can and already has been transferred to financial modelling as well.

5   Conclusions

This study presented a two dimensional implementation of a population dynamics model proposed by Jabłońska (2011) and Jabłońska and Kauranne (2011). This approach stemming from a combination of fluid dynamics and animal spatial dynamics was used previously in modelling financial time series. In this work it was used to analyze behavior of individuals on a two-dimensional plane, when a specific subgroup of the whole population heads in a deterministically set direction, while the remaining part follows only its internal dynamics.

The results have shown that which particular model setting the value of 4.85% for the size of the escaping subgroup is a minimum necessary to pull the rest of the population in the same direction and let it stay close to the subgroup. Any smaller size may influence the movement of the whole population, but is not sufficient to attract the individuals to stay constantly close to the escape. This leads to conclusion that, as the model is able to reproduce the natural behaviour of biological organisms (Mob mentality, 2009), then it may be appropriate to be used as a basis for modelling animal behaviour in more complicated settings.

One suggestion for future work is to analyze the population movement when closed in a bounded space, or having obstacles on the way. Moreover, it would be interesting to see the model's performance in a prey-predator setting, that is when the total population would contain two main subpopulations: preys and predators, whose aggregation and repulsion forces would be defined through their biological interactions. The model should also be continuously revised for its financial modelling applications.

References

Bianchi, A., Capasso, V., Morale, D., Sioli, F.: A Mathematical Model for Price Herding. Application to Prediction of Multivariate Time Series. MIRIAM report (confidential), Milan (2003)

Capasso, V., Bianchi, A., Morale, D.: Estimation and prediction of a nonlinear model for price herding. In: Provasi, C. (ed.) Complex Models and Intensive Computational Methods for Estimation and Prediction, Padova (2005)

Couzin, I.D., Krause, J., Franks, N.R., Levin, S.A.: Effective leadership and decision making in animal groups on the move. Nature 433, 513–516

Jabłońska, M.: From fluid dynamics to human psychology. What drives financial markets towards extreme events. Doctoral dissertation, Lappeenranta University of Technology, Finland (2011)

Jabłońska, M., Kauranne, T.: Multi-agent Stochastic Simulation for the Electricity Spot Market Price. In: Osinga, S., Hofstede, G.J., Verwaart, T. (eds.) Emergent Results on Artificial Economics. LNEMS, vol. 652, pp. 3–14. Springer, Heidelberg (2011)

Keynes, J.M.: The General Theory of Employment, Interest and Money. Macmillan, London (1936)

Morale, D., Capasso, V., Oelschlager, K.: An interacting particle system modelling aggregation behavior: from individuals to populations. Journal of Mathematical Biology 50(1), 49–66 (2005)

Mob mentality. Science illustrated (November/December 2009)

Autonomous Shaping via Coevolutionary Selection of Training Experience

Marcin Szubert and Krzysztof Krawiec

Institute of Computing Science, Poznan University of Technology, Poznań, Poland
{mszubert,kkrawiec}@cs.put.poznan.pl

Abstract. To acquire expert skills in a sequential decision making domain that is too vast to be explored thoroughly, an intelligent agent has to be capable of inducing crucial knowledge from the most representative parts of it. One way to shape the learning process and guide the learner in the right direction is effective selection of such parts that provide the best training experience. To realize this concept, we propose a shaping method that orchestrates the training by iteratively exposing the learner to subproblems generated autonomously from the original problem. The main novelty of the proposed approach consists in equalling the learning process with the search in subproblem space and in employing a coevolutionary algorithm to perform this search. Each individual in the population encodes a sequence of subproblems that is evaluated by confronting the learner trained on it with other learners shaped in this way by particular individuals. When applied to the game of Othello, temporal difference learning on the best found subproblem sequence yields substantially better players than learning on the entire problem at once.

Keywords: reinforcement learning, coevolutionary algorithms, shaping.

1 Introduction

Many real-world problems concern sequential decision making where every single decision changes the state of the environment and results in a *reward*. The main difficulties with handling such problems arise from the fact that rewards can be delayed in time. As a result, acting greedily is not always the best strategy and, even more importantly, it is hard to determine which actions should be credited with future rewards. Training an autonomous agent to maximize the cumulative payoff in this kind of problems is formalized as reinforcement learning (RL) [1]. This machine learning paradigm encapsulates the nature-related concept of *trial-and-error* search for optimal behavior, guided by the *interactions* between a learner and an unknown environment.

Past research shows that the most difficult RL problems are those with a long sequence of unrewarded decisions leading to a single payoff at the end. Typical examples of such scenarios are board games, where the only explicit reward is the final game outcome. One way to aid the learning process in this case is to use the idea of *shaping,* borrowed from behavioral psychology [2]. It assumes that a

C.A. Coello Coello et al. (Eds.): PPSN 2012, Part II, LNCS 7492, pp. 215–224, 2012.

learner is trained on a series of easier problems before approaching the original one. The main difficulty with shaping is that it requires very careful selection of training problems that should possibly approximate the desired behavior [3]. Such expert-driven shaping involves substantial amount of domain knowledge, and can introduce unnecessary biases into the learning process. In this context, learning from scratch remains an unbiased and thus attractive alternative.

In this paper we propose a method for autonomous shaping without giving up the above *tabula rasa* attitude. We employ competitive coevolution [4] to identify appropriate training experience for an agent that learns a game playing strategy. This leads to mapping the original problem of optimizing an agent's policy into a *dual* problem of finding the best input for the policy learning algorithm, while preserving the ultimate goal of learning — maximization of an adopted quality measure. The critical question one needs to answer to implement this form of shaping is: where can we get the simpler training problems from? In the case of games, *endgames* are the most obvious form of subproblems, as they naturally include the final rewards, which are essential to do any learning at all. Assuming that the training experience is gathered dynamically, starting from a given initial game state, our idea is to change this initial state in such way that the following interactions allow for faster and more general learning. More specifically, we consider *sequences of endgames*, represent them as *shaping vectors*, and search for the shaping vector that provides the best possible learning gradient.

We expect that learning from the pre-selected experience will converge faster and improve the final performance of the trained agents. Additionally, the dual problem definition can bring even more benefits. Firstly, the selected set of subproblems is a valuable source of knowledge about the problem structure. Indeed, shaping vector can be considered as an analog to the concept of *underlying objectives* of the problem [5], which here can be interpreted as the crucial set of skills needed for successfully operating in the given environment. Secondly, diversification of learning experience is a natural answer to the *exploration-exploitation* trade-off. Performing random moves to explore the environment (for instance, according to the so-called ε-greedy action selection scheme) could no longer be needed if the shaping vector is diverse enough.

2   Shaping by Initial State Selection

We consider sequential decision problems, which are defined by a state space S, a set of possible actions A, a default initial state $s_o \in S$, and a subset of terminal states. Additionally, the environment specifies a reward function $r : S \times A \to \mathbb{R}$ and a transition function $f : S \times A \to S$, which can be non-deterministic. The objective is to automate the process of learning agents that solve such problems, i.e., maximize the expected reward. An agent's behavior is determined by its policy $\pi : S \to A$, $\pi \in \Pi$ that for each state chooses an action leading to one of the subsequent states. The set of states traversed by an agent in a single episode is a directed path from s_0 to one of the terminal states in the transition graph that spans S. Such paths form samples of experience that can be used for improving the policy.

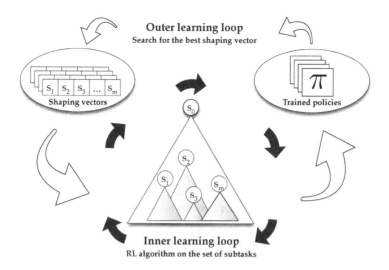

Fig. 1. The inner and outer loops of the shaping by initial state selection

We assume that an incremental learning algorithm $T : \Pi \times S \to \Pi$ is given that, provided with a current policy π_k and an initial state $s \in S$, produces an improved policy π_{k+1}. In the online variant considered here, learning occurs during exploration of the state graph: a training episode $T(s, \pi_k)$ consists in a simulation of agent's traversal through S, starting in s, with a single learning step taking place after each state transition.

It is usually assumed, particularly in the domain of board games, that the training process starts from the default initial state, i.e., T is always applied to s_0. This seems obvious, as, in the end, we want to learn a policy capable of solving the *entire* problem (e.g. playing the full game). However, for many problems the number of states that can be reached in the initial steps of problem solving is low, and grows exponentially with subsequent steps. As a result, a learner that starts from s_0 is doomed to overexplore the initial stages of problem solving while underexploring the final ones.

The main tenet of the proposed approach is that training a policy on a well-assorted, properly diversified and representative set of subproblems can be more beneficial than confronting it with the entire problem. We implement the concept of a set of subproblems by defining *shaping vector*, which is simply a vector **s** of m states $s_i \in S$, $i = \{1, \ldots, m\}$, where, in accordance with the sequential nature of considered problems, every $s_i \neq s_0$ identifies a subproblem of problem s_0 (assuming the transition graph is acyclic). A shaping vector can represent the training experience from potentially different areas of the state graph.

We orchestrate the learning process by iteratively applying the learning algorithm to consecutive elements of shaping vector: $\pi_i \leftarrow T(s_i, \pi_{i-1})$, where π_0 is an initial policy created in some arbitrary way. In this way, the experience gathered in π_i while solving subproblem s_i can be preserved when learning from subsequent subproblems. This inner learning loop (see Fig. 1) can iterate over the elements of experience multiple times, if needed. Ultimately, the obtained strategy $\pi_{\mathbf{s}}$ is expected to embody the knowledge derived form the set of subproblems embedded in the shaping vector $\mathbf{s}$.

The choice of subproblems to form the training experience is essential for the performance of the trained policy. For instance, the particular assortment of subproblems can make it impossible for the learner to visit certain states in S during learning. In absence of objective guidelines that would help making this choice, we delegate this task to evolutionary algorithm, which maintains a population of individuals, each defining a training experience. Shaping vector $\mathbf{s}$ forms then the *genotype* of an individual, while its *phenotype* is the policy $\pi_{\mathbf{s}}$ trained using the above learning procedure. Evaluation of the phenotype formed in this way consists in running $\pi_{\mathbf{s}}$ on the entire problem, starting from s_0, possibly multiple times if indeterminism is involved. The fitness of individual can be then defined as, e.g., the average reward obtained by $\pi_{\mathbf{s}}$. In this way, the evolutionary process becomes responsible for searching for the useful training experiences, forming the outer loop of the proposed approach (Fig. 1).

In this paper we apply the above recipe for autonomous shaping to *competitive environments*, in which solving sequential decision problems boils down to playing games, and subproblems correspond to endgames. Rather than maximizing the expected reward on a single problem in a static, single-agent environment, we want to maximize the expected game outcome when playing against *any* opponent, i.e., another agent that interferes at the decision making process. This objective can be naturally implemented using coevolutionary algorithms, in which the fitness of an individual depends on the outcomes of its interactions with the other individuals in the population. Technically, we implement single-population competitive coevolution [4]: in the evaluation phase, the strategies $\pi_{\mathbf{s}}$ derived from particular individuals play a round-robin tournament against each other, and the total score received determines individual's fitness.

Independently of the choice of the algorithm performing the outer learning loop, the proposed approach can be then considered dual with respect to traditional methods of policy learning. Rather than aiming at acquisition of maximum knowledge from the original problem by, e.g., tuning the parameters of the training algorithm, the focus of the method is on shaping, i.e., exposing the learner to the 'right' training experience represented by selected subproblems. In short, *what* to learn becomes here more important than *how* to learn. In this context, the choice of the actual training algorithm T is of secondary importance: its parameters, if any, remain fixed during the entire training process, and it only serves as a means to assess the usefulness of particular set of initial states.

3   Experimental Setup

In the following we apply the proposed approach of autonomous shaping in its coevolutionary variant to the problem of learning to play the board game of Othello (Fig. 2a). The experiments have been conducted using our coevolutionary algorithms library cECJ [6] built upon the Evolutionary Computation in Java framework. For each considered setup, evolutionary runs have been repeated 20 times.

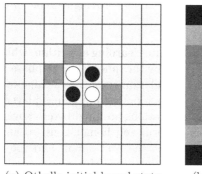

(a) Othello initial board state   (b) Heuristic WPC weights

Fig. 2. Othello board and its coloring according to heuristic player weights (darker color — greater weight)

Learner Architecture. One of the main issues to consider when learning game-playing strategy is the architecture of the learner, which is mainly determined by a strategy representation. Of many possible ways in which the strategies $\pi \in \Pi$ could be represented, we chose the simple weighted piece counter (WPC). WPC assigns a weight w_i to each board location i and uses scalar product to calculate the utility f of a board state $\mathbf{b}$:

$$f(\mathbf{b}) = \sum_{i=1}^{8 \times 8} w_i b_i, \tag{1}$$

where b_i is $+1$, -1, or 0 if, respectively, location i is occupied by a black piece, white piece, or empty. The players interpret the values of f in a complementary manner: the black player prefers moves leading to states with larger values, while smaller values are favored by the white player. Alternatively, WPC may be viewed as an artificial neural network comprising a single linear neuron with inputs connected to board locations. The standard heuristic player represented as a WPC is illustrated in the Fig. 2b. We use it also as an opponent in our experiments to measure the post-training performance of agents.

The Inner Learning Algorithm. We used the basic temporal difference method TD(0) as the learning algorithm T that improves a game-playing strategy on the basis of the training experience represented as a shaping vector (cf. Section 2). The initial strategy π_0 has all weights zeroed (see Eq. (1)). Given a state s_i from the shaping vector $\mathbf{s}$, invoking $T(s_i, \pi_k)$ consists in a single training pass of self-play TD(0) with s_i as an initial state and π_k determining the initial values of WPC weights. Game outcomes determine the rewards. The weights of WPC are modified after every move by a gradient-descent temporal difference update rule [7] with the learning rate parameter set to $\alpha = 0.01$. Each element of the shaping vector was used as an initial state for 100 learning episodes to increase the amount of experience gathered in the corresponding part of the game tree. TD(0) was previously applied for Othello [8], proving capable of producing very good players in short training times.

The Outer Learning Algorithm. The outer learning algorithm, which targets on optimizing the training experience used by the inner learning phase, is framed as coevolutionary learning. The initial population comprises 50 shaping vectors, each composed of $m = 50$ states selected randomly from games played between two random players. The subsequent generations are bred by crossover followed by mutation. The former operator is uniform and homologous, so an offspring inherits $m/2$ randomly selected states from the first parent and the rest from the second one, and the order of states is preserved. Mutation is applied to the offspring with probability 0.05 per state and consists in replacing a state with a newly generated random state. The genotype-phenotype mapping is realized by the inner learning loop, and the evaluation consists of playing a population-wide round-robin tournament between strategies created in this way. The players score 3, 1, or 0 points for winning, drawing, and losing, respectively. The total score earned in the tournament becomes individual's fitness, which is then subject to tournament selection of size 5. Thus, we evaluate shaping vectors by judging the performance of players created with their guidance.

4   The Results

The complete process of learning a game strategy using autonomous shaping involves two phases. First, the method proposed in Section 2 attempts to evolve the best shaping vector for the given learning algorithm. In the second phase, this vector is employed to train a strategy, which becomes the final outcome of the overall training process. All players in our experiments are deterministic, as well as the game of Othello itself. Thus, in order to estimate the score of a given trained player against the WPC-heuristic (Fig. 2b), we forced both players to make random moves with probability $\varepsilon = 0.1$. This provides richer repertoire of players' behaviors and makes the resulting estimates more continuous and robust.

Phase 1: Search for the Best Shaping Vector. The objective progress of this procedure was monitored by assessing the quality of the fittest player, i.e., the player that appeared the best among all the players trained with particular shaping vectors. We call this player the *best-of-generation learner*.

Figure 3 illustrates the performance of the best-of-generation learners, averaged over 20 coevolutionary runs. For reference, we plot also best-of-generation players found by standard coevolutionary search performed directly in the space of WPC strategies (for more details see [8]). Clearly, coevolution of training experience outperforms the direct approach. The level of play it attains is very similar to the best strategies obtained using CTDL, a hybrid of coevolution and TDL proposed in our previous work [8].

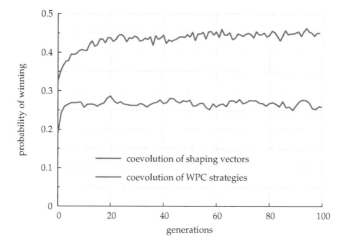

Fig. 3. Comparison of the average performance of the *best-of-generation learners* shaped by coevolved training experience and the *best-of-generation players* coevolved directly, against the WPC-heuristic opponent

Phase 2: Training the Strategy Using the Best Shaping Vector Found.

In this phase, the best shaping vector found in phase 1 is mapped to a strategy using the genotype-phenotype mapping described in Section 2. We take a deeper look at this inner learning process realized by TD(0) algorithm.

Figure 4 visualizes the learning from the training experience embodied by the best shaping vector. Every thin blue curve depicts the mean performance of a strategy trained using the best shaping vector found in one of 20 evolutionary runs. The horizontal axis corresponds to the inner learning loop shown in Fig. 1 (as opposed to Fig. 3, where it marked the iterations of the outer loop). Each learning episode corresponds to an application of the training algorithm (TD(0)) to a single initial state, $T(s_i, \pi_k)$, so the horizontal axis is simply the k axis.

The thick red line shown in Fig. 4 depicts the behavior of the the standard TDL learning process, starting always from s_0 (illustrated in Fig. 2a), which gathers experience by ε-greedy action selection scheme (with ε equal to 0.1). Standard TDL clearly stalls much earlier than the shaping approach, and attains substantially worse performance at the end of training.

Performance against the WPC-heuristic says only a little about the overall objective quality of a strategy, because in practice we typically aim at producing versatile and robust players, capable of winning against a wide range of opponents. Thus, we gauged also the *relative performance* against other players trained using different methods. To this aim, we confront the teams of best-of-run strategies obtained from 20 runs with the team of players that have been trained on full games, using TD(0) randomized self-play starting from the default initial state s_0. Table 1 presents the outcomes of that duel, with the shaping-trained strategies sorted descendingly with respect to their outcome. The teams of strategies produced using the proposed approach are clearly superior. Even the worst

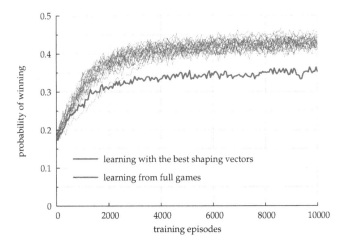

Fig. 4. The average performance of the learners trained with the best shaping vectors (blue, one plot per vector) vs. the average performance of the learners trained from full games (default experience, red), as a function of the number of TD(0) training episodes

Table 1. The outcomes of matches played between the teams of players obtained using the shaping approach with a team of strategies trained using randomized self-play TD(0)

Run #	Wins	Draws	Losses	Points	% pts.	Run #	Wins	Draws	Losses	Points	% pts.
1	508	20	272	1544	64.33	11	476	22	302	1450	60.42
2	497	36	267	1527	63.63	12	472	33	295	1449	60.38
3	497	23	280	1514	63.08	13	475	23	302	1448	60.33
4	495	22	283	1507	62.79	14	476	15	309	1443	60.13
5	492	20	288	1496	62.33	15	472	26	302	1442	60.09
6	488	28	284	1492	62.17	16	474	17	309	1439	59.96
7	486	26	288	1484	61.83	17	465	28	307	1423	59.29
8	482	32	286	1478	61.58	18	464	24	312	1416	59.00
9	478	35	287	1469	61.21	19	453	37	310	1396	58.17
10	476	27	297	1455	60.63	20	457	24	319	1395	58.13

of them wins substantially more games than it loses. Also, the performance of particular teams varies only slightly, with most of them scoring between 59 and 62% of all available points. This clearly suggests that the search for initial state vectors, though intermediated by the nontrivial genotype-phenotype mapping, repeatedly leads to producing stable and well-performing players.

5   Discussion and Related Work

The problem of selecting the training experience for a reinforcement learning agent has been addressed by several authors. Mihalkova and Mooney [9] propose a method for improving the reinforcement learning by allowing the learner to relocate, i.e., to be placed in a requested state of the environment. An agent may benefit from this possibility by omitting the already known regions of the environment (when the agent is "bored" and is not learning anything new) or escaping from the parts of the state space that are unlikely to be visited again using the optimal policy (when the agent made a wrong exploratory move and fell "in trouble"). Relocation destinations are chosen according to an uncertainty measure reflecting agent's confidence about the best action in a given state. This approach differs from ours in being inherently *active* — it is the learner who makes decisions about when and where to relocate within an online training process. In this context, our method resembles more the selective sampling used in traditional supervised active learning [10].

A complementary *passive* approach is taken by Rachelson *et al.* [11] who introduce the meta-algorithm of Optimal Sample Selection (OSS). Given a batch-mode RL policy inference algorithm, a policy evaluation method, and a generative model of the environment, OSS attempts to identify a set of one-step transitions which, when supplied to the policy learning algorithm, lead to an optimal behavior with respect to the evaluation measure. In this method the learning proceeds independently from the selection of training experience — the learner cannot affect the way the experience is gathered. Also, the policy learning algorithm is assumed to work in an offline batch manner, i.e., it exploits a fixed, prepared in advance set of training examples (sample of transitions), without a need of dynamically interacting with the environment. Our approach abstracts from the character of the policy learning algorithm and is more coarse-grained — instead of selecting single transitions, we find entire states that implicitly identify many useful paths through the environment. This allows us to represent the training experience in a more compact, illustrative, and generic way.

Finally, changing the initial state can be seen as a slight modification of the learning task itself. What we finally want to achieve is then the transfer of knowledge from a set of adjusted task to the original problem. We expect that it can improve the learning process in the same way as the transfer learning [12], including initial performance, time of learning and final performance. In a similar spirit, Konidaris and Barto [13] investigate knowledge transfer across a sequence of tasks and employ an autonomous shaping approach by augmenting reward functions – they use the knowledge about predicted rewards from one task to shape the reward function of the other one.

6   Summary

In learning game-playing strategies, it is typically the role of a *trainer* to guide the learner through the paths of the game tree from which it can learn the most. This study revolved around the observation that such guidance can take on different forms. Naturally, the trainer is embodied by an opponent, e.g., an expert player or the learner itself [14]. In the proposed method, the opponent strategy, though varying with time, is stationary in being produced by a fixed learning algorithm, while the role of guidance is delegated to a set of initial states that limit the exploration to the corresponding partial game trees. We are not looking for an ideal trainer here, but for an ideal training experience. Eventually, the goal is to shape the learning process so that it produces proficient learners prepared to perform well in every, potentially unseen before, region of environment. This goal has been attained in this study for the game of Othello: rephrasing a learning task in a way that enables autonomous shaping led to better performing and more versatile players. Applicability of this approach to other interactive and non-interactive domains is to be verified in future research.

Acknowledgment. This work has been supported by grant no. N N519 441939.

References

1. Sutton, R.S., Barto, A.G.: Reinforcement Learning: An Introduction. The MIT Press (1998)
2. Skinner, B.: The behavior of organisms: An experimental analysis. Appleton-Century (1938)
3. Randløv, J., Alstrøm, P.: Learning to drive a bicycle using reinforcement learning and shaping. In: Proceedings of the Fifteenth International Conference on Machine Learning, pp. 463–471. Morgan Kaufmann, San Francisco (1998)
4. Popovici, E., Bucci, A., Wiegand, R.P., de Jong, E.D.: Coevolutionary principles. In: Handbook of Natural Computing. Springer, Berlin (2010)
5. Jaśkowski, W., Krawiec, K.: Formal analysis, hardness and algorithms for extracting internal structure of test-based problems. Evolutionary Computation 19(4), 639–671 (2011)
6. Szubert, M.: cECJ — Coevolutionary Computation in Java (2010), http://www.cs.put.poznan.pl/mszubert/projects/cecj.html
7. Sutton, R.S.: Learning to predict by the methods of temporal differences. Machine Learning 3, 9–44 (1988)
8. Szubert, M., Jaśkowski, W., Krawiec, K.: Coevolutionary Temporal Difference Learning for Othello. In: 2009 IEEE Symposium on Computational Intelligence and Games, pp. 104–111 (2009)
9. Mihalkova, L., Mooney, R.: Using active relocation to aid reinforcement learning. In: Proceedings of the 19th International FLAIRS Conference, pp. 580–585 (2006)
10. Cohn, D., Atlas, L., Ladner, R.: Improving generalization with active learning. Machine Learning 15(2), 201–221 (1994)
11. Rachelson, E., Schnitzler, F., Wehenkel, L., Ernst, D.: Optimal sample selection for batch-mode reinforcement learning. In: Proceedings of the 3rd International Conference on Agents and Artificial Intelligence, ICAART 2011 (2011)
12. Torrey, L., Shavlik, J.: Transfer Learning. In: Handbook of Research on Machine Learning Applications and Trends: Algorithms, Methods, and Techniques, pp. 242–264. IGI Global (2009)
13. Konidaris, G., Barto, A.: Autonomous shaping: Knowledge transfer in reinforcement learning. In: Proceedings of the 23rd International Conference on Machine Learning, pp. 489–496. ACM (2006)
14. Epstein, S.: Toward an ideal trainer. Machine Learning 15(3), 251–277 (1994)

A Parallel Cooperative Co-evolutionary Genetic Algorithm for the Composite SaaS Placement Problem in Cloud Computing

Maolin Tang and Zeratul Izzah Mohd Yusoh

Queensland University of Technology,
2 George Street, Brisbane, QLD 4001, Australia

Abstract. A composite SaaS (Software as a Service) is a software that is comprised of several software components and data components. The composite SaaS placement problem is to determine where each of the components should be deployed in a cloud computing environment such that the performance of the composite SaaS is optimal. From the computational point of view, the composite SaaS placement problem is a large-scale combinatorial optimization problem. Thus, an Iterative Cooperative Co-evolutionary Genetic Algorithm (ICCGA) was proposed. The ICCGA can find reasonable quality of solutions. However, its computation time is noticeably slow. Aiming at improving the computation time, we propose an unsynchronized Parallel Cooperative Co-evolutionary Genetic Algorithm (PCCGA) in this paper. Experimental results have shown that the PCCGA not only has quicker computation time, but also generates better quality of solutions than the ICCGA.

1 Introduction

Cloud computing is a new computing paradigm in which all the resources are provided to users as a service over the Internet [1]. According to Gartner, one of the world's leading information technology research and advisory company, by 2014 the cloud computing services revenue is expected to reach 148.8 billion dollars [2]. Software as a Service (SaaS) is one of the most important configurable computing services in cloud computing [3]. It uses a software distribution model in which software is hosted by a SaaS vendor in the cloud and made it available to users as a service over the Internet. SaaS in cloud computing has three distinct characteristics that differentiate itself from a traditional on-premise software. First, it is sold on demand, typically by pay per use. Second, it is elastic - a user can have as much or as little of the service as they want at any given time. Third, the software that provides the service is fully managed by the SaaS vendor. These features allow users to obtain the same benefits of on-premise software without the associated complexity of installation, management, support, licensing, and high initial cost, and therefore make SaaS in cloud computing so compelling.

A composite SaaS is a kind of SaaS that is developed using component-based software development technologies. It usually consists of several software components and data components, such as databases. The composite SaaS placement

C.A. Coello Coello et al. (Eds.): PPSN 2012, Part II, LNCS 7492, pp. 225–234, 2012.
© Springer-Verlag Berlin Heidelberg 2012

problem is to place those software components and data components on those compute servers and storage servers, respectively, in the cloud such that the performance of the composite SaaS is optimal. The problem is similar to the task assignment problem and the terminal assignment problems addressed in [7,6]. However, the composite SaaS placement problem is more challenging than the task assignment problem as it has more constraints. For example, a software component can be only placed on a compute server and the compute server must meet the CPU and memory requirements of the software component. Thus, the algorithms for the task assignment and the terminal assignment problem cannot be immediately applied to solve the composite SaaS placement problem.

From the computational point of view, the composite SaaS placement is a large-scale combinatorial optimization problem as a cloud may contain thousands of compute severs and storage servers, and a composite SaaS may have dozens of components. Thus, a Penalty-based Genetic Algorithm (PGA) was initially developed [8]. In order to improve the quality of solutions, an Iterative Cooperative Co-evolutionary Genetic Algorithm (ICCGA) was then developed [9]. Experimental results showed that the ICCGA can produce better solutions than the PGA. However, the computation time of the ICCGA was noticeably slow. In order to improve the computation time, this paper presents a Parallel Cooperative Co-evolutionary Genetic Algorithm (PCCGA). The PCCGA has been implemented and tested. Experimental results have shown that the PCCGA not only has quicker computation time, but also produces better solutions, than the ICCGA. In addition, experimental results have shown that the PCCGA has better scalability than the ICCGA.

The remaining paper is organized as follows. Section 2 formulates the composite SaaS placement problem. Section 3 proposes a PCCGA model and entails the PCCGA. Section 4 evaluates the PCCGA. Finally, section 5 concludes this research work.

2   Problem Formulation

Let $C = \{c_1, c_2, \cdots, c_m\}$ be the entire set of m compute servers and $S = \{s_1, s_2, \cdots, s_n\}$ be the complete set of n storage servers in a cloud computing environment. The compute servers and storage servers are interconnected through a set of communication links E. The servers and the communication links together form a cloud communication network. A cloud communication network can be modeled in a graph $G = < V, E >$, where $V = C \cup S$, and if $< v_i, v_j > \in E$ if and only if there exists a communication link between v_i and v_j and $v_i, v_j \in V$.

A composite SaaS, X, consists of a set of software components $SC = \{sc_1, sc_2, \cdots, sc_p\}$ and a set of data components $SS = \{ss_1, ss_2, \cdots, ss_q\}$, where p is the number of software components and q the number of data components in X. The control dependencies and data dependencies between those software components are stored in sets CD and DD, respectively.

A SaaS component has a CPU requirement and a memory requirement. A compute server has a CPU capacity and a memory capacity. A software component

$sc_i =< sc_i^{cpu}, sc_i^{mem} >$ can be deployed on a compute server $c_j =< c_j^{cpu}, c_j^{mem} >$ only when the compute serve c_j can meet both the CPU and memory requirements of the software component sc_i, that is $sc_i^{cpu} \leq c_j^{cpu}$ and $sc_i^{mem} \leq c_j^{mem}$, where $1 \leq i \leq p$ and $1 \leq j \leq m$.

Similarly, a data component has a space requirement and a storage server has a space capacity. A data component $sd_i =< sd_i^{space} >$ can be placed on a storage server $s_j =< s_j^{space} >$ only when the storage server has enough room to hold the data component sd_i, that is $sd_i^{space} \leq s_j^{space}$, where $1 \leq i \leq q$ and $1 \leq j \leq n$.

Given a composite SaaS $X =< SC, SS, CD, DD >$ and a cloud computing communication network $G =< C \cup S, E >$, the composite SaaS placement problem is to find $f_1 : SC \rightarrow C$ and $f_2 : SS \rightarrow S$ such that the performance of the composite SaaS is optimal measured by the Estimated Execution Time of the composite SaaS, which is derived in [9].

3 Parallel Cooperative Co-evolutionary Genetic Algorithm

This section presents an unsynchronized parallel computation model and describes the algorithm of the PCCGA.

3.1 Parallel Model

The parallel model based on which the PCCGA is developed is derived from a cooperative co-evolution model proposed by Potter and de Jong [10]. In the cooperative co-evolution model, a problem is divided into several interacting subproblems. For each of the subproblems, an evolutionary algorithm, such as genetic algorithm, is used to solve it independently, and the multiple subproblems are solved concurrently using multiple independent evolutionary algorithms. The interaction between the evolutionary algorithms occurs only when evaluating the fitness value of an individual in the population of an evolutionary algorithm as the individual is only part of the solution to the problem in the domain and therefore in order to evaluate its fitness the PCCGA needs to combine the individual with a representative from each of the other evolutionary algorithms to form a complete solution. An individual is rewarded when it works well with the representative from the other evolutionary algorithms and is punished when it does not work well with the representative.

Based on the cooperative co-evolutionary model, we developed an unsynchronized parallel model as shown in Fig. 1. In the parallel model, we decompose the computation into two unsynchronized and parallel sub-computations. One of the sub-computations is the placement of the software components; another the placement of the data components. For each of the sub-computations, we use a genetic algorithm (GA) to solve it. The communication between the two GAs is asynchronous through a buffer. The buffer has two units. One unit keeps the best solution from the software component placement sub-computation; the other the best solution from the the data component placement sub-computation. At the

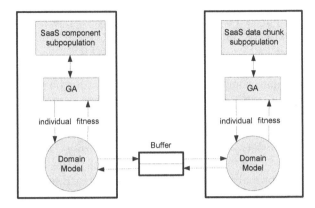

Fig. 1. Parallel Cooperative Co-evolutionary Genetic Algorithm Model

end of each generation, the GAs update their best solution in the buffer. Since the two GAs are not synchronized, one GA may update its best solution in the buffer more frequently than the other during the computation.

3.2 Algorithm Descriptions

The PCCGA invokes a classical GA for the software component placement problem and a classical GA for the data component placement problem. Thus, before giving the algorithm description of the PCCGA, we present the two classical GAs. The encoding scheme and genetic operators used in the GAs are the same with those in [9].

The GA for the Software Component Placement Problem

1. randomly generate an initial population of solutions to the software component placement problem;
2. while the termination condition is not true
   (a) get the best solution to the data component placement problem from the buffer;
   (b) for each individual in the population:
       i. combine the individual with the best solution to the data component placement problem to form a complete SaaS placement solution;
       ii. calculate the fitness value of the SaaS placement solution.
   (c) select individuals for recombination from the population based on their fitness values and pair them up;
   (d) probabilistically apply the crossover operator to each of the pairs to generate new individuals;
   (e) probabilistically use for the mutation operator to each of the new individuals;
   (f) use the new individuals to replace the old individuals in the population;
   (g) update the best software component placement solution in the buffer.

The GA for the Data Component Placement Problem

1. randomly generate an initial population of solutions to the data component placement problem;
2. while the termination condition is not true
   (a) get the best solution to the software component placement problem from the buffer;
   (b) for each individual in the population:
      i. combine the individual with the best solution to the software component placement problem to form a complete SaaS placement solution;
      ii. calculate the fitness value of the SaaS placement solution.
   (c) select individuals for recombination from the population based on their fitness values and pair them up;
   (d) probabilistically apply the crossover operator to each of the pairs to generate new individuals;
   (e) probabilistically use for the mutation operator to each of the new individuals;
   (f) use the new individuals to replace the old individuals in the population;
   (g) update the best data component placement solution in the buffer.

The Algorithm Description of the PCCGA

1. while the termination condition is not true
   (a) run the GA for the software component placement problem and the GA for the data component placement problem in parallel;
2. combine the best solution to the software component placement problem and the best solution to the data component placement problem in the buffer to form a solution to the SaaS placement and output it.

4 Evaluation

This section evaluates the performance of the PCCGA, including its computation time, quality of solution and scalability. Since there is no benchmark available for the composite SaaS placement problem, we have to use the performance of the ICCGA as a benchmark.

In order to conduct a comparative study of the PCCGA and the ICCGA, we implemented both of them in Microsoft Visual Studio C#. We also developed a C# program to randomly generate a cloud communication network of a given configuration based on the cloud model presented in [11] and another C# program to randomly create a composite SaaS placement problem of a given configuration.

Since the complexity of a composite SaaS placement problem depends on both the size of the cloud communication network and the size of the composite SaaS placement problem, we conducted two groups of experiments. In the first group of experiments, we randomly generated a cloud communication network that

Table 1. The characteristics of the five composite SaaS placement problems

Test	Test Problem Characteristics		
Problem	No. of software comp.	No. of data comp.	Total no. of comp.
1	5	5	10
2	10	10	20
3	15	15	30
4	20	20	40
5	25	25	50

has 100 compute servers and 100 storage servers, and then randomly generated five composite SaaS placement problems of different sizes. Table 1 shows the characteristics of the five composite SaaS placement problems.

This group of experiments were designed to evaluate the speed-up ratio and the quality of the solutions of the PCCGA when it is used for solving different sizes of composite SaaS placement problems and to study how the computation time of the PCCGA would increase when the size of the composite SaaS placement problem increases. We used both the PCCGA and the ICCGA to solve each of the five composite SaaS placement problems. Considering the stochastic nature of the PCCGA and the ICCGA, for each of the composite SaaS placement problems we repeated the experiment for 10 times and recorded the quality of the solutions generated by both of the algorithms and their computation times. The statistics about the computation time and the quality of solutions of the two algorithms are shown in Table 2 and Table 3 respectively.

It can be seen from Table 2 that the average computation time of the PCCGA is between 11.29% and 37.15% of the average computation time of the ICCGA for the five test problems. However, the average estimated execution time of the composite SaaS placement produced by the PCCGA is also better than that produced by the ICCGA (the average estimated execution time of the PCCGA is only between 23% and 53% of that of the ICCGA).

In addition, in the evaluation, we also compared the scalability of the PCCGA with the scalability of the ICCGA. Fig. 2 displays how the average computation times increased when the size of the composite SaaS increased. When the total number of SaaS components and data components increased from 10 to 50, the average computation time of the ICCGA increased from 278.5 seconds to 9266.8 seconds, while the computation time of the PCCGA only increased from 103.3 seconds to 1046.9 seconds linearly.

In the second group of experiments, we randomly generated a composite SaaS placement problem that has 10 software components and 10 data components, and randomly generated five cloud communication networks. Table 4 shows the characteristics of the five randomly generated cloud communication networks.

Then, we used both the PCCGA and the ICCGA to solve the composite SaaS placement problem in the five cloud communication networks of different sizes. Considering the stochastic nature of the two algorithms, for each of the experiments we repeated for 10 times and recorded the qualities of the solutions and the computation times for each run of the experiments. The statistics about

Table 2. Comparison of the computation times of the PCCGA and the ICCGA for different composite SaaS placement problems

Test	PCCGA (second)				ICCGA (second)			
Problem	Best	Worst	Ave	SD	Best	Worst	Ave	SD
1	69.6	154.1	103.5	27.4	188.4	419.4	278.5	96.2
2	132.6	371.4	231.5	80.4	1020.0	2416.8	1557.1	415.8
3	382.8	832.2	621.0	144.9	1743.6	6877.8	3859.9	1541.3
4	415.2	950.4	710.3	174.1	2466.0	8473.8	5505.0	2103.3
5	632.4	1938	1046.9	403.6	3436.2	15763.8	9266.82	3755.8

Table 3. Comparison of the qualities of solutions produced by the PCCGA and the ICCGA for different composite SaaS placement problems

Test	PCCGA (millisecond)				ICCGA (millisecond)			
Problem	Best	Worst	Ave	SD	Best	Worst	Ave	SD
1	513	89309	24958.5	29056.3	89748	129609	108866.9	12613.6
2	112	180144	36680.2	53964.5	107524	263752	140893.0	46928.2
3	22042	114729	72857.1	36736.7	130569	235881	172433.3	33817.7
4	12674	180144	129565.7	63641	174319	263752	243215.3	47358.1
5	16100	216704	123456.6	88892.8	217730	648594	316120.1	120410.2

Table 4. The characteristics of the clouds

Test	Test Problem Characteristics		
Problem	No. of compute servers	No. of storage servers	Total no. of servers
1	50	50	100
2	100	100	200
3	150	150	300
4	200	200	400
5	250	250	500

the computation times and the quality of solutions of the two algorithms are shown in Table 5 and Table 6 respectively.

It can be seen from Table 6 that the average computation time of the PCCGA is between 17.66% and 31.89% of the average computation time of the ICCGA for the five test problems. However, the average estimated execution time of the composite SaaS placement produced by the PCCGA is also better than that produced by the ICCGA (the average estimated execution of the PCCGA is only between 45% and 55% of that of the ICCGA).

In addition, in the evaluation, we also compared the scalability of the PCCGA with the scalability of the ICCGA. Fig. 3 displays how the average computation times increased when the size of the composite SaaS increased. When the total number of compute and storage servers in cloud computing increased from 100 to

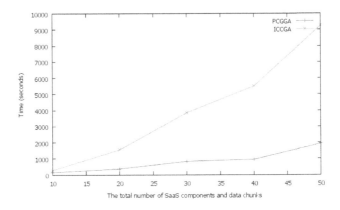

Fig. 2. The computation time increasing trend when the size of the composite SaaS increases

Table 5. Comparison of the computation times of the PCCGA and the ICCGA in different clouds

Test	PCCGA (sec)				ICCGA (sec)			
Problem	Best	Worst	Ave	SD	Best	Worst	Ave	SD
1	54.0	136.2	89.4	31.3	145.2	440.4	280.3	88.5
2	115.8	440.9	288.5	107.2	813.0	2228.4	1362.4	482.3
3	233.4	1302.0	672.4	326.3	712.8	3451.2	2456.0	809.7
4	401.4	1404.6	774.6	316.9	3024.0	8580.0	4385.4	1629.9
5	771.6	1707.6	1350.5	309.9	3885.6	10029.6	7187.6	1824.2

Table 6. Comparison of the qualities of solutions produced by the PCCGA and the ICCGA in different clouds

Test	PCCGA (millisecond)				ICCGA (millisecond)			
Problem	Best	Worst	Ave	SD	Best	Worst	Ave	SD
1	7284	53536	28519.9	15211.9	48064	79026	66425.4	9186.6
2	34357	110918	61238.9	24669.8	97370	168355	137089.4	21842.2
3	34588	254672	162969.4	65807.6	202857	379929	295677.8	47531.4
4	120114	359175	241791.2	77513.7	368345	623340	501810.4	73681.5
5	44786	509168	274609.9	136396.5	363253	623821	547364.0	80760.1

500, the average computation time of the ICCGA increased from 280.3 seconds to 7187.6 seconds, while the computation time of the PCCGA only increased from 89.4 seconds to 1350.5 seconds linearly.

In all the experiments, the subpopulation sizes for the compute server GA and the storage server GA were set at 100 in both the PCCGA ad the ICCGA. The probabilities for crossover and mutation were set at 0.95 and 0.15, respectively,

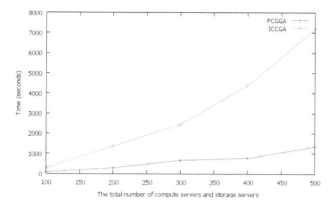

Fig. 3. The computation time increasing trend when the size of cloud computing increases

in both the PCCGA and the ICCGA. The termination condition used in both the PCCGA and the ICCGA was 'no improvement on the best solution for 25 consecutive generations'. All the experiments were carried out in a computer with 3.00 GHz Intel Core 2 Duo CPU and 4GB RAM.

5 Conclusion and Future Work

This paper has proposed an unsynchronized parallel cooperative co-evolutionary genetic algorithm (PCCGA) for the composite SaaS placement problem in cloud computing, and has evaluated the performance of the PCCGA by comparing it with an iterative cooperative co-evolutionary genetic algorithm (ICCGA). The experimental results have shown that the computation time of the PCCGA was noticeably quicker than that of the ICCGA. In addition, the experimental results have shown that on average the quality of the solutions produced by the PCCGA is much better than that of the ICCGA for those randomly generated test problems. Moreover, the experimental results have shown that the PCCGA has better scalability than the ICCGA. To the best of our knowledge, this research is the first attempt to tackle the composite SaaS placement algorithm using parallel evolutionary computation.

The SaaS placement is a large-scale complex combinatorial optimization. Thus, how to further improve its computation time by increasing its parallelism is an issue that we will investigate in the future, and one possible way to improve the parallelism is to break down the composite SaaS placement problem into more subcomponents using the random grouping techniques proposed in [12]. In addition, in the PCCGA we always select the best individual from the other population, which may not be appropriate in some cases. Thus, another work that I will do in the future is to the selection strategy.

Acknowledgment. The programs used in the evaluation were developed by Peter Wong at Queensland University of Technology. In addition, the authors would like to thank the PC members and reviewers of this papers for your valuable comments.

References

1. Foster, I., Yong, Z., Raicu, I., Lu, S.: Cloud computing and grid computing 360-degree compared. In: Proceeding of Grid Computing Environment Workshop, pp. 1–10 (2008)
2. Gartner Inc., Gartner says worldwide cloud services market to surpass \$68 Billion in 2010 (2010), http://www.gartner.com/it/page.jsp?id=1389313
3. Armbrust, M., et al.: Above the clouds: a Berkeley view of cloud computing. Tech. Rep. UCB/EECS-2009-28, EECS Department, U.C. Berkeley (2009)
4. Khare, V., Yao, X., Sendhoff, B.: Multi-network evolutionary systems and automatic problem decomposition. International Journal of General Systems 35(3), 259–274 (2006)
5. Khare, V.R., Yao, X., Sendhoff, B.: Credit Assignment Among Neurons in Co-evolving Populations. In: Yao, X., Burke, E.K., Lozano, J.A., Smith, J., Merelo-Guervós, J.J., Bullinaria, J.A., Rowe, J.E., Tiňo, P., Kabán, A., Schwefel, H.-P. (eds.) PPSN VIII. LNCS, vol. 3242, pp. 882–891. Springer, Heidelberg (2004)
6. Salcedo-Sanz, S., Xu, Y., Yao, X.: Hybrid meta-heuristics algorithms for task assignment in heterogeneous computing systems. Computers and Operations Research 33(3), 820–835 (2006)
7. Salcedo-Sanz, S., Yao, X.: A hybrid hopfield network – genetic algorithm approach for the terminal assignment problem. IEEE Transactions on Systems, Man and Cybernetics, Part B: Cybernetics 34(6), 2343–2353 (2004)
8. Yusoh, Z., Tang, M.: A penalty-based genetic algorithm for the composite SaaS placement problem in the cloud. In: Proceeding of IEEE World Congress on Computational Intelligence, pp. 600–607. IEEE, Spain (2010)
9. Yusoh, Z.I.M., Tang, M.: A Cooperative Coevolutionary Algorithm for the Composite SaaS Placement Problem in the Cloud. In: Wong, K.W., Mendis, B.S.U., Bouzerdoum, A. (eds.) ICONIP 2010, Part I. LNCS, vol. 6443, pp. 618–625. Springer, Heidelberg (2010)
10. Potter, M., de Jong, K.: A Cooperative Coevolutionary approach to Function Optimization. In: Davidor, Y., Männer, R., Schwefel, H.-P. (eds.) PPSN III. LNCS, vol. 866, pp. 249–257. Springer, Heidelberg (1994)
11. IBM System Storage, http://www-07.ibm.com/storage/au/
12. Yang, Z., Tang, K., Yao, X.: Large scale evolutionary optimization using cooperative coevolution. Information Sciences 178(15), 2985–2999 (2008)

Community Detection Using Cooperative Co-evolutionary Differential Evolution

Qiang Huang[1], Thomas White[2], Guanbo Jia[2], Mirco Musolesi[2], Nil Turan[3],
Ke Tang[4], Shan He[2,3,*], John K. Heath[3], and Xin Yao[2,4]

[1] School of Software, Sun Yat-sen University, Guangzhou, China
[2] Cercia, School of Computer Science
[3] Center for Systems Biology, School of Biological Sciences, The University of Birmingham,
Birmingham, B15 2TT, UK
[4] Nature Inspired Computation and Application Laboratory (NICAL), Department of Computer
Science, University of Science and Technology of China, Hefei, Anhui 230027, China
s.he@cs.bham.ac.uk

Abstract. In many scientific fields, from biology to sociology, community detection in complex networks has become increasingly important. This paper, for the first time, introduces Cooperative Co-evolution framework for detecting communities in complex networks. A Bias Grouping scheme is proposed to dynamically decompose a complex network into smaller subnetworks to handle large-scale networks. We adopt Differential Evolution (DE) to optimize network modularity to search for an optimal partition of a network. We also design a novel mutation operator specifically for community detection. The resulting algorithm, Cooperative Co-evolutionary DE based Community Detection (CCDECD) is evaluated on 5 small to large scale real-world social and biological networks. Experimental results show that CCDECD has very competitive performance compared with other state-of-the-art community detection algorithms.

1 Introduction

Many complex systems, such as social [11] and biological networks [3], can be naturally represented as complex networks. A complex network consists of nodes (or vertices) and edges (or links) which respectively represent the individual members and their relationships in systems. By representing complex systems as complex networks, many theories and methods in graph theory can be applied to enable us to gain insights into complex systems. Therefore, in recent years, the study of complex networks has attracted more and more attention.

Unlike simple networks such as lattices or random graphes, complex networks possess many distinctive properties, of which community structure [1] is one of the most studied. The community structure is usually considered as the division of networks into subsets of vertices within which intra-connections are dense, while between which inter-connections are sparse [1]. The identification of the community structure provides

---

* Correspondence and requests for materials should be addressed to Dr. S. He.

C.A. Coello Coello et al. (Eds.): PPSN 2012, Part II, LNCS 7492, pp. 235–244, 2012.

important information about the relationship and interaction among nodes in the complex network. Such information ultimately leads to insights into how network function and topology affect each other.

In the past few years, many algorithms have been proposed to detect the underlying community structure in complex networks [1]. These algorithms can roughly be grouped as traditional methods, such as graph partitioning, spectral methods, modularity maximization methods, and methods based on statistical inference. Among them, the most popular group is *modularity maximization* methods, because of its superior performance on real-world complex networks. For modularity maximization methods, many deterministic optimization algorithms such as greedy algorithms have been employed [1]. However, according to [4], we should also treat results from deterministic algorithms such as greedy optimization or spectral methods with "particular caution" because they only return one unique solution, which might "obscure the magnitude of the degeneracy problem and the wide range alternative solutions".

To address the above problem, we previously proposed a stochastic network community detection algorithm, Differential Evolution based Community Detection (DECD) [5], in which Differential Evolution (DE) was used to evolve a population of potential solutions for network partitions, to maximize the network modularity [8]. The results show that DECD can achieve competitive community detection results on several benchmark and real-world complex networks. However, our further investigation showed that DECD is not satisfactory on large-scale networks.

In order to achieve better scalability to handle large-scale networks, this paper proposes CCDECD (Cooperative Co-evolutionary Differential Evolution based Community Detection) by incorporating a Cooperative Co-evolution (CC) framework into our DECD algorithm. To the best of our knowledge, this is the first time CC framework has been introduced for community detection. A CC framework employs a divide and conquer strategy, which divides a large-scale problem into subcomponents and evolves those subcomponents independently and co-adaptively. Compared with traditional Evolutionary Computation, the advantages of a CC framework are: 1) it is capable of handling large scale optimization problems; and 2) it can better deal with problems with complex structure. Such a framework is very natural and attractive to community detection because of two distinctive properties of complex networks: 1) large scale, e.g., consists of thousands or even millions of nodes; and 2) highly structured, e.g., hierarchical.

Apart from introducing CC framework for community detection, the other main contributions of this paper include: 1) a Bias Grouping scheme to dynamically decompose the complex network into smaller subcomponents; 2) a novel mutation operator called global network mutation specifically designed for community detection; and 3) a thorough evaluation of the performance of CCDECD on several real-world networks, including a large scale network which consists of 6927 nodes.

The remainder of this paper is organized as follows. Section 2 introduces the details of CCDECD. In Section 3, the performance of CCDECD is tested on biological and real-world social networks and then the experimental results are discussed. Finally, Section 4 concludes this paper.

2   The Proposed Algorithm

In this paper, a new algorithm based on CCDE called CCDECD is proposed for community detection in complex networks. Similar to the random grouping framework in [15], the main idea behind our CCDECD is also to split a large network into m s-dimensional subcomponents, and then evolve each of them with standard DE. However, we found that the random grouping scheme used in [15] is not suitable for a complex network community detection problem because it will lose connectivity information of the network, which is crucial for the search performance of DE on modularity. Therefore, we introduce a novel bias grouping scheme which utilizes the connectivity information. The key steps of our CCDECD can be summarized as follows:

Step 1) Set $g = 0$ where g denotes the generation number.

Step 2) Randomly initialize population P_g.

Step 3) $g = g + 1$

Step 4) Split the n-dimensional complex network into m sub-components G_i ($i = 1, \ldots, m$), where G_i consists of s indices of nodes ($n = m \times s$) using bias grouping scheme (See Section 2.4 for details).

Step 5) Set $i = 1$.

Step 6) Construct subpopulation SP_i for G_i by extracting s genes as defined by G_i from P.

Step 7) For subpopulation SP_i, optimize the network division using a standard DE by maximizing network modularity of G_i with g_s generations (See Section 2.2 for details).

Step 8) Select the best individual SI_{best} from SP_i.

Step 9 Update population P_g by replacing the s genes as defined by G_i with SI_{best}.

Step 10 $g = g + g_s$

Step 11) If $i < m$ then $i + +$, and go to **Step 6**.

Step 12) Optimize the network division of the whole network represented by P_g using a modified DE with the global network mutation operator for g_g generations (See Section 2.5 for details).

Step 13 $g = g + g_g$

Step 14) Stop if $g > g_{\max}$ where $g_{\max}$ is the maximum number of generations and output the best individual I_{best}; otherwise go to **Step 4**.

2.1   Individual Representation

CCDECD uses the community identifier-based representation proposed in [14] to represent individuals in the population for the community detection problem. For a graph $G = (V, E)$ with n nodes modelling a network, the kth individual in the population is a vector that consists of n genes $\boldsymbol{x}_k = \{x_1, x_2, \ldots, x_n\}$ in which each gene x_i can be assigned an allele value j in the range $\{1, 2, \ldots, n\}$. The gene and allele represent the node and the community identifier (commID) of communities in G respectively. Thus, $x_i = j$ denotes that the node i belongs to the community whose commID is j, and nodes i and d belong to the same community if $x_i = x_d$. Since at most n communities exist in G and then the maximum value of commID is n.

2.2 Fitness Function

Newman and Girvan [8] proposed the network modularity to measure the strength of the community structure found by algorithms. The network modularity is a very efficient quality metric for estimating the partitioning of a network into communities and has been used by many community detection algorithms recently [1,14,7].

CCDECD also employs the network modularity which is maximized as the fitness function to evaluate individuals in the population. The network modularity is defined as follows [14].

$$Q = \sum_{j=1}^{m} \left[\frac{l_j}{L} - \left(\frac{d_j}{2L} \right)^2 \right], \tag{1}$$

where j is the commID, m is the the the total number of communities, l_j is the number of links in module j, L is the total number of edges in the network and d_j is the degree of all nodes in module j.

2.3 Initialization

At the beginning of the initialization process, CCDECD places each node into a random community by assigning a random commID and generates individuals in the initial population. However, such random generation of individuals is likely to cause some unfavorable individuals that consist of some nodes having no connectivity with each other in the original graph. Considering that nodes in the same community should connect with each other and in the simple case are neighbors, the initialization process proposed in [14] is used to overcome the above drawbacks. The process works as follows: once an individual is generated, some nodes in an individual are randomly selected and their commIDs are assigned to all of their neighbors. By this process, the space of the possible solutions is restricted and the convergence of CCDECD is improved.

2.4 Bias Grouping Scheme

Similar to the random group scheme proposed in [15], we proposed a bias grouping scheme for handling large scale networks. The idea behind this bias grouping scheme is to dynamically decompose the whole networks into smaller subcomponents which each consist of nodes that are more likely connected to each other. Therefore, the search algorithm can optimize these tightly interacting variables together, which will ultimately lead to better results than splitting variables into subcomponents with unconnected nodes. The bias grouping scheme works as follows: we randomly select s nodes in the network, where s is the size of a subcomponent. Then we find all the first neighbors of the s nodes and concatenate them to form a set TG. Finally, we select the first s nodes from TG to form a subcomponent G_i ($i = 1, \ldots, m$). If all the s nodes have no first neighbors, all the s nodes will be selected.

2.5 Mutation

There are two different mutation operators in our CCDECD. For the standard DE used in Step 3 for optimizing the division of subcomponents, the most popular "rand/1" mutation strategy is used [6] since it has no bias to any special search directions.

In Step 5, in order to optimize the division of the global network, we design a novel global network mutation operator: for each population, we randomly select one node i and find all its neighbors. For each node in its neighbors, we randomly assign a probability in the range $[0, 1]$. If the probability of nodes is larger than the mutation rate we predefined, their commIDs will be mutated to the commID of the selected node i. Otherwise, nothing will be changed. This mutation can make use of connectivity information of the network, and thus improve the search ability.

2.6 Clean-Up Step

CCDECD also adopts the clean-up operation proposed by Tasgin and Bingol [14] to correct the mistakes of putting nodes into wrong communities in both mutant and trial vectors and improves the search ability. The clean-up operation is based on the community variance $CV(i)$, which is defined as the fraction of the number of different communities among the node i and its neighbors to the degree of the node i as follows:

$$CV(i) = \frac{\sum_{(i,j) \in E} \text{neq}(i, j)}{\deg(i)}, \tag{2}$$

where $\text{neq}(i, j) = \begin{cases} 1, & \text{if commID}(i) \neq \text{commID}(j) \\ 0, & \text{otherwise} \end{cases}$, $\deg(i)$ is the degree of the ith node, E is the set of edges, and commID is the community containing ith node.

The clean-up step works as follows: Firstly some nodes are randomly selected. Then for each of these nodes i, $CV(i)$ is computed and compared with a threshold which is a predefined constant obtained by experience. If $CV(i)$ is larger than the threshold, the community ID of this node will be assigned to the one which is the most common community ID among the neighbors. Otherwise, no operation is performed on this node.

3 Experiments and Results

In this section, the performance of CCDECD is evaluated on 4 well known real-world social and biological networks. CCDECD is implemented in MATLAB 7.0 and all the experiments are performed on Windows XP SP2 with a Pentium Dual-Core 2.5GHz processor and 2.0GB RAM. The parameters in CCDECD are set as follows: the population size is 30; the maximum number of cycles is $c_{\max} = 100$ and $m = 30$; the mutation rate for the global network mutation operator is set to be 0.2; for the standard DE, the maximum of generations was 30 and for the "rand/1" mutation operator, the scaling factor is $F = 0.9$ and the threshold value is $\eta = 0.32$. The threshold for clean step is set to be 0.35 as used in [14].

For comparison, we implement DECD and another community detection algorithm based on a Genetic Algorithm (GA), named GACD. We adopt the MATLAB Genetic

Algorithm Optimization Toolbox (GAOT) to optimize the network modularity to detect communities in networks. The GA we use is real encoded GA with heuristic crossover and uniform mutation. The values of all the parameters use in the experiments are the default parameters in GAOT. Moreover, for the sake of fairness, the same initialization process and the clean-up operation in CCDECD are employed in the DECD and GACD algorithms. The number of function evaluations of DECD and GACD is set to be the same as CCDECD. We also adopt MATLAB implementations of Girvan-Newman (GN) algorithm [7] from Matlab Tools for Network Analysis (http://www.mit.edu/˜gerganaa) for comparison.

3.1 Datasets

In this paper, we selected the following 5 well known real-world social and biological networks to further verify the performance of CCDECD: 1) the Zachary's Karate Club network; 2) Dolphins network; 3) the American College Football network; 4) Protein and Protein Interaction (PPI) network and 5) Erdös collaboration network.

We selected the above 5 datasets because for small to medium scale datasets 1) to 4), their true community structures are known, which provide gold-standards, e.g., normalized mutual information, for the evaluation of our CCDECD algorithm. We also selected the Erdös collaboration network which is the largest network tested in [9]. The characteristics of the five networks are summarized in Table 1.

Table 1. The characteristics of the five networks tested in the paper. N and M stand for nodes and edges of the network, respectively. Q_{opt} is the known global optimal modularity value.

Dataset	N	M	Q_{opt}
Karate	34	78	0.41979
Dolphins	62	159	0.52852
Football	115	613	0.60457
PPI	1430	6531	–
Erdös	6927	11850	–

3.2 Small Real-World Social Networks

We first validate our algorithm on the small-scale real-wold social networks with true community structure: 1) the Zachary's Karate Club network; 2) Dolphins network; and 3) the American College Football network. As pointed out in [13], performance metrics based on network modularity Q are not always reliable. Therefore, apart from Q, we also adopt normalized mutual information (NMI) as proposed in [2] for performance evaluation.

Since CCDECD, DECD and GACD are stochastic optimization algorithms, we perform the experiments 30 times on these three networks. The average values of Q and NMI, e.g., Q_{avg} and NMI_{avg} and their best values, e.g., Q_{bst} and NMI_{bst}, are compared with that obtained by GN (a deterministic algorithm) from one run of an experiment. We also perform two sample student's t-test between the results obtained from CCDECD and those from other algorithms. The results are presented in Table 2

Table 2. Experimental results of the Zachary's Karate Club, Dolphins and the American College Football networks. N_{pr} is the average predicted number of communities; Q_{avg} and NMI_{avg} are the average values of modularity Q and NMI, respectively. Q_{bst} and NMI_{bst} are the best values of modularity Q and NMI, respectively. The results with asterisks indicate the results are significantly difference from the results obtained from CCDECD.

Network	Algorithm	N_{pr}	Q_{avg}	Q_{bst}	NMI_{avg}	NMI_{bst}
Karate	CCDECD	4.0 ± 0.0	**0.41979 ± 0.00000**	**0.41979**	0.69 ± 0.00	0.69
	DECD	4.1 ± 0.3	$0.41341 \pm 0.00446^*$	**0.41979**	$0.65 \pm 0.06^*$	0.71
	GACD	3.3 ± 0.9	$0.39552 \pm 0.01492^*$	0.41724	0.69 ± 0.10	**0.84**
	GN	2	0.35996	0.35996	**0.84**	0.84
Dolphins	CCDECD	4.1 ± 0.3	**0.52078 ± 0.00026**	**0.52162**	0.80 ± 0.04	0.93
	DECD	4.7 ± 0.8	$0.51557 \pm 0.00374^*$	0.52069	$0.83 \pm 0.05^*$	0.95
	GACD	4.9 ± 0.8	$0.50987 \pm 0.01499^*$	0.51986	**$0.87 \pm 0.07^*$**	**1.00**
	GN	4	0.50823	0.50823	0.84	0.84
Football	CCDECD	10.1 ± 0.7	**0.60382 ± 0.00089**	0.60457	0.89 ± 0.02	**0.93**
	DECD	10.1 ± 0.8	0.60363 ± 0.00071	**0.60457**	0.90 ± 0.02	0.92
	GACD	8.7 ± 1.4	$0.59044 \pm 0.01239^*$	**0.60457**	0.85 ± 0.05	0.93
	GN	12	0.59726	0.59726	**0.93**	0.93

From Table 2, it can be seen that CCDECD performed better than the other three competitors, i.e., DECD, GACD and GN on the three networks. In [9], the author proposed a novel multi-objective genetic algorithm (MOGA-Net) for community detection. The objective is not to maximize modularity but to maximizes the number of connections inside each community and minimizes the number of links between the modules. The average best Q values obtained by MOGA-Net are 0.416, 0.505 and 0.515 for Karate, Dolphin and Football networks, respectively; and the corresponding average NMI values are 0.602, 0.506 and 0.775. The best NMI obtained by MOGA-Net on the Football network is 0.795, even worse than NMI_{avg} obtained by CCDECD. Such results show that maximizing Q with our CCDECD can also achieve better NMI, a gold standard for evaluating CD algorithms, than MOGA-Net.

3.3 Biological Network: Yeast Protein-Protein Interaction Network

We apply our CCDECD algorithm to a biological network, e.g., Yeast Protein-Protein Interaction (PPI) Network [3], which contains 1430 nodes (proteins) and 6531 edges (interactions). We use CYC2008 [10], a complete and up-to-date set of yeast protein complexes (or communities) as a reference set to evaluate the predicted modules by CCDECD. We compute precision, recall and F-measure to measure the performance of CCDECD. The performance of CCDECD is compared with DECD, GACD and GN. We also adopt results from recent literature, e.g., [12] for comparison.

Similar to the experiments in [12], we use the affinity score to decide whether a predicted module is matched with a reference complex:

$$\text{Affinity}(A, B) = \frac{|A \bigcup B|^2}{|A| \times |B|}, \tag{3}$$

where A and B are two modules of proteins, e.g., one of predicted module or reference complexes. We assume a module A matches module B if and only if Affinity(A, B) is above a predefined threshold ω. Then we can define $Hit(\mathcal{A}, \mathcal{B})$ which contains all the matched modules:

$$Hit(\mathcal{A}, \mathcal{B}) = \{A_i \in \mathcal{A} | \text{Affinity}(A_i, B_j) > \omega, \exists B_j \in \mathcal{B}\}. \tag{4}$$

We define precision, recall and F-measure as follows:

$$\text{Recall} = \frac{|Hit(\mathcal{R}, \mathcal{P})|}{|\mathcal{R}|}, \tag{5}$$

$$\text{Percision} = \frac{|Hit(\mathcal{P}, \mathcal{R})|}{|\mathcal{P}|}, \tag{6}$$

$$\text{F-measure} = \frac{2 \times \text{Recall} \times \text{Percision}}{\text{Recall} + \text{Percision}}, \tag{7}$$

where $\mathcal{P}$ is the predicted module set and $\mathcal{R}$ is the reference complex set.

Following the experimental settings in [12], we set $\omega = 0.4$ and 0.5 and select the best results from 30 runs of experiments in order to compare with their algorithms fairly. We compared the results from Critical Module (CM) algorithm proposed in [12]. It is worth mentioning that, due to the large size of the PPI network, the GN algorithm in the Matlab Tools for Network Analysis failed to produce results in reasonable time. Therefore, we adopt the results of the GN algorithm from [12] for comparison.

Table 3. The best results from 30 runs of experiments of the Yeast Protein-Protein Interaction Network

ω	Algorithm	#. pred. complex	Precision	Recall	F-measure
0.4	CCDECD	108	0.5093	**0.3**	**0.3776**
	DECD	143	0.5083	0.2927	0.3715
	GACD	109	0.5046	0.2902	0.3685
	CM	65	**0.5745**	0.0667	0.1195
	GN	65	0.383	0.042	0.0757
0.5	CCDECD	94	0.4681	**0.2683**	**0.3411**
	DECD	115	0.4696	0.2390	0.3168
	GACD	106	0.4340	0.2220	0.2937
	CM	65	**0.6154**	0.0691	0.1241
	GN	65	0.5231	0.0568	0.1025

From Table 3, we can see that compared with other algorithms, CCDECD has better performance. It is interesting to see that, the difference of performance among CCDECD, DECD and GACD is not as significant as those between CCDECD and other non-population-based algorithms, e.g., CM. Such results indicate that, at least for medium size networks, which are commonly seen in biology, population-based algorithms are preferred because of their better search performance.

3.4   Large-Scale Network: Erdös Collaboration Network

In this section, we further evaluate the performance of CCDECD with a large-scale network: Erdös collaboration network. We report the results, e.g., average number of communities and average values of modularity obtained by our CCDECD in comparison with those of DECD, GACD, GN and MOGA-Net [9] in Table 4.

Table 4. Experimental results of Erdös collaboration network. N_{pr} is the average number of communities; Q_{avg} are the average values of modularity Q. The results with asterisks indicate the results are significantly difference from the results obtained by CCDECD.

Algorithm	N_{pr}	Q_{avg}
CCDECD	194.8 ± 17.89	0.6390 ± 0.0042
DECD	407.5 ± 44.92	$0.5598 \pm 0.0095^*$
GACD	277.4 ± 22.47	$0.6070 \pm 0.0108^*$
MOGA-Net	302	0.5502
GN	57	0.6723

Table 4 clearly show that in terms of Q_{avg}, CCDECD performed much better than the other three population-based algorithms. More specifically, in contrast to the results on small scale networks presented in Section 3.2, the performance of CCDECD in terms of Q_{avg} is much better than DECD and GACD, which indicates that CCDECD is more scalable to handle large-scale networks. However, we should notice that, compared with the greedy based GN algorithm, the results of our CCDECD is still not competitive.

4   Conclusion

This paper, for the first time, introduces the Cooperative Co-evolutionary algorithm to detect community structure in complex networks. We have proposed the Bias Grouping scheme to dynamically decompose the complex network into smaller subcomponents for independent and co-adaptive evolution. We have also designed the global network mutation operator specifically for community detection problems which exploits the network connectivity information. We have tested our CCDECD on several benchmark real-world social and biological networks, including the Erdös collaboration network which consists of 6927 nodes, in comparison with DECD, GACD, GN and MOGA-Net algorithms. Apart from the modularity value, for the small scale real-world networks, we have also employed NMI based on true community structure as the performance metric [13]. Compared with other state-of-the-art EACD algorithms, the experimental results have demonstrated that CCDECD is very effective for community detection in complex networks. Compared with greedy based CD algorithms, e.g., GN algorithm, our CCDECD generates more accurate results on small to medium scale networks. However, although it is a step forward, it is still not competitive to handle large-scale network. It will be our future work to incorporate local search algorithm into our CC framework to further improve CCDECD's scalability.

Acknowledgment. Mr Qiang Huang and Dr Shan He would like to thank Paul and Yuanbi Ramsay for their financial support. Mr Thomas White and Dr Shan He are supported by EPSRC (EP/J01446X/1). Dr Ke Tang, Dr Shan He and Professor Xin Yao are supported by an EU FP7-PEOPLE-2009-IRSES project under Nature Inspired Computation and its Applications (NICaiA) (247619).

References

1. Clauset, A., Newman, M.E.J., Moore, C.: Finding community structure in very large networks. Physical Review E 70, 066111 (2004)
2. Danon, L., Guilera, A.D., Duch, J., Arenas, A.: Comparing community structure identification. J. Stat. Mech. (2005)
3. Gavin, A.C., et al.: Proteome survey reveals modularity of the yeast cell machinery. Nature 440, 631–636 (2006)
4. Good, B.H., Montjoye, Y., Clauset, A.: Performance of modularity maximization in practical contexts. Physical Review E 81, 046106 (2010)
5. Jia, G., Cai, Z., Musolesi, M., Wang, Y., Tennant, D.A., Weber, R., Heath, J.K., He, S.: Community detection in social and biological networks using differential evolution. In: Learing and Intelligent OptimizatioN Conference (2012)
6. Neri, F., Tirronen, V.: Recent advances in differential evolution: a survey and experimental analysis. Artificial Intelligence Review 33, 61–106 (2010)
7. Newman, M.E.J.: Fast algorithm for detecting community structure in networks. Physical Review E 69, 026113 (2004)
8. Newman, M.E.J., Girvan, M.: Finding and evaluating community structure in networks. Physical Review E 69, 026113 (2004)
9. Pizzuti, C.: A multiobjective genetic algorithm to find communities in complex network. IEEE Transactions on Evolutionary Computation (2011)
10. Pu, S., Wong, J., Turner, B., Cho, E., Wodak, S.J.: Up-to-date catalogues of yeast protein complexes. Nucleic Acids Res. 37, 825–831 (2009)
11. Scott, J.: Social network analysis: A Handbook. Sage Publications, London (2000)
12. Sohaee, N., Forst, C.V.: Modular clustering of protein-protein interaction networks. In: 2010 IEEE Symposium on Computational Intelligence in Bioinformatics and Computational Biology, CIBCB (2010)
13. Steinhaeuser, K., Chawla, N.V.: Identifying and evaluating community structure in complex networks. Pattern Recognition Letters 31, 413–421 (2009)
14. Tasgin, M., Bingol, H.: Community detection in complex networks using genetic algorithm. In: Proceedings of the European Conference on Complex Systems (2006)
15. Yang, Z., Tang, K., Yao, X.: Large scale evolutionary optimization using cooperative coevolution. Information Sciences 178, 2985–2999 (2008)

On-Line Evolution of Controllers for Aggregating Swarm Robots in Changing Environments

Berend Weel[1], Mark Hoogendoorn[1], and A.E. Eiben[1]

Vrije Universiteit Amsterdam, The Netherlands
{b.weel,m.hoogendoorn,a.e.eiben}@vu.nl

Abstract. One of the grand challenges in self-configurable robotics is to enable robots to change their configuration, autonomously, and in parallel, depending on changes in the environment. In this paper we investigate, in simulation, if this is possible through evolutionary algorithms (EA). To this end, we implement an unconventional on-line, on-board EA that works inside the robots, adapting their controllers to a given environment on-line. This adaptive robot swarm is then exposed to changing circumstances that require that robots aggregate into "organisms" or dis-aggregate into swarm mode again to improve their fitness. The experimental results clearly demonstrate that this EA is capable of adapting the system in real time, without human intervention.

1 Introduction

Within the domain of self-configurable robotics, Stoy and Kurokawa [24] have identified a number of grand challenges, including one that a self-configurable robot *should be able to change its configuration, autonomously, and in parallel, depending on changes in the environment.* This is exactly the problem we address in this paper.

The main assumption and working hypothesis of the present study is that this problem can be solved by using evolutionary algorithms. Therefore, this paper falls in the area of evolutionary robotics, to be more specific in evolutionary swarm robotics, since we consider a swarm of robotic units that can physically aggregate and form a so-called organism, as envisioned by the Symbrion research project [16]. A specific feature of our system, that distinguishes it from the huge majority of related work, is that we use on-line evolution. In most evolutionary robotics systems the robot controllers are evolved off-line, *before* deploying the robots in some operational environment, cf. [20]. In contrast, we apply on-line evolution, *after* deployment, during the operational period of the robots. This feature is essential for robotic systems that are requested to operate long periods without direct human intervention, possibly in unforeseen and dynamically changing environments [19]. Our previous work has addressed the issue of self-driven aggregation and we have shown that even light environmental pressure is sufficient for the on-line evolution of aggregated organisms [27]. In this paper we switch from a static environment, as used in [27], to a dynamically changing one. The main research question is:

C.A. Coello Coello et al. (Eds.): PPSN 2012, Part II, LNCS 7492, pp. 245–254, 2012.
© Springer-Verlag Berlin Heidelberg 2012

Is our on-line evolutionary capable of repeatedly re-adapting the robot controllers if the circumstances change?

To find an answer to this question we design three different environments. One where aggregated organisms have an advantage, one where they have a disadvantage, and one that is neutral from this perspective. Then we expose a group of 50 robots to a scenario where the environment repeatedly changes and try to find out whether the organisms can adapt their sizes appropriately. To this end, there are two important things to note. Firstly, that the behaviour of organisms is the result of the behaviour of the individual robots that form their cells. Secondly, that robot controllers can only change through evolution and we do NOT use any specific fitness function to reward aggregation or disaggregation, only environmental selection.

2   Related Work

A seemingly related area of existing work is that of evolutionary optimisation in dynamic environments [4,18]. Our kind of on-line on-board evolutionary algorithms are similar to this because the actual (on-line) performance is more important than the end result (off-line performance). However, we are working with robots whose controllers need to be evolved on-the-fly (in vivo). Here lies a big difference: in our application one cannot afford bad candidate solutions, because they could "kill" the given robot, while in a usual EA bad individuals merely slow down the search.

Our work is related to both swarm robotics and self-reconfigurable modular robot systems. Swarm Robotics [17] is a field that stems from Swarm Intelligence [3], where swarm-robots often have the ability for physical self-assembly. Swarm-bots were created in order to provide a system which was robust towards hardware failures, versatile in performing different tasks, and navigating different environments. Similarly, self-reconfigurable modular robot systems were designed with three key motivations: versatility, robustness and low cost. The first two are identical to motivations for swarm-robots, while low cost can be achieved through economy of scales and mass production as these systems use many identical modules. Yim gives an overview of self-reconfigurable modular robot systems in [29], the research is mainly on creation of modules in hardware and showcasing their abilities to reconfigure and lift other modules. For our research, we assume a self-reconfigurable robot system which is independently mobile, as reported in [12,14,28]. The task of multiple robots connecting autonomously is usually called self-assembly, and has been demonstrated in several cases: [7,21,28,30]. Most of these however, are limited to pre-programmed control sequences without any evolution. In self-reconfigurable robots, self-assembly is restricted to the docking of two modules as demonstrated in [14,28].

On-Line On-Board evolution is a relatively new field in evolutionary robotics, initiated by the seminal paper of Watson *et al.* [26] who present a system where a population of physical robots (i.e. their controllers) autonomously evolves while situated in their task environment. Since then the area of on-line on-board evolution in swarm-robotics, as classified in [6], has gained a lot of momentum [5,10,12,13,22,25].

The work in this paper is part of the SYMBRION/REPLICATOR projects in which robots are being developed and used that are independently mobile and can operate as a swarm, but also have a mechanical docking mechanism allowing the modules to form and control a multi-robot organism [12]. The most closely related existing work is that of [2,7,8] that explores self-assembly of swarm robots. The controllers of the so-called s-bots (Recurrent Neural Networks) were evolved off-line in simulation, and deployed and tested in real s-bots afterwards. That research shows it is possible to evolve controllers which create organisms. Our present work is to demonstrate that it can be done through on-line evolution as a response to environmental changes.

3   System Description and Experimental Setup

As explained in the Introduction, we design three different environments. One where aggregated organisms have an advantage, one where they have a disadvantage, and one that is neutral from this perspective. Then we expose a group of 50 robots to a scenario where the environment repeatedly changes and watch whether they can adapt appropriately. In this section we describe the details.

Fig. 1. Overview of the final arena, consisting of neutral (white), organism-friendly (light-blue), and organism-unfriendly terrains

Arena. The main idea behind our implementation is to relate the environmental (dis)advantage of organisms to their ability to move and to use different terrains. To be specific, we add a "basic instinct" to the robots to move eastwards (from left to right in our arena) by defining their fitness through their positions: the more they move to the right during evaluation, the higher. Then we create three terrains that differ in their organism-friendliness. In the organism friendly terrain single robots cannot progress to the right and the speed of a larger organism is higher. Metaphorically speaking we have river with a west-bound current here, where only multi-cellular organism have the strength to swim eastwards. We make the organism unfriendly terrain by laying out narrow pathways where big organisms get stuck. The neutral terrain imposes no minimum nor maximum organism size. These three terrains are laid side by side and the resulting composed field is repeated three times in order to increase the number of environmental changes in one run, the resulting arena is shown in Fig. 1. Note that this arena is suited to test the populations response to changes, because robots are driven to move to the right by the fitness function. However, this fitness is certainly does not provide a specific reward for aggregating behaviour, thus it does not represent "cheating". In the meanwhile, it provides a well defined measure to assess success of robot behaviours: the more to the right at the end of an evaluation, the better.

Robots. We conduct our experiments with simulated e-puck robots in a simple 2D simulator: RoboRobo[1]. The robots can steer by setting their desired left and right wheel speeds. Each robot has 8 sensors to detect obstacles (static obstacles as well as other robots), as well as 8 sensors to detect the the river-like zones.

Connections. In our experiments robots can create new organisms, join an already existing organism, and two existing organisms can merge into a larger organism. When working with real robots, creating a physical connection between two robots can be challenging, and movements of joints are noisy because of actuator idiosyncrasies, flexibility of materials used, and sensor noise. We choose to disregard these issues and create a very simple connection mechanism which is rigid the moment a connection is made. The connection is modelled as a magnetic slip-ring, which a robot can set to 'positive', 'negative' or 'neutral'. When robots are close enough, they automatically create a rigid connection if both have their ring on the 'positive' setting. The connection remains in place as long a neither sets its slip-ring to 'negative'. Thus, a positive-neutral combination is not sufficient to establish a new connection, but it is sufficient to maintain an existing one. The neutral setting is important in this experiment to allow for organisms to maintain a certain size, as it allows connections to be maintained without creating new ones.

Controller. The controller is a feed-forward artificial neural network that selects one of 5 pre-programmed strategies based on sensory inputs. The neural net has 20 inputs (cf. Table 1), 8 outputs and no hidden nodes. It uses a *tanh* activation function. The inputs are normalised between 0 and 1.

The output of the neural network, as described in Table 1, is interpreted as follows: the first five outputs each vote for an action, the action with the highest activation level is selected. The sixth output describes the desired organism size which is used when the 'form organism' strategy is chosen. The seventh output describes the direction the robot should move in when performing the 'move' strategy. The eighth output is the desired speed the controller wants to move in, and is used in all strategies except 'halt' (which sets speed to 0).

Evolutionary Algorithm and Runs. We use an on-line on-board hybrid evolutionary mechanism. The first constituent of the hybrid is the $(\mu + 1)$ ON-LINE [9] method, where each robot is an island with a population of μ individuals (genotypes encoding possible controllers) that undergo evolution locally [1]. The other component is the peer-to-peer protocol based EVAG method [15]. The hybridised algorithm as described in detail in [11] also allows recombination across all robots in a panmictic overlay topology.

---

[1] http://www.lri.fr/~bredeche/roborobo/

Table 1. Neural Network inputs (left) and outputs (right)

8x Obstacle distance sensors	Vote for Form Organism
8x Zone distance sensors	Vote for Leave Organism
1x Size of the organism	Vote for Halt
1x Angle to the end	Vote for Avoid Obstacles
1x Distance to the end	Vote for Move
1x Bias node	Desired organism size
	Desired direction for Move
	Desired speed for Move

To represent robot controllers we use a genome which directly encodes the N weights of the neural net using a real-valued vector of length N. This genome is extended to include N mutation step sizes ($\sigma's$) for these N genes. Mutation is a standard Gaussian perturbation with noise drawn from $N(0, \sigma)$ using self-adaptation of $\sigma's$ through the standard formula's. For recombination we use averaging crossover. As for selection, we have a mixed system of global parent selection and local survivor selection. That is, parents are selected using a binary tournament over all genomes in all robots. Once the parents create a newborn controller its fitness is assessed by allowing it to control the robot for 1000 time steps: first a 'free' phase of 200 time steps to allow it to get out of bad situations, followed by an evaluation period for 800 time steps. Each 1000 time steps therefore constitutes 1 generation. At the start of a generation a choice is randomly made between creating a new controller as described above, or choosing an existing controller for re-evaluation, the chance of re-evaluating is controlled by the re-evaluation rate. At the end of the evaluation cycle the given controller is compared to the local population of μ others and replaces the worst one if it is better.

We ran the experiment using 50 robots, we used this number to have a relatively large amount of robots, while not over-crowding the starting area. Too many robots in the start area could lead to an inability of a controller to perform its otherwise good behaviour by getting stuck behind bad controllers.

We used the parameters shown in Table 2 for our evolutionary algorithm and repeated the experiment 50 times, each run lasting 2000 generations. The parameter settings are based on parameters found in our earlier paper [27] in which we used the BONESA toolbox[2] [23] to optimise settings for crossover rate, mutation rate, initial mutation step size, re-

Table 2. Parameters

Parameter	Value
Local population size	3
Mutation chance	0.4
Crossover chance	0.05
Re-evaluation rate	0.5
Initial mutation step-size	0.1
Generations	2000

evaluation rate, and population size. Our experiments are fully repeatable, as the source code is available via the web-page of the first author[3].

---

[2] http://sourceforge.net/projects/tuning/

[3] A zip-file can be found at http://www.few.vu.nl/~bwl400/papers/parcours.zip

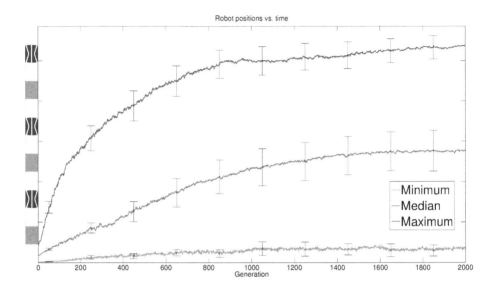

Fig. 2. Robot positions

4   Results and Analysis

This section presents the results of the experiment that have been performed. We essentially want to investigate (1) whether the robots are able to find their way through to the end the obstacle course, (2) analyse *how* they find their way through the obstacle course with respect to the formation of organisms. We will address both questions below.

4.1   Are They Able to Find Their Way?

In order to answer the first question, we have studied the positions of the robots within the obstacle course over time. Hereby, we have taken the position of the best performing robot during each run (i.e. the robot which came closest to the end of the obstacle course), and also recorded the position of the robot closest to the beginning of the course (the worst performing robot). Furthermore, we have taken the position of the median robot. The results averaged over 50 runs are shown in Figure 2. The layout of the obstacle course is shown on the y-axis whereas the x-axis show time (by means of the number of generations).

In the figure, it can be seen that the best robot on average is almost able to complete the entire obstacle course, meaning that it manages to pass the river three times, and ends up in the last narrow passageway. The reason why the best robots on average do not make it all the way to the end is due to the fact that there are some incidental bad runs where the best robot does not even pass the first obstacle.

When considering the worst individual, it can be seen that the worst performing robots hardly progresses within the obstacle course. On average the robots are not able to get beyond the first river which they encounter. In fact, they do

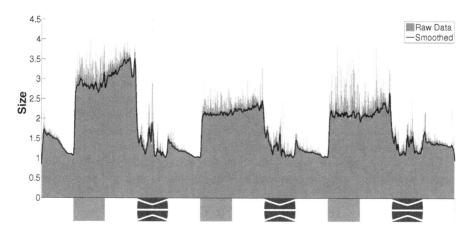

Fig. 3. Organism Size per Position

not even end up at the beginning of the river. The median robots manage to pass the first set of obstacles (the river and the narrow passageway) and are also able to pass the second river.

Overall, it can be concluded that on average a majority of the robots manage to find a way passed the river twice as well as a single passage of the narrow passageway. Some are able to do this three consecutive times. A minority of the robots is hindered too much by the obstacles, resulting in them never passing the first obstacle, namely the river.

4.2   How Do They Find Their Way?

It is interesting to see that the robots learn how to deal with the obstacles, but the question that remains is: how do they achieve it? Do they form organisms? And do they leave organisms? We will try to obtain some insights by studying the behaviour of the robots on a more detailed level. Therefore, we investigate the size of the organisms over the obstacle course to see whether they learn to form an organism and leave it at the appropriate locations.

Figure 3 shows the position in the obstacle course on the x-axis and the average organism size (a one denotes a single agent that is not part of an organism) on the y-axis.

In Fig. 3 can be seen that the average size of the organisms at the river area is a lot higher compared to the narrow passageway. In the neutral territories, the robots tend to continue with the organism size required by the obstacle they encountered last (i.e. after a river they remain within an

Table 3. Mean organism size in different zones. Zones are defined as the start and end of the river/narrow passage.

Zone	X region	Mean	Std
River 1	900–1800	2.76	0.45
Corridors 1	2700–3600	1.31	0.33
River 2	4500–5400	2.07	0.27
Corridors 2	6300–7200	1.28	0.27
River 3	8100–9000	2.07	0.35
Corridors 3	9900–10800	1.30	0.23

organism, and after the narrow passageway they remain single). When looking closer at the behaviour of the robots during the narrow passageway, a spike in the centre of the passageway can be seen. We assume that this is due to the fact that the curve in the passageway is difficult to pass for the robots, and therefore one option for them is to try and form an organism. In the trend of the organism size during the river passage it can be seen that the average size of the individuals is declining a bit after the first river. This because there are simply fewer robots around with which an organism can be formed, resulting in a disadvantage for robots that want to form large organisms.

Table 3 shows more detailed data on the average organism size at the various regions within the obstacle course. The standard deviations are also included.

5   Concluding Remarks and Further Research

In this paper we addressed the challenge of enabling a group of self-configurable robots to adapt their controllers to changing circumstances autonomously, without human intervention. The basic idea behind our approach is to equip the robots with evolutionary operators that keep working, during the operational period of the robots. Our algorithmic solution combines ideas from island-based EAs [1] and peer-to-peer EAs [15], offering –in principle– the best of both worlds.

Our experiments have provided convincing evidence that this approach is capable of evolving the robot controllers in real time and respond to environmental changes, without using a problem-tailored fitness function to "push" some targeted behaviour. Inevitably, we used a number of simplifying assumptions and design decisions in our experimental setup (e.g., using distance from the origin as an abstract measure of fitness), but these did not include any specific bias either. The emerging system behaviour was rooted in the interplay of the evolutionary mechanism and the environmental pressure.

Further work will be carried out in two directions. Firstly, we will explore the niche of applicability of our approach, by testing it in a number of different cases, i.e., in different (changing) environments, with various tasks for aggregated robots and measures of viability (fitness). One of the most interesting questions here concerns the combination of environmental selection (open-ended evolution for pure survival) and human-defined tasks (directed evolution with quantifiable performance measures). Secondly, in close cooperation with roboticists, we will port the whole machinery to real robots to validate its working *in vivo*.

Acknowledgements. This work was made possible by the European Union FET SYMBRION project. The authors would like to thank Evert Haasdijk and our partners in the SYMBRION consortium for many inspirational discussions.

References

1. Araujo, L., Merelo, J.: Diversity through multiculturality: Assessing migrant choice policies in an island model. IEEE Transactions on Evolutionary Computation 15(4), 456 – 469 (2011)

2. Bianco, R., Nolfi, S.: Toward open-ended evolutionary robotics: evolving elementary robotic units able to self-assemble and self-reproduce. Connection Science 16(4), 227–248 (2004)
3. Bonabeau, E., Dorigo, M., Theraulaz, G.: Swarm Intelligence: From Natural to Artificial Systems. Oxford University Press (1999)
4. Branke, J., Kirby, S.: Evolutionary Optimization in Dynamic Environments. Kluwer Academic Publishers, Boston (2001)
5. Bredeche, N., Montanier, J.-M.: Environment-Driven Embodied Evolution in a Population of Autonomous Agents. In: Schaefer, R., Cotta, C., Kołodziej, J., Rudolph, G. (eds.) PPSN XI. LNCS, vol. 6239, pp. 290–299. Springer, Heidelberg (2010)
6. Eiben, A.E., Haasdijk, E., Bredeche, N.: Embodied, on-line, on-board evolution for autonomous robotics. In: Levi, P., Kernbach, S. (eds.) Symbiotic Multi-Robot Organisms: Reliability, Adaptability, Evolution,, ch. 5.2, pp. 361–382. Springer (May 2010)
7. Groß, R., Bonani, M., Mondada, F., Dorigo, M.: Autonomous self-assembly in swarm-bots. IEEE Transactions on Robotics 22, 1115–1130 (2006)
8. Groß, R., Dorigo, M.: Evolution of solitary and group transport behaviors for autonomous robots capable of self-assembling. Adaptive Behavior 16(5), 285 (2008)
9. Haasdijk, E., Eiben, A.E., Karafotias, G.: On-line evolution of robot controllers by an encapsulated evolution strategy. In: Proceedings of the 2010 IEEE Congress on Evolutionary Computation. IEEE Computational Intelligence Society. IEEE Press, Barcelona (2010)
10. Hettiarachchi, S.: Distributed evolution for swarm robotics. ProQuest (2007)
11. Huijsman, R.J., Haasdijk, E., Eiben, A.: An On-line On-board Distributed Algorithm for Evolutionary Robotics. In: Proceedings of the 10th International Conference, Evolution Artificielle, EA 2011. LNCS. Springer (2011)
12. Kernbach, S., Meister, E., Scholz, O., Humza, R., Liedke, J., Rico, L., Jemai, J., Havlik, J., Liu, W.: Evolutionary robotics: The next-generation-platform for online and on-board artificial evolution. In: 2009 IEEE Congress on Evolutionary Computation, pp. 1079–1086 (2009)
13. König, L., Mostaghim, S., Schmeck, H.: Online and onboard evolution of robotic behavior using finite state machines. In: Proceedings of the 8th International Conference on Autonomous Agents and Multiagent Systems, vol. 2, pp. 1325–1326. International Foundation for Autonomous Agents and Multiagent Systems (2009)
14. Kutzer, M.D.M., Moses, M.S., Brown, C.Y., Scheidt, D.H., Chirikjian, G.S., Armand, M.: Design of a new independently-mobile reconfigurable modular robot. In: 2010 IEEE International Conference on Robotics and Automation (ICRA), pp. 2758–2764. IEEE (2010)
15. Laredo, J.L.J., Eiben, A.E., Steen, M., Merelo, J.J.: EvAg: a scalable peer-to-peer evolutionary algorithm. Genetic Programming and Evolvable Machines 11(2), 227–246 (2009)
16. Levi, P., Kernbach, S. (eds.): Symbiotic Multi-Robot Organisms, Cognitive Systems Monographs, vol. 7. Springer, Heidelberg (2010)
17. Mondada, F., Pettinaro, G.C., Guignard, A., Kwee, I.W., Floreano, D., Deneubourg, J.L., Nolfi, S., Gambardella, L.M., Dorigo, M.: Swarm-bot: A new distributed robotic concept. Autonomous Robots 17(2/3), 193–221 (2004)
18. Morrison, R.W.: Designing Evolutionary Algorithms for Dynamic Environments. Springer (2004)
19. Nelson, A.L., Barlow, G.J., Doitsidis, L.: Fitness functions in evolutionary robotics: A survey and analysis. Robotics and Autonomous Systems 57(4), 345–370 (2009)

20. Nolfi, S., Floreano, D.: Evolutionary Robotics: The Biology, Intelligence, and Technology of Self-Organizing Machines. MIT Press, Cambridge (2000)
21. O'Grady, R., Christensen, A.L., Dorigo, M.: Autonomous Reconfiguration in a Self-assembling Multi-robot System. In: Dorigo, M., Birattari, M., Blum, C., Clerc, M., Stützle, T., Winfield, A.F.T. (eds.) ANTS 2008. LNCS, vol. 5217, pp. 259–266. Springer, Heidelberg (2008)
22. Schwarzer, C., Schlachter, F., Michiels, N.: Online evolution in dynamic environments using neural networks in autonomous robots. International Journal On Advances in Intelligent Systems 4(3-4), 288–298 (2012)
23. Smit, S.K., Eiben, A.E.: Multi-problem parameter tuning using BONESA. In: Hao, J., Legrand, P., Collet, P., Monmarché, N., Lutton, E., Schoenauer, M. (eds.) Proceedings of Artificial Evolution, 10th International Conference, Evolution Artificielle (EA 2011), pp. 222–233 (2011)
24. Stoy, K., Kurokawa, H.: Current topics in classic self-reconfigurable robot research. In: Proceedings of the 2011 IROS Workshop (SW9), Reconfigurable Modular Robotics: Challenges of Mechatronic and Bio-Chemo-Hybrid Systems (2011)
25. Szymanski, M., Winkler, L., Laneri, D., Schlachter, F., Van Rossum, A., Schmickl, T., Thenius, R.: Symbricatorrtos: a flexible and dynamic framework for bio-inspired robot control systems and evolution. In: IEEE Congress on Evolutionary Computation, CEC 2009, pp. 3314–3321. IEEE (2009)
26. Watson, R.A., Ficici, S.G., Pollack, J.B.: Embodied evolution: Distributing an evolutionary algorithm in a population of robots. Robotics and Autonomous Systems 39(1), 1–18 (2002)
27. Weel, B., Haasdijk, E., Eiben, A.E.: The Emergence of Multi-cellular Robot Organisms through On-Line On-Board Evolution. In: Di Chio, C., et al. (eds.) EvoApplications 2012. LNCS, vol. 7248, pp. 124–134. Springer, Heidelberg (2012)
28. Wei, H., Cai, Y., Li, H., Li, D., Wang, T.: Sambot: A self-assembly modular robot for swarm robot. In: 2010 IEEE International Conference on Robotics and Automation (ICRA), pp. 66–71. IEEE (2010)
29. Yim, M., Shen, W.M., Salemi, B., Rus, D., Moll, M., Lipson, H., Klavins, E., Chirikjian, G.S.: Modular self-reconfigurable robot systems [grand challenges of robotics]. IEEE Robotics & Automation Magazine 14(1), 43–52 (2007)
30. Yim, M., Shirmohammadi, B., Sastra, J., Park, M., Dugan, M., Taylor, C.: Towards robotic self-reassembly after explosion. In: 2007 IEEE/RSJ International Conference on Intelligent Robots and Systems, pp. 2767–2772. IEEE (2007)

Buildable Objects Revisited

Martin Waßmann and Karsten Weicker

HTWK Leipzig, IMN, Postfach 301166, 04251 Leipzig, Germany
http://portal.imn.htwk-leipzig.de/fakultaet/weicker/

Abstract. Funes and Pollack introduced the The buildable objects experiments where LEGO® structures are evolved that may carry loads. This paper re-evaluates and extends the approach. We propose a new evaluation scheme using maximum network flow algorithms and a graph representation. The obtained structures excel previous results. Furthermore, we are now able to address problems that require more than one bearing.

Keywords: Buildable objects, physics simulation, evolutionary design.

1 Introduction

The evolution of structures with a load-carrying capacity, e.g., structural frame design or whole buildings, is an appealing part of evolutionary design. Since the work of Funes and Pollack [3–5], the evolution of LEGO® structures serves as an interesting application domain and research playground.

Although the problem is present for 15 years, there are still two major challenges:

- The stability of LEGO® structures is difficult to determine because the physics of structural frame design cannot be applied. And dynamic physics engine apply the various forces subsequently and isolated from each other which often leads to inexactness. A more precise approximation of the forces within a structure could improve the results considerably.
- Funes and Pollack [4] call their representation of the structure, a tree, underconstrained because it induces several problems, e.g., the overlapping of tree branches.

Also, if we consider problems with multiple bearings, the tree is not suited and it is unclear how well the existing evaluation of the stability works.

As a consequence, we present a graph-based representation together with a new technique for approximating the stability using flow networks based on the findings in the master's thesis [9]. However, the focus is on the representation and the results for a crane and a bridge problem with multiple bearings.

2 Related Work

There are two application domains of EAs using LEGO® bricks. First, Funes and Pollack [3–5] construct structures like cranes that are required to be stable in the

C.A. Coello Coello et al. (Eds.): PPSN 2012, Part II, LNCS 7492, pp. 255–265, 2012.
© Springer-Verlag Berlin Heidelberg 2012

presence of external forces. They use a tree-based representation and a repair function. However, since certain connections between bricks are not represented by the tree, the operators cannot use this information. Moreover, the approach is limited to single bearings. In [5], 3D models are constructed. Second, Petrovic [8] constructs models that mimic the shape of real-world objects. He uses a direct representation in form of a list of bricks (with the respective coordinates) – external forces are not considered.

Another relevant publication is the work of Devert et al. [2] who construct structures using "toy bricks" without considering pinches for connecting bricks.

The problem at hand appears to be closely related to structural frame design for which a comprehensive survey is available [7]. However, the very specific connection mechanism of LEGO® bricks inhibit re-using ideas from those approaches for both constructing and evaluating structures.

Related to our representation are research projects that evolve graphs, e.g. [6]. However, as we will see later, the graphs in our representation are very restricted such that knowledge transfer is difficult.

3   Evaluating Structures of Buildable Objects

The quality of a LEGO® structure depends on the evaluation mechanism, i.e., the computation of the forces within the structure. A graph model of the structure, e.g. in Fig. 1, enables the evaluation similarly to [3, 4].

Fig. 2 shows the forces that affect the stability of a LEGO® structure.

Mere compressive forces do not affect the stability of structures. Tensile forces might excess the adhesion of two connected bricks. But the moments are the primary cause for instable LEGO® structures when they exceed the capacity of a joint of two bricks. Basically, the forces are caused by the weight of each single brick as well as external loads.

Funes and Pollack [3–5] computed the stability by isolating forces and moments. For each force, the resulting moment is computed at each joint which leads to a flow network of moments and counteracting moments that are propagated towardsthe bearings. At each joint, the sum of all resulting moments for all forces must respect the moment capacity of the joint, i.e. it is a multi-commodity network flow problem. However, this approach overestimates the moments within the structure since the isolation of forces (and resulting moments) does not consider compensation of opposing moments.

Fig. 1. Modelling a LEGO® structure: the placement of the bricks (left) and the corresponding graph where in each vertex the first two values are the position and the third value is the size of the brick (right)

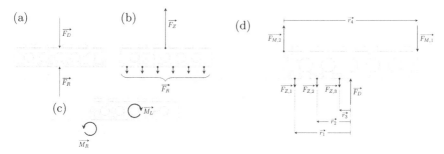

Fig. 2. Forces between bricks of a structure: (a) compressive force, (b) tensile force, and (c) moment. Subfigure (d) shows the moments from (c) as pairs of forces.

Our method for evaluating the stability of a structure follows the approach of classical statics. A system of equations is constructed, where the equilibrium of forces and moments for each brick is described by equations. Additional inequalities guarantee that no connection between two bricks is stressed beyond it's capacity. This system of equations can be solved if for each brick the external loads, e.g. gravitational forces or resulting moments, are counteracted by an equal and opposite reaction.

In general, such a system cannot be solved trivially as it may be underdetermined. We have investigated two techniques to solve such a system. First, constraint satisfaction solvers have been tried. However, the asymptotic runtime is exponential with the number of forces which leads to an unacceptable runtime. Instead, the system of equations is turned into maximum flow problems which can be solved in polynomial time using the push-relabel algorithm [1].

In a first phase, we consider only the compressive and tensile forces without moments – as if each brick is fixed in its orientation and cannot be tilted. For each brick the forces and counteracting forces are considered and a flow network is constructed that guarantees that all forces and counteracting forces are balanced. The maximum flow represents a possible distribution of the forces in the LEGO® structure – resulting in the effective forces for each single brick. If the tensile force exceeds the respective capacity of a joint the structure is not stable.

In a second phase, for each brick, the moments are computed from the effective forces. The moments need to be balanced too – modeled as a flow network with moments and counteracting moments. If the resulting effective moments exceed the capacity of a joint the structure is not stable.

The two phased approach allows us to use single commodity flow networks. However the approach has the disadvantage, that forces and moments are distributed successively and only one force distribution is calculated. Hence a *bad* force distribution may lead to an impossible moment distribution. The structure is declared not stable, even though there might exist a force distribution for which a moment distribution is possible. The approach is decribed in more detail in [10].

Exemplarily, Fig. 3 shows a simple LEGO® structure and the flow network model for distributing vertical forces. The source vertex s supplies the forces due

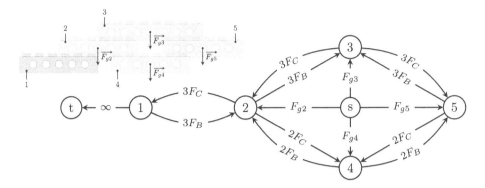

Fig. 3. A simple structure and the flow network modelling the forces with maximal compressive force F_C and maximal tensile force F_B

to the bricks' weight and the target vertex t is connected to the bearing. Each vertex represents a brick and each pair of edges between two vertices a joint of two bricks. Throughout the experiments, we use the maximal compressive force $F_c = 5.44$ kN and the maximal tensile force $F_B = 2.5$ N.

This approach has the advantage that we can consider problems with an arbitrary number of bearings. However, the computation of the moments is still an approximation since the distribution of forces might be unrealistic. Moreover, additional stabilizing factors, e.g., if bricks are placed side by side, are not considered. The maximal possible load for a structure is determined by binary search and iterative solution of the networks.

4   Concepts of the Evolutionary Algorithm

Given the expensive evaluation function described in the previous section, we decided to use a steady-state genetic algorithm to make the newly generated individual immediately available (similarly to [3]). In each generation one individual is produced with the following procedure:

1. select two parents with rank-based fitness-proportional selection
2. apply the recombination operator; if the new individual is invalid, use the first parent
3. apply the mutation operator; if the new individual is invalid, use the recombinant

The new individual replaces the worst individual in the population.

The LEGO® structures are represented directly as graphs like in Fig. 1. Such a graph $G = (B \cup L, E)$ contains the vertices B (placeable bricks) and L (bearings). Both are annotated by the position and the size of the brick. The edges in E reflect the connections between bricks. We consider an individual to be *legal* iff the bricks do not overlap pairwise and the represented structure is stable.

For recombination, one of the following operations is applied to the parents' brick sets B_1 and B_2.

horizontal cut: A brick $b \in B_1$ is chosen uniformly. The new individual contains all bricks $\{a \in B_1 | a_y \leq b_y\} \cup \{a \in B_2 | a_y > b_y\}$ where c_y denotes the y-coordinate of a brick c.

vertical cut: A brick $b \in B_1$ is chosen uniformly. The new individuals contains all bricks $a \in B_1$ with $a_x \leq b_x$ where c_x denotes the x-coordinate of the left end of brick c. Furthermore, all bricks $a \in B_2$ are added that fulfil $a_x > b_x$ and do not overlap with the bricks selected from B_1.

mixing: A random set of bricks $C \subseteq B_1$ is chosen. Let $D \subseteq B_2$ be the set of all bricks, that do not overlap with the bricks in C. The new individual contains $C \cup D$.

Because mixing is more disruptive, we use the probability 0.2 for mixing and 0.4 for the other two operators.

The mutation operator applies one of the following operations to the individual's brick set B.

addition: A brick $a \in B$ with size $(t, 1)$ and a new brick b with size $(t', 1)$ is chosen. The new brick is placed at position $(a_x + z_1, a_y + z_2)$ with random integer numbers $1 - t' \leq z_1 < t$ and $z_2 \in \{-1, 1\}$.

deletion: A brick $a \in B$ is removed.

shifting: A brick $a \in B$ with size $(t, 1)$ is moved to the new position $(a_x + z, a_y)$ with random number $1 - t \leq z < t$.

replacement: A brick $a \in B$ with size $(t, 1)$ is replaced by a new brick b of size $(t', 1)$ where the centre of mass is kept unchanged, i.e., b is positioned at $(a_x + \lfloor \frac{t-t'}{2} \rfloor, a_y)$.

exchange: Two bricks $a, b \in B$ are chosen and exchanged aligned according to the left end of the bricks.

shifting partial structures: A random brick $a \in B$ is chosen as well as a random number $z \sim \mathcal{N}(0, 1)$. All bricks $b \in B$ with $b_y > a_y$ are moved by $\lfloor z + \frac{1}{2} \rfloor$ along the x-axis.

The operations are applied with probability 0.1 (first two) and 0.2 (rest).

In our experiments, we start with a population of empty individuals that contain only the given bearings. As a consequence, the evolution process focuses on growing functional structures during the early phases. More sophisticated initialisation procedures have not been investigated in detail.

5 Resulting Structures

Using our algorithm, we investigated two problems classes – the classical crane problem as in [3] and a new bridge problem, which is distinct from the bridge in [3] because we consider more than one bearing.

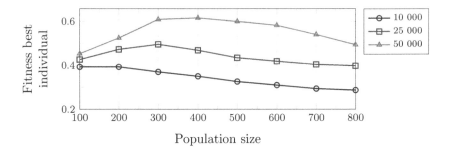

Fig. 4. Calibrating the population size: final fitness after 10^4, $2.5 \cdot 10^4$, and $5 \cdot 10^4$ generations

5.1 Crane Problem

In the crane problem, a fixed bearing is given and the aim is to produce a structure that is able to carry a weight $load^* \in \mathbb{R}^+$ at a distance $length^* \in \mathbb{R}^+$ from the bearing. The tuple $(length^*, load^*)$ describes an instance of the problem class with which we can adjust whether we want short structures that can carry heavy loads or more outreaching structures.

The fitness of such a LEGO® structure M is measured using the function

$$f_{\text{crane}}(M) = \left(\min \left\{ 1, \frac{length(M)}{length^*} \right\} \right)^s \cdot \min \left\{ 1, \frac{load(M)}{load^*} \right\} \cdot (1 - c \cdot size(M))$$

where $length(M)$ is the actual length of M, $size(M)$ the number of bricks, and $load(M)$ is the resulting maximum load using the physics simulation. Moreover, s is a parameter to put more emphasis on the length of the structures (which seems to be more difficult to evolve); the parameter c controls to what extent the number of bricks should be minimised.

All parameters have been investigated thoroughly for the crane problem. Fig. 4 shows how the final fitness changes with varying population size. A too big population size exhibits problems due to missing convergence. As a consequence, we used a population with 400 individuals and 50,000 generations.

Even more interesting are the parameters s and c to modify the fitness function. Fig. 5 shows how length and load capacity change for three different values of s. The inverse direction of the curves in the two subfigures shows that s is a proper means to control the focus of the optimisation. However, as Fig. 6 demonstrates, there is not a unique scale for the different problem instances.

Parameter c controls the impact of the number of bricks onto the fitness function. Fig. 7 shows that values between 10^{-4} and 10^{-3} decrease the number of bricks considerably with only little loss in length and load capacity.

Two of our results are shown in Figures 8 and 9. The crane for $(1\,m, 0.25\,kg)$ was created using $s = 1.4$, $c = 10^{-9}$, and 50,000 generations. Remarkable about these crane arms is, besides achieved length and load-carrying capacity, their quality. The graph overlay in Fig. 8 shows the regular structure of the load arm, that allows the distribution of the applied load along as much as possible

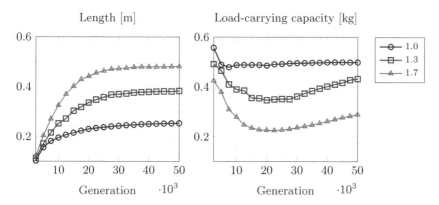

Fig. 5. Calibrating the factor s for the problem instance $(0.5m, 0.5kg)$: length (left) and load capacity (right) for three values for s

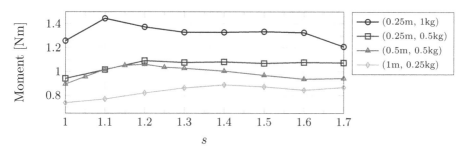

Fig. 6. Calibrating the factor s for other problem instances: the maximum momentum for four problem instances and various values for s

bricks, reducing the strain per joint. Furthermore a counter balance is used to counteract the moment created by the load. These two features are reoccurring – depending on the parameter c, which inhibits the creation of a counter balance – and distinguish our solutions for the crane problem from those in the literature. The structure left of the bearing is randomly shaped since it only serves as a counter balance. The crane for $(0.5 \text{ m}, 0.5 \text{ kg})$ is the result of an experiment using $c = 10^{-3.5}$.

Table 1 compares the results of the Funes/Pollack approach and our results in load and length. Although a comparison is difficult, the results of the new approach appear to outperform the older results in both load and length.

5.2   Bridge Problem

The bridge problem requires the evolution to create a supporting structure between the bearings and the surface of the bridge where each brick is required to support a given load.

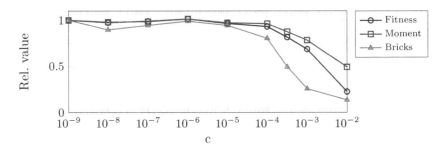

Fig. 7. Calibrating the factor c: normalised values for the number of bricks, the length, and the load capacity after 10^4 generations

Table 1. Comparison of the constructed structures

Funes/Pollack		Waßmann/Weicker		
length	load	length	load	problem instance
0.5 m	0.148 kg	0.496 m	0.482 kg	(0.5 m, 0.5 kg)
0.16 m	0.5 kg	0.256 m	1.0 kg	(0.25 m, 1 kg)
		0.664 m	0.236 kg	(1 m, 0.25 kg)
		0.296 m	0.41 kg	(0.5 m, 0.5 kg)

The fitness function is designed in such a way that the evolution focusses on connecting all parts of the bridge first:

$$f_{\text{bridge}}(M) = \begin{cases} \frac{1}{1+f_{\text{con}}(M')}, & \text{if } M' \text{ is not connected} \\ 1 + p_{\max}(M'')^{\frac{1}{1+\delta}} \cdot c^{size(M)}, & \text{otherwise} \end{cases}$$

where M' is the structure extended by the road surface as mounting, $f_{\text{con}}(M')$ is the minimal distance between the connected components of the graph including the mounting bricks. M'' is constructed by adding iteratively those bricks of the surface for which the forces are supported by the LEGO® structure – the value δ corresponds to the number of unsupported bricks. The factor c controls again the number of involved bricks – in the following experiments $c = 0.995$ was used.

A first scenario with three bearings is shown in Fig. 10 together with one of our first results. However, the bridge problem has not been analysed in the same depth as the crane problem. The structure in the lower left region serves as a counter balance and moves the centre of mass above the central bearing.

The second scenario (Fig. 11) has four bearings and requires a free space between the two central supports.

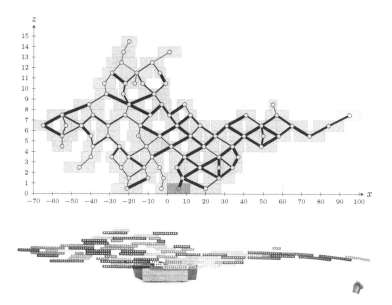

Fig. 8. Result for problem instance (1 m, 0.25 kg): length 0.664 m and load 0.236 kg

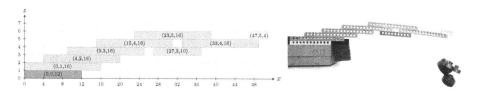

Fig. 9. Result for problem instance (0.5 m, 0.5 kg): length 0.296 m and load 0.41 kg

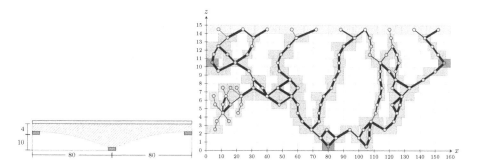

Fig. 10. First bridge problem

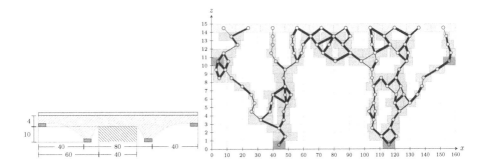

Fig. 11. Second bridge problem

6   Conclusion

The new algorithm convinces in two respects, the quality of the exemplary structures as well as the reliability with which competitive structures are produced – as the calibration results demonstrate. This is due to (a) a representation and respective operators that are designed specifically for the problem at hand and (b) a more exact evaluation mechanism for the structures.[1]

Where the crane problem has been investigated thoroughly, we presented only few preliminary results for the bridge problem with multiple bearings. Future research should focus on this problem class as well as 3D structures for which we tested our approach already successfully (not reported here). Furthermore, the computation of the stability could be extended to handle external horizontal forces like it would be necessary for tower structures.

References

1. Cherkassky, B.V., Goldberg, A.V.: On implementing the push-relabel method for the maximum flow problem. Algorithmica 19, 390–410 (1997)
2. Devert, A., Bredeche, N., Schoenauer, M.: Evolution design of buildable objects with blind builder: an empirical study. In: Pham, T.L., Le, H.K., Nguyen, X.H. (eds.) Proc. of the Third Asian-Pacific Workshop on Genetic Programming, pp. 98–109. Military Technical Academy, Hanoi (2006)
3. Funes, P., Pollack, J.B.: Computer evolution of buildable objects. In: Husbands, P., Harvey, I. (eds.) Fourth European Conf. on Artificial Life, pp. 358–367. MIT Press, Cambridge (1997)
4. Funes, P., Pollack, J.B.: Computer evolution of buildable objects. In: Bentley, P.J. (ed.) Evolutionary Design by Computers, pp. 387–403. Morgan Kaufmann, San Francisco (1999)
5. Funes, P.J.: Buildable evolution. SIGEVOlution 2(3), 6–19 (2007)
6. Globus, A., Lawton, J., Wipke, T.: Automatic molecular design using evolutionary techniques. Tech. Rep. NAS-99-005, NASA Ames Research Center (1999)

---

[1] The source code is available for download at
http://portal.imn.htwk-leipzig.de/fakultaet/weicker/forschung/download

7. Kicinger, R., Arciszewski, T., De Jong, K.A.: Evolutionary computation and structural design: a survey of the state of the art. Computers & Structures 83(23-24), 1943–1978 (2005)
8. Petrovic, P.: Solving lego brick layout problem using evolutionary algorithms. In: Norsk Informatikkonferanse (NIK 2001), pp. 87–97 (2001)
9. Waßmann, M.: Physiksimulation und Evolution von LEGO®-Strukturen. Master's thesis, HTWK Leipzig, Leipzig, Germany (2011)
10. Waßmann, M., Weicker, K.: Maximum flow networks for stability analysis of LEGO® structures (2012), under submission (ESA 2012)

Collective Robot Navigation
Using Diffusion Limited Aggregation

Jonathan Mullins[1], Bernd Meyer[1], and Aiguo Patrick Hu[2]

[1] Centre for Research in Intelligent Systems
Monash University, Australia
[2] Department of Electrical and Computer Engineering
University of Auckland, New Zealand

Abstract. Many applications of swarm robotics require autonomous navigation in unknown environments. We describe a new collective navigation strategy based on diffusion limited aggregation and bacterial foraging behaviour. Both methods are suitable for typical swarm robots as they require only minimal sensory and control capabilities. We demonstrate the usefulness of the strategy with a swarm that is capable of autonomously finding charging stations and show that the collective search can be significantly more effective than individual-based search.

1 Introduction

Swarm robotics is becoming an increasingly active field of research. This is unsurprising, as deploying a swarm of simple, small, and inexpensive robots can present an extremely attractive alternative to the use of a single complex and costly robot in a significant number of application scenarios. Clearly, there are situations where a larger robot may not be able to operate effectively at all, for example in space constrained areas under collapsed buildings in the aftermath of an earthquake. Here, a swarm of small robots would be far more effective in sifting through the rubble and exploring every small cavity. Robot swarms are also generally thought to be more resistant to damage and disruption, and to be more resilient in the face of changing environment conditions.

For robots used in tasks such as disaster response, space exploration, or environmental tracking it is immediately obvious that robustness and adaptivity are core requirements. A case in point is the proposed NASA mission PAM (Prospecting Asteroid Mission) [6]: it aims to deploy a swarm of approximately $1,000$ pico-spacecraft to explore the asteroid belt. In the asteroid belt it is not unlikely for a spacecraft to be hit by another object, so that a mission relying on a single complex spacecraft could easily fail. A self-organising swarm could be more resilient and also help to address the challenge of delayed communication by performing time-critical behaviour changes autonomously.

While swarm robotics has become a very active field of research, its real-world applications have been limited so far. Limited battery power, poor communication facilities, minimal sensory equipment and low computational processing capacity pose significant challenges, as does the design of distributed algorithms

C.A. Coello Coello et al. (Eds.): PPSN 2012, Part II, LNCS 7492, pp. 266–276, 2012.

that reliably produce a desired collective behaviour. Typical academic test tasks for self-organised collective control are clustering robots onto a particular position (such as a light source) [5], optimal dispersal of robots to cover a target range [11] [8], collective transport [9], and various forms of structure formations, such a assembling a chain of robots [4].

In this paper, we tackle a collective navigation task that arises in many realistic applications: the guidance of swarm members to a common target. There are uncountable scenarios where finding a particular target is an important sub-task of the swarm's mission. A prime example is a clean-up mission after an (industrial) accident. Typically, more swarm members will have to be guided to the source of contamination once it has been located by one of the swarm's scouts to assist with its removal and clean-up. This is akin to the recruitment mechanisms in social insects that allow them effective exploitation of food sources through collective transport [7].

This task is also required for the construction of our own experimental test bed for swarm robotics. Our aim is to build a robot swarm that will autonomously roam the building of Monash's Computer Science Department to perform continuous inspection. One of the many challenges with this is the limited battery capacity and thus operating time of individual swarm robots.

The e-puck robot that we are using in our experiments [12] is typically not able to operate for more than three hours before having to be recharged. As we are aiming for fully autonomous swarm operation, the first challenge to address was the e-puck's reliance on human intervention for recharging. We overcame this by modifying the e-puck's hardware to allow contact-less inductive charging. With this modification it is sufficient for a robot to drive onto a specifically constructed wireless charging platform and rest there until the batteries are fully recharged. The details of this modification are beyond the scope of this paper and described elsewhere [2].

However, even with autonomous charging being physically possible, the challenge remains for the robot to locate the charging platform before it runs out of batteries. In the absence of perfect knowledge of the environment this clearly requires the swarm to search for the charger.

A strategy for returning the e-puck to the charging platform autonomously was developed in two parts: Firstly, a simple gradient search was implemented on the physical e-puck robot after fitting its charging platform with an audible beacon (Section 3.1). Audio was chosen as a gradient medium both for its relative ease of experimentation and its imperfect gradient field (due to reflections and interference), providing a test-case for other forms of perturbed fields such as chemical gradients. Secondly, a collective navigation strategy was implemented, such that an agent can be guided to the charging platform from beyond audible range with the assistance of the swarm (Section 3.2). To do so, the swarm constructs a space-filling beacon structure in the environment which the searching agent traverses toward the charging platform.[1]

---

[1] For demos see http://www.csse.monash.edu.au/~berndm/autonomous_epuck/

Since we are interested in swarm robots with extremely limited sensory capabilities and processing power, we decided to use control algorithms that are so simple that they can in theory be implemented on devices without any digital computation capabilities whatsoever. To achieve this, we only used two types of nature-inspired behaviours that could be implemented even with just analog control circuits: bacterial gradient search [13] for the individual navigation and Diffusion Limited Aggregation (DLA) [14] for the formation of the collective navigation structure. The conclusion from our preliminary experiments is that even with such extremely simple behaviours effective collective search is possible.

2   Experiment Setup

To conduct the e-puck robot experiments, we constructed a rectangular environment measuring 2.4m x 3.6m with 5cm high walls from MDF and pine (Figure 1). An omni-directional audio beacon was placed midway down the long edge of the environment, 30cm from the wall. Pink noise was emitted from the audio beacon for detection by the robot, which averaged the volume of its three microphones during experiments to minimise rotationally induced bias.

For development of the collective navigation strategy, 60 virtual e-pucks were dispersed in a 10m x 10m virtual environment using the ASEBA Framework [10]. Simulation was used as no e-puck swarm of adequate size was available and a real e-puck swarm of similar size would cost in excess of $50,000. ASEBA provided physically realistic simulation of the e-puck swarm and charging platform, the e-puck model simulating all sensors and motors with the exception of the speaker and microphones (see Figure 5). We ported the Swis2D audio plugin from Webots [3] to ASEBA, which provides 2D audio simulation, to overcome this limitation. Other physical robotics simulators (such as Webots) were considered for the simulation component, however ASEBA was determined to be the most flexible in terms of software customisation and the sharing of control scripts between the real-world and virtual e-pucks.

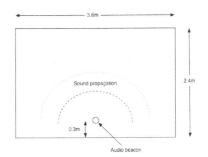

Fig. 1. Layout of the physical environment and audio beacon for conducting e-puck experiments

3   Algorithms

3.1   E-coli Inspired Gradient Search

We developed an audio-based search strategy by adapting the foraging behaviour of *E. coli* bacteria to the capabilities of the e-puck robot. *E. coli* perform a gradient search on the nutrient in their environment in order to move to the most favourable location by alternating between two states: *tumble* and *run* (Figure

2). During a tumble, the bacterium briefly rotates on the spot, randomly picking a new direction to start moving, slightly biased toward the current direction of travel. During a run, the bacterium moves in a relatively straight line for an amount of time, the length of which increases when the bacterium detects that it is moving toward a more favourable nutrient source, or decreases if moving toward a noxious substance. Eventually, convergence on the most favourable location in the environment occurs. The foraging behaviour has previously been applied to distributed function minimisation problems, and identified as a potential search strategy for mobile agents [13]. Note that whilst *E. coli* occur in groups, this foraging behaviour is conducted individually without real interactions, so that this is not a true collective swarming mechanism.

Whereas the *E. coli* bacterium constantly measures the improvement or deterioration in nutrient level while moving through its environment, the motor noise of the e-puck is such that ambient audio can be accurately

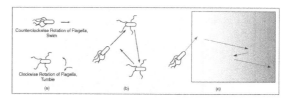

Fig. 2. *E. coli* bacterium foraging behaviour [13]

sampled only when the robot is stationary. Consequently, rather than modulating the length of the current run phase based on the measured gradient, we instead modulate the length of the following run. Also, following a run that results in a volume increase, we select a new direction randomly from a normal distribution around the current heading, otherwise the new direction is selected uniformly random over all directions. The search terminates when both volume and proximity measurements indicate that the target is reached.

Although initial experiments demonstrated the search strategy to be effective, we observed instances where an unfortunate combination of tumbles would result in the robot passing within a few centimetres of the target without acquiring it. To improve the search performance at close proximity to the audio beacon, two further behaviours were activated when the robot measured a volume v greater than some pre-determined thresholds. This threshold V_{warm} is selected sufficiently high to guarantee the robot is near the beacon and not in some local maximum elsewhere. The area in which the robot measures a volume level above this threshold is defined as the *warm zone*. Once the robot enters the warm zone, if a subsequent run phase results in a measurement below this threshold, the robot backtracks to the previous location (Fig. 3a, 3b). The second threshold, V_{hot}, is selected sufficiently high to guarantee the robot is within approximately 5cm of the beacon, allowing for the proximity sensors to be used to steer the robot directly to the target (Fig. 3c, 3d).

Algorithm 3.1 describes the bacterial search adapted for the e-puck robot. $N(a, b)$ is defined as a random number taken from a normal distribution with mean a and standard deviation b. $U(a, b)$ is a random number taken from a uniform distribution between a and b. L_{min} and L_{max} are constants representing minimum and maximum drive lengths, whilst V_{min} and V_{max} are constants representing the noise floor volume of the e-puck microphone and the volume it measures when at the target. v_{last} is the volume measured at the e-puck's previous location. A series of experiments were conducted to quantify the performance of this simple algorithm compared to a random walk baseline. The results and analysis are given in Section 4.

Algorithm 3.1. BACTERIALSEARCH()

---

while target not found

$$\text{do} \begin{cases} \textbf{if } v < V_{warm} < v_{last} \\ \textbf{then } \text{backtrack} \\ \\ \textbf{else} \begin{cases} \textbf{if } V_{min} < v > v_{last} \\ \quad \textbf{then} \begin{cases} \text{rotate } N(0, 60) \\ \text{drive length} = \frac{(L_{max} - L_{min})(v - v_{last})}{V_{max} - V_{min}} \end{cases} \\ \quad \textbf{else} \begin{cases} \text{rotate } U(0, 360) \\ \text{drive length} = V_{min} \end{cases} \\ \textbf{if } v > V_{hot} \\ \quad \textbf{then } \text{drive towards closest object} \\ \quad \textbf{else } \text{drive forward} \end{cases} \end{cases}$$

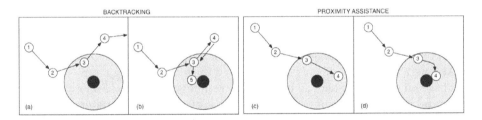

Fig. 3. E-puck searching without backtracking (a) and with backtracking (b); without proximity assistance (c), and with proximity assistance (d)

3.2   DLA Inspired Navigation

The usable area of the single-agent bacterial search is constrained to the area in which the beacon is audible. To extend this area when operating as part of an e-puck swarm, a collective navigation strategy based on DLA was implemented. DLA is a natural process where particles aggregate in a random manner, forming fractal-like tree structures rooted at the starting particle (Figure 4). The process was first described in [14], and examples in nature include dust and snowflake formation, coral growth and the path taken by lightning. The simple, self organised process generates a space-filling structure from a fixed point, making it a suitable approach for our e-puck swarm to construct a traversable structure rooted at the charging platform.

3.3   Implementation

In contrast to the real-world audio beacon which emitted pink noise, the simulated audio beacon emits a single tone, such that it is identifiable as the root of the DLA structure. The collective navigation strategy is initiated when an agent beyond the audible range of the charging platform requires a recharge, and sounds a call for help using a specific frequency. The assisting agents retransmit the call for help throughout the swarm and begin random walking.

Each assistant random walks until in range of a DLA tone (either the signal of the charging platform itself or that of an aggregated agent), at which point it stops (aggregates) and retransmits the detected DLA tone at a slightly higher frequency. The result is a tree of audible nodes with the lowest frequency at the charging platform. Each node remains in this state until the DLA tone it initially detected (its parent) is disabled, at which time the agent disables its own tone and returns to its primary task.

Fig. 4. Simulated DLA structure [1]

The agent requiring recharging continuously performs the bacterial search on the lowest audible DLA tone. As it approaches each non-root node in the structure, the node's parent becomes audible and becomes the new target. When the charging platform detects the agent has boarded, the root DLA tone is disabled, releasing the entire swarm back to its primary task. As the agent traverses the structure, node agents that detect it on its way past (using proximity sensors) disconnect from the structure, as they are no longer required. This causes disconnection of all the node's children, dramatically reducing the amount of total agent time committed to the process. Figure 7 depicts the states and decisions that each agent implements as part of the strategy.

3.4   Synchronisation

Scenarios were observed where one or more assisting agents broke away from the swarm during the *random walk* state, but never aggregated onto the DLA structure before the recharge request was completed. In such cases, once the disconnected agents(s) rejoined, the swarm was incorrectly commanded back into the *random walk* state, even though no agent required recharging.

With no guarantee of a fully connected swarm with respect to audio communication (and thus no inherent temporal ordering of events), this issue was resolved by having each agent maintain a clock, synchronised with the swarm. Alert signals are no longer transmitted on one single frequency, but can fall anywhere in the range A_{lo} to A_{hi}. The frequency of an alert signal is defined as $A_{lo} + t_{alert}$, where t_{alert} is the time the alert was initiated. Each agent also

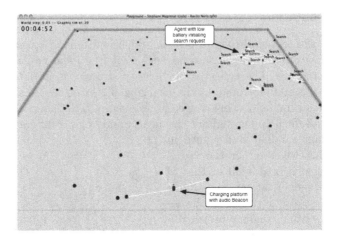

Fig. 5. Alert signal from initiating agent (labelled "low battery") is being relayed through the swarm. The red cylinder (centre bottom) is the charging platform beacon; yellow lines indicate audio communication.

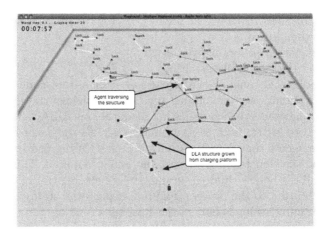

Fig. 6. DLA structure has been constructed (red/orange lines), agent is traversing it toward the charging platform. See also http://www.csse.monash.edu.au/ ∼berndm/autonomous_epuck/.

maintains a record of the most recent time they transitioned back to the primary task state, $t_{release}$. An agent that detects an alert signal a, first confirms that $a - A_{lo} > t_{release}$. If it is, it relays the alert as normal. If not, the alert is ignored and the agent with the more recent $t_{release}$ time updates the transmitting agent's stale release time. To do so, we define another frequency range R_{lo} to R_{hi}, where $R_{lo} > A_{hi}$. The updating agent transmits a tone on $R_{low} + t_{release}$, and upon reception, the stale agent updates its own $t_{release}$ and returns to the *primary task* state.

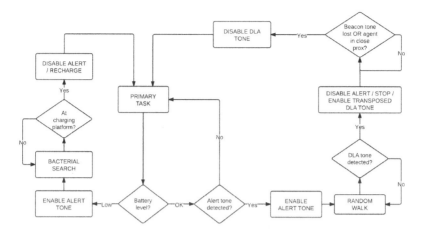

Fig. 7. Statechart of DLA Collective Navigation Strategy

The above modifications ensured that break-away agents were brought up to date with the state of the swarm upon their return, however it did so with the added cost of requiring agent's clocks to be synchronised. It is hoped that our navigation strategy can be improved by solving the synchronisation problem through some decentralised means.

4 Results and Discussion

The single-agent bacterial search was compared to a pure random walk using the target acquisitions completed over 150 minutes (hence variation in n acquisitions). Albeit a low baseline, we were interested in demonstrating an effective search strategy with minimal processing and memory requirements. Between each target acquisition, the robot performed a random walk to a new starting position in the environment. The results indicated that the bacterial algorithm performed significantly better than the random walk, even without the backtracking and proximity assistance at close range. Table 1 shows a comparison between the algorithms.

We compared the DLA collective navigation strategy to an individual agent search in identically configured simulation environments, such that the performance improvement could be quantified and assessed with respect to the time cost to the swarm. To measure the individual (unassisted) search performance, an agent was positioned toward the opposing wall of the environment from the charging platform, well beyond audible range. Unsurprisingly, the time to complete a search from outside the audible range is extremely large, as it was simply random walking until coming within range. A total of 21 experiments were completed, with a mean search time of 01:37:27 ($hh{:}mm{:}ss$), and a standard deviation of 01:12:58. The high variance stems from the fact that the search starts outside of the reach of the audio beacon so that initially a pure random walk is performed.

Table 1. Comparison of e-puck search algorithms (150 min. observation time)

Algorithm	n acquisitions	μ	σ
Random walk	14	8m 56s	6m 38s
Bacterial search	37	3m 05s	2m 50s
Bact. search with backtracking / prox. assist	47	2m 29s	1m 52s

Performance of the collective navigation strategy was measured by deploying a swarm of 60 agents randomly in the environment, one agent initiating an alert from the same location that the individual search was measured from. As evident in the dramatically improved search time, the collective navigation strategy proved very effective, however it did so with a substantial time-cost to the swarm. Table 2 shows the summary of results, where t_{search} is the total search time from when the agent sounded an alert until arriving at the charging platform, and t_{cost} is the total time committed to the process by assisting agents. $t_{traversal}$ is the search time from the moment the searching agent makes contact with the DLA structure through to search completion, provided to distinguish the DLA assembly time from the structure traversal time.

Table 2. Performance of collective search over 22 experiments ($hh{:}mm{:}ss$)

	μ	σ
t_{search}	00:15:06	00:06:46
$t_{traversal}$	00:09:13	00:05:33
t_{cost}	10:58:19	05:40:20

The single-agent bacterial search proved to be an effective means of performing a gradient search with minimal processing requirements on an audio source. It is conceivable that the search strategy could be implemented on miniaturised robots with simple analog circuitry, making it potentially very useful for swarm robotics applications requiring basic localisation capability with minimal hardware. Whilst our experiments were limited to a simple rectangular environment, it is anticipated that the search may also perform well in more complex environments, as audio does not require a sightline for detection, and effectively provides a profile of the physical environment through the propagation and reflection of sound waves.

The DLA collective navigation strategy was demonstrated in simulation to successfully guide an agent toward its charging platform from outside audible range. If only a single agent needs to be guided, the experiment results indicate that the time cost to the swarm outweighs the gain in single-agent search time. However, in applications where multiple agents need to home in on the target, such as where collective transport is required, the net time-cost amortises quickly. In fact, the strategy would break even in total time cost with just 7 of the 60 agents homing in on the target. At the same time the actual duration of the homing phase is dramatically reduced. Similarly, applications where loss of a single agent is unacceptable would also warrant this approach. The experiments conducted on both the e-puck and the simulated e-puck swarm were limited to specific environment configurations, and we consider them proof-of-concept only. With performance data for more environment configurations and specifically different swarm densities, a more complete analysis of both search strategies' characteristics could be ascertained.

Variations on the strategy could be applied to other behavioural requirements, for example the navigation of an entire swarm to a single location (collective homing) or the retrieval of some object to a pre-defined point (collaborative search-and-retrieve). In a more general sense, such a strategy may be useful for any application requiring navigation toward a single target from anywhere in an environment. Our research demonstrates the usefulness of two very simple nature-inspired strategies, single-agent bacterial search and DLA-based collaborative search, as the basis of such applications.

References

1. Bourke, P.: DLA - Diffusion Limited Aggregation (June 2004), http://paulbourke.net/fractals/dla/ (accessed May 26, 2010)
2. Chen, L.J., Simon, W.I., Hu, A.P., Meyer, B.: A contactless charging platform for swarm robots. In: Annual Conf. of the IEEE Industrial Electronics Society (IECON), Melbourne (November 2011)
3. Cyberbotics: Webots user guide, http://www.cyberbotics.com/cdrom/common/doc/webots/guide/section6.6.html (accessed March 22, 2010)
4. Dorigo, M., Tuci, E., Groß, R., Trianni, V., Labella, T.H., Nouyan, S., Ampatzis, C., Deneubourg, J.-L., Baldassarre, G., Nolfi, S., Mondada, F., Floreano, D., Gambardella, L.M.: The SWARM-BOTS Project. In: Şahin, E., Spears, W.M. (eds.) Swarm Robotics 2004. LNCS, vol. 3342, pp. 31–44. Springer, Heidelberg (2005)
5. Hamann, H., Meyer, B., Schmickl, T., Crailsheim, K.: A model of symmetry breaking in collective decision-making. In: Simulation of Adaptive Behavior (SAB), Paris (August 2010)
6. Hinchey, M., Sterritt, R., Rouff, C., Rash, J., Truszkowski, W.: Swarm-based space exploration. ECRIM News (65), 26 (2006)
7. Hölldobler, B., Wilson, E.: The Ants. Springer, Berlin (1990)
8. Hsiang, T.R., Arkin, E., Bender, M., Fekete, S., Mitchell, J.: Algorithms for rapidly dispersing robot swarms in unknown environments. In: Algorithmic Foundations of Robotics V, vol. 7, pp. 77–94. Springer, Berlin (2004)
9. Kube, C., Bonabeau, E.: Cooperative transport by ants and robots. Robotics and Autonomous Systems 30(1-2), 85–101 (2000)
10. Magnenat, S., Rtornaz, P., Bonani, M., Longchamp, V., Mondada, F.: ASEBA: A Modular Architecture for Event-Based Control of Complex Robots. IEEE/ASME Transactions on Mechatronics 16(2), 321–329 (2011)
11. McLurkin, J., Smith, J.: Distributed algorithms for dispersion in indoor environments using a swarm of autonomous mobile robots. In: Alami, R., Chatila, R., Asama, H. (eds.) Distributed Autonomous Robotic Systems 6, pp. 399–408. Springer, Japan (2007)

12. Mondada, F., Bonani, M., Raemy, X., Pugh, J., Cianci, C., Klaptocz, A., Magnenat, S., christophe Zufferey, J., Floreano, D., Martinoli, A.: The e-puck, a robot designed for education in engineering. In: Autonomous Robot Systems and Competitions, Castelo Branco (May 2009)
13. Passino, K.: Cooperative foraging and search. In: Biomimicry for Optimization, Control, and Automation, pp. 765–828. Springer, London (2005)
14. Witten, T.A., Meakin, P.: Diffusion-limited aggregation at multiple growth sites. Phys. Rev. B 28(10), 5632–5642 (1983)

Global Equilibrium Search Algorithms for Combinatorial Optimization Problems

Oleg Shylo[1], Dmytro Korenkevych[2], and Panos M. Pardalos[2,3,*]

[1] Department of Industrial Engineering, University of Pittsburgh,
Pittsburgh, PA 15261
[2] Department of Industrial and Systems Engineering, University of Florida,
303 Weil Hall, Gainesville, FL, 32611, USA
[3] National Research University Higher School of Economics, Laboratory
of Algorithms and Technologies for Network Analysis (LATNA), 136 Rodionova St.,
Nizhny Novgorod 603093, Russia

Abstract. Global Equilibrium Search (GES) is a meta-heuristic framework that shares similar ideas with the simulated annealing method. GES accumulates a compact set of information about the search space to generate promising initial solutions for the techniques that require a starting solution, such as the simple local search method. GES has been successful for many classic discrete optimization problems: the unconstrained quadratic programming problem, the maximum satisfiability problem, the max-cut problem, the multidimensional knapsack problem and the job-shop scheduling problem. GES provides state-of-the-art performance on all of these domains when compared to the current best known algorithms from the literature. GES algorithm can be naturally extended for parallel computing as it performs search simultaneously in distinct areas of the solution space. In this talk, we provide an overview of Global Equilibrium Search and discuss some successful applications.

Keywords: discrete optimization, meta-heuristics, global equilibrium search.

1 Method Description

Simulated annealing (SA) [1] is a randomized metaheuristic approach that was successfully applied to a variety of discrete optimization problems. Usually, the search process under SA consists of a sequence of transitions between feasible solutions that is guided by certain probabilistic rules. Standard simulated annealing is a memoryless optimization approach – the transitions between solutions are independent from the previous search states. Global equilibrium search (GES) [2] shares similar ideas, but, unlike simulated annealing, GES uses adaptive memory structures to collect information about visited solutions, and actively uses this knowledge to control future transitions.

---

* The author is partially supported by LATNA Laboratory, NRU HSE, RF government grant, ag. 11.G34.31.0057.

C.A. Coello Coello et al. (Eds.): PPSN 2012, Part II, LNCS 7492, pp. 277–286, 2012.

Consider a general formulation of a combinatorial optimization problem with binary variables:

$$\min\{f(x)|x \in D \subset B^n\}, \tag{1}$$

where B^n is a set of all n-dimensional vectors, whose components are either 1 or 0, and $f(x)$ is an objective function. At each stage of the simulated annealing method, a set of solutions $N(x) \in D$ that belong to a user-defined neighborhood of x is generated according to some predefined rule. The method sequentially evaluates $N(x)$, and moves from the current solution to one of the solutions in $N(x)$ based on the Metropolis acceptance criterion [3], which describes the change of states in thermodynamic systems. The transition from x to $y \in N(x)$ happens with the probability $P(x \to y)$, which depends on a temperature parameter μ:

$$P(x \to y) = \begin{cases} \exp(-\mu[f(y) - f(x)]), & \text{if } f(x) \le f(y); \\ 1 & \text{if } f(x) > f(y). \end{cases}$$

Given a sufficient number of iterations with constant temperature parameter μ, the SA method will converge to an equilibrium state. If we decide to terminate it after reaching the equilibrium, then the final solution – which can be modeled by a random vector $\xi(\mu, \omega)$ – will follow a Boltzmann distribution [4]:

$$P\{\xi(\mu, \omega) = x\} = \begin{cases} \dfrac{\exp(-\mu f(x))}{\sum\limits_{x \in D} \exp(-\mu f(x))}, & x \in D \\ 0, & x \notin D. \end{cases} \tag{2}$$

The set of feasible solutions D can be represented as a union of two disjoint subsets: $D_j^1 = \{x \mid x \in D, x_j = 1\}$, and $D_j^0 = \{x \mid x \in D, x_j = 0\}$, representing feasible solutions with the j-th component equal to 1 or 0 respectively. The probability that the j-th component of random vector $\xi(\mu)$ is equal to 1 can be expressed as

$$\pi_j(\mu) \equiv P\{\xi_j(\mu) = 1\} = \frac{\sum_{x \in D_j^1} \exp(-\mu f(x))}{\sum_{x \in D} \exp(-\mu f(x))}. \tag{3}$$

The stationary probabilities $\pi_j(\mu)$ can be used to generate random vectors that have approximately Boltzmann distribution provided by (2). Usually, $\xi_j(\mu)$ and $\xi_i(\mu)$ are not independent for all $i \ne j$, therefore we can only achieve an approximation of (2) when generating random solutions by fixing their components independently according to (3). GES collects information about some solutions from D and uses it to approximate the equilibrium distribution (2).

Let S be a subset of D; $S_j^1 = \{x \mid x \in S, x_j = 1\}$, and $S_j^0 = \{x \mid x \in S, x_j = 0\}$. For example, this set can contain all local optima discovered in the past search stages. Instead of explicitly storing the solutions in S, it is sufficient to store only the values that are required to approximate (3). Specifically, we set:

$$Z(\mu) = \sum_{x \in S} \exp(-\mu[f(x) - \min_{x \in S} f(x)]) \qquad (4)$$

$$Z_j^1(\mu) = \sum_{x \in S_j^1} \exp(-\mu[f(x) - \min_{x \in S} f(x)]) \qquad (5)$$

Every time a new solution is included in S, these memory structures can be updated by adding the corresponding terms to (4) and (5), without storing complete solutions with all attributes. These values are scaled using the value of the best objective function in S, $\min_{x \in S} f(x)$. Whenever the best objective is improved by a newly found solution, $Z(\mu)$ and $Z_j^1(\mu)$ are rescaled by multiplying each value by $\exp(x^{oldbest} - x^{newbest})$, where $x^{oldbest}$ and $x^{newbest}$ are the old best objective value and the new best objective value, respectively.

Using these memory structures, the stationary probabilities $\pi_j(\mu)$ from (3) can be approximated as

$$p_j(\mu) = \frac{Z_j^1(\mu)}{Z(\mu)}. \qquad (6)$$

1.1   Intensification and Diversification

GES generates solutions according to the distribution defined by (6), or applies a sequence of perturbations to the components of a given solution guided by (6), which is usually more efficient in practice. The latter approach to generating new solutions should be used when a single perturbation might lead to an infeasible solution, in which case such perturbations are simply prohibited.

By increasing the value of the temperature parameter we can generate solutions that more closely resemble the best solution in the set of known solutions S. Let $x^{min} \in S$ denote the best solution in S: $f(x^{min}) = \min_{x \in S} f(x)$. If all other solutions in S have larger objective functions, then

$$\lim_{\mu \to \infty} p_j(\mu) = x_j^{min}.$$

This follows from the definition of $p_j(\mu)$ and the fact that $\lim_{\mu \to \infty} Z(\mu) = 1$. This property is used in GES to alternate between diversification and intensification stages using a monotonically increasing sequence of temperature parameters: $\mu_1, \mu_2, \ldots, \mu_K$. In order to calculate $p_j(\mu_k)$ for all temperature values from this sequence, we need to store $(n+1) \cdot K$ values corresponding to $Z_j^1(\mu_k)$ and $Z(\mu_k)$, where n is the number of binary variables in the problem definition.

The specific values for the temperature parameters are usually calculated using simple recursive formulas: $\mu_0 = 0$, $\mu_{k+1} = \alpha \mu_k$ for $k = 1, \ldots, K-1$. The parameters involved in this recursion (μ_1, α and K) are chosen to guarantee the convergence to the best solution in the set S:

$$\|x_j^{min} - p_j^K\| \approx 0$$

Below, we will address the dynamic adjustment of the temperature schedule that can be easily implemented in practice.

1.2 Alternative Memory Implementations

Even for small values of the temperature parameter μ, the expression given by (6) can provide biased probabilities. For example, if we generate solutions with temperature equal to zero and the set S contains only solutions with jth component equal to 1, then the probability $p_j(0) = 1$. To deal with this bias, one can use alternative memory implementations.

For example, the probabilities can be approximated using the best objective values corresponding to each component:

$$f_j^0 = \begin{cases} \min\{f(x) : x \in S_j^0\} & \text{if } |S_j^0| \neq 0 \\ \infty & \text{otherwise} \end{cases}$$

$$f_j^1 = \begin{cases} \min\{f(x) : x \in S_j^1\} & \text{if } |S_j^1| \neq 0 \\ \infty & \text{otherwise} \end{cases}$$

Let x^{min} be a solution from the set S with a minimum value: $f(x^{min}) = \min\{f(x) : x \in S\}$. The probability $p_j(\mu)$ can be approximated as:

$$p_j(\mu) = \frac{\exp(-\mu[f_j^1 - f(x^{min})])}{\exp(-\mu[f_j^1 - f(x^{min})]) + \exp(-\mu[f_j^0 - f(x^{min})])} \tag{7}$$

In Figure 1, we present a pseudo-code of the procedure that calculates the generation probabilities using f_j^1 and f_j^0. To achieve convergence to the best solution in the set S when using Formula (7), we penalize solutions that have the same objective value as x^{min} (Figure 1, lines 10–13; 20–23). In addition, if S_j^1 (or S_j^0) is empty, one can use the maximum absolute difference between objective values instead of $f_j^1 - f(x^{min})$ ($f_j^0 - f(x^{min})$) to avoid premature convergence (Figure 1, lines 5, 15).

The general scheme of the GES method is presented in Figure 2. In the beginning memory structures are initialized by a randomly generated solution. The temperature cycle is repeated until $nfail^{max}$ cycles without improvement to the best found solution, x^{best}. Generation probabilities are recalculated at every temperature stage using the corresponding temperature parameter. These probabilities are used to generate $ngen$ solutions that are used as initial points for local search procedure. Locally-optimal solutions are used to update the adaptive memory structures, which will affect the future generation probabilities.

1.3 Parallel Implementations

GES algorithm can be naturally extended for parallel computing as it performs search simultaneously in distinct areas of the solution space. One can trivially accelerate GES by initiating a set of copies of GES procedures with different random seeds. Due to the randomness, each copy will follow a distinct search trajectory, which often leads to significant parallel acceleration in practice [5]. Further

Require: μ – temperature parameter μ, x^{min} – current best solution, f^0 and f^1 – vectors of the best object values corresponding to each component, n – number of solution components;

1: $maxdif = \max\limits_{j}\{abs(f_j^0 - f_j^1) : f_j^1 < \infty; f_j^0 < \infty; \}$
2: $sum_0 = sum_1 = 0$
3: **for** $j = 1$ to n **do**
4:   **if** $f_j^0 = \infty$ **then**
5:     $sum_0 = sum_0 + \exp(-\mu \cdot maxdif)$
6:   **else**
7:     **if** $f_j^0 > f(x^{min})$ **then**
8:       $sum_0 = sum_0 + \exp(-\mu[f_j^0 - f(x^{min})])$
9:     **else**
10:       **if** $x_j^{min} = 0$ **then**
11:         $sum_0 = sum_0 + 1$
12:       **else**
13:         $sum_0 = sum_0 + \exp(-\mu)$
14:   **if** $f_j^1 = \infty$ **then**
15:     $sum_1 = sum_1 + \exp(-\mu \cdot maxdif)$
16:   **else**
17:     **if** $f_j^1 > f(x^{min})$ **then**
18:       $sum_1 = sum_1 + \exp(-\mu[f_j^1 - f(x^{min})])$
19:     **else**
20:       **if** $x_j^{min} = 1$ **then**
21:         $sum_1 = sum_1 + 1$
22:       **else**
23:         $sum_1 = sum_1 + \exp(-\mu)$
24:   $p_i(\mu) = \frac{sum_1}{sum_1 + sum_0}$
25: **return** $p(\mu)$

Fig. 1. Calculation of the transition probabilities

improvements can be achieved by sharing the best found solutions and/or the adaptive memory structures (for example, by sending the updates to vectors f^0 and f_j^1) between different copies. Such sharing is equivalent to increasing the number of solutions, $ngen$, generated at each temperature stage of GES algorithm.

2 Applications

Global equilibrium search has a number of advantages when compared to the simulated annealing method. Firstly, the presence of adaptive memory allows GES to outperform SA in terms of solution quality and computational speed. Its performance was tested on classic optimization problems that capture the complexities of modern optimization applications. Here we mention some of these applications and provide detailed results for the quadratic assignment problem.

Require: μ – vector of temperature values, K – number of temperature stages, $maxnfail$ – restart parameter, $ngen$ – # of solutions generated during each stage
Ensure:
1: x^{best} = construct random solution; $S=\{x^{best}\}$
2: **while** stopping criterion = FALSE **do**
3:     $x \leftarrow$ construct random solution
4:     $x^{min} = x$
5:     $S = \{x^{min}\}$ (set of known solutions)
6:     **for** $nfail = 0$ to $nfail^{max}$ **do**
7:         $x^{old} = x^{min}$
8:         **for** $k = 0$ to K **do**
9:             $p(\mu_k)$ = calculate generation probabilities(S, μ_k)
10:            **for** $g = 0$ to $ngen$ **do**
11:                x = generate solution$(x, p(\mu_k))$
12:                x = local search method(x)
13:                $S = S \cup x$
14:                $x^{min} = \arg\min\{f(x) : x \in S\}$
15:                **if** $f(x^{min}) < f(x^{best})$ **then**
16:                    $x^{best} = x^{min}$
17:        **if** $f(x^{old}) > f(x^{min})$ **then**
18:            $nfail = 0$
19: **return** x^{best}

Fig. 2. Pseudo-code of the GES method

2.1   Unconstrained Binary Quadratic Problem

One of the well-known and most interesting classes of integer optimization problems is the maximization of the quadratic 0–1 function:

$$\max_{\mathbf{x} \in \{0,1\}^n} f(\mathbf{x}) = \sum_{i=1}^{n} \sum_{j=1}^{n} q_{ij} x_i x_j, \tag{8}$$

where q_{ij} are elements of an $n \times n$ symmetric real matrix $Q \in \mathbb{R}^{n \times n}$. This problem is referred to as an unconstrained binary quadratic programming problem. Many fundamental problems in science, engineering, finance, medicine and other diverse areas can be formulated as quadratic binary programming problems. Quadratic functions with binary variables naturally arise in modeling selections and interactions.

GES was applied to a wide spectrum of large-scale instances of the unconstrained binary quadratic programming problem [6]. The computational experiments revealed favorable performance compared to the best known heuristics on a set of publicly available benchmark instances [7,8,9].

2.2   Maximum Satisfiability Problem

Maximum satisfiability problem consists of finding an assignment of boolean variables that satisfies as many given logical clauses as possible. In the weighted maximum satisfiability problem each logical clause has a predetermined positive weight, and the goal is to search for an assignment, which maximizes the total weight of the satisfied clauses.

Among various heuristic approaches for solving this problem – such as, algorithms based on reactive tabu search [10], simulated annealing [11], GRASP [12,13,14,15,16], iterated local search [17] and guided local search [18] – GES provides the state-of-the-art performance [19] on many benchmark instances [20,21,22,23].

2.3   Quadratic Assignment Problem

The Quadratic Assignment Problem (QAP) stated for the first time by Kooper and Beckman in 1957 is a well known combinatorial optimization problem and remains one of the greatest challenges in the field [24], [25]. Firstly it was formulated in the context of the plant location problem. Given n facilities with some physical products flow between them and n locations with known pairwise distances, one should determine to which location each facility must be assigned in order to minimize total distance $\times$ flow.

Mathematically QAP can be formulated as follows: let $A_{n \times n} = a(i, j)$ be a matrix, where $a_{i,j} R^+$ represents product flow between facilities f_i and f_j, let $B_{n \times n} = (b_{i,j})$ be a matrix, where $b_{i,j} \in R^+$ represents the distance between locations l_i and l_j. Let $p : \{1, \ldots, n\} \to \{1, \ldots, n\}$ be a permutation of integers. The cost of a permutation is defined as follows:

$$c(p) = \sum_{i=1}^{n} \sum_{j=1}^{n} a_{ij} b_{p(i)p(j)}.$$

The goal is to find a permutation $p*$ with a minimal cost. The QAP is well known to be strongly NP-hard, and even small instances may require long computational time. A number of practical problems can be formulated as QAP. Among those are problems dealing with backboard wiring, scheduling, manufacturing, statistical data analysis, typewriter keyboard design, image processing, turbine balancing and so on [25], [26].

We performed a series of computational experiments on well known benchmark problems from the QAPLIB library [27] (online version is available at http://www.opt.math.tu-graz.ac.at/qaplib/). In our implementation we used Tabu Search algorithm as a local search method [28]. The neighborhood of a given permutation $\hat{p}$ was defined as $N(\hat{p}) = \{p : \|p - \hat{p}\| = 2\}$, where $\| \cdot \|$ denotes Hamming distance.

We compared our algorithm to Robust Tabu search (RoTabu), Ant Colony and Simulated Annealing (SA) algorithms (the codes for these algorithms were obtained from QAPLIB resource as well). Each algorithm was executed on each instance 10 times. The experiments were performed on a 3GHz AMD computer. In Table 1 and Table 2 we report the average deviation (in %) between the best solution found by the algorithm and the best known solution.

Table 1. Experiments on real world instances. *Instance* - problem instance name, n - size of the instance, *BKS* - best known solution value, *time* - maximum allowed computational time in seconds.

Instance	n	BKS	RoTabu	SA	Ant Colony	GES	time, s
bur26a	26	5426670	0.03	0.52	**0.02**	0.03	0.1
bur26b	26	3817852	0.09	0.32	**0.03**	0.07	0.1
bur26c	26	5426795	0.03	0.29	0	0	0.1
bur26d	26	3821225	0.05	0.10	0	0	0.1
bur26e	26	5386879	0.01	0.18	0	0	0.1
bur26f	26	3782044	0.01	0.06	0	0	0.1
bur26g	26	10117172	0.01	0.30	0	0	0.1
bur26h	26	7098658	0.13	0.13	0	0	0.1
chr25a	25	3796	4.23	45.5	1.24	0	2
nug30	30	6124	0.04	5.29	0.15	0	2
kra30a	30	88900	0	3.77	0.70	0	4
kra30b	30	91420	**0.01**	3.59	0.04	**0.01**	4
tai64c	64	1855928	0.37	1.28	0	0	1
tai20b	20	122455319	0.05	8.75	0.09	0	0.1
tai25b	25	344355646	0.02	2.82	0	0	0.5
tai30b	30	637117113	0.04	2.66	0	0	1
tai35b	35	283315445	0.1	3.09	0	0	2
tai40b	40	637117113	0.43	2.12	0.11	0	2
tai50b	50	458821517	1.58	0.57	0.26	0	8
tai60b	60	608215054	1.05	0.66	0.32	0	20
tai80b	80	818415043	0.84	1.43	0.94	**0.12**	40

Table 2. Experiments on randomly generated instances. *Instance* - problem instance name, n - size of the instance, *BKS* - best known solution value, *time* - maximum allowed computational time in seconds.

Instance	n	BKS	RoTabu	SA	Ant Colony	GES	time, s
tai20a	20	703482	**0.05**	0.55	0.44	0.06	2.5
tai25a	25	1167256	0	0.93	1.5	0	5
tai30a	30	1818146	0.29	0.47	0.93	**0.03**	7.5
tai35a	35	2422002	0.63	0.86	1.14	**0.16**	10
tai40a	40	3139370	0.78	1.01	1.43	**0.30**	30
tai50a	50	4941410	1.04	1.37	1.75	**0.64**	45
tai60a	60	7205962	1.17	1.25	1.94	**0.84**	60
tai80a	80	13546960	1.28	1.07	1.39	**0.62**	120

3   Conclusions

One of the most important qualities of GES is its ability to process and utilize the solutions that are obtained by different search techniques. GES offers a mechanism of information processing that can be used to organize an intelligent multi-start search that involve different optimization techniques. This collaborative functionality can be effectively used in parallel implementations. Numerous successful applications on a wide range of combinatorial optimization problems corroborate the efficiency of the GES method.

References

1. Kirkpatrick, S., Gelatt, C.D., Vecchi, M.P.: Optimization by simulated annealing. Science 220(4598), 671–680 (1983)
2. Shylo, V.: A global equilibrium search method (in russian). Kybernetika i Systemniy Analys 1, 74–80 (1999)
3. Schuur, P.C.: Classification of acceptance criteria for the simulated annealing algorithm. Mathematics of Operations Research 22(2), 266–275 (1997)
4. Henderson, D., Jacobson, S., Johnson, A.: The theory and practice of simulated annealing. In: Glover, F., Kochenberger, G. (eds.) Handbook of Metaheuristics, pp. 287–319. Kluwer Academic Publishers, Boston (2003)
5. Shylo, O.V., Middelkoop, T., Pardalos, P.M.: Restart strategies in optimization: parallel and serial cases. Parallel Comput. 37, 60–68 (2011)
6. Pardalos, P.M., Prokopyev, O.A., Shylo, O.V., Shylo, V.P.: Global equilibrium search applied to the unconstrained binary quadratic optimization problem. Optimization Methods Software 23(1), 129–140 (2008)
7. Palubeckis, G.: Unconstrained binary quadratic optimization, http://www.soften.ktu.lt/~gintaras/binqopt.html (last accessed on October 2008)
8. Beasley, J.E.: OR-library webpage, http://people.brunel.ac.uk/ mastjjb/jeb/info.html (last updated on June 2008)
9. Sloane, N.: Challenge problems: Independent sets in graphs, http://www.research.att.com/~njas/doc/graphs.html (last accessed on October 2008)
10. Battiti, R., Protasi, M.: Reactive search, a history-sensitive heuristic for max-sat. J. Exp. Algorithmics 2 (January 1997)
11. Spears, W.M.: Simulated annealing for hard satisfiability problems. In: Johnson, D., Trick, M. (eds.) The Second DIMACS Implementation Challenge. DIMACS Series on Discrete Mathematics and Theoretical Computer Science, vol. 26, pp. 533–558. American Mathematical Society (1996)
12. Du, D., Gu, J., Pardalos, P.: Satisfiability problem: theory and applications: DIMACS Workshop, March 11-13, vol. 35. Amer. Mathematical Society (1997)
13. Festa, P., Pardalos, P.M., Pitsoulis, L.S., Resende, M.G.C.: Grasp with path relinking for the weighted maxsat problem. J. Exp. Algorithmics 11, Article No. 2.4 (2006)
14. Festa, P., Pardalos, P.M., Pitsoulis, L.S., Resende, M.G.C.: Grasp with path relinking for the weighted maxsat problem. J. Exp. Algorithmics 11 (February 2007)
15. Resende, M., Feo, T.: A grasp for satisfiability. In: Johnson, D., Trick, M. (eds.) The Second DIMACS Implementation Challenge. DIMACS Series on Discrete Mathematics and Theoretical Computer Science, vol. 26, pp. 499–520. American Mathematical Society (1996)
16. Resende, M., Pitsoulis, L., Pardalos, P.: Approximate solution of weighted max-sat problems using grasp. In: Gu, J., Pardalos, P. (eds.) Satisfiability Problem: Theory and Applications. DIMACS Series on Discrete Mathematics and Theoretical Computer Science, vol. 35, pp. 393–405. American Mathematical Society (1997)

17. Yagiura, M., Ibaraki, T.: Analyses on the 2 and 3-flip neighborhoods for the max sat. Journal of Combinatorial Optimization 3, 95–114 (1999)
18. Mills, P., Tsang, E.: Guided local search for solving sat and weighted max-sat problems. J. Autom. Reason. 24(1-2), 205–223 (2000)
19. Shylo, O.V., Prokopyev, O.A., Shylo, V.: Solving weighted max-sat via global equilibrium search. Oper. Res. Lett., 434–438 (2008)
20. Resende, M.: Algorithm source code distribution, http://research.att.com/~mgcr/src (last accessed on October 2008)
21. Resende, M.: Test problem distribution, http://research.att.com/~mgcr/data (last accessed on October 2008)
22. Yagiura, M.: Benchmark instances and program codes for max sat, http://www.al.cm.is.nagoya-u.ac.jp/~yagiura/msat.html (last accessed on October 2008)
23. Mills, P.: Constraint programming and optimization laboratory, university of essex, gls solver, algorithm: Extended gls, http://cswww.essex.ac.uk/CSP/glsdemo.html (last accessed on October 2008)
24. Pardalos, P., Wolkowicz, H.: Quadratic assignment and related problems: DIMACS Workshop, May 20-21, vol. 16. Amer. Mathematical Society (1994)
25. Burkard, R., Cela, E., Pitsoulis, L., Pardalos, P.M.: The quadratic assignment problem. In: Handbook of Combinatorial Optimization, vol. 3, pp. 241–337. Springer (1998)
26. Pardalos, P., Pitsoulis, L.: Nonlinear assignment problems: algorithms and applications, vol. 7. Kluwer Academic Pub. (2000)
27. Burkard, R.E., Karisch, S.E., Rendl, F.: Qaplib–a quadratic assignment problem library. Journal of Global Optimization 10(4), 391–403 (1997)
28. Misevicius, A.: A tabu search algorithm for the quadratic assignment problem. Computational Optimization and Applications 30(1), 95–111 (2005)

A Genetic Programming Approach for Evolving Highly-Competitive General Algorithms for Envelope Reduction in Sparse Matrices

Behrooz Koohestani and Riccardo Poli

School of Computer Science and Electronic Engineering,
University of Essex, CO4 3SQ, UK
{bkoohe,rpoli}@essex.ac.uk

Abstract. Sparse matrices emerge in a number of problems in science and engineering. Typically the efficiency of solvers for such problems depends crucially on the distances between the first non-zero element in each row and the main diagonal of the problem's matrix — a property assessed by a quantity called the *size of the envelope* of the matrix. This depends on the ordering of the variables (i.e., the order of the rows and columns in the matrix). So, some permutations of the variables may reduce the envelope size which in turn makes a problem easier to solve. However, finding the permutation that minimises the envelope size is an NP-complete problem. In this paper, we introduce a hyper-heuristic approach based on genetic programming for evolving envelope reduction algorithms. We evaluate the best of such evolved algorithms on a large set of standard benchmarks against two state-of-the-art algorithms from the literature and the best algorithm produced by a modified version of a previous hyper-heuristic introduced for a related problem. The new algorithm outperforms these methods by a wide margin, and it is also extremely efficient.

Keywords: Hyper-Heuristic, Genetic Programming, Envelope Reduction Problem, Graph Labelling, Sparse Matrices.

1   Background

A substantial number of problems in science and engineering require the solution of large systems of linear equations. The effectiveness of methods designed to handle such systems depends critically on finding an ordering for the variables for which the distances between the first non-zero element in each row and the main diagonal of the problem's matrix is small [15]. This property is typically assessed by a quantity called the *size of the envelope* of the matrix. Let us start by providing a formal definition of it.

Let A be an $N \times N$ symmetric matrix with entries a_{ij}. The *row bandwidth* of the ith row of A is defined as follows: $b_i(A) = i - \min\{j : a_{ij} \neq 0\}$. In other words, the row bandwidth is the distance (in columns) from the first non-zero entry in a row to the diagonal [7]. The *envelope* of matrix A is directly related

C.A. Coello Coello et al. (Eds.): PPSN 2012, Part II, LNCS 7492, pp. 287–296, 2012.

to its row bandwidths, and can be thought of as a function $e(i, A) = b_i(A) + 1$, which returns the number of elements between the first non-zero entry in a row and the main diagonal (inclusive). Then the *size of the envelope* of the matrix is defined as [20]:

$$|Env(A)| = \sum_{i=1}^{N} e(i, A).$$

Finding a permutation of rows and columns of A which minimises the envelope size $|Env(A)|$ — a problem known as the *Envelope Reduction Problem* (ERP) — is the focus of this paper. Since there are $N!$ possible permutations for an $N \times N$ matrix, the ERP is considered, in general, a very difficult combinatorial optimisation problem. Indeed, ERP was shown to be NP-complete [3].

1.1   Envelope Reduction Algorithms

A variety of methods have been proposed in order to address the ERP. One of the earliest heuristic approaches for reducing the bandwidth and envelope size of sparse matrices was introduced by Cuthill and McKee [6]. Their algorithm (CM) is still one of the most widely used algorithms to (approximately) solve these problems. In this method, the nodes in the graph representation of a matrix are partitioned into equivalence classes based on their distance from a given root node. The partition is known as *level structure* for the given node. In CM, the root node for the level structure is normally chosen from the nodes of minimum degree in the graph. George [8] observed that renumbering the CM ordering in a reverse way (RCM) often yielded a result superior to the original ordering. The GPS algorithm, introduced by Gibbs, Poole and Stockmeyer [10], also uses level structures, and it is comparable with RCM in terms of solution quality, while being several times faster. The GK (Gibbs-King) algorithm [9], which is a variation of GPS, provides considerably better reduction of the envelope in comparison with the original GPS, but it is often much slower in execution. The Sloan algorithm [20] offered a significant improvement over the methods mentioned earlier by introducing a new step in which the ordering obtained from a variant of the GPS algorithm was locally refined. Adopting a very different approach Barnard *et al.* [2] proposed the use of spectral analysis of the Laplacian matrix associated with the graph representing the non-zero elements in a sparse matrix as an effective method for the reduction of the envelope of a sparse matrix. Recently, also a new variation of the GPS algorithm has been presented [21].

1.2   Hyper-Heuristics

The term *hyper-heuristic* was first introduced by Cowling *et al.* [5]. According to their definition, a hyper-heuristic manages the choice of which lower-level heuristic method should be applied at any given time, depending upon the characteristics of the heuristics and the region of the solution space currently under exploration. Here, a *heuristic* is a rule-of-thumb or "educated guess" that reduces the search required to find a solution. More generally a hyper-heuristic could be defined as "heuristics to choose other heuristics" [4]. Here, we embrace a slightly different definition and see a *hyper-heuristic as a search algorithm that explores the space of problem solvers*. Genetic Programming (GP) [13,16] has been very successfully used as a hyperheuristic. For example, GP has evolved competitive SAT solvers [1], state-of-the-art or better than state-of-the-art bin packing algorithms [19], particle swarm optimisers [18], evolutionary algorithms [14], and TSP solvers [11].

In this paper, a hyper-heuristic approach based on GP is introduced for evolving graph-theoretic envelope reduction algorithms. Our approach is to adopt the basic ideas of some of the best algorithms for ERP, in particular their use of level structures, but to evolve the strategy the algorithm uses to construct permutations.[1] The paper is organised as follows: in Sec. 2, we describe the proposed hyper-heuristic for the solution of the ERP in detail; in Sec. 3, we report the results of our experiments; and finally, our conclusions are given in Sec. 4.

2   Proposed Hyper-Heuristic

In our method for addressing the ERP, which we call *Genetic Hyper-Heuristic* or GHH for brevity, GP is given a training set of matrices as input, and it produces a novel solver for ERPs as its output. To cope with such a complex task, following the strategy adopted in previous work [19,12], we provide GHH with the "skeleton" of a generic level-structure-based ERP solver and we ask GP to evolve the "brain" of that solver, that is the decision-making element of the system which prioritises nodes for insertion into a permutation.

A description of GHH is given in Algorithm 1. For efficiency, GHH computes the fitness of all individuals in a new generation incrementally, by testing the whole population on a problem in the training set before moving to the next (Step 4). For the same reason, we operate on the graph representation of sparse matrices instead of directly acting on the matrices. Note also that, unlike previous solvers (including our method [12]) which prioritise nodes at each level in a level structure independently, GHH is capable of exploring and sorting vertices located beyond a specific level. More details on Algorithm 1 are provided below.

2.1   Our GP System

We used a tree-based GP system with some additional decoding steps required for the ERP. The initial population was generated randomly using a modified version of the ramped half-and-half method [13,16] using the functions and terminals shown in Table 1 (more on these below). As shown in Algorithm 1, the

---

[1] To the best of our knowledge, no prior attempt to use a hyper-heuristic to evolve ERP solvers has been reported in the literature. However, we conducted previous research with a hyper-heuristic for the related bandwidth minimisation problem where the objective is to minimise $\max_i b_i(A)$ [12]. We will compare our new envelope reduction approach against an envelope-reduction version of such hyper-heuristic in Sec. 3.

Algorithm 1. GHH for ERP

1: Randomly generate an initial population of programs from the available primitives.
2: **repeat**
3:    Initialise the fitness of each program $p \in$ population to 0.
4:    **for** each instance $G_i \in$ training set of ERPs **do**
5:       Select a starting vertex s and construct a level structure rooted at s.
6:       **for** each program $p \in$ population **do**
7:          $l \leftarrow$ empty list
8:          **for** each vertex $v \in V(G_i)$ **do**
9:             Insert s into array $perm[1...n]$ and update l.
10:            Scan l and if $l.count = 0$, then break.
11:            **for** each vertex $v' \in l$ **do**
12:               Execute p.
13:            **end for**
14:            Create permutation σ represented by p.
15:            Sort vertices in l in order given by σ.
16:            $s \leftarrow$ first element of the ordered list l; $l.remove(s)$.
17:         **end for**
18:         Apply $perm$ to the adjacency list of the graph G_i.
19:         Compute the $envelope$.
20:         $fitness[p] = fitness[p] + envelope(G_i, p)$.
21:      **end for**
22:   **end for**
23:   Apply selection.
24:   Produce a new generation of individual programs.
25: **until** the termination condition is met.
26: **return** the best program tree.

Table 1. The functions and terminals used in our GP system

Primitive set	Arity	Description
+	2	Adds two inputs
-	2	Subtracts second input from first input
*	2	Multiplies two inputs
ED	0	Returns the number of unvisited vertices connected to each vertex
DFSV	0	Returns the distance from starting vertex for each vertex
Constants	0	Uniformly-distributed random constants in the interval $[-1.0, +1.0]$

fitness of a program tree (to be minimised) is the sum of the envelopes of the solutions that it creates when run on each problem instance in the training set. The parameters of our GP runs are given in Table 2.[2] Tournament selection was used. New individuals were created by applying reproduction, sub-tree crossover and point mutation. We also used elitism to preserve the overall best found solution. Also, to control excessive code growth, the Tarpeian method [17] was utilised in the system. The termination criterion used was based on the predetermined maximum number of generations to be run.

2.2 Specialised Primitives

To make it possible for GHH to exploit the new possibilities offered by its ability to explore and prioritise vertices located at different depths in the level structure, we provided two special primitives, ED and DFSV (see Table 1).

---

[2] Parameters were selected after conducting a number of preliminary experiments, considering both the quality of solutions and run times.

Table 2. Parameters used for our runs

Parameter	Value
Maximum Number of Generations	100
Maximum Depth of Initial Programs	3
Population Size	2000
Tournament Size	4
Elitism Rate	0.1%
Reproduction Rate	0.9%
Crossover Rate	70%
Mutation Rate	29%
Mutation Per Node	0.05%

The primitive ED, which stands for *Effective Degree*, is motivated by the common method of sorting vertices in a level structure based on their degree in classical ERP solvers. The degree of a vertex is the number of vertices connected to that vertex. There is no doubt that this is of fundamental importance in node ordering algorithms for ERP. However, we found that prioritising using the primitive ED, which does not include the vertices already visited when counting the number of vertices connected to a vertex, provides more accurate guidance.

In a level structure, all vertices in a level are located at the same distances from the root vertex. Since in traditional ERP solvers nodes are sorted only within each level before moving to the next, node distance from the root is an irrelevant feature for such algorithms. However, in GHH, after the first step of the algorithm, nodes from different levels will be present in the list l. These nodes will thus have different distances from the root node. The primitive DFSV captures this information. This may help prioritise vertices and break ties.[3]

2.3   Vertex Selection

Let us analyse Algorithm 1 from the vertex selection point of view. First, a level structure rooted at a suitable starting vertex s (vertex of minimum degree or a *pseudo-peripheral vertex*) is constructed (Step 5). Next, an empty list l is formed for each program p in the population (Step 7). The vertex s is then inserted into the first position of array *perm*, and l is updated (Step 9). The update process includes finding all unvisited vertices connected to s and inserting them into l. Note that further vertices will sequentially be assigned to s and inserted in the second, third, etc. positions in *perm*.

Next, the GP interpreter is called k times, where k is the number of vertices in l (Step 12). Each call of the interpreter executes the selected program with respect to the different values returned by ED and DFSV. The outputs obtained from each execution of the given program are stored in a one dimensional array. This array is then sorted in ascending order while also recording the position that each element originally had in the unsorted array. Reading such positions sequentially from the sorted array produces a permutation associated with the

---

[3] Sloan [20] also uses a distance quantity in his algorithm, but he computes distances from the end node of a pseudo-diameter.

original program (Step 14). The vertices located in l are then ordered based on the permutation generated (Step 15).

In Step 16, the first element of l is then removed and considered as a new starting vertex. This process is repeated for each vertex in $V(G_i)$ until all the vertices of graph G_i have been numbered. Finally, $perm$ is applied to the adjacency list of the initial graph (or matrix), a new adjacency list is generated (Step 18), and its envelope is computed (Step 19).

2.4 Training and Test Sets

We used a training set of 25 benchmark instances G_i from the Harwell-Boeing sparse matrix collection. This is a collection of standard test matrices arising from problems in FEM grids, linear systems, least squares, and eigenvalue calculations from a wide variety of scientific and engineering disciplines. The benchmark matrices were selected from 5 different sets in this collection, namely BCSSTRUC1, BCSSTRUC3, CANNES, LANPRO and LSHAPE with sizes ranging from 24×24 to 960×960. This training set was used only to evolve the heuristics. The performance of the evolved heuristics was then evaluated using a completely separate test set of 30 matrices taken from Everstine's collection (DWT) and BCSPWR, both included in the Harwell-Boeing database. DWT set is closely related to CANNES and LSHAPE sets used in our training set in terms of its discipline and the class of problems. We also picked the six largest instances from the BCSPWR set, which is in a totally different class compared to the training set used. We did this to assess how well the generated heuristics generalised in unseen situations.

3   Results

Ten independent runs of GHH with the training set specified above were performed, and the corresponding best-of-run individual in each was recorded. We then selected as our overall best evolved heuristic the best program tree from these ten best-of-run results.[4] The simplified version of the best heuristic evolved by GHH is as follows:

```
(((((((DFSV + ED) + (ED * ED)) * (((DFSV * ((0.616301555473498 + (DFSV * (DFSV - -
0.156489470580821))) + ((DFSV - -0.778113556456805) * ((DFSV * 0.680593788009413) + (DFSV *
ED))))) - (DFSV - -0.778113556456805)) - 0.273723254573403)) - 0.616301555473498) + (((ED - ED) +
DFSV) - -0.163010843639733)) + ((DFSV + (0.316761550175381 * DFSV)) - 0.889497244679135)) + -
0.00709300954225148)
```

This function is shown graphically in Figure 1. The function is monotonic in both ED and DFSV. For small values of ED, nodes closer to the root are preferred

---

[4] Due to the high computational load involved in the use of hyper-heuristics one can normally only perform a very small number of runs. However, this is normally considered acceptable since whenever focusing on human-competitive results one is more interested in the algorithms resulting from the application of a hyper-heuristic than on the analysis of the hyper-heuristic itself.

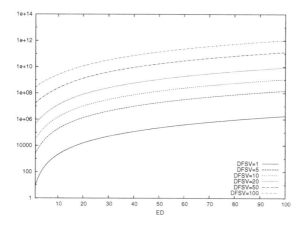

Fig. 1. Plot of the best GHH heuristic

over nodes further away. So, in a highly sparse matrix the algorithm behaves similarly to CM. However, if there are significant differences in ED values, the algorithm looks ahead and may prefer a deeper node with a lower ED to a closer one with a higher ED, thus exhibiting a previously totally unexplored strategy.

We incorporated this heuristic into a level structure system and carried out experiments with the test set. In order to assess the performance of the heuristic generated, we compared it against two well-known and high-performance algorithms: RCM and GK. In practice, both algorithms are still among the best and most widely used methods for envelope reduction. We also tested this heuristic against GP-HH, which is the best algorithm produced by an envelope-minimising version of our previous hyper-heuristic method for evolving *bandwidth* reduction heuristics [12]. Unlike GHH's heuristics, GP-HH is constrained to operating at only one level of a level structure at a time.

Table 3 shows a performance comparison of the algorithms under test. All results associated with RCM and GK on the DWT set were taken from [20]. Because there were no results available in the literature for the BCSPWR set, we used the highly enhanced version of the RCM algorithm contained in the MATLAB library to compute the related envelopes. We do not report the results of GK on the BCSPWR problems as we did not have access to the original code, or a reliable software package.

As shown in the table, the results of GHH are extremely encouraging with respect to the mean of the envelope values and the number of the best results obtained (shown in the "Wins/Draws" rows). GHH's best evolved program outperforms RCM, GK and GP-HH's best evolved program by a significant margin, and produces extremely good results for the BCSPWR set.

Our system was implemented in C#, and all the experiments were performed on an AMD Athlon(tm) Dual-core Processor 2.20 GHz. We measured the time required for our method to solve each problem instance on this computer. The running times for DWT 59 (the smallest instance) and BCSPWR10 (the largest instance) were 0.0307 and 1.0094 seconds, respectively, while the average running

Table 3. Comparison of GHH's best evolved program against the RCM and GK algorithms as well as GP-HH's best evolved program

Instance	Dimension	Envelope			
		RCM	GK	GP-HH	GHH
DWT 59	59 × 59	314	314	327	**297**
DWT 66	66 × 66	217	**193**	**193**	194
DWT 72	72 × 72	**244**	**244**	355	291
DWT 87	87 × 87	696	682	685	**556**
DWT 162	162 × 162	1641	**1579**	1611	1610
DWT 193	193 × 193	5505	**4609**	4851	5196
DWT 209	209 × 209	3819	4032	3851	**3580**
DWT 221	221 × 221	2225	2154	2335	**2053**
DWT 245	245 × 245	4179	3813	4884	**3081**
DWT 307	307 × 307	8132	8132	8644	**7693**
DWT 310	310 × 310	3006	3006	3045	**2974**
DWT 361	361 × 361	5075	**5060**	**5060**	**5060**
DWT 419	419 × 419	8649	8073	8635	**7411**
DWT 503	503 × 503	15319	15042	15139	**13759**
DWT 592	592 × 592	11440	**10925**	11933	11160
DWT 758	758 × 758	8580	**8175**	8479	8250
DWT 869	869 × 869	19293	15728	16942	**15296**
DWT 878	878 × 878	22391	**19696**	22074	21572
DWT 918	918 × 918	23105	**20498**	22032	22471
DWT 992	992 × 992	38128	**34068**	37288	37288
DWT 1005	1005 × 1005	43068	40141	41525	**38107**
DWT 1007	1007 × 1007	24703	**22465**	24692	24156
DWT 1242	1242 × 1242	50052	52952	50515	**44666**
DWT 2680	2680 × 2680	105663	99271	105967	**92500**
Mean		16893.50	15868.83	16710.92	**15384.20**
Wins/Draws		0/1	8/3	0/2	**13/1**
BCSPWR05	443 × 443	11227	*NA*	10246	**5377**
BCSPWR06	1454 × 1454	64636	*NA*	55897	**29499**
BCSPWR07	1612 × 1612	75956	*NA*	65675	**32664**
BCSPWR08	1624 × 1624	79811	*NA*	80057	**33045**
BCSPWR09	1723 × 1723	80983	*NA*	76222	**42477**
BCSPWR10	5300 × 5300	672545	*NA*	655482	**296313**
Mean		164193.00	*NA*	157263.20	**73229.16**
Wins/Draws		0/0	*NA*	0/0	**6/0**

Numbers in bold face are the best results.

time was 0.1605 seconds. This reveals that our evolved algorithm is not only very effective but also extremely efficient.

4 Conclusions

We have proposed a hyper-heuristic approach (GHH) based on genetic programming for evolving envelope reduction algorithms. The algorithm is novel not only from the point of view of being the first to use GP on this problem but also because it incorporates new ideas for using a level structure system without its conventional constraints. Also, we have employed two novel features in the process of prioritising nodes for the construction of permutations.

The best heuristic generated by GHH were compared against two well-known and high-performance algorithms, i.e., the RCM and GK, as well as the best heuristic evolved by a hyper-heuristic method we previously developed, on a large set of standard benchmarks from the Harwell-Boeing sparse matrix collection. GHH's best evolved heuristic showed remarkable performance, both on benchmark instances from the same class as the training set and also on large problem instances from a totally different class, confirming the efficacy of our approach. The evolved heuristic was also extremely efficient.

References

1. Bader-El-Den, M.B., Poli, R.: A GP-based hyper-heuristic framework for evolving 3-SAT heuristics. In: Thierens, D., Beyer, H.G., Bongard, J., Branke, J., Clark, J.A., Cliff, D., Congdon, C.B., Deb, K., Doerr, B., Kovacs, T., Kumar, S., Miller, J.F., Moore, J., Neumann, F., Pelikan, M., Poli, R., Sastry, K., Stanley, K.O., Stutzle, T., Watson, R.A., Wegener, I. (eds.) GECCO 2007: Proceedings of the 9th Annual Conference on Genetic and Evolutionary Computation, vol. 2, pp. 1749–1749. ACM Press, London (2007)
2. Barnard, S.T., Pothen, A., Simon, H.D.: A spectral algorithm for envelope reduction of sparse matrices. In: Supercomputing 1993: Proceedings of the 1993 ACM/IEEE Conference on Supercomputing, pp. 493–502. ACM, New York (1993)
3. Barnard, S.T., Pothen, A., Simon, H.D.: A spectral algorithm for envelope reduction of sparse matrices, dedicated to william kahan and beresford parlett (1993), http://citeseer.ist.psu.edu/64928.html
4. Burke, E., Kendall, G., Newall, J., Hart, E., Ross, P., Schulenburg, S.: Hyper-Heuristics: An Emerging Direction in Modern Search Technology. In: Handbook of Metaheuristics. International Series in Operations Research & Management Science, ch. 16, pp. 457–474 (2003)
5. Cowling, P.I., Kendall, G., Soubeiga, E.: A Hyperheuristic Approach to Scheduling a Sales Summit. In: Burke, E., Erben, W. (eds.) PATAT 2000. LNCS, vol. 2079, pp. 176–190. Springer, Heidelberg (2001)
6. Cuthill, E., McKee, J.: Reducing the bandwidth of sparse symmetric matrices. In: ACM National Conference, pp. 157–172. Association for Computing Machinery, New York (1969)
7. Everstine, G.C.: A comparison of three resequencing algorithms for the reduction of matrix profile and wavefront. International Journal for Numerical Methods in Engineering 14, 837–853 (1979)

8. George, J.A.: Computer implementation of the finite element method. Ph.D. thesis, Stanford, CA, USA (1971)
9. Gibbs, N.E.: A hybrid profile reduction algorithm. ACM Transactions on Mathematical Software 2(4), 378–387 (1976)
10. Gibbs, N.E., Poole, W.G., Stockmeyer, P.K.: An algorithm for reducing the bandwidth and profile of a sparse matrix. SIAM Journal on Numerical Analysis 13(2), 236–250 (1976)
11. Keller, R.E., Poli, R.: Linear genetic programming of parsimonious metaheuristics. In: Srinivasan, D., Wang, L. (eds.) 2007 IEEE Congress on Evolutionary Computation, September 25-28, pp. 4508–4515. IEEE Computational Intelligence Society, IEEE Press, Singapore (2007)
12. Koohestani, B., Poli, R.: A hyper-heuristic approach to evolving algorithms for bandwidth reduction based on genetic programming. In: Bramer, M., Petridis, M., Nolle, L. (eds.) Research and Development in Intelligent Systems XXVIII, pp. 93–106. Springer, London (2011)
13. Koza, J.R.G.P.: On the Programming of Computers by Means of Natural Selection. MIT Press, Cambridge (1992)
14. Oltean, M.: Evolving evolutionary algorithms using linear genetic programming. Evolutionary Computation 13(3), 387–410 (Fall 2005)
15. Pissanetskey, S.: Sparse Matrix Technology. Academic Press, London (1984)
16. Poli, R., Langdon, W.B., McPhee, N.F.: A Field Guide to Genetic Programming (2008), published via http://lulu.com, with contributions by J. R. Koza
17. Poli, R.: Covariant tarpeian method for bloat control in genetic programming. In: Riolo, R., McConaghy, T., Vladislavleva, E. (eds.) Genetic Programming Theory and Practice VIII, Genetic and Evolutionary Computation, May 20-22, vol. 8, ch.5, pp. 71–90. Springer, Ann Arbor (2010)
18. Poli, R., Langdon, W.B., Holland, O.: Extending Particle Swarm Optimisation via Genetic Programming. In: Keijzer, M., Tettamanzi, A.G.B., Collet, P., van Hemert, J., Tomassini, M. (eds.) EuroGP 2005. LNCS, vol. 3447, pp. 291–300. Springer, Heidelberg (2005)
19. Poli, R., Woodward, J., Burke, E.K.: A histogram-matching approach to the evolution of bin-packing strategies. In: Srinivasan, D., Wang, L. (eds.) IEEE Congress on Evolutionary Computation, September 25-28, pp. 3500–3507. IEEE Computational Intelligence Society, IEEE Press, Singapore (2007)
20. Sloan, S.W.: A FORTRAN program for profile and wavefront reduction. International Journal for Numerical Methods in Engineering 28(11), 2651–2679 (1989)
21. Wang, Q., Shi, X.W.: An improved algorithm for matrix bandwidth and profile reduction in finite element analysis. Progress In Electromagnetics Research Letters 9, 29–38 (2009)

A Memetic Approach for the Max-Cut Problem

Qinghua Wu and Jin-Kao Hao*

LERIA, Université d'Angers, 2 Boulevard Lavoisier, 49045 Angers Cedex 01, France
{wu,hao}@info.univ-angers.fr

Abstract. The max-cut problem is to partition the vertices of a weighted graph $G = (V, E)$ into two subsets such that the weight sum of the edges crossing the two subsets is maximized. This paper presents a memetic max-cut algorithm (MACUT) that relies on a dedicated multi-parent crossover operator and a perturbation-based tabu search procedure. Experiments on 30 G-set benchmark instances show that MACUT competes favorably with 6 state-of-the-art max-cut algorithms, and for 10 instances improves on the best known results ever reported in the literature.

Keywords: Multi-parent crossover, memetic algorithm, local search, graph partitioning.

1 Introduction

Consider an undirected graph $G = (V, E)$ with vertex set $V = \{1, ..., n\}$ and edge set $E \subset V \times V$. Let $w_{ij} \in Z$ be the weight associated with edge $\{i, j\} \in E$. The well-known max-cut problem is to seek a partition of the vertex set V into two disjoint subsets $S_1 \subset V$ and $S_2 = V \setminus S_1$, such that the weight of the cut, defined as the sum of the weights on the edges connecting the two subsets, is maximized, i.e., $max \sum_{u \in S_1, v \in S_2} w_{uv}$. The max-cut problem, more precisely its weighted version, is one of Karp's 21 NP-complete problems [10].

The computational challenge of the max-cut problem has motivated a large number of solution procedures including approximation algorithms, exact methods and metaheuristics. The approximation approach (see for example [5,9,11]) provides a guaranteed performance, but is generally outperformed by other methods in computational testing. Recent examples on exact methods include the cut and price approach [14] and the branch and bound approach [18]. For large instances, various metaheuristic algorithms have been extensively used to find high-quality solutions in an acceptable time. Some representative examples include GRASP [4], ant colony [8], hybrid genetic algorithm [12], tabu search [16,13,21], scatter search [15], global equilibrium search [19] and maximum neural network [20].

In this paper, we present a memetic algorithm for the max-cut problem which is inspired by a very recent algorithm initially designed for the balanced max-bisection problem [22]. Experiments on a set of 30 well-known benchmark instances show that our memetic approach performs very well compared with state of the art algorithms.

---

* Corresponding author.

C.A. Coello Coello et al. (Eds.): PPSN 2012, Part II, LNCS 7492, pp. 297–306, 2012.

2   A Memetic Algorithm for the Max-Cut Problem

2.1   Outline of the Memetic Algorithm

Memetic algorithms are hybrid search methods that typically blend population-based search and neighborhood-based local search framework. The basic idea behind memetic approaches is to combine advantages of the crossover that discovers unexplored promising regions of the search space, and local optimization that finds good solutions by concentrating the search around these regions. The general architecture of our memetic algorithm for the max-cut problem is summarized in Algorithm 1. From an initial population of solutions which are first improved by a tabu search procedure, the algorithm carries out a number of evolution cycles. At each cycle, which is also called a generation, m ($m \geq 2$) parents are randomly selected to serve as parents and the crossover operator is applied to create an offspring solution, which is further optimized by tabu search. Subsequently, the population updating rule decides whether the improved offspring should be inserted into the population and which existing individual should be replaced. We describe below the main components of our memetic algorithm.

Algorithm 1. Memetic algorithm for the max-cut problem

Require: A weighted graph $G = (V, E, \omega)$, population size p
Ensure: The best solution I^* found
1: $Pop = \{I_1, ..., I_p\} \leftarrow GeneratePopulation(p)$ /* Section 2.3 */
2: $I^* \leftarrow Best(Pop)$
3: **while** Stop condition is not verified **do**
4:     $(I^1, ..., I^m) \leftarrow ChooseParents(Pop)$ /*Randomly select $m \geq 2$ parents */
5:     $I_0 = Recombination(I^1, ..., I^m)$ /* Section 2.5 */
6:     $I_0 \leftarrow Tabu\_Search(I_0)$ /* Section 2.4 */
7:     **if** $f(I_0) > f(I^*)$ **then**
8:         $I^* \leftarrow I_0$ /* Update the best solution found so far */
9:     **end if**
10:    $Pop \leftarrow Pool\_Updating(I_0, Pop)$ /* Section 2.6 */
11: **end while**

2.2   Search Space and Fitness Function

Given a graph $G = (V, E)$ where each edge $\{i, j\} \in E$ is assigned a weight w_{ij}, the search space explored by our memetic algorithm is defined as the set of all the partitions of V into 2 disjoint subsets, i.e., $\Omega = \{\{S_1, S_2\} : S_1 \cap S_2 = \emptyset, S_1 \cup S_2 = V\}$. For a given partition or cut $I = \{S_1, S_2\}$, its fitness $f(I)$ is the weight of the cutting edges crossing S_1 and S_2, i.e.,

$$f(I) = \sum_{i \in S_1, j \in S_2} w_{ij} \tag{1}$$

2.3   Initial Population

The initial population of size p is constructed as follows. For each individual, we first assign randomly the vertices of the graph to the two vertex subsets S_1 and S_2 to produce a starting solution, and then apply the tabu search improvement procedure (see section 2.4) to obtain a local optimum. The resulting solution is added to the population if the solution does not duplicate any solution in the population. This procedure is repeated until $2 \times p$ solutions are obtained from which we retain the p best ones to form the initial population.

2.4   Perturbation-Based Tabu Search Improvement

To improve the newly generated offspring created by the crossover, we apply a perturbation-based tabu search procedure which integrates a periodic perturbation mechanism to bring diversification into the search. The general procedure of our tabu search method is described in Algorithm 2. Starting from an given solution, the tabu search procedure is first used to optimize the solution as far as possible until the best solution found so far cannot be improved within a certain number of iterations (lines 6–12), then the perturbation mechanism is applied to the current solution to generate a new starting solution (line 13–15), whereupon a new round of tabu search is launched. This process is repeated until a maximum allowed number ($MaxIter$) of iterations is reached.

Algorithm 2. Perturbation-based tabu search for the max-cut problem

Require: A weighted graph $G = (V, E, \omega)$, initial solution $I = \{S_1, S_2\}$, number
$\quad Piter$ of consecutive iterations eclipsed before triggering a perturbation, number
$\quad MaxIter$ of tabu search iterations
Ensure: The best solution I^* found and $f(I^*)$
1: $I^* \leftarrow I$ /* Records the best solution found so far */
2: $Iter \leftarrow 0$ /* Iteration counter */
3: Compute the move gain Δ_v according to Eq. 2 for each vertex $v \in V$.
4: Initiate the tabu list and tabu tenure
5: **while** $Iter < MaxIter$ **do**
6:    Select an overall best allowed vertex $v \in V$ with the maximal move gain (ties
    are broken randomly)
7:    Move v from its original subset to the opposite set
8:    Update the tabu list and the move gain Δ_v for each $v \in V$
9:    **if** $f(I) > f(I^*)$ **then**
10:      $I^* \leftarrow I$ /* Update the best solution found so far */
11:    **end if**
12:    $Iter \leftarrow Iter + 1$
13:    **if** I^* not improved after $Piter$ iterations **then**
14:      $I \leftarrow Perturb(I)$ /* Apply perturbations to I */
15:    **end if**
16: **end while**

Our tabu search procedure employs a neighborhood defined by the simple one-flip move, which consists of moving a vertex $v \in V$ from its original subset to the opposite set. Notice that this neighborhood is larger than the neighborhood used in [22] where the move operator displaces consecutively two vertices between the two subsets of the current solution to keep the partition balance.

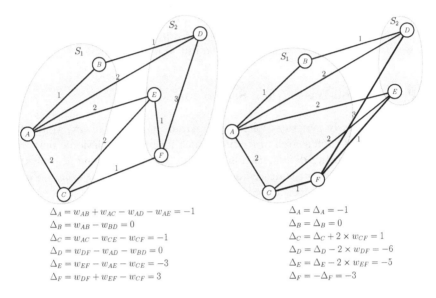

$$\Delta_A = w_{AB} + w_{AC} - w_{AD} - w_{AE} = -1 \qquad \Delta_A = \Delta_A = -1$$
$$\Delta_B = w_{AB} - w_{BD} = 0 \qquad \Delta_B = \Delta_B = 0$$
$$\Delta_C = w_{AC} - w_{CE} - w_{CF} = -1 \qquad \Delta_C = \Delta_C + 2 \times w_{CF} = 1$$
$$\Delta_D = w_{DF} - w_{AD} - w_{BD} = 0 \qquad \Delta_D = \Delta_D - 2 \times w_{DF} = -6$$
$$\Delta_E = w_{EF} - w_{AE} - w_{CE} = -3 \qquad \Delta_E = \Delta_E - 2 \times w_{EF} = -5$$
$$\Delta_F = w_{DF} + w_{EF} - w_{CF} = 3 \qquad \Delta_F = -\Delta_F = -3$$

Fig. 1. An example of the initialization (left) and update (right) of the move gain

The concept of move gain is used to represent the change in the fitness function f (Eq. 1, Section 2.2). It expresses how much a cut could be improved if a vertex v is moved from its subset to the other subset. In our implementation, we employ a streamlined incremental technique for fast evaluation of move gains. Specifically, let Δ_v be the move gain of moving vertex v to the other subset. Then initially, each move value can be calculated in linear time using the following formula (see Figure 1 (left)).

$$\Delta_v = \begin{cases} \sum_{x \in S_1, x \neq v} w_{vx} - \sum_{y \in S_2} w_{vy}, & \text{if } v \in S_1 \\ \sum_{y \in S_2, y \neq v} w_{vy} - \sum_{x \in S_1} w_{vx}, & \text{otherwise.} \end{cases} \qquad (2)$$

Each time one displaces a vertex v from its set to the other set, one just needs to update a subset of move gains affected by this move by applying the following abbreviated calculation (see Figure 1 (right)):

1. $\Delta_v = -\Delta_v$
2. for each $u \in V - \{v\}$,

$$\Delta_u = \begin{cases} \Delta_u - 2 \times w_{uv}, & \text{if } u \text{ is in the same set as } v \text{ before moving } v \\ \Delta_u + 2 \times w_{uv}, & \text{otherwise.} \end{cases}$$

Then each iteration of our tabu search procedure selects a move with the largest Δ value (breaking ties randomly) whcih is not forbidden by the tabu list. Each time a vertex v is moved from its original subset to the opposite subset, v is forbidden to go back to its original set for a certain number tt of iterations (tt is called the tabu tenure). The tabu tenure is tuned dynamically according to the mechanism described in [6]. Finally, a simple aspiration criterion is applied which allows a move to be performed in spite of being tabu if it leads to a solution better than the current best solution.

When the best solution cannot be further improved by tabu search, a perturbation operator is triggered to vary the local optimum solution from which a new round of tabu search is launched. The perturbation consists in randomly moving γ vertices from their original subsets to the opposite subsets where γ is a parameter which indicates the strength of the perturbation.

2.5   The Multi-parent Crossover

It is commonly admitted that, in order to be efficient, a crossover operator should be adapted to the problem being solved and should integrate useful problem-specific knowledge of the given problem.

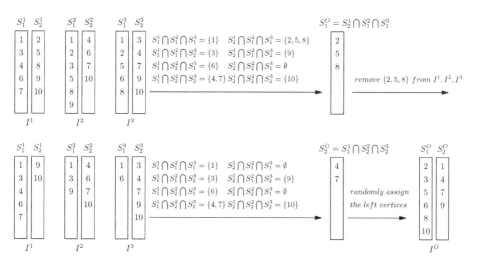

Fig. 2. An example of the multi-parent crossover operator

The max-cut problem is a grouping problem [3], i.e., a cut is composed of two distinct groups of vertices. An important principle in crossover design for grouping or partitioning problems is to manipulate promising groups of objects rather than individual objects. Such an approach for designing crossover operators has been successfully applied to solve a number of grouping problems such as graph coloring [7,17], bin packing [3] and graph partitioning [1,6].

We propose a grouping-based multi-parent crossover for the max-cut problem, the proposed crossover tries to preserve subsets (or grouping vertices) of the vertex partitions which are common to all parent individuals. More formally, given m chosen parents $\{I^1, ..., I^m\}$ ($m \geq 2$ is chosen randomly from a given range, fixed at $\{2,3,4\}$ in this paper), each cut I^i can be represented as $I^i = \{S_1^i, S_2^i\}$. Then we produce an offspring solution $I^O = \{S_1^O, S_2^O\}$ using these m parent individuals as follows.

We first select one subset from each of the m parents such that the cardinality of intersection of these chosen subsets is maximal. Then we build S_1^O as the intersection of these m selected subsets, i.e., $S_1^O = arg\ max\{|S_{x_1}^1 \cap ... \cap S_{x_m}^m| : x_1, ..., x_m \in \{1, 2\}\}$. When S_1^O is built, for each $v \in S_1^O$, v is removed from all the parent individual subsets in which it occurs. Then, S_2^O is constructed in the same way as for building S_1^O such that $S_2^O = arg\ max\{|S_{x_1}^1 \cap ... \cap S_{x_m}^m| : x_1, ..., x_m \in \{1, 2\}\}$. If a vertex v is left unassigned after this procedure, v is placed either to S_1^O or S_2^O at random. Figure 2 shows an example with 3 parents.

Notice that this crossover differs from that of [22] for at least two reasons. It operates on multi-parents (instead of 2 parents) and its offspring is not required to be a balanced cut.

2.6   The Population Updating Rule

The updating procedure of Pop is invoked each time an offspring solution is created by the crossover operator and then improved by tabu search. Specifically, the improved solution I^O is added into Pop if I^O is distinct from any solution in Pop and the fitness $f(I^O)$ is higher (better) than the worst solution I^w in Pop. Under this circumstance, we update Pop by replacing I^w with I^O.

3   Computational Results

3.1   Experimental Protocol and Benchmark Instances

Our MACUT algorithm is coded in C and compiled using GNU GCC on a PC (Pentium 2.83GHz CPU and 8G RAM). We show our results on a selection of 30 well-known G-set benchmark graphs (see Table 1)[1] [4,13,15,16,19,21]. The first 24 instances (with at most 3000 variables) are the most popular and we include 6 additional larger instances with 5000 to 10000 variables. The edge weights of these graphs take values in the set $\{-1, 0, 1\}$.

The parameters of our algorithm are determined by a preliminary experiment on a selection of problem instances and are fixed as follows: population size $p = 10$, non-improvement tabu search iterations before perturbation $Piter = 500$, perturbation strength $\gamma = 150$, number of tabu search iterations applied to each offspring $MaxIter = 10^6$, number of parents for crossover $m \in \{2, 3, 4\}$. Given the stochastic nature of MACUT, each instance is independently solved 20 times, each run being limited to 30 minutes for graphs with $|V| < 5000$ and 120 minutes

---

[1] Available at http://www.stanford.edu/~yyye/yyye/Gset/

for graphs with $|V| \geq 5000$. These timeout limits are comparable with the stop conditions used in [15,21].

3.2   Comparisons with the Best Known Results

Table 1 presents the detailed computational results of our MACUT algorithm as well as its underlying perturbation-based tabu search (PTS). The first two columns in the table indicate the name and the number of vertices of the graph. Column 3 presents the best-known objective value f_{pre} in the literature [4,13,15,16,19,21]. Columns 4 to 7 show MACUT's results including the best objective value (f_{best}), the averaged objective value (f_{avg}) over the 20 runs, the success rate (hit) for reaching f_{best} and the average CPU time in seconds ($time$) over the 20 runs for which the f_{best} value is reached. The last 4 columns present the results of its underlying perturbation-based tabu search.

Table 1. Computational results of MACUT and its underlying PTS on 30 G-set max-cut instances

Instance	$|V|$	f_{pre}	MACUT				PTS			
			f_{best}	f_{avg}	hit	$time(s)$	f_{best}	f_{avg}	hit	$time(s)$
G_1	800	11624	11624	11624	20/20	8.0	11624	11624	20/20	6.9
G_2	800	11620	11620	11620	20/20	6.3	11620	11620	20/20	7.5
G_3	800	11622	11622	11622	20/20	3.0	11622	11622	20/20	2.8
G_{11}	800	564	564	564	20/20	2.5	564	564	20/20	1.9
G_{12}	800	556	556	556	20/20	2.5	556	556	20/20	4.0
G_{13}	800	582	582	582	20/20	3.4	582	582	20/20	4.1
G_{14}	800	3064	3064	3063.95	19/20	450.0	3064	3063.9	18/20	661.2
G_{15}	800	3050	3050	3050	20/20	22.1	3050	3050	20/20	24.9
G_{16}	800	3052	3052	3052	20/20	11.6	3052	3052	20/20	10.7
G_{22}	2000	13359	13359	13359	20/20	74.8	13359	13359	20/20	206.2
G_{23}	2000	13342	**13344**	13344	20/20	280.0	13344	13343.2	12/20	651.7
G_{24}	2000	13337	13337	13337	20/20	252.2	13337	13335.6	20/20	815.6
G_{32}	2000	1410	1410	1410	20/20	349.2	1410	1408.3	3/20	844.6
G_{33}	2000	1382	1382	1382	20/20	391.4	1380	1379.6	18/20	667.6
G_{34}	2000	1384	1384	1384	20/20	220.7	1384	1381.6	2/20	512.4
G_{35}	2000	7685	**7686**	7685.9	18/20	895.7	7676	7674.2	2/20	1400.9
G_{36}	2000	7677	**7679**	7676.3	6/20	1395.4	7671	7669.3	1/20	1024.7
G_{37}	2000	7689	**7690**	7689.65	16/20	903.7	7678	7675.8	1/20	1175.3
G_{43}	1000	6660	6660	6660	20/20	3.6	6660	6660	20/20	3.7
G_{44}	1000	6650	6650	6650	20/20	3.7	6650	6650	20/20	3.0
G_{45}	1000	6654	6654	6654	20/20	18.2	6654	6654	20/20	20.1
G_{48}	3000	6000	6000	6000	20/20	0.2	6000	6000	20/20	0.2
G_{49}	3000	6000	6000	6000	20/20	0.4	6000	6000	20/20	0.4
G_{50}	3000	5880	5880	5880	20/20	15.0	5880	5880	20/20	13.6
G_{55}	5000	10236	**10299**	10290.8	2/20	2496.0	10235	10221	1/20	1807.3
G_{56}	5000	3934	**4016**	4006.9	2/20	2897.2	3954	3941.7	1/20	2108.9
G_{60}	7000	14057	**14186**	14171.1	1/20	5827.1	14065	14048.4	1/20	789.3
G_{65}	8000	5518	**5550**	5538.7	1/20	5879.6	5488	5479.1	1/20	1476.5
G_{66}	9000	6304	**6352**	6331.9	1/20	6203.8	6266	6255.2	1/20	2748.2
G_{67}	10000	6894	**6934**	6922.4	1/20	6761.3	6901	6892.1	1/20	1142.0

From Table 1, we observe that MACUT attains the best-known result for each of the 30 graphs. More importantly, MACUT improves on the best known results for 10 instances (indicated in bold). The average computing time required for MACUT to reach its best results f_{pre} varies from 3 seconds to 1.8 hours. It is clear that the required time to attain the current best-known objective value of column f_{pre} is shorter for the 10 graphs where MACUT finds improved solutions.

When comparing MACUT with its underlying PTS, one observes that MACUT outperforms PTS in terms of the best and average objective values. Indeed, for 10 instances, MACUT is able to achieve better solution with much larger cut values. For 15 instances, MACUT reaches larger average objective values than PTS. In particular, it is remarkable that for each of the instances with at least 5000 vertices, MACUT performs far better than PTS. These comparative results demonstrate that the crossover operator is essential for the success of our MACUT algorithm and help MACUT to discover better solutions that are not attainable by our tabu search algorithm alone.

3.3   Comparisons with State-of-Art Max-Cut Algorithm

To further assess the performance of our MACUT approach, we now compare the results of our MACUT algorithm with the most effective heuristic algorithms in the literature. Due to the differences among the programming languages, data structures, compiler options and computers, we do not focus on computing time. Instead, we are mainly interested in solution quality for this experiment. We just mention that the timeout limits we used are quite similar to those adopted by some recent references like [15,21].

Table 2. Comparison with 6 state-of-the-art algorithms in terms of the best results obtained

Instance	f_{pre}	f_{best}	Best results of 6 reference max-cut algorithms					
			GES[19]	SS[15]	TS-UBQP[13]	VNSPR[4]	CirCut[2]	GRASP-TS/PM[21]
G_1	11624	11624	11624	11624	11624	11621	11624	11624
G_2	11620	11620	11620	11620	11620	11615	11617	11620
G_3	11622	11622	11622	11622	11620	11622	11622	11620
G_{11}	564	564	564	562	564	564	560	564
G_{12}	556	556	556	552	556	556	552	556
G_{13}	582	582	582	578	580	580	574	582
G_{14}	3064	3064	3064	3060	3061	3055	3058	3063
G_{15}	3050	3050	3050	3049	3050	3043	3049	3050
G_{16}	3052	3052	3052	3045	3052	3043	3045	3052
G_{22}	13359	13359	13359	13346	13359	13295	13346	13349
G_{23}	13342	13344	13342	13317	13342	13290	13317	13332
G_{24}	13337	13337	13337	13303	13337	13276	13314	13324
G_{32}	1410	1410	1410	1398	1406	1396	1390	1406
G_{33}	1382	1382	1382	1362	1378	1376	1360	1374
G_{34}	1384	1384	1384	1364	1378	1372	1368	1376
G_{35}	7685	7686	7685	7668	7678	7635	7670	7661
G_{36}	7677	7679	7677	7660	7660	7632	7660	7660
G_{37}	7689	7690	7689	7664	7664	7643	7666	7670
G_{43}	6660	6660	6660	6656	6660	6659	6656	6660
G_{44}	6650	6650	6650	6648	6639	6642	6643	6649
G_{45}	6654	6654	6654	6642	6652	6646	6652	6654
G_{48}	6000	6000	6000	6000	6000	6000	6000	6000
G_{49}	6000	6000	6000	6000	6000	6000	6000	6000
G_{50}	5800	5800	5880	5880	5880	5880	5880	5880
G_{55}	10236	10299	-	-	10236	-	-	-
G_{56}	3934	4016	-	-	3934	-	-	-
G_{60}	14057	14186	-	-	14057	-	-	-
G_{65}	5518	5550	-	-	5518	-	-	-
G_{66}	6304	6352	-	-	6304	-	-	-
G_{67}	6894	6934	-	-	6894	-	-	-
Better			4	18	18	18	19	12
Equal			20	6	12	6	5	12
Worse			0	0	0	0	0	0

Table 2 compares our MACUT algorithm with 6 state-of-the-art algorithms, which cover the best known results for the tested instances. Columns 2 and 3 recall the previous best known results (f_{pre}) and the best results found by MACUT (f_{best}). Columns 4 to 9 present the best results obtained by these reference algorithms. The last three rows show the summary of the comparison between our MACUT algorithm and these reference algorithms. The rows 'Better', 'Equal' and 'Worse' respectively denotes the number of instances for which our MACUT algorithm gets better, equal and worse results than the corresponding reference algorithm. From the last three rows of Table 2, it is observed that our MACUT algorithm outperforms the 6 reference algorithms in terms of the quality of the best solution found. In comparison with each of these 6 algorithm, MACUT achieves at least 4 better solutions and in no case, MACUT's result is worse than that of these reference algorithms. This experiment confirms thus the effectiveness of the proposed memetic approach to deliver high quality solutions for the tested 30 benchmark max-cut instances.

4 Conclusions

We presented an effective memetic algorithm for the NP-hard max-cut problem. The proposed MACUT algorithm integrates a grouping-based multi-parent crossover which tries to preserve groups of the vertex shared by the parent solutions and a dedicated perturbation-based tabu search procedure. The design of our crossover operator is motivated by an experimental observation (not shown in the paper due to the page limit) that groups of vertices are always shared by high quality solutions. Experimental results confirmed that the crossover operator boosts the performance of the algorithm and helps the search to discover high quality solutions unachievable by a local search algorithm alone. The experiments of MACUT on 30 well-known G-set benchmark instances demonstrated, by providing new best results for 10 instances, its competitiveness compared to 6 state-of-the-art algorithms. Additional studies are needed to better understand the proposed algorithm.

Acknowledgment. We are grateful to the referees for their comments and questions which helped us to improve the paper. The work is partially supported by the RaDaPop (2009-2013) and LigeRO (2010-2013) projects (Pays de la Loire Region, France).

References

1. Benlic, U., Hao, J.K.: A multilevel memetic approach for improving graph k-partitions. IEEE Transactions on Evolutionary Computation 15(5), 624–642 (2011)
2. Burer, S., Monteiro, R.D.C., Zhang, Y.: Rank-two relaxation heuristics for max-cut and other binary quadratic programs. SIAM Journal on Optimization 12, 503–521 (2001)
3. Falkenauer, E.: Genetic algorithms and grouping problems. Wiley, New York (1998)

4. Festa, P., Pardalos, P.M., Resende, M.G.C., Ribeiro, C.C.: Randomized heuristics for the max-cut problem. Optimization Methods and Software 7, 1033–1058 (2002)
5. Frieze, A., Jerrum, M.: Improved approximation algorithm for max k-cut and max-bisection. Algorithmica 18, 67–81 (1997)
6. Galinier, P., Boujbel, Z., Fernandes, M.C.: An efficient memetic algorithm for the graph partitioning problem. Annals of Operations Research 191(1), 1–22 (2011)
7. Galinier, P., Hao, J.K.: Hybrid evolutionary algorithms for graph coloring. Journal of Combinatorial Optimization 3(4), 379–397 (1999)
8. Gao, L., Zeng, Y., Dong, A.: An ant colony algorithm for solving Max-cut problem. Progress in Natural Science 18(9), 1173–1178 (2008)
9. Goemans, M.X., Williamson, D.P.: Improved approximation algorithms for maximum cut and satisfiability problems using semidefinite programming. Journal of the Association for Computing Machinery 42(6), 1115–1145 (1995)
10. Karp, R.M.: Reducibility among combinatorial problems. In: Miller, R.E., Thacher, J.W. (eds.) Complexity of Computer Computation, pp. 85–103. Plenum Press (1972)
11. Karish, S., Rendl, F., Clausen, J.: Solving graph bisection problems with semidefinite programming. SIAM Journal on Computing 12, 177–191 (2000)
12. Kim, S.H., Kim, Y.H., Moon, B.Y.: A Hybrid Genetic Algorithm for the MAX CUT Problem. In: Genetic and Evolutionary Computation Conference, pp. 416–423 (2001)
13. Kochenberger, G., Hao, J.K., Lü, Z., Wang, H., Glover, F.: Solving large scale max cut problems via tabu search. Accepted to Journal of Heuristics (2012)
14. Krishnan, K., Mitchell, J.: A semidefinite programming based polyhedral cut and price approach for the Max-Cut problem. Computational Optimization and Applications 33, 51–71 (2006)
15. Marti, R., Duarte, A., Laguna, M.: Advanced scatter search for the max-cut problem. INFORMS Journal on Computing 21(1), 26–38 (2009)
16. Palubeckis, G.: Application of multistart tabu search to the MaxCut problem. Information Technology and Control 2(31), 29–35 (2004)
17. Porumbel, D.C., Hao, J.K., Kuntz, P.: An evolutionary approach with diversity guarantee and well-informed grouping recombination for graph coloring. Computers and Operations Research 37(10), 1822–1832 (2010)
18. Rendl, F., Rinaldi, G., Wiegele, A.: Solving Max-Cut to optimality by intersecting semidefinite and polyhedral relaxations. Mathematical Programming 121, 307–335 (2008)
19. Shylo, V.P., Shylo, O.V.: Solving the maxcut problem by the global equilibrium search. Cybernetics and Systems Analysis 46(5), 744–754 (2010)
20. Wang, J.: An Improved Maximum Neural Network Algorithm for Maximum Cut Problem. Neural Information Processing 10(2), 27–34 (2006)
21. Wang, Y., Lü, Z., Glover, F., Hao, J.K.: Probabilistic GRASP-tabu search algorithms for the UBQP problem. Accepted to Computers and Operations Research (2012), http://dx.doi.org/10.1016/j.cor.2011.12.006
22. Wu, Q., Hao, J.K.: Memetic search for the max-bisection problem. Accepted to Computers and Operations Research (2012), http://dx.doi.org/10.1016/j.cor.2012.06.001

An Improved Choice Function Heuristic Selection for Cross Domain Heuristic Search

John H. Drake[1], Ender Özcan[1], and Edmund K. Burke[2]

[1] School of Computer Science, University of Nottingham
Jubilee Campus, Wollaton Road, Nottingham, NG8 1BB, UK
{jqd,exo}@cs.nott.ac.uk
[2] Computing Science and Mathematics, School of Natural Sciences
University of Stirling, Stirling, FK9 4LA, Scotland
e.k.burke@stir.ac.uk

Abstract. Hyper-heuristics are a class of high-level search technologies to solve computationally difficult problems which operate on a search space of low-level heuristics rather than solutions directly. A iterative selection hyper-heuristic framework based on single-point search relies on two key components, a heuristic selection method and a move acceptance criteria. The Choice Function is an elegant heuristic selection method which scores heuristics based on a combination of three different measures and applies the heuristic with the highest rank at each given step. Each measure is weighted appropriately to provide balance between intensification and diversification during the heuristic search process. Choosing the right parameter values to weight these measures is not a trivial process and a small number of methods have been proposed in the literature. In this study we describe a new method, inspired by reinforcement learning, which controls these parameters automatically. The proposed method is tested and compared to previous approaches over a standard benchmark across six problem domains.

Keywords: Hyper-heuristics, Choice Function, Heuristic Selection, Cross-domain Optimisation, Combinatorial Optimization.

1 Introduction

The term 'hyper-heuristic' was first used in the field of combinatorial optimisation by Cowling et al. [1] and was defined as '*heuristics to choose heuristics*'. This paper investigated the application of a number of random, greedy and Choice Function-based hyper-heuristic approaches to a real-world sales summit scheduling problem using two deterministic move acceptance criteria, all moves (AM) and only improving (OI). Although the term hyper-heuristic was first used at this time, ideas which exhibited hyper-heuristic behaviour can be traced back as early as 1961 [2] in the field of job shop scheduling where combining scheduling rules was shown to perform better than taking any of the rules individually. In the first journal article to appear using the term Burke et al. [3] presented a tabu-search-based hyper-heuristic. In this system a set of low-level heuristics are

C.A. Coello Coello et al. (Eds.): PPSN 2012, Part II, LNCS 7492, pp. 307–316, 2012.
© Springer-Verlag Berlin Heidelberg 2012

ranked using rules based on reinforcement learning and compete against each other for selection. The hyper-heuristic selects the highest ranked heuristic not present in the tabu list. If an improvement is made after applying the selected heuristic its rank is increased, if not, its rank is decreased and it is placed in the tabu list until the current solution has changed. This hyper-heuristic was applied to nurse scheduling and university course timetabling problems obtaining competitive results. Hyper-heuristics have since been applied successfully to a wide range of problems such as examination timetabling [3–7], production scheduling [2], nurse scheduling [3, 8], bin packing [8, 9], sports scheduling [10], dynamic environments [11] and vehicle routing [8, 12].

Research trends have lead to a number of different hyper-heuristics approaches being developed, particularly those concerned with automatically generating new heuristics, for which the original definition of a hyper-heuristic is too limited to cover. A more general definition is offered by Burke et al. [13, 14]:

'A hyper-heuristic is a search method or learning mechanism for selecting or generating heuristics to solve computational search problems.'

This more general terminology includes systems which use high level strategies other than heuristics within the definition of hyper-heuristics and covers the two main classes of hyper-heuristics, those concerned with heuristic selection and those with heuristic generation. Here, our concern will be those methodologies which are used to select heuristics.

2   Selection Hyper-Heuristics and the Choice Function

Traditional single-point based search hyper-heuristics rely on two key components, a heuristic selection method and a move acceptance criteria as decomposed by Özcan et al. [15] and depicted in Figure 1. Such hyper-heuristics will sometimes be labelled *selection method-acceptance criteria* in this paper. Hyper-heuristics using this framework operate on a single solution and repeatedly select and apply low-level heuristics to this solution. At each stage a decision made as to whether to accept the move until some termination criteria is met.

Cowling et al. [1] experimented with a number of heuristic selection mechanisms including Simple Random and Choice Function using accept All Moves and accept Only Improving moves as acceptance criteria. Simple Random selects a heuristic to apply randomly from the set of low-level heuristics at each point in the search. The Choice Function is an elegant selection method which scores heuristics based on a combination of three different measures. The heuristic to apply is then be chosen by a strategy based on these scores. The first measure (f_1) records the previous performance of each individual heuristic, with more recent executions carrying larger weight. The value of f_1 for each low-level heuristic $h_1, h_2, ..., h_j$ is calculated as:

$$f_1(h_j) = \sum_n \alpha^{n-1} \frac{I_n(h_j)}{T_n(h_j)} \tag{1}$$

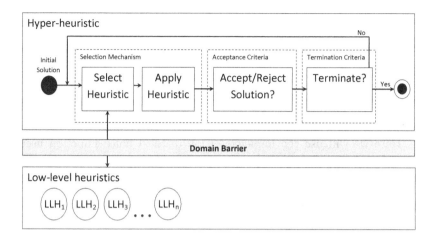

Fig. 1. Classic single-point search hyper-heuristic framework

where $I_n(h_j)$ is the change in evaluation function, $T_n(h_j)$ is the time taken to call the heuristic for each previous invocation n of heuristic h_j and α is a value between 0 and 1 giving greater importance to recent performance.

The second measure (f_2) attempts to capture any pair-wise dependencies between heuristics. Values of f_2 are calculated for each heuristic h_j when invoked immediately following h_k using the formula in Equation 6:

$$f_2(h_k, h_j) = \sum_n \beta^{n-1} \frac{I_n(h_k, h_j)}{T_n(h_k, h_j)} \tag{2}$$

where $I_n(h_k, h_j)$ is the change in evaluation function, $T_n(h_k, h_j)$ is the time taken to call the heuristic for each previous invocation n of heuristic h_j following h_k and β is a value between 0 and 1 which also gives greater importance to recent performance.

The third measure (f_3) is the time elapsed ($\tau(h_j)$) since the heuristic was last selected by the Choice Function. This allows all heuristics at least a small chance of selection.

$$f_3(h_j) = \tau(h_j) \tag{3}$$

In order to rank heuristics a score is given to each heuristic with Choice Function F calculated as:

$$F(h_j) = \alpha f_1(h_j) + \beta f_2(h_k, h_j) + \delta f_3(h_j) \tag{4}$$

where α and β as defined previously weight f_1 and f_2 respectively to provide intensification of the heuristic search process whilst δ weights f_3 to provide sufficient diversification. In this initial work these parameters were set as static values based on the authors experimental insight. This study showed the Choice Function selection combined with All Moves acceptance worked well.

Further to this work, Cowling et al. [16] described a method to adaptively change these parameters. A mechanism is proposed which increases the weights of α or β when using a heuristic selected by the Choice Function results in an improvement in the objective value. Although no specific implementation details are provided, this reward is said to be proportional to the size of improvement over the previous solution. Conversely, if a decrease in solution quality is obtained these weights are penalised proportionally to the change in objective value. Using this mechanism lead to an improved performance compared to the original results. Here we will use the implementation of the Choice Function used by Bilgin et al. [17] for benchmark function optimisation and Özcan et al. [4] and Burke et al. [6] for Examination Timetabling. This implementation increases α and β and reduces δ by the same value if an improvement is made and reduces α and β and increases δ if no improvement is made.

3   Modified Choice Function

Recently the HyFlex framework [8] was proposed and developed in order to support the first Cross-domain Heuristic Search Challenge, CHeSC 2011 [18]. HyFlex was designed with the goal of providing a common framework to test and compare different cross domain algorithms. Currently HyFlex contains six problem domains for algorithms to be tested on; maximum satisfiability (MAX-SAT), one-dimensional bin packing, personnel scheduling, permutation flow shop, the travelling salesman problem (TSP) and the vehicle routing problem (VRP). Using this framework allows us to directly compare our approach with previously proposed algorithms.

Using the classic version of the Choice Function has some limitations when applied to the HyFlex framework. Firstly, we are often not interested in the proportional improvement gained by a given heuristic but rather whether there has been any improvement at all. In the early stages of a search, a relatively poor heuristic could gain a large reward if it obtains a large improvement in objective value from a poor starting position. Later on in the search, a heuristic may yield a small improvement which is much more significant in the context of the optimisation process but will not receive such a large reward for this improvement. Secondly, if no improving solutions are found for a period of time, the Choice Function can very quickly descend into random search if the weighting is dominated by the diversification component. This can be a useful trait however the rate at which the diversification increases in significance must be controlled. Özcan et al. [4] observed that Simple Random heuristic selection with Late Acceptance Strategy move acceptance performed very well on a set of Examination Timetabling instances. In this particular case very few (4) perturbative low-level heuristics were implemented.

We propose a modified version of the Choice Function which aims to address these issues through the management of the parameters weighting f_1, f_2 and f_3 inspired by reinforcement learning [19]. This mechanism will rely on a system of reward and punishment in order to tune these parameters. Our Modified Choice

Function does not make a distinction between the values of α or β which weight f_1 and f_2 respectively and considers them as a single intensification parameter which we will refer to as simply ϕ. This value will also be used to give greater importance to recent performance as with the original Choice Function of Cowling et al. [1]. The parameter to weight f_3 is used to control the level of diversification of heuristic search as before and will still be referred to as δ. In the Modified Choice Function the score F_t for each heuristic h_j is now calculated as:

$$F_t(h_j) = \phi_t f_1(h_j) + \phi_t f_2(h_k, h_j) + \delta_t f_3(h_j) \tag{5}$$

where t is the number of invocations of h_j since an improvement was made using this heuristic. At each stage, if an improvement in objective value is made ϕ is rewarded and set to a static maximum value close to the upper limit of the interval $(0,1)$ whilst δ is concurrently reduced to a static minimum value close to the bottom end of this interval. This leads to a greater emphasis on intensification and greatly reduces the level of diversity of heuristic selection choice each time an improvement is obtained. If no improvement in objective value is made the level of intensification is decreased by linearly reducing ϕ and the weighting of diversification is increased at the same rate. This gives the intensification component of the Choice Function more time as the dominating factor in the calculation of F. For the experiments in this paper we define the parameters ϕ_t and δ_t as:

$$\phi_t(h_j) = \begin{cases} 0.99, & \text{if an improving move is made} \\ \max\{\phi_{t-1} - 0.01, 0.01\}, & \text{if a non-improving move is made} \end{cases} \tag{6}$$

and

$$\delta_t(h_j) = 1 - \phi_t(h_j) \tag{7}$$

4   Computational Results

Prior to the original competition, the results of eight hyper-heuristics were provided by the organisers to assess an algorithms performance [18]. These hyper-heuristics were inspired by state-of-the-art techniques from the hyper-heuristic literature. Each hyper-heuristic performs a single run on 10 instances for each of 4 problem domains; maximum satisfiability (MAX-SAT), one-dimensional bin packing, personnel scheduling and permutation flow shop. They are then ranked using a system based on the Formula One scoring system, the best performing hyper-heuristic for each instance is awarded 10 points, the second 8 points and then each further hyper-heuristic awarded 6, 5, 4, 3, 2, 1 and 0 points respectively. As this ranking system is based on relative performance, the Choice Function and Modified Choice Function are compared to the competition entries independently. All experiments were carried out on machines allowing a hyper-heuristic 576 seconds running time for each instance by the benchmarking tool provided by the competition organisers. In order for a fair comparison to be made crossover heuristics are ignored as the original Choice Function provides no details of how to manage operators which require more than one argument. Figure 2(a) shows the results of the Modified Choice Function using accept All Moves as an acceptance criteria when compared with the eight hyper-heuristics (HH1-HH8) provided for the competition. Figure 2(b) shows the results of the same experiments using the original Choice Function and accept All Moves acceptance as implemented by Bilgin et al. [17] and Burke et al. [6].

	HH1	HH2	HH3	HH4	HH5	HH6	HH7	HH8	ModCF-AM
MAX-SAT	55.25	73.25	36.5	24.5	1	46	51.75	14.5	**87.25**
Bin Packing	59	61	**76**	71	15	51	39	1	17
Personnel Scheduling	64	57.5	22	50.5	50	0	49.5	31	**65.5**
Flow Shop	30	21	26.5	86	19.5	**77.5**	21	69	39.5
Overall	208.25	212.75	161	**232**	85.5	174.5	161.25	115.5	209.25

(a)

	HH1	HH2	HH3	HH4	HH5	HH6	HH7	HH8	CF-AM
MAX-SAT	58	**78**	39.5	25.5	1	49	54.5	15.5	69
Bin Packing	59	61	**78**	71	19	51	38	7	6
Personnel Scheduling	**65.5**	64.5	23	52.5	53	0	52	31	48.5
Flow Shop	38	27.5	32	**86**	25.5	78.5	26.5	70	6
Overall	220.5	231	172.5	**235**	98.5	178.5	171	123.5	129.5

(b)

Fig. 2. Formula One scores for a single run of Modified Choice Function - All Moves hyper-heuristic and the CHeSC default hyper-heuristics (a) and a single run of classic Choice Function - All Moves hyper-heuristic and the CHeSC default hyper-heuristics

From these tables we see that the Modified Choice Function outperforms all of the CHeSC default hyper-heuristics in MAX-SAT and Personnel Scheduling. More importantly the Modified Choice Function outperforms the original Choice Function in all four problem domains although both versions seem to struggle more on the Bin Packing and Flow Shop instances. This could be due to the omission of crossover operators if such operators perform well in these problem domain. The best performing hyper-heuristic in this set (HH4) is based on iterative local search, this supports the work of Özcan et al. [20] which showed that the F_C selection hyper-heuristic framework performed well compared to other hyper-heuristic frameworks.

Following the competition the results were provided for the competition entries over a subset of the problems of all six problem domains. These results were taken as the median of 31 runs of each hyper-heuristic on each instance. Our results are also taken as the median of 31 runs in order to maintain consistency and allow direct comparison to the competition entries. Figure 3(a) shows the results of the classic Choice Function and All Moves acceptance criteria compared to the 20 competition entries using the Formula One scoring system. Figure 3(b) shows the results of the same experiments using the Modified Choice Function and accept All Moves as an acceptance criteria.

Rank	Name	Score
1	AdapHH	181
2	VNS-TW	134
3	ML	131.5
4	PHunter	93.25
5	EPH	89.25
6	HAHA	75.75
7	NAHH	75
8	ISEA	71
9	KSATS-HH	66
10	HAEA	53.5
11	ACO-HH	39
12	GenHive	36.5
13	DynILS	27
14	SA-ILS	24.25
15	XCJ	22.5
16	AVEG-Nep	21
17	GISS	16.75
18	SelfSearch	7
19	MCHH-S	4.75
20	**Classic CF - AM**	**1**
21	Ant-Q	0

(a)

Rank	Name	Score
1	AdapHH	177.1
2	VNS-TW	131.6
3	ML	127.5
4	PHunter	90.25
5	EPH	88.75
6	NAHH	72.5
7	HAHA	71.85
8	ISEA	68.5
9	KSATS-HH	61.35
10	HAEA	52
11	ACO-HH	39
12	**Modified CF - AM**	**38.85**
13	GenHive	36.5
14	DynILS	27
15	SA-ILS	22.75
16	XCJ	20.5
17	AVEG-Nep	19.5
18	GISS	16.25
19	SelfSearch	5
20	MCHH-S	3.25
21	Ant-Q	0

(b)

Fig. 3. Results of the median of 31 runs of the classic Choice Function - All Moves hyper-heuristic (a) and the Modified Choice Function - All Moves hyper-heuristic (b), compared to CHeSC competitors using Formula One scores

Since the competition results were made available, Di Gaspero and Urli [21] described variations of their original method (AVEG-Nep), which are also based on Reinforcement Learning. The best of the variants included in this paper ranked 13th overall compared to the original competitors. Here we see that managing the parameter settings of a Choice Function using Reinforcement Learning inspired techniques can outperform such methods, ranking 12th overall. For this hyper-heuristic points are only scored in two problem domains, Personnel Scheduling and MAX-SAT leaving room for improvement in the other four domains. The vast majority (32.85) of these points were scored in MAX-SAT where our method excels. When compared to the competition entries, the Modified Choice Function outperforms all other competitors. It is likely that a very small number of heuristics are providing improvement in this problem domain and the increased focus on intensification is providing the gain in performance. Figure 4 shows a breakdown of the number of points awarded to each technique over the MAX-SAT competition instances.

Using the classic choice function performs particularly badly against the other competition entries ranking 20th out of 21 overall only obtaining a single point in Personnel Scheduling. The Formula One scoring system is limited in that

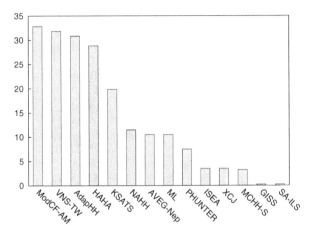

Fig. 4. Number of points scored in the MAX-SAT domain using the Formula One system for each CHeSC competitor

it only measures relative performance against a set of previous approaches. As such a more direct comparison between the two experiments can be performed on the objective values achieved by both hyper-heuristics. Table 1 shows the results of an independent student's t-test on the values of each of the 31 runs for each instance in the competition within a 95% confidence interval. These results show the Modified Choice Function statistically significantly outperforming the classic Choice Function completely in 3 of the 6 problem domains. In many cases there is no statistically significant difference in performance. In only 3 of the 30 problem instances the classic choice function performs statistically significantly better than the Modified Choice Function.

Table 1. Pairwise comparison between MCF-AM and CF-AM using independent T-Test. In this table s+ (s-) denotes that using MCF-AM (CF-AM) is performing statistically significantly better than using CF-AM (MCF-AM), while =+ (=−) denotes that there is no statistically significant performance variation between MCF-AM and CF-AM however MCF-AM (CF-AM) performs slightly better (worse) on average.

Problem Domain	Instance 1	Instance 2	Instance 3	Instance 4	Instance 5
MAX-SAT	s+	s+	s+	s+	s+
Bin Packing	=−	s+	s+	=−	s−
Personnel Scheduling	=−	=+	=+	=−	=+
Flow Shop	s+	s+	s+	s+	s+
TSP	s−	=−	s−	s+	=+
VRP	s+	s+	s+	s+	s+

5   Concluding Remarks

In this work we have described a modified version of the Choice Function heuristic selection method which manages the parameters which weight the intensification and diversification components of Choice Function scores through methods inspired by reinforcement learning. This Modified Choice Function aggressively rewards the intensification weighting and heavily punishes the diversification component each time an improvement is made. We have shown that managing these parameters in such a way provides great benefits compared to a classic implementation of the Choice Function. So far, this work has been limited to improving the Choice Function selection mechanism itself. Previous work in the literature has suggested that performance can be improved by reducing the search space of heuristics [22, 23]. We plan to include the method proposed by Özcan and Kheiri [24] to reduce the set of active heuristics in combination with the Modified Choice Function heuristic selection method and apply it to the problem instances available in HyFlex. We restricted this study to focus on only the selection mechanism component of a traditional hyper-heuristic, Özcan et al. [15, 20] tested a number of hyper-heuristics over a set of benchmark functions and observed that the acceptance criteria used can have a more significant impact on the performance of a hyper-heuristic than selection mechanism. We would like to extend this work to analyse the effect of using different move acceptance criteria in conjunction with the Modified Choice Function. In this paper we have not made use of any operators in the HyFlex framework which require more than one argument such as crossover. Drake et al. [25] described a number of methods for managing potential second arguments for crossover and other n-ary operators. As future work we will include the multiple argument management techniques from this study to analyse whether including crossover operators can benefit hyper-heuristics based on the Modified Choice Function.

References

1. Cowling, P.I., Kendall, G., Soubeiga, E.: A Hyperheuristic Approach to Scheduling a Sales Summit. In: Burke, E., Erben, W. (eds.) PATAT 2000. LNCS, vol. 2079, pp. 176–190. Springer, Heidelberg (2001)
2. Fisher, M., Thompson, G.: Probabilistic learning combinations of local job-shop scheduling rules. In: Factory Scheduling Conference (1961)
3. Burke, E.K., Kendall, G., Soubeiga, E.: A tabu-search hyperheuristic for timetabling and rostering. Journal of Heuristics 9, 451–470 (2003)
4. Özcan, E., Bykov, Y., Birben, M., Burke, E.K.: Examination timetabling using late acceptance hyper-heuristics. In: CEC 2009, pp. 997–1004 (2009)
5. Özcan, E., Misir, M., Ochoa, G., Burke, E.K.: A reinforcement learning - great-deluge hyper-heuristic for examination timetabling. Int. Journal of Applied Meta-heuristic Computing 1, 39–59 (2010)
6. Burke, E.K., Kendall, G., Misir, M., Özcan, E.: Monte carlo hyper-heuristics for examination timetabling. Annals of Operations Research (in press)
7. Sabar, N.R., Ayob, M., Qu, R., Kendall, G.: A graph coloring constructive hyper-heuristic for examination timetabling problems. Applied Intelligence (in press)
8. Ochoa, G., Hyde, M., Curtois, T., Vazquez-Rodriguez, J.A., Walker, J., Gendreau, M., Kendall, G., McCollum, B., Parkes, A.J., Petrovic, S., Burke, E.K.: HyFlex: A Benchmark Framework for Cross-Domain Heuristic Search. In: Hao, J.-K., Middendorf, M. (eds.) EvoCOP 2012. LNCS, vol. 7245, pp. 136–147. Springer, Heidelberg (2012)

9. López-Camacho, E., Terashima-Marín, H., Ross, P.: A hyper-heuristic for solving one and two-dimensional bin packing problems. In: GECCO 2011, pp. 257–258. ACM, New York (2011)
10. Gibbs, J., Kendall, G., Özcan, E.: Scheduling English Football Fixtures over the Holiday Period Using Hyper-heuristics. In: Schaefer, R., Cotta, C., Kołodziej, J., Rudolph, G. (eds.) PPSN XI. LNCS, vol. 6238, pp. 496–505. Springer, Heidelberg (2010)
11. Kiraz, B., Uyar, A.Ş., Özcan, E.: An Investigation of Selection Hyper-heuristics in Dynamic Environments. In: Di Chio, C., Cagnoni, S., Cotta, C., Ebner, M., Ekárt, A., Esparcia-Alcázar, A.I., Merelo, J.J., Neri, F., Preuss, M., Richter, H., Togelius, J., Yannakakis, G.N. (eds.) EvoApplications 2011, Part I. LNCS, vol. 6624, pp. 314–323. Springer, Heidelberg (2011)
12. Garrido, P., Castro, C.: Stable solving of cvrps using hyperheuristics. In: GECCO 2009, pp. 255–262. ACM (2009)
13. Burke, E.K., Hyde, M., Kendall, G., Ochoa, G., Özcan, E., Woodward, J.: A Classification of Hyper-heuristics Approaches. In: Handbook of Metaheuristics, 2nd edn., pp. 449–468. Springer (2010)
14. Burke, E.K., Hyde, M., Kendall, G., Ochoa, G., Özcan, E., Qu, R.: Hyperheuristics: A survey of the state of the art. Technical Report No. NOTTCS-TR-SUB-0906241418-2747, School of Comp. Sci., University of Nottingham (2010)
15. Özcan, E., Bilgin, B., Korkmaz, E.E.: Hill Climbers and Mutational Heuristics in Hyperheuristics. In: Runarsson, T.P., Beyer, H.-G., Burke, E.K., Merelo-Guervós, J.J., Whitley, L.D., Yao, X. (eds.) PPSN IX. LNCS, vol. 4193, pp. 202–211. Springer, Heidelberg (2006)
16. Cowling, P., Kendall, G., Soubeiga, E.: A parameter-free hyperheuristic for scheduling a sales summit. In: MIC 2001, pp. 127–131 (2001)
17. Bilgin, B., Özcan, E., Korkmaz, E.E.: An Experimental Study on Hyper-Heuristics and Exam Timetabling. In: Burke, E.K., Rudová, H. (eds.) PATAT 2006. LNCS, vol. 3867, pp. 394–412. Springer, Heidelberg (2007)
18. Ochoa, G., Hyde, M.: The cross-domain heuristic search challenge (CHeSC 2011) (2011), http://www.asap.cs.nott.ac.uk/chesc2011/
19. Kaelbling, L.P., Littman, M.L., Moore, A.W.: Reinforcement learning: a survey. Journal of Artificial Intelligence Research 4, 237–285 (1996)
20. Özcan, E., Bilgin, B., Korkmaz, E.E.: A comprehensive analysis of hyper-heuristics. Intelligent Data Analysis 12(1), 3–23 (2008)
21. Di Gaspero, L., Urli, T.: Evaluation of a family of reinforcement learning cross-domain heuristics for optimization. In: LION 6 (2012)
22. Özcan, E., Basaran, C.: A case study of memetic algorithms for constraint optimization. Soft Computing 13(8-9), 871–882 (2009)
23. Chakhlevitch, K., Cowling, P.I.: Choosing the Fittest Subset of Low Level Heuristics in a Hyperheuristic Framework. In: Raidl, G.R., Gottlieb, J. (eds.) EvoCOP 2005. LNCS, vol. 3448, pp. 23–33. Springer, Heidelberg (2005)
24. Özcan, E., Kheiri, A.: A hyper-heuristic based on random gradient, greedy and dominance. In: ISCIS 2011, pp. 404–409 (2011)
25. Drake, J.H., Özcan, E., Burke, E.K.: Controlling crossover in a selection hyper-heuristic framework. Technical Report No. NOTTCS-TR-SUB-1104181638-4244, School of Computer Science, University of Nottingham (2011)

Optimizing Cellular Automata through a Meta-model Assisted Memetic Algorithm

Donato D'Ambrosio[1], Rocco Rongo[2],
William Spataro[1], and Giuseppe A. Trunfio[3]

[1] Department of Mathematics, University of Calabria, 87036 Rende (CS), Italy
[2] Department of Earth Sciences, University of Calabria, 87036 Rende (CS), Italy
[3] DADU, University of Sassari, 07041 Alghero (SS), Italy

Abstract. This paper investigates the advantages provided by a Meta-model Assisted Memetic Algorithm (MAMA) for the calibration of a Cellular Automata (CA) model. The proposed approach is based on the synergy between a global meta-model, based on a radial basis function network, and a local quadratic approximation of the fitness landscape. The calibration exercise presented here refers to SCIARA, a well-established CA for the simulation of lava flows. Compared with a standard Genetic Algorithm, the adopted MAMA provided much better results within the assigned computational budget.

Keywords: Cellular Automata, Model Calibration, Meta-modelling, Memetic Algorithms.

1 Introduction

Most applications of Cellular Automata (CA) models to the simulation of real complex systems, require a preliminary calibration process [1,2]. The latter consists of finding the unknown values of the model parameters in such a way that the outcomes of the model itself better correspond to the observed dynamics of the system under consideration. For such purpose, automated methods have been developed by defining calibration as a global optimization problem in which the solution in terms of parameter values must maximize a fitness measure [2,3]. Because of the size of the search space, such a process usually requires a large number of fitness evaluations, which consist of computationally expensive CA simulations. Hence, in dealing with CA calibration the use of parallel computing is often mandatory [2]. As shown in [4], an additional strategy for increasing the search efficiency may consists of the so-called meta-model assisted (or surrogate assisted) optimization [5], which is based on inexpensive surrogate functions able to approximate the fitness corresponding to the CA simulations.

However, in most cases the global search struggles to provide an accurate solution. This is often because search heuristics, for example based on the Genetic Algorithms (GA) operators, are more effective at exploring the search space

C.A. Coello Coello et al. (Eds.): PPSN 2012, Part II, LNCS 7492, pp. 317–326, 2012.

rather than at the fine-tuning of a particular solution candidate. Therefore, a further enhancement of the classical calibration based on a pure global search approach, may be obtained introducing a local search (LS) phase, which in many applications proved to be capable of providing much more efficient and accurate global optimization processes. For example, in the hybrid GAs known as Memetic Algorithms [6] a sub-process of LS is introduced to refine individuals by more or less standard hill climbing procedures. As in the case of the meta-modelling approach, such hybridization has the main aim of increasing the overall efficiency of the optimization process (i.e., leading to better solutions within an assigned computational budget). In some recent applications, also Meta-model Assisted Memetic Algorithms (MAMAs) have been described and successfully applied to optimization problems [7,8]. However, to our knowledge the advantages provided by a MAMA for the calibration of CAs have not been explored. Some preliminary results in this direction are the object of this paper, in which a MAMA has been applied to the calibration of a well-established CA for the simulation of lava flows, namely the SCIARA model [2,9].

The paper is organized as follows. In section 2 the CA calibration problem is formalized. Section 3 describes in detail the tested MAMA. Section 4 illustrates the results of the numerical experiments and section 5 concludes the paper outlining possible future work.

2   Optimization of Cellular Automata

In many applications of the CA modelling approach the cells' transition function depends on a vector of constant parameters $\mathbf{p} = [p_1, \ldots, p_n]^T$, which belongs to a set Λ (e.g. [1,2]). In particular, the overall transition function Φ gives the global configuration $\Omega^{(t+1)}$ (i.e. the set of all cell states) at the step $t + 1$ as:

$$\Omega^{(t+1)} = \Phi(\Omega^{(t)}, \mathbf{p}) \tag{1}$$

The iterative application of Φ, starting from an initial configuration $\Omega^{(0)}$, leads to the CA simulation:

$$\Omega^{(0)} \xrightarrow{\Phi} \Omega^{(1)} \xrightarrow{\Phi} \cdots \xrightarrow{\Phi} \Omega^{(t)} \implies \Omega^{(t)} = \Phi^t(\Omega^{(0)}, \mathbf{p}) \tag{2}$$

where the dependence of the automaton configuration at the time step t on both the initial configuration and the parameters is explicit, with the other automaton characteristics (i.e. the model structure) being fixed.

The CA model can be optimized with respect to $\mathbf{p}$ to maximise the agreement between the simulated patterns and those belonging to a spatio-temporal dataset $\bar{\mathcal{V}}$, which come from an experiment of the real system behaviour. In particular, let $\bar{\mathcal{V}}$ be composed by a sequence of q configurations:

$$\bar{\mathcal{V}} = \left\{ \bar{\Omega}^{(k)} : k \in \{0, \tau_1, \ldots, \tau_q\} \right\} \tag{3}$$

where $\tau_i \in \mathbb{N}$ indicates the time step in which a configuration is known. Starting from $\bar{\Omega}^{(0)}$, and given a vector $\mathbf{p}$ of parameters, the process (2) can be executed for the computation of the $q - 1$ configurations:

$$\mathcal{V} = \left\{ \Omega^{(k)} : k \in \{\tau_1, \ldots, \tau_q\} \right\} \tag{4}$$

where $\Omega^{(j)} = \Phi^j(\bar{\Omega}^{(0)}, \mathbf{p})$. The agreement θ between the real and simulated processes is usually measured by a suitable fitness function:

$$\theta = \Theta\left(\bar{\mathcal{V}}, \mathcal{V}\right) = \Theta\left(\bar{\mathcal{V}}, \mathbf{p}\right) \tag{5}$$

Therefore, the calibration consists of the following maximisation problem:

$$\max_{\mathbf{p} \in \Lambda} \Theta\left(\bar{\mathcal{V}}, \mathcal{V}\right) \tag{6}$$

which involves finding a proper value of $\mathbf{p}$ that leads to the best agreement between the real and simulated spatio-temporal sequences.

Different heuristics have been used to tackle the automatic solution of problem (6) [2,3]. In this paper a MAMA has been adopted, which is designed according to some of the basic ideas described in [7]. The optimization process consists of a GA assisted by a fitness approximation model and endowed with a LS phase. The MAMA is used to evolve a population, whose generic *chromosome* is a n-dimensional vector $\mathbf{p}$. In the latter, the i-th element is obtained as the binary encoding of the parameter p_i. Each chromosome can be decoded back in a vector of parameters $\mathbf{p}$ and, through performing a CA simulation, the corresponding fitness can be computed.

3   A Meta-Model Assisted Memetic Algorithm

In the MAMA object of this paper the original fitness evaluations are partly replaced by the fitness estimates provided by an inexpensive model. This allows to reduce the number of CA simulations needed to evaluate the individuals generated by the genetic operators during the search. As detailed later, the CA simulations carried out during the optimization provide a training set $\mathcal{T}$:

$$\mathcal{T} = \left\{ \langle \theta^{(1)}, \mathbf{p}^{(1)} \rangle, \langle \theta^{(2)}, \mathbf{p}^{(2)} \rangle, \ldots, \langle \theta^{(n_t)}, \mathbf{p}^{(n_t)} \rangle \right\} \tag{7}$$

where each fitness value $\theta^{(i)}$ corresponds to a parameter vector $\mathbf{p}^{(i)}$. Thus, on the basis of the patterns in $\mathcal{T}$ a meta-model $\hat{\theta}$ is dynamically built for evaluating each candidate solution $\mathbf{p}$ through the estimated fitness value $\hat{\theta}(\mathbf{p})$.

It is worth noting that, given the patterns in $\mathcal{T}$, either a global meta-model or a local one can be trained [5,7]. For example, an *ad-hoc* surrogate of the real fitness can be constructed for each individual $\mathbf{p}$ to be evaluated using only the k nearest neighbours of $\mathbf{p}$ in $\mathcal{T}$. However, even if a local meta-model can potentially be more accurate than a global one, the cost of training a number of local surrogates should be compared with the cost of the true fitness evaluation.

Since the meta-model only provides a more or less accurate approximation of the fitness landscape, to avoid convergence to false optima the surrogate-assisted optimization should also use, in some way, the true fitness function

[5]. On the other hand, the involvement of the latter should be minimized due to its high computational cost. A trade-off is provided by a suitable *evolution control* strategy. In particular, the approach adopted in this paper is the so-called *individual-based control*, which consists of using the true fitness function to evaluate at each generation some of the offspring individuals (i.e. the *controlled individuals*). The latter are chosen according to the so-called *best strategy*, in which the exact fitness value is assigned to some of the individuals that are the best according to the meta-model. For a detailed discussion about the commonly adopted evolution control strategies, the reader is referred to [5].

The pseudo-code of the corresponding MAMA is outlined in Figure 1. As in other elitist GAs, the optimization procedure begins with the initialization and exact evaluation of a population of individuals encoding CA vectors of parameters. The evolutionary search is iterated until the assigned budget n_{sim} of CA evaluations is exhausted. During the search, each CA simulation leads to a new element for the archive $\mathcal{T}$ of the training patterns. Usually, in the first GA generations the set $\mathcal{T}$ does not contains enough elements to build a reliable meta-model. Thus, while the current number of elements in $\mathcal{T}$ is less than the threshold $\rho_1 n_{\text{sim}}$ (see line 6), where $\rho_1 \in [0, 1]$, the search consists of a standard GA, in which the fitness evaluations are carried out through CA simulations (see line 8). Subsequently, when $|\mathcal{T}| \geq \rho_1 n_{\text{sim}}$ the meta-model $\hat{\theta}$ is built/updated at each generation (see line 10) in order to estimate the fitness of each individual belonging to the set of offspring $\mathcal{S}$. The adopted global surrogate is a Radial

```
1    Q ← populationInit();
2    for each q in Q do
3        simulateCAAndUpdateArchive(q, T);
4    while ( |T| < n_sim )
5        S ← crossoverAndMutation(Q);
6        if ( |T| < ρ1 n_sim )
7            for each q ∈ S do
8                simulateCAsAndUpdateArchive(q, T);
9        else
10            θ̂ ← createRBFN(T);
11            for each q ∈ S do
12                surrogateFitnessEvaluation(q, θ̂);
13            κ ← π |S|;
14            controlTheBestAndUpdateArchive(S, κ, T);
15            if ( |T| > ρ2 n_sim )
16                for i = 0 to κ do
17                    S[i] ←localSearchAndUpdateArchive(S[i], T);
18            end if
19        end if
20        Q ← elitistSelection(Q, S);
21    end while
```

Fig. 1. Outline of the meta-model assisted memetic CA optimization. The variable n_{sim} indicates the assigned budget of CA evaluations for the optimization process.

Basis Function Network (RBFN), a special type of artificial neural network that uses radial basis functions as activation functions [10]. The RBFN is often used as surrogate to assist optimizations because of its good generalization ability and because of its simpler topology compared to other networks [5,7]. Formally, the adopted RBFN can be expressed as:

$$\theta(\mathbf{p}) = \sum_{i=1}^{n_h} w_i \delta(\mathbf{p} - \mathbf{c}_i) \tag{8}$$

where n_h is the number of hidden neurons, $\delta(\mathbf{x})$ is the kernel function, $\mathbf{c}_i$ is the i-th center and w_i are the weights. The adopted kernel function is the Gaussian:

$$\delta(\mathbf{p} - \mathbf{c}_i) = \exp\left(-\frac{\|\mathbf{p} - \mathbf{c}_i\|^2}{2\sigma_i^2}\right) \tag{9}$$

where σ_i is the bandwidth assumed for the centre $\mathbf{c}_i$. In particular, the RBFN implementation has been based on the SHARK C++ library, a machine learning framework for regression and classification tasks including neural networks and kernel methods [11]. The first stage for building $\hat{\theta}$ consists of a fully unsupervised learning in which the centres and the corresponding bandwidths are determined. In particular: (i) first the RBFN centres $\mathbf{c}_i$ are obtained by few iterations of a k-means clustering algorithm on the set $\mathcal{T}$; (ii) then the value of σ_i for a given cluster centre $\mathbf{c}_i$ is set to the average Euclidean distance between $\mathbf{c}_i$ and the training vectors which belong to that cluster. Subsequently, the weights w_i need to be trained to achieve good generalization. In this work, the weights of $\hat{\theta}$ are trained using the iRprop algorithm [12] implemented in the SHARK library, which is quite fast and efficient.

Once all the offspring are evaluated through $\hat{\theta}$, in order to avoid convergence towards false optima, the control strategy mentioned above is applied at line 14 by invoking the function *controlTheBestAndUpdateArchive*. In particular, the latter ensures that the fraction π of the individuals in $\mathcal{S}$ which are the best according to $\hat{\theta}$ are re-evaluated through CA simulations. As a further result of the function *controlTheBestAndUpdateArchive*, the first $\kappa = \pi |\mathcal{S}|$ offspring in $\mathcal{S}$ are sorted in descending order according to their fitness. Also, each CA simulation carried out during the control process contributes to the enrichment of the archive $\mathcal{T}$, which is used for future meta-model buildings/updates.

In an advanced stage of the optimization, in particular when $|\mathcal{T}| > \rho_2 n_{\text{sim}}$, with $\rho_2 \in [0,1]$ and $\rho_2 > \rho_1$, the κ controlled individuals are taken as starting point of the LS. The latter starts from each controlled individual $\mathbf{p}$ and is conducted on the local Quadratic Polynomial Approximation (QPA) defined as:

$$\theta(\mathbf{p}) = \beta_0 + \sum_{1 \le i \le n} \beta_i \, p_i + \sum_{1 \le i \le j \le n} \beta_{(n-1+i+j)} \, p_i \, p_j = \boldsymbol{\beta}^T \, \tilde{\mathbf{p}} \tag{10}$$

where:

$$\boldsymbol{\beta} = [\beta_0, \beta_1, \ldots, \beta_{n_v - 1}]^T \tag{11}$$

is the vector collecting the $n_v = (n+1)(n+2)/2$ model coefficients and:

$$\tilde{\mathbf{p}} = [1, p_1, p_2, \ldots, p_1 p_2, \ldots, p_n^2]^T \tag{12}$$

is the vector of the CA parameters mapped into the polynomial model. In particular, in order to improve the LS reliability, an ad-hoc local QPA is built for each individual $\mathbf{p}$ on the basis of its $n_s \geq n_v$ nearest neighbours in $\mathcal{T}$. In this study, the model coefficients $\boldsymbol{\beta}$ are estimated using the least square method.

Even using an accurate QPA, the LS procedure may converge towards a point that does not represent an actual improvement of the starting individual $\mathbf{p}$. Hence, at the cost of some more CA simulations, the LS has been based on a trust-region approach [13]. In the latter, the LS iteratively operates on a region in which the accuracy of the QPA is verified by executing *ad-hoc* CA simulations. In particular, if the QPA accuracy is satisfying then the region is expanded; conversely, if the QPA accuracy is poor then the region is contracted. In practice, following the classical trust-region approach, the LS is structured in a sequence of subproblems as follows:

$$\max \hat{\psi}(\mathbf{p}^{(j)} + \mathbf{d}), \qquad j = 0, 1, 2, \ldots, \lambda$$
$$\text{subject to } \|\mathbf{d}\| \leq r^{(j)} \tag{13}$$

where $\hat{\psi}(\mathbf{x})$ is the QPA meta-model, $\mathbf{p}^{(j)}$ is the starting point of the j-th iteration (i.e. $\mathbf{p}^{(0)}$ is the individual to optimize), $\mathbf{p}^{(j)} + \mathbf{d}$ represents a point within the current trust-region radius $r^{(j)}$. In this paper, the BLG code for solving an optimization problem with bound constraints through a gradient method, described in [14], is used for the trust-region subproblems. At the first sub-problem of the LS, the radius $r^{(0)}$ is initialized as the average of all the n_s nearest neighbours of $\mathbf{p}^{(0)}$ in $\mathcal{T}$. Then, the value of $r^{(j)}$ is determined for each of the following sub-problems on the basis of a parameter $\omega^{(j)}$, which is computed at the end of each subproblem as follows:

$$\omega^{(j)} = \frac{\theta(p^{(j)}) - \theta(p_{\text{opt}}^{(j)})}{\hat{\theta}(p^{(j)}) - \hat{\theta}(p_{\text{opt}}^{(j)})} \tag{14}$$

where each evaluation of the function $\theta(\mathbf{x})$ requires a CA simulation. Then, the trust region is contracted or expanded for low or high values of $\omega^{(j)}$ respectively, according to the empirical rule described in [7].

The LS process terminates when the maximum number of subproblems λ is reached. The latter parameter represents the individual learning intensity, that is the amount of computational budget in terms of CA simulations devoted on improving a single solution. At the end of each LS, any locally optimized vector of CA parameters is encoded back into the offspring according to a Lamarckian evolutionary approach [15].

4   Calibration Tests and Discussion

A Master-slaves parallel version of the MAMA described above has been developed and applied to the last release of SCIARA, a CA model for lava flows simulation. In the current implementation, based on the Message Passing paradigm, a

Table 1. Parameters object of calibration, explored ranges and target values

Parameter	Explored range for calibration	Target value
r_s	[0, 1]	0.096
r_v	[0, 1]	0.853
h_s [m]	[1, 50]	13.67
h_v [m]	[1, 50]	1.920
p_c	[0, 100]	8.460

master process executes the algorithm outlined in Figure 1, while the remaining processes carry out all the required CA simulations.

In the SCIARA model, which is described in detail in [9], a specific component of the transition function computes lava outflows from the central cell towards its neighbouring ones on the basis of the altitudes, lava thickness and temperatures in the neighbourhood. In the model, lava can flow out when its thickness overcomes a *critical height*, so that the basal stress exceeds the yield strength. The critical height mainly depends on the lava temperature according to a power law. Moreover, viscosity is accounted in terms of flow relaxation rate, being this latter the parameter of the distribution algorithm that influences the amount of lava that actually leaves the cell. At each time-step the new cell temperature is updated according to the mass and energy exchange between neighbouring cells and also by considering thermal energy loss due to lava surface irradiation. The temperature variation, besides the change of critical height, may lead to the lava solidification which, in turn, determines a change in the morphology. In SCIARA the transition function depends on the following scalar parameters: r_s, the relaxation rate at the temperature of solidification; r_v, the relaxation rate at the temperature of extrusion; h_s, the critical height at the temperature of solidification; h_v, the critical height at the temperature of extrusion; p_c, the "cooling parameter", which regulates the thermal energy loss due to lava surface irradiation. Once that the input to the model has been provided, such as parameter values, terrain topography, vents and the effusion rates as a function of time, SCIARA can simulate the lava flow. The simulation stops when the fluxes fall below a small threshold value. However, before using the model for predictive applications, the parameters must be optimized for a specific area and lava type. To this end, the following fitness measure was defined:

$$\theta = \frac{|R \cap S|}{|R \cup S|} \tag{15}$$

where R and S represent the areas affected by the real and simulated event, respectively. Note that $\theta \in [0,1]$; its value is 0 if the real and simulated events are completely disjoint, being $|R \cap S| = 0$; it is 1 in case of perfect overlap, being $|R \cap S| = |R \cup S|$.

For the calibration task the MAMA was compared with the corresponding standard GA (SGA). In both algorithms a population of 100 bit-strings, each encoding a candidate solution $\mathbf{p} = [r_s, r_v, h_s, h_v, p_c]$, was evolved. In particular,

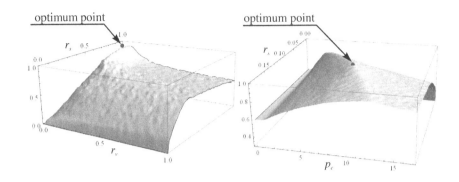

Fig. 2. The rugged fitness landscape generated by SCIARA

each of the SCIARA parameters was encoded on a string of 12 bits using the intervals shown in Table 1. As for the genetic operators, the standard 1-point crossover applied with probability $p_c = 1.0$ was adopted, while the mutation consisted of a bit flipping with probability $p_m = 1/n_b$, being n_b the number of bits per individual. Also, the standard Roulette Wheel Selection was applied together with an elitist replacement scheme.

The calibration exercise concerns a real event occurred on Mt. Etna (Sicily, Italy) in 2001 which is described in details in [2]. However, the target final configuration was obtained with SCIARA itself, using the set of parameters shown in Table 1. This guarantees the existence of a zero-error solution of the calibration problem, thus allowing for a more objective evaluation of the calibration procedures. In Figure 2 the landscape generated by the fitness defined in Equation (15) is depicted. In particular, the two surfaces were obtained executing a number of SCIARA simulations on a grid covering the whole search space, with a refinement in a neighbourhood of the target point shown in Table 1. The ruggedness of the fitness landscape, which can be observed in Figure 2, is known as one of the causes of slow convergence when using most optimization heuristics. In these cases, it is known that using global meta-models can help on smoothing the fitness landscape, thus speeding-up the optimization convergence. In the preliminary experiments presented here, besides the overall effectiveness of the

Table 2. Overview of the calibration results obtained assigning to each search algorithm a budget of 1000 SCIARA evaluations. The statistics were computed on 10 independent run of each algorithm.

	λ	Average	Min	Max	Std. Dev.
SGA	-	0.821	0.740	0.910	0.048
	0	0.918	0.901	0.950	0.015
	2	0.894	0.872	0.925	0.017
MAMA	4	0.910	0.862	0.966	0.035
	6	0.939	0.912	0.971	0.019
	10	0.901	0.860	0.921	0.019

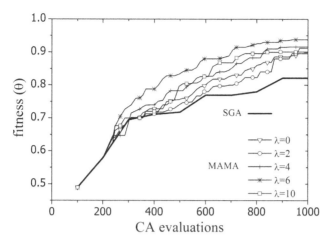

Fig. 3. Average behaviour of the optimization heuristics SGA and MAMA. In the latter, different learning intensities were tested.

MAMA, also the influence of the individual learning intensity (i.e., the parameter λ) was investigated. In particular, five different values of λ were considered, namely 0, 2, 4, 6 and 10. To all runs, a budget of $n_{sim} = 1000$ CA simulations was assigned. In the MAMA, the remaining parameters were $\rho_1 = 0.2$, $\rho_2 = 0.3$ and $\pi = 0.1$. Therefore, up to 200 CA simulations the MAMA worked as the SGA. Starting from 200 CA simulations, the MAMA operated exploiting the RBFN as fitness surrogate. Only after the first 300 CA simulations, the LS was applied to about 10 individuals per generation. For each type of heuristic search, 10 independent runs were carried out. In Table 2 an overview of the results is shown. Within the limited budget of 1000 CA evaluations, the SGA achieved an average fitness value of $\theta \approx 0.82$ and a maximum of $\theta \approx 0.91$. As expected, the MAMA outperformed the SGA providing the best result for $\lambda = 6$, that is a final average $\theta \approx 0.94$ and a maximum $\theta \approx 0.97$. Figure 3-a shows the average behaviour of the algorithms during the search process. Interestingly, for any number of SCIARA simulations and regardless of the learning intensity, the MAMA attained an average fitness significantly higher than that of the SGA. In particular, the MAMA with $\lambda = 6$ reached a significant average speed of convergence, by requiring only about one half of the computational budget to achieve the same fitness given by the SGA at the end of the process. Since each CA evaluation takes several minutes on a standard PC, the MAMA can thus provide the same results of a SGA saving a few hours of computation.

As can be seen on Table 2, in the present application the trade-off between exploration and exploitation regulated by λ had a limited influence (i.e. about 5% at most) on the achieved optimum. Probably, in this case the beneficial smoothing effects provided by the global meta-model plays a major role on speeding-up the optimization. However, it is important to remark that a small gain in the fitness defined by Equation (15) corresponds to a significant difference in the final map of the lava invasion.

5   Conclusions and Future Work

The preliminary results of this study indicate that the automatic optimization of CA models can greatly benefit by the use of a MAMA. Future work will focus on more sophisticated strategies for choosing the individuals on which it is worth investing CA simulations for a Lamarckian learning. In particular, an interesting direction to explore is that proposed in [7], where a pre-selection criterion based on a *probability of improvement* was adopted to rank the promising individuals.

References

1. Di Gregorio, S., Serra, R.: An empirical method for modelling and simulating some complex macroscopic phenomena by cellular automata. Future Generation Computer Systems 16, 259–271 (1999)
2. Rongo, R., Spataro, W., D'Ambrosio, D., Avolio, M.V., Trunfio, G.A., Gregorio, S.D.: Lava flow hazard evaluation through cellular automata and genetic algorithms: an application to mt etna volcano. Fundam. Inform. 87, 247–267 (2008)
3. Blecic, I., Cecchini, A., Trunfio, G.A.: A Comparison of Evolutionary Algorithms for Automatic Calibration of Constrained Cellular Automata. In: Taniar, D., Gervasi, O., Murgante, B., Pardede, E., Apduhan, B.O. (eds.) ICCSA 2010, Part I. LNCS, vol. 6016, pp. 166–181. Springer, Heidelberg (2010)
4. D'Ambrosio, D., Rongo, R., Spataro, W., Trunfio, G.A.: Meta-model Assisted Evolutionary Optimization of Cellular Automata: An Application to the SCIARA Model. In: Wyrzykowski, R., Dongarra, J., Karczewski, K., Waśniewski, J. (eds.) PPAM 2011, Part II. LNCS, vol. 7204, pp. 533–542. Springer, Heidelberg (2012)
5. Jin, Y.: A comprehensive survey of fitness approximation in evolutionary computation. Soft Comput. 9, 3–12 (2005)
6. Moscato, P.: On evolution, search, optimization, genetic algorithms and martial arts - towards memetic algorithms (1989)
7. Zhou, Z., Ong, Y.S., Nair, P.B., Keane, A.J., Lum, K.Y.: Combining global and local surrogate models to accelerate evolutionary optimization. IEEE Transactions on Systems, Man, and Cybernetics, Part C 37, 66–76 (2007)
8. Lim, D., Jin, Y., Ong, Y.S., Sendhoff, B.: Generalizing surrogate-assisted evolutionary computation. IEEE Trans. Evolutionary Computation 14, 329–355 (2010)
9. Spataro, W., Avolio, M.V., Lupiano, V., Trunfio, G.A., Rongo, R., D'Ambrosio, D.: The latest release of the lava flows simulation model sciara: First application to Mt Etna (Italy) and solution of the anisotropic flow direction problem on an ideal surface. Procedia CS 1, 17–26 (2010)
10. Park, J., Sandberg, I.W.: Universal approximation using radial-basis-function networks. Neural Comput. 3, 246–257 (1991)
11. Igel, C., Heidrich-Meisner, V., Glasmachers, T.: Shark. Journal of Machine Learning Research 9, 993–996 (2008)
12. Igel, C., Hüsken, M.: Empirical evaluation of the improved Rprop learning algorithm. Neurocomputing 50 (2003)
13. Celis, M.R., Dennis Jr., J.E., Tapia, R.A.: A trust region strategy for nonlinear equality constrained optimization. In: Boggs, P., Byrd, R., Schnabel, R. (eds.) Numerical Optimization 1984, pp. 71–82. SIAM Publications (1985)
14. Hager, W.W., Zhang, H.: A new active set algorithm for box constrained optimization. SIAM Journal on Optimization 17, 526–557 (2006)
15. Ong, Y., Keane, A.J.: Meta-lamarckian learning in memetic algorithms. IEEE Transactions on Evolutionary Computation 8, 99–110 (2004)

A Memetic Algorithm for Community Detection in Complex Networks

Olivier Gach[1,2] and Jin-Kao Hao[2,⋆]

[1] LIUM & IUT, Université du Maine, Av. O. Messiaen, 72085 Le Mans, France
olivier.gach@univ-lemans.fr
[2] LERIA, Université d'Angers, 2 Bd Lavoisier, 49045 Angers Cedex 01, France
hao@info.univ-angers.fr

Abstract. Community detection is an important issue in the field of complex networks. Modularity is the most popular partition-based measure for community detection of networks represented as graphs. We present a hybrid algorithm mixing a dedicated crossover operator and a multi-level local optimization procedure. Experimental evaluations on a set of 11 well-known benchmark graphs show that the proposed algorithm attains easily all the current best solutions and even improves 6 of them in terms of maximum modularity.

Keywords: heuristic, community detection, complex networks, graph partitioning, modularity, combinatorial optimization.

1 Introduction

Complex networks are a graph-based model which is very useful to represent connections and interactions of the underlying entities in a real networked system [19]. A vertex of the complex network represents an object of the real system while an edge symbolizes an interaction between two objects. A typical example is social network where each vertex corresponds to a particular member of the network while the edges incident to the vertex represent the relationships between this member and other members. Other prominent complex networks include biological networks, citation networks, and the World Wide Web.

Complex networks typically display non-trivial topological features and special patterns which characterize its connectivity and impact the dynamics of processes applied to the network [17]. Discovering these particular features and patterns helps understand the dynamics of the networks and represents a real challenge for research [6].

In particular, complex networks may contain specific groups of highly interconnected vertices which are loosely associated with other groups. Such a group is commonly called community, cluster or still module [19] and all the communities of a network form a clustering. In terms of graph theory, a clustering can be defined as a partition of the vertices of the underlying graph into disjoint subsets,

---

⋆ Corresponding author.

C.A. Coello Coello et al. (Eds.): PPSN 2012, Part II, LNCS 7492, pp. 327–336, 2012.

each subset representing a community. A community is typically characterized by two basic factors: intra-cluster density and inter-cluster density. Intuitively, a community is a cohesive group of vertices that are connected more "densely" to each other than to the vertices in other communities. To quantify the quality of a given community and more generally a clustering, *modularity* is certainly the most popular measure [18]. Under this quality measure, the problem of community detection is a pure combinatorial optimization problem. Formally, the modularity measure can be stated as follows.

Given a weighted graph $G = (V, E, w)$ where w is a weighting function, i.e., $w : V \times V \longmapsto \mathbb{R}$ such that for all $\{u, v\} \in E, w(\{u, v\}) \neq 0$, and for all $\{u, v\} \notin E, w(\{u, v\}) = 0$. Let $C \subseteq V$ and $C' \subseteq V$ be two vertex subsets, $W(C, C')$ the weight sum of the edges linking C and C', i.e., $W(C, C') = \sum_{u \in C, v \in C'} w(\{u, v\})$ (in this formula, each edge is counted twice). The modularity of a clustering with K communities $I = \{C_1, C_2, ..., C_K\}$ ($\forall i \in \{1, 2, ..., K\}, C_i \subset V$ and $C_i \neq \varnothing$; $\cup_{i=1}^{K} C_i = V$; $\forall i, j \in \{1, 2, ..., K\}, C_i \cap C_j = \varnothing$) is given by:

$$Q(I) = \sum_{i=1}^{K} \left[\frac{W(C_i, C_i)}{W(V, V)} - \left(\frac{d_i}{W(V, V)} \right)^2 \right] \tag{1}$$

where d_i is the sum of the degrees of the vertices of community C_i, i.e., $d_i = \sum_{v \in C_i} deg(v)$ with $deg(v)$ being the degree of vertex v.

It is easy to show that Q belongs to the interval [-0.5,1]. A clustering with a small Q value close to -0.5 implies the absence of real communities. A large Q value close to 1 indicates a good clustering containing highly cohesive communities. The trivial clustering with a single cluster has a Q value of 0.

Given the modularity measure Q, the community detection problem aims to find, among the space of all possible clusterings (partitions) of a given graph, a particular clustering with the maximal modularity Q. This is thus a highly combinatorial optimization problem and known to be NP-hard [3]. Consequently, heuristic algorithms are a natural choice to handle this problem. The heuristic algorithms proposed recently for community detection with the modularity measure belong to three general approaches: fast greedy agglomeration like [4], local search [22,14] and hybrid algorithms like [1,13] as some examples.

In this paper, we introduce a memetic algorithm for community detection (MA-COM). MA-COM combines a dedicated crossover operator and a multilevel optimization procedure. MA-COM uses a quality-and-distance based population updating strategy to maintain population diversity. Tested on a set of 11 well-known complex networks, MA-COM attains improved solutions (with a larger Q value) for 6 cases with respect to the best-known values of the literature.

2 Hybrid Evolutionary Algorithm

2.1 Main Scheme

Memetic algorithms are known to be highly effective for solving a number of hard combinatorial optimization problems [16]. A memetic algorithm typically combines a recombination (or crossover) operator and a local optimization operator. The recombination operator generates new solutions which are hopefully located in new promising regions in the search space while the local optimization operator searches around the newly generated solutions in order to discover solutions of good quality.

The general scheme of our MA-COM algorithm for community detection is summarized in Algorithm 1. Basically, MA-COM begins with an initial population of solutions (line 1, Section 2.2) and then repeats an iterative process for a number of times (generations) (lines 3–11). At each generation, two solutions are randomly selected to serve as parents (line 4). The recombination operator is applied to the parents to generate a new offspring solution which is further improved by the local optimization procedure (lines 5–6, see Section 2.3). Finally, we apply a quality-and-distance based rule to decide whether the improved offspring solution can be inserted into the population (line 10, Section 2.4). The solution with the highest modularity discovered during the search is always recorded (line 7-8). The whole algorithm stops if during g consecutive generations, the modularity improvement is inferior to a given threshold ϵ. In the following subsections, we give more details on the components of our algorithm.

Algorithm 1. Pseudo-code of memetic algorithm for community detection

Require: Graph $G = (V, E)$.
Ensure: A clustering I^* of G with a maximal modularity.
1: $P = \{I^1, I^2, ..., I^p\} \leftarrow$ Initialize_Population()   /* Sect. 2.2*/
2: $I^* = \arg\max_{I \in P}\{Q(I)\}$ /* Record the best clustering found so far */
3: **repeat**
4:     $(I^i, I^j) \leftarrow$ Choose_Parents(P)
5:     $I \leftarrow$ Recombine_Parents(I^i, I^j)   /* Sect. 2.3 */
6:     $I \leftarrow$ Improve(I)   /* Sect. 2.2 and 2.3 */
7:     **if** $Q(I) > Q(I^*)$ **then**
8:         $I^* \leftarrow I$
9:     **end if**
10:    $P \leftarrow$ Update_Population(I, P)   /* Sect. 2.4 */
11: **until** end_criterion

2.2   Initial Population

Each solution (clustering) is represented by a n-vector C where n is the order of the graph and $C[i] \in \{1, ..., K\}$ is the community label of vertex i. To generate the initial population P, we employ a *randomized* multi-level algorithm due to Blondel et al. (named BGLL) [1] which uses the vertex mover (VM) heuristic [22] as its refinement procedure. Each VM application displaces a vertex from its current community to another community if the move increase the modularity.

Specifically, we begin with the initial graph G_0 (called it the lowest level graph) where each vertex forms a community and iteratively apply the VM heuristic to improve the modularity of the clustering C of graph G_0 until no improvement is possible for C. From this point, we transform G_0 into a new (a higher-level) graph G_1 where each vertex is a community of the clustering C and an edge links two vertices in G_1 if they represent two neighboring communities in C. Now we apply the VM heuristic to the new graph G_1 to obtain another clustering and then use the clustering to transform G_1 to a new graph G_2 of higher level. This coarsening phase stops when the last graph cannot be further improved by the VM heuristic.

At this point, a second phase (uncoarsening) unfolds the hierarchy of graphs starting from the highest level. At each uncoarsening step, the communities represented by the vertices of the current graph are recovered. The uncoarsening phase stops when the lowest level is reached to recover the initial graph G_0. The corresponding clustering of G_0 constitutes an individual of the initial population of our memetic algorithm.

Experiments show that this initialization procedure is able to provide the memetic algorithm with diversified initial solutions of good quality.

2.3 A Priority-Based Crossover Operator

Crossover is a key element for the effectiveness of the memetic approach [16]. We develop a crossover operator which is dedicated to the clustering problem, named priority-based crossover operator. Our crossover uses two parents (which are selected at random from the population) to generate a new offspring clustering. Random selection suffices in our context because 1) all the individuals of the population are generally of good quality (since they are improved by local optimization) and 2) they are sufficiently distanced in terms of community structure due to the pool updating strategy used in Section 2.4.

The key idea of this operator is to take communities as genetic material and try to preserve some communities from the parents. Specifically, let (I^1, I^2) be two parent clusterings and $\mathbf{p}$ a priority vector. Let s and r be respectively the number of communities of clusterings I^1 and I^2. The vector $\mathbf{p}$, indexed from 1 to $s + r$, is defined by a random permutation of $\{1, 2, ..., s + r\}$. The indices between 1 and s of $\mathbf{p}$ denotes the communities of one parent and those between $s + 1$ and $s + r$ the communities of the other parent. Thus each community of the parents is designated by a unique number from 1 to $s + r$. For each community $C_i, i \in \{1, 2, ..., s + r\}$, the corresponding value in $\mathbf{p}$ (i.e., $\mathbf{p}[i]$) gives the priority of C_i. By convention, a smaller $\mathbf{p}$ value indicates a higher priority for the community and vise versa.

The crossover procedure generates from (I^1, I^2) its offspring clustering I^o as follows. We go through one by one all the communities by following the priority order given by the vector $\mathbf{p}$. We begin by selecting the highest priority community C according to $\mathbf{p}$ and transfer all the vertices of the community to form a community of the offspring I^o. We then pick the community C' with the second highest priority, remove the vertices already in I^o and use the remaining vertices of C' to form a new community of I^o (empty community is discarded). We repeat this process until the community with the lowest priority is handled. Finally, the communities of I^o are re-labeled from one to the number of communities contained in the offspring.

Figure 1 illustrates the crossover procedure applied to a small graph. Among the 7 communities of the two parents, the one with the highest priority 1 (labeled 5 in parent 2 with vertices {1,2,8,10,13,17}) is transfered to the off-spring. The second selected community is the one labeled 2 from parent 1 (i.e., {3,7,9,13,16}). After removing vertex 13 which appears already in the offspring, we use {3,7,9,16} to form another community of the offspring. The next selected community is labeled 1 from parent 1 ({1,2,8,10,17}), removing the shared vertices leads to an empty community which is discarded. This process continues until all the 7 communities are examined. The resulting offspring is composed of 5 communities originating from both parents. This crossover operator leads generally to an offspring with more communities than in the parents, deteriorating thus the modularity objective. To improve the quality of the offspring, we apply the BGLL algorithm described in Section 2.2 by taking the offspring as its initial solution. The improved offspring is then considered for inclusion in the population according to the quality-and-distance strategy explained in Section 2.4.

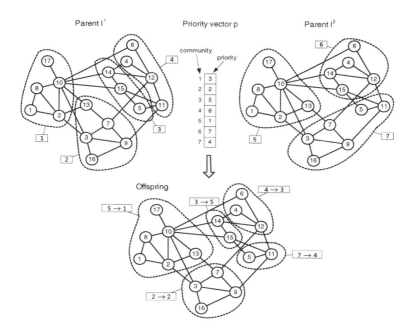

Fig. 1. Illustration of the crossover operator. Five new communities in the offspring are created from seven communities of two parents.

The time complexity of the crossover operator is $O(n)$. With appropriate data structures, the crossover operator can be implemented in one pass of the vertices of the graph.

Finally, we notice the the priority associated to each community can be defined by considering other factors like the modularity and size of the community. Due to space limitations, we do not explore these possibilities in this paper. Yet, as shown in the experimental evaluation section, our memetic algorithm equipped with the crossover operator using random priorities works well for the set of the test graphs.

2.4 Population Updating Strategy

Population diversity is another critical issue in a memetic algorithm to avoid premature convergence [16]. Our experiments show that this particularly holds in our case due to the small size of the population used (typically several tens of solutions). For this reason, we employ a population updating strategy which considers not only the quality of the offspring, but also its distance to the solutions of the population.

Distance Function. Let $X = \{X_1, X_2...X_K\}$ and $Y = \{Y_1, Y_2...Y_{K'}\}$ be two clusterings of graph $G = (V, E)$. For an edge $e = \{u, v\} \in E$ and a community C of X or Y, we use $e \in C$ to state the fact that the vertices u and v of e are in the same community. Then we use the Rand Index [21] to define our distance d between X and Y as follows:

$$d(X, Y) = \frac{\sum_{e \in E} d_e(X, Y)}{m} \qquad (2)$$

where $d_e(X, Y)$ of edge $e = \{u, v\}$ is defined by:

$$d_e(X, Y) = \begin{cases} 0 \text{ if } \exists X_i \in X, \exists Y_j \in Y \text{ s.t. } e \in X_i \text{ and } e \in Y_j \text{ OR} \\ \quad \text{if } \forall X_i \in X, \neg(e \in X_i) \text{ and } \forall Y_i \in Y, \neg(e \in Y_j) \\ 1 \text{ otherwise.} \end{cases} \qquad (3)$$

We can show that d (called Edge Rand Index - ERI) satisfies the conditions of a mathematical distance and its values belong to $[0,1]$. Intuitively, this distance measures the edge disagreements between two clusterings.

Updating Procedure. Let P be the current population and I^o be the offspring to be considered for inclusion in P. Let $I_c \in P$ be the closest clustering to I^o according to the above distance and $I_w \in P$ the worst clustering (with the smallest modularity). Let δ_{min} is a fixed distance threshold. We apply the following replacement rule: if $d(I^o, I_c) < \delta_{min}$ and $Q(I^o) \geq Q(I_c)$, then I^o replaces I_c in P; otherwise, if $Q(I^o) \geq Q(I_w)$ then I^o replaces I_w in P.

By taking into account both quality and distance, this updating strategy reinforces the population diversity when the search progresses.

3   Computational Results

3.1   Experimental Setup

This section is dedicated to a performance assessment of our MA-COM algorithm which is coded in Pascal. We carry out extensive experiments on a set of 11 networks (with 34 to 27519 vertices) commonly used for community detection (Table 1). Directed graphs are transformed into undirected graphs and loops are removed. Our algorithm also takes into account weighted graphs (*Condmat2003*). We run the program 20 times on each graph and report the maximal modularity, the average modularity and the average computing time, based on a PC equipped with a Pentium Core i7 870 of 2.93 GHz and of 8 GB of RAM. The algorithm stops after 500 consecutive generations without an improvement of modularity greater than 10^{-4}. The values for the other parameters are the following: population size (30), distance threshold δ_{min} used for population management (0.01). These same values are used to report all the results of this section, though better results could probably be obtained by fine tuning some parameters. Experiments show that population size and distance threshold have an important influence on MA-COM's performance. In Section 3.2, we show our results in terms of the modularity criterion while in Section 3.3 we analyze some structural features of the solutions found.

3.2   Results in Terms of Modularity

Table 1 shows the results of the proposed memetic algorithm (MA-COM) compared to the current best-known results (BKR) ever reported in the literature in terms of the modularity values. We also include the results of the BGLL algorithm which is used to generate the initial population of our memetic algorithm. From Table 1, we observe that the proposed MA-COM algorithm obtains clusterings of equal or greater modularity for all the tested graphs. In particular, for the 6 largest graphs (from *C. elegans* to the last network), MA-COM improves the current best-known results by finding solutions with a larger modularity. For the first 5 graphs which are also the smallest ones (with no more 200 vertices and 3000 edges), even BGLL alone attains the current best-known modularity values during the population initialization phase.

We also observe that the average modularity of our MA-COM algorithm is very closed to the maximum and, for all the graphs, is always equal to or better than the best-known result. This shows that MA-COM is quite stable, despite of its stochastic nature. The computing time grows more than linearly with respect to the number of edges m. Experimental statistics show that the time complexity could be approximated by $O(m^{\alpha})$ with $\alpha \approx 1.3$.

3.3   Structural Changes in Clusterings

In the last section, we show that MA-COM improves the solutions of the BGLL algorithm in terms of modularity. Now we turn our attention to structural transformations of solutions achieved by MA-COM from solutions given by BGLL.

Table 1. Results on 20 runs of the proposed MA-COM algorithm on 11 commonly used real graphs (sources in brackets). The BKR column shows the best known result with its sources in brackets. The other columns give the average and maximum modularity of the best solutions in the initial population (BGLL) and the final population of MA-COM. The number of communities of the best solution is indicated between parenthesis. Improved results are highlighted in bold.

Graph	BKR	BGLL [1]		MA-COM		
		Avg Q	Max Q (K)	Avg Q	Max Q (K)	Time(s)
Karate Club [23]	0.4198 [13,20,14]	0.4198	0.4198 (4)	0.4198	*0.4198 (4)*	0.3
Dolphins [15]	0.529 [13]	0.5281	0.5286 (5)	0.5286	*0.5286 (5)*	0.5
Political Books [12]	0.527[13]	0.5273	0.5273 (5)	0.5273	*0.5273 (5)*	1.0
College Football [7]	0.605 [13]	0.6046	0.6046 (10)	0.6046	*0.6046 (10)*	1.4
Jazz [8]	0.4452 [14]	0.4452	0.4452 (4)	0.4452	*0.4452 (4)*	5.2
C. elegans [5]	0.452 [13]	0.4457	0.4497 (11)	0.4531	**0.4533 (10)**	8.3
E-mail [10]	0.582 [13]	0.5748	0.5772 (10)	0.5828	**0.5829 (10)**	23.1
Erdos [9]	0.7162 [20]	0.6993	0.7021 (32)	0.7184	**0.7188 (34)**	88.4
Arxiv [11]	0.813 [1]	0.8166	0.8181 (60)	0.8246	**0.8254 (56)**	197.2
PGP [2]	0.8841 [13,20]	0.8841	0.8850 (95)	0.8865	**0.8867 (94)**	156.7
Condmat2003	0.8146 [20]	0.8112	0.8116 (77)	0.8165	**0.8170 (73)**	1369.7

For this purpose, we consider, for each of the 11 graphs and each of the 20 runs of MA-COM, the best solution I^*_{init} (i.e., the clustering with the largest modularity) from the initial population (generated by BGLL) and the best solution I^*_{final} from the final population (generated by MA-COM). We compute then the distance between I^*_{init} and I^*_{final} using two distance measures: the well-known Normalized Mutual Information (NMI) and the Edge Rand Index (ERI) which is defined in Section 2.4 for population management. While NMI measures the information shared by I^*_{init} and I^*_{final}, ERI indicates the percentage of edges which disagree in the clusterings I^*_{init} and I^*_{final}. Table 2 show the statistics of these measures averaged over the 20 runs for each graph. Additionally, we indicate the averaged number of communities (indicator K) in the initial and final population. Finally, we present the averaged sizes of the smallest and the largest communities in the initial and final best solutions.

Table 2 shows that for the small graphs except *Dolphins*, the memetic algorithm has a limited effect on the best BGLL clustering. On the contrary, structural changes for other graphs are more or less important because an edges difference of 2.7% to 13.1% are observed in the initial best and the final best solutions. Some graphs have probably a simple structure with few local optima, for instance *PGP* (with a high NMI). Some smaller graphs like *C. elegans* seem to have a more complexe modularity landscape (13.1% of edges of the initial best solutions are changed in final best solutions).

The indicator K confirms the well-known propensity of modularity based methods to reduce the number of communities. However, the reduction is moderate, indicating that the changes revealed by the ERI distance are mainly due to moves of vertices rather than merges of communities. The good surprise comes with the smallest and largest communities. The memetic algorithm has a clear trend to help discover small communities (which are known to be difficult to detect). More generally, we believe that the crossover operator of the algorithm

Table 2. Several structural measures to compare the best solution in the initial population and the best solution in the final population: NMI (Normalized Mutual Information), ERI (Edge Rand Index), K (number of communities), average sizes of the smallest and largest community over 20 runs.

Graph	NMI	ERI	K Initial	K Final	Smallest com. size Initial	Smallest com. size Final	Largest com. size Initial	Largest com. size Final
Karate Club	1.000	0.0%	4.0	4.0	5.0	5.0	12.0	12.0
Dolphins	0.976	1.9%	5.0	5.0	5.0	5.0	19.9	20.0
Political Books	0.982	0.4%	5.0	5.0	3.0	3.0	40.6	40.0
College Football	1.000	0.0%	10.0	10.0	9.0	9.0	16.0	16.0
Jazz	0.999	0.1%	4.0	4.0	21.9	22.0	62.1	62.0
C. elegans	0.733	13.1%	10.2	9.2	7.5	5.0	92.0	82.2
E-mail	0.780	9.1%	10.8	10.1	43.2	36.2	185.8	168.5
Erdos	0.771	12.0%	32.5	33.9	23.8	9.7	622.5	619.6
Arxiv	0.795	7.6%	59.6	55.5	4.5	4.5	920.5	812.2
PGP	0.915	2.7%	98.0	95.0	5.9	6.0	668.5	641.7
Condmat2003	0.758	8.9%	75.8	70.6	18.5	6.6	2478.7	2266.6
Total	0.883	5.1%	28.6	27.5	13.4	10.2	465.3	431.0

acts mainly on the ambiguous vertices which are attached to several communities and help discover the right community for these vertices.

4   Conclusion and Perspectives

This paper deals with the community detection problem in complex networks with the popular modularity criterion. To approximate this hard combinatorial problem, we proposed a memetic algorithm mixing a dedicated crossover operator and a multi-level local optimization procedure. The proposed crossover operator blends the communities of two clusterings (parents) according to a priority rule. Offspring solutions are improved with the multi-level local optimizer. To maintain a healthy population diversity, we introduce a Rand Index based distance and consider for population management both the quality of the offspring and its distance to the solutions of the population. Experimental results on a set of 11 popular networks showed that the proposed approach can easily match the best known results in 5 cases and discover improved solutions for the 6 other largest networks. The analysis of initial solutions and final solutions showed the benefit of memetic approach in discovering communities of small size that are difficult to find. This work demonstrated that the memetic approach is a very promising method for modularity maximization. The proposed algorithm could also be used to devise more powerful methods. One possible way would be to embed the memetic approach into the multi-level approach in order to handle very large networks.

Acknowledgment. We are grateful to the referees for their comments and questions which helped us to improve the paper. The work is partially supported by the Pays de la Loire Region (France) within the RaDaPop (2009-2013) and LigeRO (2010-2013) projects.

References

1. Blondel, V.D., Guillaume, J.-L., Lambiotte, R., Lefebvre, E.: Fast unfolding of communities in large networks. J. Stat. Mech.: Theory Exp., P10008 (October 2008), doi:10.1088/1742-5468/2008/10/P10008
2. Boguñá, M., Pastor-Satorras, R., Díaz-Guilera, A., Arenas, A.: Models of social networks based on social distance attachment. Phys. Rev. E 70(5), 056122 (2004)
3. Brandes, U., Delling, D., Gaertler, M., Gorke, R., Hoefer, M., Nikoloski, Z., Wagner, D.: On modularity clustering. IEEE Trans. Knowl. Data Eng. 20(2), 172–188 (2008)
4. Clauset, A., Newman, M.E.J., Moore, C.: Finding community structure in very large networks. Phys. Rev. E 70(6), 066111 (2004)
5. Duch, J., Arenas, A.: Community detection in complex networks using extremal optimization. Phys. Rev. E 72(2), 027104 (2005)
6. Fortunato, S.: Community detection in graphs. Physics Reports 486, 75–174 (2010)
7. Girvan, M., Newman, M.E.J.: Community structure in social and biological networks. Proc. Natl. Acad. Sci. USA 99(12), 7821–7826 (2002)
8. Gleiser, P., Danon, L.: Community structure in social and biological networks. Advances in Complex Systems 6, 565–573 (2003)
9. Grossman, J.: The Erdös number project (2007), http://www.oakland.edu/enp/
10. Guimerà, R., Danon, L., Díaz-Guilera, A., Giralt, F., Arenas, A.: Self-similar community structure in a network of human interactions. Phys. Rev. E 68(6), 065103 (2003)
11. KDD. Cornell kdd cup (2003), http://www.cs.cornell.edu/projects/kddcup/
12. Krebs, V.: A network of books about recent us politics sold by the online bookseller amazon.com. (2008), http://www.orgnet.com
13. Liu, X., Murata, T.: Advanced modularity-specialized label propagation algorithm for detecting communities in networks. Phys. A 389(7), 1493–1500 (2009)
14. Lü, Z., Huang, W.: Iterated tabu search for identifying community structure in complex networks. Phys. Rev. E 80(2), 026130 (2009)
15. Lusseau, D., Schneider, K., Boisseau, O.J., Haase, P., Slooten, E., Dawson, S.M.: The bottlenose dolphin community of Doubtful Sound features a large proportion of long-lasting associations. Behav. Ecol. Sociobiol. 54(4), 396–405 (2003)
16. Neri, F., Cotta, C., Moscato, P. (eds.): Handbook of Memetic Algorithms. SCI, vol. 379. Springer (2011)
17. Newman, M.E.J.: The structure of scientific collaboration networks. Proc. Natl. Acad. Sci. USA 98(2), 404–409 (2001)
18. Newman, M.E.J., Girvan, M.: Finding and evaluating community structure in networks. Phys. Rev. E 69(2), 026113 (2004)
19. Newman, M.E.J.: Networks: An Introduction. Oxford University Press (2010)
20. Noack, A., Rotta, R.: Multi-level Algorithms for Modularity Clustering. In: Vahrenhold, J. (ed.) SEA 2009. LNCS, vol. 5526, pp. 257–268. Springer, Heidelberg (2009)
21. Rand, W.M.: Objective criteria for the evaluation of clustering methods. J. Amer. Statistical Assoc. 66(336), 846–850 (1971)
22. Schuetz, P., Caflisch, A.: Efficient modularity optimization by multistep greedy algorithm and vertex mover refinement. Phys. Rev. E 77(4), 046112 (2008)
23. Zachary, W.W.: An information flow model for conflict and fission in small groups. J. Anthropol. Res. 33, 452–473 (1977)

Local Optima Networks, Landscape Autocorrelation and Heuristic Search Performance

Francisco Chicano[1], Fabio Daolio[2], Gabriela Ochoa[3], Sébastien Vérel[4],
Marco Tomassini[2], and Enrique Alba[1]

[1] E.T.S. Ingeniería Informática, University of Málaga, Spain
[2] Information Systems Department, University of Lausanne, Lausanne, Switzerland
[3] Inst. of Computing Sciences and Mathematics, University of Stirling, Scotland, UK
[4] INRIA Lille - Nord Europe and University of Nice Sophia-Antipolis, France

Abstract. Recent developments in fitness landscape analysis include the study of Local Optima Networks (LON) and applications of the Elementary Landscapes theory. This paper represents a first step at combining these two tools to explore their ability to forecast the performance of search algorithms. We base our analysis on the Quadratic Assignment Problem (QAP) and conduct a large statistical study over 600 generated instances of different types. Our results reveal interesting links between the network measures, the autocorrelation measures and the performance of heuristic search algorithms.

1 Introduction

An improved understanding of the structure of combinatorial fitness landscapes can facilitate the design and further successful application of heuristic search methods to solve hard computational problems. This article brings together two recent developments in fitness landscape analysis for combinatorial optimisation, namely, local optima networks (LONs) and elementary landscape decomposition. LONs represent a new model of combinatorial landscapes based on the idea of compressing the information given by the whole problem configuration space into a smaller mathematical object that is the graph having as vertices the local optima and as edges the possible transitions between them [15,16]. This characterization of landscapes as complex networks enables the use of tools and metrics of the complex networks domain [4] and has brought new insights into the global structure of the landscapes studied in the past [9].

The QAP has been recently analysed using this model [9] and the clustering structure of the local optima networks of two classes of QAP instances was studied in [8]. The study revealed that the so-called "real-like" instances have significantly more optima cluster (or modular) structure than the class of random uniform instances of the QAP. Using the theory of elementary landscapes [3] the QAP has been analysed in [6] and the elementary landscape decomposition has been computed. This decomposition can then be used to exactly compute the

C.A. Coello Coello et al. (Eds.): PPSN 2012, Part II, LNCS 7492, pp. 337–347, 2012.

autocorrelation coefficient and the autocorrelation length of any arbitrary QAP instance [7].

In this article, the expression in [7] is used to calculate the autocorrelation length of the two classes of QAP instances studied in [8]. Since for those instances the LONs were exhaustively computed, the exact number of local optima are known in all cases. This will allow us to support the autocorrelation length conjecture [13], which links the autocorrelation length to the number of local optima of a landscape. We also conduct a correlation study among several network metrics calculated on the extracted LONs and the success rate of two heuristic search algorithms: simulated annealing and genetic algorithms. Our goal is to discover relationships between fitness landscape features and the performance of heuristic search methods.

The article is structured as follows. Section 2 includes the relevant definitions, methodologies and metrics used in this article. Section 3 presents the correlation study and Section 4 discusses our main findings and suggests directions for future work.

2   Background

In this section we introduce all the background concepts required in the rest of the paper. We define the QAP, describe the LONs, introduce the network metrics, the autocorrelation length and describe the heuristic search algorithms used in the experimental section.

2.1   The Quadratic Assignment Problem

The QAP is a combinatorial problem in which a set of facilities with given flows have to be assigned to a set of locations with given distances in such a way that the sum of the product of flows and distances is minimized. A solution to the QAP is generally written as a permutation π of the set $\{1, 2, ..., n\}$. The cost associated with a permutation π is: $C(\pi) = \sum_{i=1}^{n} \sum_{j=1}^{n} a_{ij} b_{\pi_i \pi_j}$, where n denotes the number of facilities/locations and $A = (a_{ij})$ and $B = (b_{ij})$ are referred to as the distance and flow matrices, respectively. The contents of these two matrices characterize the class of instances of the QAP.

For the statistical analysis conducted here, the two instance generators proposed in [12] for the multi-objective QAP were adapted for the single-objective QAP. The first generator produces uniformly random instances where all flows and distances are integers sampled from uniform distributions. The second generator produces flow entries that are non-uniform random values. The instances produced have the so-called "real-like" structure since they resemble the structure of QAP instances found in practical applications. These instance generators are based on the procedures described by Taillard in [14]. In particular, uniform instances are similar to the TaiXXXa instances of QAPLIB [5] and real-like instances are similar to the TaiXXXb instances. We consider here these two types of instances and three problem dimensions: 9, 10 and 11. Therefore, we have six different instance groups. For each group, 100 instances were generated for a total of 600 QAP instances that will be used in our study.

2.2 Local Optima Networks

In order to define the local optima network of the QAP instances, we need to provide the definitions for the nodes and edges of the network. The vertices of the graph can be straightforwardly defined as the local minima of the landscape. In this work, we select small QAP instances such that it is feasible to obtain the nodes exhaustively by running a best-improvement local search algorithm from every configuration (permutation) of the search space. The neighborhood of a configuration is defined by the *pairwise exchange* or *swap* operation, which is the most basic operation used by many metaheuristics for QAP. This operator simply exchanges any two positions in a permutation, thus transforming it into another permutation. The neighborhood size is thus $|V(s)| = n(n-1)/2$. Given a local optima s, its *basin of attraction* is defined as the set of solutions s' from which s can be reached using a hill-climbing algorithm [9].

The edges account for the transition probability between basins of attraction of the local optima. More formally, the edges reflect the probability of going from basin b_i to basin b_j, which is computed as the average over all $s \in b_i$ of the transition probabilities to solutions $s' \in b_j$. The reader is referred to [9] for a more detailed exposition.

We define a *Local Optima Network* (LON) as being the graph $G = (S^*, E)$ where the set of vertices S^* contains all the local optima, and there is an edge $e_{ij} \in E$ with weight $w_{ij} = p(b_i \to b_j)$ between two nodes i and j if and only if $p(b_i \to b_j) > 0$, where $p(b_i \to b_j)$ is the probability of moving from basin b_i to basin b_j in one step. Notice that since each optimum has its associated basin, G also describes the interconnection of basins.

2.3 Network Metrics

We describe below the six network metrics considered in our analysis.

Number of vertices, N_v : The number of nodes of a LON is simply the number of local optima in the fitness landscape. It is exhaustively computed running a best-improvement hill-climbing algorithm from each solution of the search space.

Clustering coefficient, Cc : Measures the probability that two neighbors of a given node are also neighbors of each other [4]. In other words, it accounts for the ratio of connected triples in the graph. In the language of social networks, it measures how likely it is that the friend of your friend is also your friend.

Shortest path length to the optimum, L_{opt}: A standard metric to characterize the structure of networks is the shortest path length (number of link hobs) between two nodes in the network. In order to compute this measure on the LONs, we considered the expected number of moves (in the case of

QAP swap moves) to pass from one basin to the other. This expected number can be computed by considering the inverse of the transition probabilities between basins: $1/w_{ij}$. We use this to calculate the average shortest paths leading to the global optimum.

Disparity, Y_2: Measures the local heterogeneity introduced by edge weights [4]. It indicates whether the outgoing links from a given node have mostly the same weights (transition probabilities) or there is one outweighing the others. Disparity for a vertex i is computed as $Y_2(i) = \sum_{j \neq i}(w_{ij}/s_i)^2$, where $s_i = \sum_{j \neq i} w_{ij}$ is the so-called *strength* of vertex i.

Fitness-fitness correlation, F_{nn}: Measures the correlation between the fitness values of adjacent local optima. More precisely, we estimate the Spearman rank correlation coefficient between the fitness value f_i of vertex i and its weighted-average nearest-neighbors fitness, defined as $F_{nn}^w(i) = 1/s_i \sum_{j \neq i} w_{ij} f_j$.

Modularity, Q: Clusters or communities in networks can be loosely defined as groups of nodes that are strongly connected between them and poorly connected with the rest of the graph. To calculate the level of community structure, also known as modularity, we consider a graph clustering algorithm that is based on the simulation of network flow [10], as in [8].

2.4   Calculation of the Autocorrelation Length

Let us consider an infinite random walk $\{x_0, x_1, \ldots\}$ on the solution space such that $x_{i+1} \in N(x_i)$. The *random walk autocorrelation function* $r : \mathbb{N} \to \mathbb{R}$ is defined as [17]:

$$r(s) = \frac{\langle f(x_t)f(x_{t+s})\rangle_{x_0,t} - \langle f(x_t)\rangle_{x_0,t}^2}{\langle f(x_t)^2\rangle_{x_0,t} - \langle f(x_t)\rangle_{x_0,t}^2} \quad (1)$$

where the subindices x_0 and t indicate that the averages are computed over all the starting solutions x_0 and along the complete random walk. The *autocorrelation length* ℓ [11] is defined as $\ell = \sum_{s=0}^{\infty} r(s)$. Using the landscape decomposition of the QAP in [7] the authors provide a closed-form formula for ℓ based on the matrices (a_{ij}) and (b_{ij}) of the QAP instance. We will use in the present article this formula to efficiently compute the autocorrelation length of all the instances in our experimental study.

2.5   Heuristic Search Algorithms and the Performance Metric

We considered two well-known heuristic search algorithms: simulated annealing (SA) and genetic algorithms (GA). The SA uses a cooling factor of 0.9983 and an initial temperature of 10^7. The neighborhood move is the same used for generating the LONs, namely, the pairwise exchange or swap operation in permutation space. The GA is a steady-state GA with a population size of 100, where one solution is computed at a time and inserted in the population using elitist replacement. The individuals are selected using a binary tournament. The genetic operators are the partially mapped crossover (PMX) [2] and the pairwise exchange mutation operation applied with probability 0.3. We perform 100 independent runs for each algorithm and instance.

In order to measure the performance of a search algorithm solving the QAP instances we use the success (hit) rate, defined as the fraction of the 100 independent runs that found the global optimum. Both algorithms stop when they reach 10,000 function evaluations.

3 Correlation Study

Our statistical analysis considers the pair-wise correlation among the six network metrics, the autocorrelation length, and the SA and GA success rates. As mentioned above, six classes of instances are considered, including two types of instances ('uniform' and 'real-like') as described in Section 2.1, and 3 problem sizes 9, 10 and 11. Each of the 6 instance classes is considered separately, and 100 instances conform the sample for the statistical analysis in each class.

The main goal of our study is to discover whether some of the studied metrics can predict the performance of a heuristic search algorithm on a given instance class. We start, then, by showing the performance of the two selected search algorithms: SA and GA. Figure 1 illustrates the range and distribution of the hit rates for each algorithm and instance class, while Table 1 contains hit rate values, number of local optima and the average shortest path length for the instance classes. We show the values of these two network metrics because they seem to have an impact on the hit rate, as we will see later.

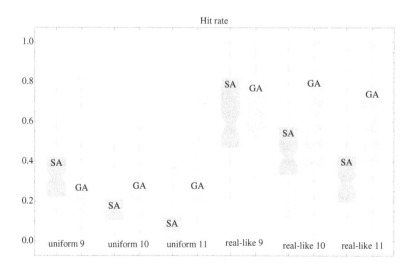

Fig. 1. Hit rate of the GA and SA for the 6 classes of instances considered. The boxplots are computed on the results of the algorithms over the 100 random instances per class.

F. Chicano et al.

The results suggest that, for each instance type (uniform or real-like), the hit rates for both algorithms decrease as the instance size increases, with the only exception of GA in the real-like 11 class. This is expected as the size of the search space increases, and so locating the global optimum is harder. However, the hit rates for real-like instances are much higher in all cases, which confirms that these instances are easier to solve for both algorithms [9]. In Table 1 we can see that real-like instances have a lower number of local optima compared to uniform instances, which explains why real-like instances are easier to solve than the uniform ones. A second observation is that the hit rate of the GA, for a given instance type, does not change much when the size of the instances is increased. However, in SA the hit rate is clearly reduced when the size increases. That is, SA seems to be quite sensitive to the size of the instance (in addition to the type) while GA is clearly sensitive to the type of the instance (uniform or real-like) but little sensitive to the size.

Table 1. Number of local optima (N_v), shortest path to the optimum (L_{opt}) and hit rate of the GA and SA for the 6 instance classes. We show the average and the standard deviation.

Class		N_v Avg.	Std. dev.	L_{opt} Avg.	Std. dev.	GA Avg.	Std. dev.	SA Avg.	Std. dev.
uni	$n = 9$	131.220	51.268	25.761	11.231	0.220	0.170	0.320	0.200
	$n = 10$	399.840	153.097	45.217	17.120	0.185	0.190	0.155	0.105
	$n = 11$	1337.300	453.520	76.815	26.698	0.210	0.175	0.090	0.070
rl	$n = 9$	14.300	7.473	8.564	4.343	0.675	0.265	0.585	0.350
	$n = 10$	26.720	17.775	13.588	7.228	0.610	0.465	0.420	0.240
	$n = 11$	64.420	47.410	22.508	11.563	0.585	0.370	0.295	0.235

Let us consider the correlations between the network metrics, the autocorrelation length and the algorithms' performance. These are shown qualitatively in Figure 2 for the instance sizes 9 and 11. The figure shows that the GA is not correlated to any measure in the real-like instances of large size. Only for the uniform instances and the ones of size 9 there are some significant correlations. We can thus, conjecture that the measures used in this study are not useful to predict the performance of the GA. A possible explanation is the presence of the crossover operator, which introduces an additional neighborhood not used for generating the LONs. On the contrary, the SA algorithm only uses a single move operator (pairwise exchange or swap) which is the same used to generate the LONs. In this case, the figure reveals correlation with some metrics. In particular, the correlation between the performance of SA and L_{opt} is the highest, which suggests that L_{opt} is the measure that better predicts the behavior of SA.

In Figure 3 we plot the hit rate against some selected measures for SA and GA in the real-like instances of size 11. The plots of the SA and GA are interleaved in order to compare the results of the regression analysis (the regression line is superimposed on the plot). We can observe how the line has a smaller slope in the case of the GA for all the plots, what explains the low correlation for the GA

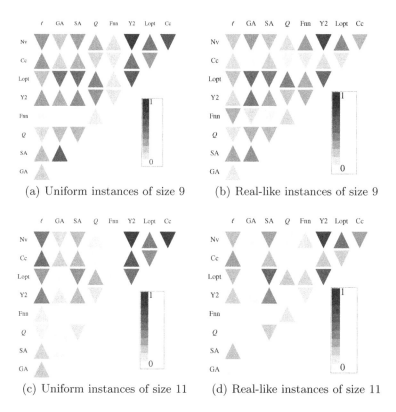

(a) Uniform instances of size 9   (b) Real-like instances of size 9

(c) Uniform instances of size 11   (d) Real-like instances of size 11

Fig. 2. Correlations between the measures. An arrow pointing up means positive correlation whereas an arrow pointing down means negative correlation. The absolute value of the correlation is shown in grey scale (the darker the higher).

hit rate and the measures. But we can also observe how L_{opt} is a good predictor of the SA performance.

An interesting observation that contributes to explain the robustness of the GA over the problem size is that while hit rate for the SA is correlated with the number of local optima, this is not the case for the GA. This suggests that the global search characteristic of a population in GA makes it more robust to the presence of larger number of local optima.

On the uniform instances, we can observe a positive correlation between the performances of GA and SA, which suggests that the search difficulty is similar for both algorithms in this case. This is observed for all instance sizes although the correlation decreases as the size increases. On the real-like instances, the observation is different. In particular, for the real-like instance of size 11, there is no correlation whatsoever between the performance of both algorithms, which suggests that the search difficulty depends on the algorithm for these instances. In other words, the hard instances for the GA are not the same as the

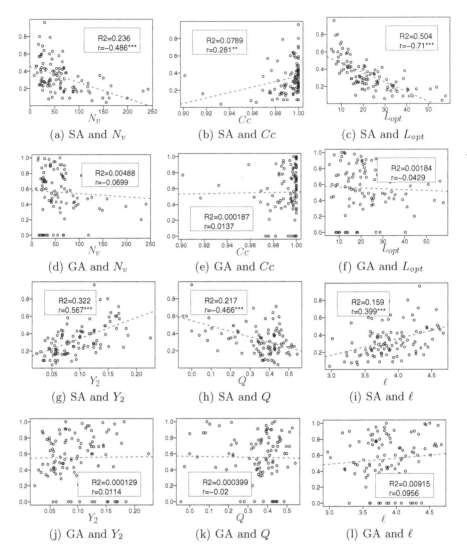

Fig. 3. Regression analysis for the hit rate of SA and GA against some selected measures for the real-like 11 instances

hard instances for the SA and vice versa. We may also speculate that the GA is more efficient at exploiting the more modular structure of the real-like instance, which makes its search dynamic different than that of the SA on these instances.

Regarding the rest of the measures, in general, the correlations are higher in the uniform instances than in the real-like instances. F_{nn} and Q seem to be the less correlated LON measures (this is particularly true for the real-like instances). The higher correlation coefficients appear between the clustering coefficient, the disparity, the number of nodes and the path to the global optimum. The autocorrelation length seems to be correlated with these measures and with the performance of SA. This is specially interesting, since the autocorrelation length can be computed from the instance data, without the need to exhaustively generate the complete search space (like it happens with the LON measures). The correlation between ℓ and the performance of SA suggests that we can use ℓ as a measure of problem difficulty, when a trajectory-based search algorithm is used. This idea is also supported by the results of Angel and Zissimopoulos [1], which provided a positive correlation between an autocorrelation measure and the performance of an SA. The correlation between ℓ and the performance of the GA is much smaller, which again indicates that it is harder to predict the performance of the GA using this type of landscape metrics.

Finally, the correlation analysis also provides evidence of the autocorrelation length conjecture. This conjecture claims that the number of local optima is inversely correlated to the autocorrelation length ℓ [13]. In [7] some results were presented that supported the autocorrelation length conjecture. In that work, the correlation between the number of local optima and ℓ was between -0.1640 and -0.3256. In our case the correlation is higher (in absolute value), in the range from -0.3729 to -0.7187. This support of the conjecture is higher in the uniform instances than in the real-like instances of the same size.

4   Discussion

We conducted a large statistical correlation study considering QAP instances of different types and sizes, a number of landscape metrics and the performance of two widely known search heuristics. Our study also brings together two recent developments in combinatorial landscape analysis, with the aim of shedding new light on the relationships between the landscape structure and the performance of heuristic search algorithms. Our study confirms that the real-like instances are easier to solve by heuristic search algorithms. Clearly, in these problems, the number of local optima in the landscape is a much better predictor of search difficulty than the size of the search space.

Overall, the GA was a stronger algorithm to solve all the studied classes of QAP instances. Moreover, the GA is more robust to the increase in problem size. Interestingly, the performance of SA and GA is correlated for the uniform instances, but this is not the case for the real-like instances. Which suggests that the GA is better at exploiting the more clustered structure of the real-like instances. However, predicting the performance of the GA seems to be a harder task than predicting the performance of SA. GAs are more complex algorithms as they incorporate a population and a recombination operator. In particular, we found some network metrics such as the average distance to the global optima and the number of local optima, which are good predictors of the SA performance, but less so for the GA. The question is still open for a better understanding and prediction of the GA performance.

Finally, our study provides supporting evidence of the correlation length conjecture indicating that the number of local optima is inversely correlated to the correlation length. This is an interesting contribution, as using elementary landscape decomposition the autocorrelation length for QAP instances can be exactly calculated from the instance data [7]. More detailed studies, additional metrics, sampling approaches to extract the LONs and larger landscapes are required to better understand and predict search difficulty in combinatorial optimization. Our study, however, is a first step that incorporates new landscape metrics coming from the field of complex networks, and try to correlate them with both previously studied landscape metrics and the performance of heuristic search methods.

Acknowledgements. This work was partially funded by the Spanish Ministry of Science and Innovation and FEDER under contract TIN2011-28194 and the Andalusian Government under contract P07-TIC-03044. Fabio Daolio and Marco Tomassini gratefully acknowledge the Swiss National Science Foundation for financial support under grant number 200021-124578. The authors would also like to thank the organizers and partipants of the seminar on Theory of Evolutionary Algorithms (10361) at Schloss Dagstuhl - Leibniz Center for Informatics.

References

1. Angel, E., Zissimopoulos, V.: On the landscape ruggedness of the quadratic assignment problem. Theoretical Computer Sciences 263, 159–172 (2000)
2. Bäck, T., Fogel, D.B., Michalewicz, Z. (eds.): Evolutionary Computation 1. Basic Algorithms and Operators. IOP Publishing Lt. (2000)
3. Barnes, J.W., Dokov, S.P., Acevedo, R., Solomon, A.: A note on distance matrices yielding elementary landscapes for the TSP. Journal of Mathematical Chemistry 31(2), 233–235 (2002)
4. Barrat, A., Barthélemy, M., Vespignani, A.: Dynamical processes on complex networks. Cambridge University Press (2008)
5. Burkard, R.E., Karisch, S.E., Rendl, F.: QAPLIB - a quadratic assignment problem library. Journal of Global Optimization 10, 391–403 (1997)
6. Chicano, F., Luque, G., Alba, E.: Elementary landscape decomposition of the quadratic assignment problem. In: Proceedings of GECCO, pp. 1425–1432. ACM, New York (2010)
7. Chicano, F., Luque, G., Alba, E.: Autocorrelation measures for the quadratic assignment problem. Applied Mathematics Letters 25(4), 698–705 (2012)
8. Daolio, F., Tomassini, M., Vérel, S., Ochoa, G.: Communities of minima in local optima networks of combinatorial spaces. Physica A: Statistical Mechanics and its Applications 390(9), 1684–1694 (2011)
9. Daolio, F., Vérel, S., Ochoa, G., Tomassini, M.: Local optima networks of the quadratic assignment problem. In: IEEE Congress on Evolutionary Computation, CEC 2010, pp. 3145–3152. IEEE Press (2010)
10. Enright, A.J., Van Dongen, S., Ouzounis, C.A.: An efficient algorithm for large-scale detection of protein families. Nucleic Acids Research 30(7), 1575–1584 (2002)
11. García-Pelayo, R., Stadler, P.: Correlation length, isotropy and meta-stable states. Physica D: Nonlinear Phenomena 107(2-4), 240–254 (1997)
12. Knowles, J.D., Corne, D.W.: Instance Generators and Test Suites for the Multiobjective Quadratic Assignment Problem. In: Fonseca, C.M., Fleming, P.J., Zitzler, E., Deb, K., Thiele, L. (eds.) EMO 2003. LNCS, vol. 2632, pp. 295–310. Springer, Heidelberg (2003)

13. Stadler, P.F.: Fitness Landscapes. In: Biological Evolution and Statistical Physics, pp. 183–204. Springer (2002)
14. Taillard, E.D.: Comparison of iterative searches for the quadratic assignment problem. Location Science 3, 87–105 (1995)
15. Tomassini, M., Vérel, S., Ochoa, G.: Complex-network analysis of combinatorial spaces: The NK landscape case. Phys. Rev. E 78(6), 066114 (2008)
16. Verel, S., Ochoa, G., Tomassini, M.: Local optima networks of NK landscapes with neutrality. IEEE Trans. on Evolutionary Computation 15(6), 783–797 (2011)
17. Weinberger, E.: Correlated and uncorrelated fitness landscapes and how to tell the difference. Biological Cybernetics 63(5), 325–336 (1990)

A Hyper-Heuristic Classifier for One Dimensional Bin Packing Problems: Improving Classification Accuracy by Attribute Evolution

Kevin Sim, Emma Hart, and Ben Paechter

Institute for Informatics and Digital Innovation, Edinburgh Napier University
Merchiston Campus, Edinburgh, EH10 5DT
{k.sim,e.hart,b.paechter}@napier.ac.uk

Abstract. A hyper-heuristic for the one dimensional bin packing problem is presented that uses an Evolutionary Algorithm (EA) to evolve a set of attributes that characterise a problem instance. The EA evolves divisions of variable quantity and dimension that represent ranges of a bin's capacity and are used to train a k-nearest neighbour algorithm. Once trained the classifier selects a single deterministic heuristic to solve each one of a large set of unseen problem instances. The evolved classifier is shown to achieve results significantly better than are obtained by any of the constituent heuristics when used in isolation.

Keywords: Hyper-heuristics, one dimensional bin packing, classifier systems, attribute evolution.

1 Introduction

The one dimensional bin packing problem (BPP) is a well researched NP-hard problem which has been tackled using a diverse range of techniques including mathematically complete procedures[16], deterministic heuristics[11], biologically inspired metaheuristics [8] as well as by the field of hyper-heuristics [15]. The plethora of research and benchmark problem instances available combined with the fact that the problem constitutes an integral part of many other more complex problems makes it an ideal domain for investigating new techniques.

This paper presents a hyper-heuristic which attempts to predict which heuristic, from an available pool, will perform best on a given problem instance. The system incorporates a classification algorithm within an EA in an attempt to generate predictor attributes that improve upon the classification accuracy obtained using predetermined characteristics. The system, once trained using half of 1370 benchmark problem instances, achieves results substantially better than any individual heuristic on the unseen problem instances.

The remainder of this paper is organised as follows. The field of hyper-heuristics and related work are introduced in section 2 with the one dimensional bin packing problem domain, the benchmark problem instances and the deterministic heuristics used in this study covered in section 3. The experimental framework is described in section 4 with the results from those experiments presented in section 5. The paper finishes with section 6 where conclusions are drawn and potential for future research is suggested.

C.A. Coello Coello et al. (Eds.): PPSN 2012, Part II, LNCS 7492, pp. 348–357, 2012.
© Springer-Verlag Berlin Heidelberg 2012

2 Hyper-Heuristics

The term hyper-heuristics (HH) first appeared in relation to combinatorial optimisation (CO) problems in [5] although the term was first coined in [6] to describe
an amalgamation of artificial intelligence techniques in the domain of automated
theorem proving. However, the concept can be traced back to the 1960's when
Fisher & Thompson [9] used machine learning techniques to select combinations
of simple heuristics to produce solutions to local job-shop scheduling problems.
Originally described as *"heuristics to select heuristics"* [2] the field has evolved
to encompass techniques including *"heuristics to generate heuristics"* using genetic programming to create new heuristics from constituent component parts
[3,4]. All hyper-heuristics, no matter the approach, have the commonality that
they search over a landscape defined by a set of heuristics, or their component
parts, for a procedure to solve a problem rather than searching directly over the
space defined by the problem itself. A more concise review can be found in [2,1].

In [15] Ross et al., proposed a hyper-heuristic approach to bin-packing that
introduced the notion of describing the state of a problem instance according to
the percentage of items that fall into 4 pre-defined "natural" categories relating
to item size, given as a ratio of the bin capacity[1]. A Michigan style Learning
Classifier System (LCS) was used to evolve a set of rules mapping problem states
to suitable heuristics. Each iteration the chosen heuristic packs a single bin with
the potential of a filler process being invoked that attempts to fill a partially filled
bin further. The remaining items are then reclassified using the new problem
state resulting in a deterministic selection of a sequence of heuristics for solving
each problem instance.

The approach presented here differs in that it does not use pre-defined categories to describe an instance's state. Using a variable-length evolutionary algorithm a set of categories is evolved that when used in conjunction with a classifier
algorithm, map the description of an instance to a suitable simple heuristic. In
contrast to [15], problem instances are only categorised once and solved using a
single heuristic. The motivation behind this is to determine whether it is possible to find an appropriate method of describing a set of problem instances such
that each instance can be mapped to the single heuristic that best solves it.
The authors of [15] showed this task to be non-trivial and were unable to find a
relationship using a perceptron. Whilst the ranges they used to describe a problem appeared "natural" choices they disregard potential relationships between
different item sizes that when combined allow for optimal bin packings.

The system presented here, conceptualised in Figure 1 , uses a heuristic selection strategy to choose which from a set of deterministic constructive heuristics
to apply to a problem instance based on knowledge of the problem domain obtained during an off-line training phase. This is achieved using a classification
algorithm that attempts to match an unseen problem instance to a procedure
for solving it based on the problem instance's characteristics. The characteristics used are the percentages of the items with weights within a number of
ranges, expressed as ratios of the bin capacity. The divisions used are not fixed
in number or dimension but are evolved by the EA during a training phase.

---

[1] Described in Section 4, Figure 2.

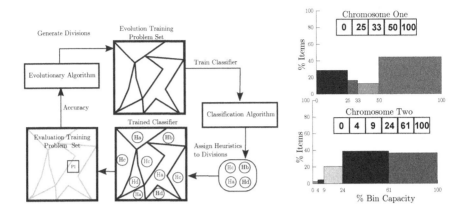

Fig. 1. During off-line training, the EA generates problem divisions, of varying dimension and number, that the classifier assigns the best known heuristic to. The classifier's accuracy in predicting which is the best heuristic for a set of unseen problem instances is used as feedback to the EA. The two graphs show the *same* problem instance encoded by the two different chromosomes shown. The x-axis depicts the evolved ranges expressed as a percentage of the bin capacity whilst the y-axis depicts the percentage of the instances' items with sizes falling within each range.

The system is described in more detail in Section 4 after introducing the BPP domain, benchmark problem instances and heuristics used during this study.

3   One Dimensional Bin Packing Problem

The objective of the one dimensional bin packing problem is to find the optimal number of bins, $OPT(I)$, of fixed capacity c required to accommodate a set of n items, $J = \{\omega_1 \ldots \omega_n\}$ with weights $\omega_j : j \in \{1 \ldots n\}$ falling in the range $1 \leq \omega_j \leq c$ whilst enforcing the constraint that the sum of weights in any bin does not exceed the bin capacity c (Scholl, et al., 1997). For any instance $OPT(I)$ must lie between the lower and upper bounds shown in Equation 1 with the upper bound occurring when all items are greater than half the bin capacity and the lower bound achieved when the total free space summed across all bins is less than the capacity of one bin.

$$\lceil (\sum\nolimits_{j=1}^{n} \omega_j) \div c \rceil \leq OPT(I) \leq n \tag{1}$$

Table 1 shows the parameters from which the benchmark data sets used in this study were generated. Data sets $ds1, ds2$ & $ds3$, introduced by Scholl et al., in [16] all have optimal solutions that vary from the lower bound given by Equation 1. However all are known and have been solved since their introduction [17]. All of the instances from $FalU$ and $FalT$, introduced by Falkenauer in [8], have optimal solutions at the lower bound except for one [12].

Four heuristics, three re-created and a fourth introduced here, were included in the system. All pre-sort and select items in decreasing weight order.

Table 1. Data sets $ds1$, $ds3$ and $FalU$ were created by generating n items with weights randomly sampled from a uniform distribution between the bounds given by ω. Those in $FalT$ were generated in a way[8] so that the optimal solution has exactly 3 items in each bin with no free space. Scholl's $ds2$ was created by randomly generating weights from a uniform distribution in the range given by $\varpi \pm \delta$. The final column gives the number of instances generated for each parameter combination.

Data Set	capacity (c)	n	ω	#Problems
$ds1$	100,120,150	50,100,200,500	[1,100],[20,100],[30,100]	$36 \times 20 = 720$
$ds3$	100000	200	[20000,30000]	10
$FalU$	150	120,250,500,1000	[20,100]	$4 \times 20 = 80$
$FalT$	1	60,120,249,501	[0.25,0.5]	$4 \times 20 = 80$

Data Set	c	n	ϖ (avg weight)	$\delta(\%)$	# Problems
$ds2$	1000	50,100,200,500	$\frac{c}{3}, \frac{c}{5}, \frac{c}{7}, \frac{c}{9}$	20,50,90	$48 \times 10 = 480$

- *First Fit Descending* (FFD) packs each item into the first bin that will accommodate it. If no bin is available a new bin is opened. All bins remain open for the duration of the procedure.
- *Djang and Finch* [7] (DJD) and an extension *DJD more Tuples* (DJT) introduced in [15] both pack items into a bin until it is at least a third full. Combinations of up to three (or five for DJT) items are then searched for that best fill the remaining space with preference given to sets that use the largest items. The bin is then closed and the procedure repeats.
- *Adaptive DJD* (ADJD), introduced here, packs items into a bin in descending order until the *free space* in the bin is less than or equal to three times the average size of the items remaining to be packed. It then operates like DJD looking for the set of up to three items that best fills the remaining capacity.

It has been noted [12] that many so called "hard" benchmark problem instances can be solved easily by simple procedures. Often benchmark instances are introduced in the literature alongside procedures specifically designed to solve them, such as those from Falkenauer whose Hybrid Grouping Genetic Algorithm (HGGA) utilises a local search heuristic inspired by Martello and Toth's Reduction Procedure (MTRP) [14] tailored for finding optimal sets of three items. It has been shown for FFD and MTRP[17], and thus DJD and HGGA which both use searches inspired by MTRP, that instances with average weights, $\varpi_j \to \frac{c}{3}$ are the most complex with those where $\varpi_j \to \frac{c}{4}, \frac{c}{5}, \frac{c}{6} \ldots$ proving difficult also.[2] All of the problems used here, except for those in ds2, have an average item weight of around $\frac{c}{3}$.

In [15] the authors showed DJT to be the most successful heuristic when used in isolation solving 73% of instances to the known optimum. The study however omitted $ds2$, on which DJT finds only 45% of the optimal solutions.[3] ADJD,

---

[2] If a solution exists at the lower bound given in Equation 1 then the total free space $\varpi_{free} \to 0$ as $\varpi_j \to \frac{c}{i} : i \in \mathbb{N} : i \geq 3$.

[3] DJT will perform best where $\varpi \geq \frac{2}{15}c$ as once the initial filling procedure has filled $\frac{1}{3}c$ the remaining $\frac{2}{3}c$ can be filled by at most five items.

introduced here, whilst the worst performer on the complete set of problems achieves significantly better results on the problem instances from $ds2$. This is accomplished by first packing items in descending order of size until the *free space* in the bin is less than or equal to average size of the items remaining to be packed thus improving the chance of finding a combination of items to fill the remaining capacity for problems with smaller average item weights when compared to DJD or DJT.

In order to get a better indication of a heuristic's performance than can be deduced solely from the number of optimal solutions found, Falkenauer's fitness function, given in Equation 2, is used with k set to 2 in order to reward solutions where any free capacity is restricted to as few bins as possible allowing for a distinction to be made between different solutions that use an equal numbers of bins as well as a measure of a non-optimal solution's quality.

$$f(x) = \sum\nolimits_{j=1}^{n} \left(\frac{fill_j}{c} \right)^k \div n \qquad (2)$$

A third metric used that gives a measure of a heuristic's ability to generalise over a diverse range of problem instances is the number of extra bins required over the optimal number. Table 2 shows the results obtained, for each heuristic, using these three metrics. It is interesting to note for instance, that whilst FFD rates highly if ranked in terms of the number of optimal solutions found, it achieves this using the second largest number of bins. In contrast ADJD, which comes 4th in terms of the number of optimal solutions found, achieves 2nd best position if ranked by either of the other two metrics.

Table 2. The table shows the results obtained by each heuristic on different data sets using three metrics; The percentages of problems solved using the optimum number of bins, the ratio for which the best fitness was attained and the percentage of extra bins required over the optimal. The headings in row 2 depict the data sets as described in Table 1 with Tr and Te depicting the training and test sets used during the experiments described here in section 4 and *All* representing the complete set of 1370 instances. None of the heuristics used here are able to find optimal solutions to any of the instance from $FalT$ or $ds3$.

Metric	Optimal						Fitness		Bins	
Heuristic	$ds1$	$ds2$	$FalU$	Tr	Te	All	All	Te	All	Te
# Problems	720	480	80	685	685	1370	1370	685	1370	685
FFD	75.83	49.17	7.5	57.66	57.37	57.52	27.52	27.45	1.78	1.81
DJD	79.03	24.05	57.5	52.55	51.97	52.26	47.74	47.74	2.00	2.02
DJT	83.75	44.58	57.5	63.21	62.77	62.99	54.96	55.47	0.73	0.75
ADJD	35.83	80.21	53.75	51.09	49.05	50.07	53.80	52.26	1.12	1.13

The following section describes the system implemented in an attempt to harness the combined abilities of the heuristics used.

4   Experimental Framework

The system, described in Figure 3, comprises of a database containing the problem instances and corresponding solutions attained by each heuristic along with a classification algorithm and an EA. The classifier predicts which heuristic will perform best on an unseen problem instance whilst the EA attempts to increase classification accuracy by evolving the predictor attributes used. Unlike other applications in which classifiers and EAs have been combined to select which predetermined predictor attributes should be used, the approach here uses the EA to evolve combinations of problem characteristics not known *a priori*. A comprehensive review of EAs combined use with classification algorithms is outwith the scope of this paper for which the reader is directed to [10].

The chromosome representation used by the EA is based on that used in [15] in which an instances' state is described by characteristics which included the percentage of each instances items with weights within certain predetermined ranges, measured as ratios of the bin capacity. The ranges used, and adopted here as a benchmark, are shown in the chromosome representation depicted in Figure 2. These were deemed "natural" choices by the authors as at most one *Huge*, two *Large* or three *Medium* items can be placed in any individual bin. These ranges, or divisions, are used as the classifiers predictor attributes with the best heuristic being the goal, or class attribute.

In this study the EA evolves variable length chromosomes which are deliberately constrained to a maximum length that was incrementally increased for each experiment conducted. A chromosome encodes each instance from the *evolution* training set by determining the percentage of items with weights in each range which along with the known best heuristic [4] for each instance is used to train the classifier. The ratio of *evaluation* training problems correctly classified is then used as the objective fitness value. Each data used in this study was created by generating either ten or twenty problems for each parameter combination as described in Table 1. The partitioning of these sets used here ensures an even distribution of instances from each parameter combination between the training and test sets and also the subdivisions of the training set described by Figure 3.

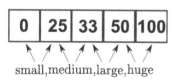

| 0 | 25 | 33 | 50 | 100 |

small,medium,large,huge

Huge: $\frac{C}{2} < \omega_j$

Large: $\frac{C}{3} < \omega_j \le \frac{C}{2}$

Medium: $\frac{C}{4} < \omega_j \le \frac{C}{3}$

Small: $\omega_j < \frac{C}{4}$

Fig. 2. For a chromosome with n genes numbered from left to right the percentage of items p_i falling into each range $r_i < p_i \le r_{i+1}$ $\forall i = 1, \ldots, n-1$ is encoded and passed to the classifier as predictor attributes. The terminal alleles, 0 & 100 were inferred.

---

[4] Determined using Equation 2 with ties awarded to the computationally simplest heuristic in the order FFD, DJD, DJT and ADJD.

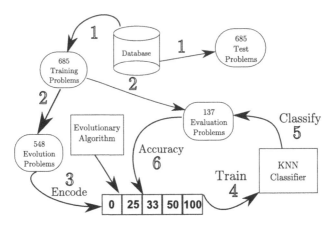

1. Separate every alternate problem into training and test sets.
2. Split the training set into evolution and evaluation sets with every
   5th problem put into the evaluation set.
3. Using the best chromosome encode the evolution set and use
   as predictor attributes for the classifier.
4. Train the classifier using the predictor attributes with the goal
   attribute being the best heuristic for each instance.
5. Use the classifier to predict the best heuristic for each problem in
   the evaluation set.
6. Measure the classification accuracy and use this figure as the
   fitness measure for that chromosome.
7. After 1000 iterations use the best chromosome and the complete
   training set to train the classifier.
8. The results, presented in section 5, show the ability of the classifier
   to select the best heuristic for the as yet unseen test set.

Fig. 3. The system elements and numbered steps explained by the pseudo code

The classification algorithm used was taken from the Waikato Environment
for Knowledge Analysis (WEKA) package [13]. After some initial observations
a K-Nearest Neighbour Classifier was chosen and used with all parameter settings as default with the exception of the variable k which was set to 2. The EA
employed, uses a steady state population, of size 40, with crossover performed
to generate one offspring each iteration with a probability of 60%. Each parent
is selected by means of a tournament between two randomly chosen competitors. Crossover takes the first parent and selects all alleles up to and including
a random position, placing these into the offspring. The second parent is then
searched sequentially until an allele value is found greater than has been introduced from the first parent. This and subsequent genes are appended to the
offspring. Mutation occurs with a probability of 2% and simply adds or removes,
with equal probability, one random value to the chromosome. In order to limit
the chromosome length a trimming process is employed. Should the chromosome

produced exceed the maximum length stipulated then the closest two allele values are merged, taking on the average value of the two. This trimming procedure is repeated as necessary until the chromosome is at most the maximum length allowed for that experiment. Each iteration the worst member of the population is replaced by the child if its fitness is better and an identical chromosome does not already exist in the population. Seven experiments were conducted, each consisting of thirty runs with each run terminated after 1000 iterations. For each experiment, the only parameter modified was the maximum allowed chromosome length, l. The values used were $l = \{3, 5, 10, 20, 50, 100, 200\}$. A chromosome length of l corresponds to $l + 1$ ranges once the terminal alleles representing 0 & 100 were added.

5 Results

The results obtained are shown in Figure 4. The best single individual heuristic, when ranked by the number of optimal solutions found, was DJT which solved 62.77% (430) of the instances in the test set using an extra 0.75% more bins (452) than the optimum. In comparison the hyper-heuristic presented here found 521

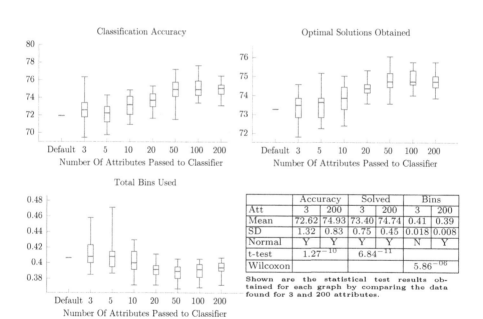

Fig. 4. The three plots, taken over 30 runs show, for the unseen 685 test problems, the percentages of problems correctly classified and solved to the known optimal along with the percentage of extra bins over the optimal of 60257 required. The default values show the results obtained when using benchmark attributes (0.25,0.33,0.5). The results of two unpaired two tailed t-tests with no assumption of equal sample variance are given for the data sets that a Shapiro-Wilk Normality test reported as being normally distributed with a non-parametric Wilcoxon Mann-Witney test used for the other.

(76.06%) optimal solutions using only 0.37% (223) more bins. A ten fold cross-validation was also conducted using the complete set of 1370 problems and the best set of evolved predictor attributes achieving 72.99% accuracy in comparison to 68.90% using the non-evolved default attributes.

Unlike in [15], the system described here is unable to solve any instances to the optimum that are unsolved by any of the constituent heuristics. As different heuristics, methodologies and problem instances are used a direct comparison is not entirely possible. However for comparison, when trained using the evolved characteristics that gave the best result in terms of the number of optimal solutions obtained along with the truncated training set of problems used in [15] the system presented here was able to find optimal solutions to 172 of the 223 test problems used in [15] as opposed to 166 reported by the papers authors.

6   Conclusions and Future Work

By combining heuristics the number of optimal solutions found is increased substantially over the number found by any individual heuristic. Furthermore by evolving *relevant* predictor attributes for use by the classifier the goal of generating a problem description that maps individual instances to an appropriate heuristic for solving it was achieved. The system developed is able to better generalise over a wider range of problem instances with varying characteristics than can be addressed by any of the heuristics when used in isolation. The new heuristic introduced, ADJD, has been shown to perform better on problem instances with certain characteristics than any of the other heuristics investigated and although the single worst heuristic over the complete set of benchmark instances it is shown to increase the generality of the hyper-heuristic system presented.

It is intended to investigate expanding the work presented here in a number of directions. Separate classifiers, one or more for each heuristic, could be combined with each attempting to predict the fitness that its associated heuristic would achieve when presented with an unseen problem instance. This would allow for multiple classifiers, even of different types, to compete potentially giving rise to improved accuracy in a similar way to ensemble classification techniques.

Another possible direction for further study is to closer emulate the research that inspired this work, where rather than using one heuristic to completely solve a problem instance, a sequence of different heuristics is used which are predicted after each new bin has been packed. All of the heuristics work in this manner already with the exception of FFD which is easily adapted, as in [15], to exhibit the same behaviour. Initial investigations into increasing the number and variety of heuristics used suggests that whilst the classification task increases in complexity with the number of heuristics used, the potential for solving more instances increases also. Although other heuristics were investigated initially many were deemed too similar, such as BFD which found only one optimal solution that FFD did not. The use of Genetic Programming techniques to generate new heuristics could potentially allow for a broader set of simple heuristics with a more diverse range of abilities to be incorporated such as has been investigated in [4] albeit using a considerably smaller set of ninety benchmark instances.

References

1. Burke, E.K., Hyde, M., Kendall, G., Ochoa, G., Özcan, E., Qu, R.: Hyper-heuristics: A survey of the state of the art. School of Computer Science and Information Technology, University of Nottingham, Computer Science Technical Report No. NOTTCS-TR-SUB-0906241418-2747 (2010)
2. Burke, E., Kendall, G., Newall, J., Hart, E., Ross, P., Schulenburg, S.: Hyper-heuristics: An emerging direction in modern search technology. In: Handbook of Metaheuristics. International Series in Operations Research & Management Science, ch. 16, pp. 457–474. Kluwer (2003)
3. Burke, E.K., Hyde, M., Kendall, G., Ochoa, G., Özcan, E., Woodward, J.R.: A classification of hyper-heuristic approaches. In: Gendreau, M., Potvin, J.Y. (eds.) Handbook of Metaheuristics, International Series in Operations Research & Management Science, vol. 146, pp. 449–468. Springer US (2010)
4. Burke, E.K., Hyde, M.R., Kendall, G., Woodward, J.: Automating the packing heuristic design process with genetic programming. Evol. Comput. 20(1), 63–89 (2012)
5. Cowling, P.I., Kendall, G., Soubeiga, E.: A Hyperheuristic Approach to Scheduling a Sales Summit. In: Burke, E., Erben, W. (eds.) PATAT 2000. LNCS, vol. 2079, pp. 176–190. Springer, Heidelberg (2001)
6. Denzinger, J., Fuchs, M.: High performance atp systems by combining several ai methods. In: Proceedings Fifteenth International Joint Conference on Artificial Intelligence, IJCAI 1997, pp. 102–107. Morgan Kaufmann (1997)
7. Djang, P.A., Finch, P.R.: Solving one dimensional bin packing problems (1998)
8. Falkenauer, E.: A hybrid grouping genetic algorithm for bin packing. Journal of Heuristics 2, 5–30 (1996)
9. Fisher, H., Thompson, G.L.: Probabilistic learning combinations of local job-shop scheduling rules. In: Muth, J., Thompson, G.L. (eds.) Industrial Scheduling, pp. 225–251. Prentice Hall, Englewood Cliffs (1963)
10. Freitas, A.A.: Data Mining and Knowledge Discovery with Evolutionary Algorithms. Springer (2002)
11. Garey, M.R., Johnson, D.S.: Computers and intractability: a guide to the theory of NP-completeness. A Series of books in the mathematical sciences. W.H. Freeman, San Francisco (1979)
12. Gent, I.P.: Heuristic solution of open bin packing problems. Journal of Heuristics 3(4), 299–304 (1998)
13. Hall, M., Frank, E., Holmes, G., Pfahringer, B., Reutemann, P., Witten, I.H.: The weka data mining software: an update. SIGKDD Explorations 11(1) (2009)
14. Martello, S., Toth, P.: Lower bounds and reduction procedures for the bin packing problem. Discrete Applied Mathematics 28(1), 59–70 (1990)
15. Ross, P., Schulenburg, S., Marín-Blázquez, J.G., Hart, E.: Hyper-heuristics: Learning to combine simple heuristics in bin-packing problems. In: Proceedings of the Genetic and Evolutionary Computation Conference, GECCO 2002, pp. 942–948. Morgan Kaufmann Publishers Inc., San Francisco (2002)
16. Scholl, A., Klein, R., Jürgens, C.: Bison: a fast hybrid procedure for exactly solving the one-dimensional bin packing problem. Comput. Oper. Res. 24(7), 627–645 (1997)
17. Schwerin, P., Wäscher, G.: The bin-packing problem: A problem generator and some numerical experiments with ffd packing and mtp. International Transactions in Operational Research 4(5-6), 377–389 (1997)

A Framework to Hybridize PBIL and a Hyper-heuristic for Dynamic Environments

Gönül Uludağ[1], Berna Kiraz[1], A. Şima Etaner-Uyar[1], and Ender Özcan[2]

[1] Istanbul Technical University, Turkey
{uludagg,etaner}@itu.edu.tr, berna.kiraz@marmara.edu.tr
[2] University of Nottingham, UK
Ender.Ozcan@nottingham.ac.uk

Abstract. Selection hyper-heuristic methodologies explore the space of heuristics which in turn explore the space of candidate solutions for solving hard computational problems. This study investigates the performance of approaches based on a framework that hybridizes selection hyper-heuristics and population based incremental learning (PBIL), mixing offline and online learning mechanisms for solving dynamic environment problems. The experimental results over well known benchmark instances show that the approach is generalized enough to provide a good average performance over different types of dynamic environments.

Keywords: hyper-heuristics, dynamic environments, multiple populations, incremental learning.

1 Introduction

Many real world optimization problems are dynamic in nature. When solving a problem in such environments, it is better to take the dynamism into account and choose an appropriate optimisation approach which is able to adapt and track the moving optima. Different types of changes may occur in the environment over time. The dynamism in the environment can be classified based on its severity, frequency, predictability, cycle length and cycle accuracy [2]. There are many techniques proposed in literature to solve dynamic optimization problems. A recent survey can be found in [5].

Recently, there has been a growing interest in Estimation of Distribution Algorithms (EDAs). The performance improvement of EDAs via the development of different algorithmic frameworks, such as multi-population approaches, inclusion of mechanisms addressing issues, such as hyper-mutation to deal with diversity loss and other mechanisms are of interest for many researchers and practitioners to solve dynamic environment problems [1,6,14,16,13,18].

There is an emerging field of research in the semi-automated design of search methodologies: *hyper-heuristics*. Burke et al. [3] defined hyper-heuristics as methodologies that search the space of heuristics by *selecting* or *generating* them

C.A. Coello Coello et al. (Eds.): PPSN 2012, Part II, LNCS 7492, pp. 358–367, 2012.
© Springer-Verlag Berlin Heidelberg 2012

to solve difficult problems. The focus of this study is selection hyper-heuristics which attempt to improve an initially generated candidate solution iteratively through *heuristic selection* and *move acceptance* stages [4,10]. In this paper, we will use *hyper-heuristics* to denote *selection hyper-heuristics*. Özcan et al. [11] proposed a hyper-heuristic framework for dynamic environments for the first time, to the best of the authors' knowledge. Empirical evidence suggests that hyper-heuristics are effective solvers in dynamic environments for real valued optimisation [7] as well as combinatorial optimisation [8].

Although variants of EDAs have been proposed to solve dynamic environment problems, it has been observed that there is almost no single approach that performs consistently well across different types of dynamic environments. This is mostly because different types of methods are capable of handling particular types of changes relatively better than others in such environments.

In this study, inspired from previous studies, we investigate the performance of a general framework which is based on a bi-population approach hybridizing a variant of EDA, in particular PBIL, and a selection hyper-heuristic across some well known benchmark functions. The goal of the study is to enhance the performance of PBIL enabling this approach to handle any given type of change dynamic and hence, raise its level of generality. The framework can combine any EDA based approach with any type of selection hyper-heuristic. We utilize an offline learning mechanism to detect the useful operators (or operator components) for different environments and then use an online learning selection hyper-heuristic to select the best operator at a given time during the search process while solving an unseen instance. The following sections discuss the details of the proposed framework.

2 Proposed Framework

In this study, we propose a new framework exploiting the advantages of hyper-heuristics and multi-population approaches. The framework can combine any multi-population EDA with selection hyper-heuristics. Here we propose *hyper-heuristic based multi-population PBIL* (HH-PBIL2), which is based on SPBIL2 introduced in [18]. SPBIL2 is a bi-population standard PBIL (SPBIL) algorithm.

Kiraz et al. [7], show that heuristic selection methods with learning, namely choice function and reinforcement learning (see [10] for details) outperform others. Both incorporate some form of a scoring mechanism. In choice function, when scoring a heuristic, the difference between the fitness values of the offspring and the current candidate solution is taken into account. In a dynamic environment setting, this means that whenever a change occurs, the current candidate solution has to be re-evaluated in the new environment. For the proposed approach this involves re-evaluating all the candidate solutions in the current population, which is computationally ineffective. Therefore, we do not use choice function as a heuristic selection method. In reinforcement learning (RL) [9] heuristic selection method, each low-level heuristic has a utility score. The scores of each heuristic are initialized to the same value and updated during the search process based on its performance.

At each step, the low-level heuristic with the maximum score is selected. If the selected heuristic produces a better solution than the previous one, it is rewarded by increasing its score, otherwise it is penalized by decreasing it. The scores are restricted to vary between predetermined lower and upper bounds.

In SPBIL, a posterior probability distribution model of promising solutions is built using statistical information obtained from the population of solution candidates. This is termed as a *probability vector*, which is used to create a population of solutions through sampling at each iteration. In SPBIL2, the population is divided into two sub-populations. Each sub-population is sampled from its own probability vector. The two probability vectors are evolved in parallel for a given maximum number of generations. As in SPBIL, the first probability vector $\vec{P}_1$ is initialized with the central probability vector, and the second probability vector $\vec{P}_2$ is initialized randomly. The size of the initial sub-populations are equal. After all candidate solutions are evaluated, sub-population sample sizes are slightly adjusted within the range of $[0.3 * n, \ 0.7 * n]$. Then, each probability vector is learnt towards the best solution(s) in the relevant sub-population. Similar to SPBIL, mutation is applied to both probability vectors before sampling. Details of SPBIL2 can be found in [18,17].

The approach proposed in this paper (HH-PBIL2) consists of two phases. In the first phase, probability vectors corresponding to a set of different environments are learned offline, using SPBIL. Then, those learned probability vectors are stored for later use. In the second phase, the probability vectors serve as low-level heuristics for the RL based hyper-heuristic.

HH-PBIL2 is proposed to enhance the performance of SPBIL2 in dynamic environments. As in SPBIL2, the population is divided into two sub-populations and two probability vectors are used in parallel. The first probability vector $\vec{P}_1$ is again initialized with the central probability vector, but the second probability vector $\vec{P}_2$ is selected randomly from the previously stored probability vectors. Each sub-population is sampled independently using the relevant probability vector. The first probability vector $\vec{P}_1$ is learned towards the best solution candidate(s) in the first population. There is no online learning step for the second probability vector $\vec{P}_2$. At each iteration, RL heuristic selection mechanism selects the probability vector with the largest score from among the previously stored probability vectors and this probability vector is assigned as $\vec{P}_2$. The score update scheme for the RL heuristic selection method is explained above.

We used two variants of HH-PBIL2 which differ in the information used to update a low-level heuristic's score. In the first variant, RL-PF, the best performing candidate solution(s) from the two populations combined, are used to update the score. In the second variant, RL-P2, the best performing solution candidate(s) from only the second population is used to update the score.

Similar to SPBIL2, after the candidate solutions are evaluated, the next population sizes are slightly adjusted. However, mutation is applied only to $\vec{P}_1$. Then, two sub-populations are sampled based on the relevant probability vectors. The approach repeats the cycle until some termination criteria are met. The pseudocode of HH-PBIL2 is shown in Algorithm 1.

Algorithm 1. Pseudocode of the proposed approach HH-PBIL2.

1: $t := 0$
2: initialize $\vec{P}_1(0) := \vec{0.5}$
3: $\vec{P}_2(0)$ is selected from RL randomly
4: $S_1(0) := sample(\vec{P}_1(0))$ and $S_2(0) := sample(\vec{P}_2(0))$
5: **while** (*termination criteria not fulfilled*) **do**
6:      evaluate $S_1(t)$ and evaluate $S_2(t)$
7:      adjust next population sizes for $\vec{P}_1(t)$ and $\vec{P}_2(t)$ respectively
8:      place k best samples from $S_1(t)$ and $S_2(t)$ into $\vec{B}(t)$
9:      **if** *(RL-PF)* **then**
10:         send the best fitness value from the whole population to RL
11:     **end if**
12:     **if** *(RL-P2)* **then**
13:         send the best fitness value from the second population to RL
14:     **end if**
15:     learn $\vec{P}_1(t)$ toward $\vec{B}(t)$
16:     mutate $\vec{P}_1(t)$
17:     $\vec{P}_2(t)$ is selected with maximum score from RL
18:     $S_1(t) := sample(\vec{P}_1(t))$ and $S_2(t) := sample(\vec{P}_2(t))$
19:     $t := t + 1$
20: **end while**

3 Experimental Design

In the experiments, the proposed approaches RL-PF and RL-P2 are compared with SPBIL and SPBIL2. The original source codes which we compared are taken from Yang's web site[1]. Our approaches are implemented based on SPBIL2. These two techniques are briefly explained in section 2.

All approaches are applied to three Decomposable Unitation-Based Functions (DUFs). All DUFs are composed of 25 copies of 4-bit building blocks. Each building block is denoted as a unitation-based function $u(x)$ which gives the number of ones in the corresponding building block. Its maximum value is 4. The fitness of a bit string is calculated as the sum of the $u(x)$ values of the building blocks. The optimum fitness value for all DUFs is 100. The DUFs can be formulated as follows [13].

$$f_{DUF1} = u(x) \qquad f_{DUF2} = \begin{cases} 4 \text{ , if } u(x) = 4 \\ 2 \text{ , if } u(x) = 3 \\ 0 \text{ , if } u(x) < 3 \end{cases} \qquad f_{DUF3} = \begin{cases} 4 & \text{ , if } u(x) = 4 \\ 3 - u(x) \text{ , if } u(x) < 4 \end{cases}$$

$DUF1$ is the *OneMax* problem whose objective is to maximize the number of ones in a bit string. $DUF2$ has a unique optimal solution surrounded by four local optima and a wide plateau with eleven points having a fitness of zero. $DUF2$ is more difficult than $DUF1$. $DUF3$ is fully deceptive [18].

The XOR dynamic problem generator [15,17] is applied to the three DUFs to obtain dynamic test problems. The XOR generator can create a dynamic environment problem with varying degrees of difficulty from any binary-encoded stationary problem using a bitwise exclusive-or (XOR) operator. Given a function $f(x)$ in a stationary environment and $x \in \{0,1\}^l$, the fitness value of the x at a given generation g is calculated as $f(x, g) = f(x \oplus m_k)$, where m_k is a binary mask for k^{th} stationary environment and $\oplus$ is the XOR operator. Firstly,

---

[1] http://www.brunel.ac.uk/~csstssy/publications.html

the mask m is initialized with a zero vector. Then, every τ generations, the mask m_k is changed as $m_k = m_{k-1} \oplus t_k$, where t_k is a binary template.

SPBIL, SPBIL2 and HH-PBIL2 share some common settings which are used as suggested in [18]. The problem consists of 25 building blocks, therefore solution candidates are of length 100. Mutation rate is 0.02 and mutation shift is 0.05. The learning rate α is taken as 0.25 and 3 best candidate solutions are used in the online learning of probability vectors. The population size for SPBIL is set to 100. For SPBIL2 and HH-PBIL2, each sub-population size is initialized as 50 and they are allowed to vary between 30 and 70. In RL, score of each heuristic is initialized to 15 and allowed to vary between 0 and 30. If the selected heuristic yields a solution with an improved fitness, its score is increased by 1, otherwise decreased by 1. The RL settings are taken as recommended in [12].

In the first phase of HH-PBIL2, probability vectors corresponding to a set of different environments are learned offline using SPBIL. To generate different environments using the XOR generator, a set of M XOR masks are randomly generated. Then, for each mask (i.e. environment), SPBIL is executed for 100 independent runs where each run consists of 10, 000 generations. During offline learning, each environment is stationary. At the end, for each environment, the probability vector producing the best solution found so far over all runs is stored. These vectors are used in in all the rest of the experiments.

This study considers the frequency of changes τ, severity of changes ρ and cycle length CL as the type of changes in the environment. In the cyclic environments, we assume that environments return to their previous locations exactly. None of the tested methods require that the time of a change is known.

As a result of some preliminary experiments, we determined the change periods as 50 generations for low frequency (LF), 25 generations for medium frequency (MF) and 5 generations for high frequency (HF) for DUF1 and DUF2. The change periods for DUF3 are determined as 100 generations for LF, 35 generations for MF and 10 generations for HF. In convergence plots, these settings for LF, MF and HF correspond respectively to stages where the algorithm has been converged for some time, where it has not yet fully converged and where it is very early on in the search. In addition, the severity of changes are chosen as 0.1 for low severity (LS), 0.2 for medium severity (MS), 0.5 for high severity (HS), and 0.75 for very high severity (VHS) for random dynamic environments. These are determined based on the definition of the XOR generator. For cyclic dynamic environments, the cycle lengths CL are selected as 2, 4 and 8. To construct cyclic environments, the masks representing the environments are selected among the randomly generated M masks used in the offline learning phase of HH-PBIL2. For each run of the algorithms, 128 changes occur after the initial environment. Therefore, the maximum number of generations is calculated as $maxGenerations = changeFrequency * changeCount$. We performed experiments to explore the effects of the severity and the frequency of the changes on the performance of the approaches for randomly changing environments, and the effects of the cycle length and the frequency of the changes on the performance of the approaches for cyclic environments.

In order to compare the performance of the algorithms, the results are reported in terms of the offline error [2], calculated as the cumulative average of the errors of the best candidate solutions found so far. The error of a candidate solution is calculated as the difference of its fitness value from the fitness value of the optimum solution at each time step. Fitness values are calculated using the corresponding DUF definitions given above. In all our experiments, while the location of the global optimum may change, its fitness value remains the same and is 100 for all time steps. An algorithm solving a dynamic environment problem aims to achieve the least overall offline error value obtained at the end of a run. All reported results are averages over 100 independent runs. Anova and Tukey's HSD tests are applied to the results at a significance level of 95% to test for statistically significant differences.

4   Results and Discussion

All test results are summarized in Table 2 for randomly changing environments and in Table 3 for cyclic environments on all DUFs. The values in the tables show the offline errors achieved at the end of a run, averaged over 100 runs. Due to lack of space, the statistical significance comparison tables are not given in the paper[2].

Firstly, we analyze the effects of the learned probability vector counts (M) on HH-PBIL2 in both randomly changing and cyclic environments. We experimented with M values of $8, 16, 32, 64$. The results of the ANOVA and Tukey's HSD tests for statistical significance at a 95% confidence level are reported in Table 1. In the table, each entry shows the total number of times the approach achieves the corresponding significance state ($s+$, $s-$, $\geq$ and $\leq$) over the others on the three DUFs for different change severity and frequency settings in randomly changing environments and for different cycle length and change frequency settings in cyclic environments. Here, the following notation is used: Given A vs B, $s+$ ($s-$) denote that A (B) is performing statistically better than B (A), while $A \geq B$ ($A \leq B$) indicates that A (B) performs slightly better than B (A) and this performance difference is not statistically significant. From the table, we can see that $M = 8$ is better overall for the tested environments under all change settings. Therefore, in the tables 2 and 3, we only report the results for $M = 8$. The statistical significance tests show that the number of learned probability vectors does not significantly affect the performance of HH-PBIL2 variants for all change frequency-severity settings. However, for cyclic environments smaller M values give better offline error values.

Secondly, we analyze the performance of HH-PBIL2 in dynamic environments showing different change properties. Both for the randomly changing environments and the cyclic environments in all DUFs, SPBIL2 is significantly better than SPBIL, except for HF cyclic changes in DUF1 and DUF2 where the performance difference is not statistically significant. In the cyclic environments,

---

[2] Statistical significance comparison tables can be download from
http://web.itu.edu.tr/etaner/ppsn2012_analysis.zip

RL-P2 is significantly better than RL-PF on average, SPBIL and SPBIL2 for all
M values and change frequencies. In the randomly changing environments, for all
HS and VHS severity settings RL-P2 is significantly better than RL-PF in all the
DUFs. For the LS and MS severity settings, their performance differences are not
statistically significant. However, statistically significant performance differences
appear in favor of RL-P2 for these severity settings in MF and HF settings in
all DUFs. In the randomly changing environments for all change severities (LS,
MS, HS, VHS) at the HF frequency setting, RL-P2 is better than SPBIL2 and
this performance difference is statistically significant. The same result is also
observed for all change frequencies (LF, MF, HF) at the HS and VHS severity
settings. For change frequency-severity combinations of LF and MF with LS and
MS, SPBIL2 is significantly better than RL-P2.

To illustrate the tracking behavior of the approaches, in Figure 1 and
Figure 2, sample plots for the error values of the generation best solution candidates versus the number of generations, for four consecutive environments after
the third change on DUF2 are given. The plots show that for randomly changing
environments, increased change severities result in significant differences between
the algorithms in favor of HH-PBIL2 variants. Increased cycle lengths in cyclic
environments have a similar effect on the algorithms. But as the frequency increases, the differences get smaller. Increased change frequencies have a similar
effect on all approaches in the cyclic environments too.

Table 1. Overall $(s+, s-, \geq$ and $\leq)$ counts for $M = 8, 16, 32, 64$ in RL-PF and RL-P2

	RL-P2-8	RL-P2-16	RL-P2-32	RL-P2-64	RL-PF-8	RL-PF-16	RL-PF-32	RL-PF-64
$s+$	376	356	342	283	202	167	137	104
$s-$	64	73	88	139	218	236	273	320
$\geq$	44	81	80	70	70	90	76	71
$\leq$	83	57	57	75	77	74	81	72

Table 2. Offline errors averaged over 100 runs, on the three DUFs for different change
severity and frequency settings in randomly changing environments

Alg.	LF				MF				HF			
	LS	MS	HS	VHS	LS	MS	HS	VHS	LS	MS	HS	VHS
DUF1												
RL-PF	4.22	8.12	9.02	9.22	9.91	16.74	20.58	22.06	27.91	32.13	36.65	38.73
RL-P2	4.24	8.15	**7.23**	**4.25**	9.95	16.73	**14.39**	**12.55**	**26.95**	**29.67**	**33.17**	**35.12**
SBIL	4.11	7.91	16.72	21.76	9.55	16.08	26.73	30.45	28.00	33.84	38.05	38.73
SPBIL2	**3.46**	**7.21**	16.21	20.72	**9.05**	**15.81**	26.00	29.18	27.75	33.35	37.23	38.12
DUF2												
RL-PF	9.28	19.26	19.14	15.05	21.23	34.53	39.88	42.37	50.03	56.56	63.43	64.61
RL-P2	9.28	19.16	**18.52**	**13.39**	21.40	34.37	**30.48**	**29.00**	49.04	53.19	**58.64**	**60.67**
SPBIL	9.00	18.42	38.86	45.88	20.43	34.48	51.51	54.83	52.30	60.45	65.21	65.72
SPBIL2	**7.63**	**17.21**	37.06	43.15	**19.53**	**33.71**	49.71	52.59	51.79	59.35	64.07	64.57
DUF3												
RL-PF	25.57	25.92	18.95	**17.79**	30.44	32.22	29.41	27.24	39.77	41.90	44.42	42.90
RL-P2	25.55	25.89	**19.02**	17.83	30.41	**32.00**	**25.03**	**25.30**	**38.94**	**39.78**	**41.78**	**40.00**
SPBIL	25.46	25.81	23.98	19.46	30.12	33.17	35.29	31.53	40.18	44.51	47.18	45.94
SPBIL2	**25.00**	**25.26**	23.19	18.52	**29.44**	32.38	34.36	30.71	39.48	43.65	46.35	45.09

Table 3. Offline errors averaged over 100 runs, on the three DUFs for different cycle length and change frequency settings in cyclic environments

Alg.	LF			MF			HF		
	CL=2	CL=4	CL=8	CL=2	CL=4	CL=8	CL=2	CL=4	CL=8
DUF1									
RL-PF	3.50	4.02	3.84	15.17	17.66	13.92	16.71	19.37	27.99
RL-P2	**0.17**	**0.17**	**0.17**	**1.80**	**2.13**	**1.93**	**8.95**	**14.76**	**19.33**
SPBIL	10.19	16.51	15.84	13.20	22.43	24.13	15.79	26.23	28.42
SPBIL2	9.08	15.73	15.29	10.73	21.05	23.08	16.24	26.20	28.19
DUF2									
RL-PF	2.18	2.06	2.58	14.38	22.72	19.81	27.29	36.27	47.93
RL-P2	**0.27**	**0.29**	**0.27**	**2.85**	**3.40**	**3.40**	**15.74**	**27.59**	**32.95**
SPBIL	20.67	36.15	36.73	24.29	43.07	46.89	27.69	45.83	51.23
SPBIL2	17.79	33.71	34.63	20.91	40.40	44.58	28.60	45.82	50.77
DUF3									
RL-PF	10.94	11.93	11.91	18.65	26.95	20.39	24.16	34.31	36.35
RL-P2	**10.53**	**11.58**	**11.57**	**12.99**	**14.35**	**14.21**	**17.51**	**29.35**	**27.79**
SPBIL	25.72	24.25	23.88	31.52	34.77	34.86	28.60	37.24	42.66
SPBIL2	25.00	23.48	23.07	30.35	33.49	33.95	28.44	36.38	41.49

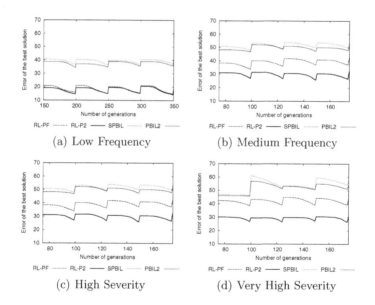

(a) Low Frequency   (b) Medium Frequency

(c) High Severity   (d) Very High Severity

Fig. 1. Sample plots for the error values of the generation best solution candidates versus the number of generations for randomly changing environments based on fixed severity - HS- (first row) and based on fixed frequency -MF- (second row) settings.

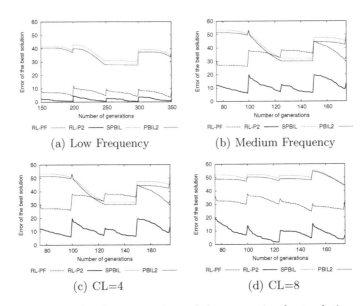

(a) Low Frequency   (b) Medium Frequency

(c) CL=4   (d) CL=8

Fig. 2. Sample plots for the error values of the generation best solution candidates versus the number of generations for cyclic changing environments based on fixed CL=4 (first row) and based on fixed frequency -MF- (second row) settings.

5   Conclusion

In this study, we investigated the performance of a bi-population framework that hybridizes a variant of population based incremental learning with a selection hyper-heuristic. The framework can combine any EDA based technique with any heuristic selection mechanism. In this study, a standard PBIL is hybridized with a reinforcement learning selection hyper-heuristic. To explore the generality of the proposed approach we performed experiments across environments exhibiting a range of different change dynamics on some well known benchmark functions for two generic approaches and our proposed approach. Previous studies indicate that stand-alone generic approaches are not sufficient to deal with different change dynamics. The results of the experiments in this study confirm this and show that the proposed approach exhibits good performance in all the tested change scenarios. This makes the proposed approach a solver which is generalized enough to provide a good average performance over different types of dynamic environments. As future work, we will experiment with hybridizing other types of EDA based methods and heuristic selection mechanisms, as well as incorporate our approach into memory based techniques. The results are promising which promote further study.

Acknowledgements. B. Kiraz is supported by TÜBİTAK 2211-National Scholarship Program for PhD students. The study is supported in part by EPSRC, grant EP/F033214/1 (The LANCS Initiative Postdoctoral Training Scheme).

References

1. Barlow, G.J., Smith, S.F.: Using memory models to improve adaptive efficiency in dynamic problems. In: IEEE Symposium on Computational Intelligence in Scheduling, CISCHED, pp. 7–14 (2009)
2. Branke, J.: Evolutionary optimization in dynamic environments. Kluwer (2002)
3. Burke, E.K., Gendreau, M., Hyde, M.R., Kendall, G., Ochoa, G., Özcan, E., Qu, R.: Hyper-heuristics: A survey of the state of the art. To appear in the Journal of the Operational Research Society (2012)
4. Cowling, P.I., Kendall, G., Soubeiga, E.: A Hyperheuristic Approach to Scheduling a Sales Summit. In: Burke, E., Erben, W. (eds.) PATAT 2000. LNCS, vol. 2079, pp. 176–190. Springer, Heidelberg (2001)
5. Cruz, C., Gonzalez, J., Pelta, D.: Optimization in dynamic environments: a survey on problems, methods and measures. Soft Computing - A Fusion of Foundations, Methodologies and Applications 15, 1427–1448 (2011)
6. Fernandes, C.M., Lima, C., Rosa, A.C.: Umdas for dynamic optimization problems. In: Proc. of the 10th Conference on Genetic and Evolutionary Computation, GECCO 2008, pp. 399–406. ACM (2008)
7. Kiraz, B., Uyar, A.Ş., Özcan, E.: An Investigation of Selection Hyper-heuristics in Dynamic Environments. In: Di Chio, C., Cagnoni, S., Cotta, C., Ebner, M., Ekárt, A., Esparcia-Alcázar, A.I., Merelo, J.J., Neri, F., Preuss, M., Richter, H., Togelius, J., Yannakakis, G.N. (eds.) EvoApplications 2011, Part I. LNCS, vol. 6624, pp. 314–323. Springer, Heidelberg (2011)
8. Kiraz, B., Topcuoglu, H.R.: Hyper-heuristic approaches for the dynamic generalized assignment problem. In: 2010 10th International Conference on Intelligent Systems Design and Applications (ISDA), pp. 1487–1492 (2010)
9. Nareyek, A.: Metaheuristics, pp. 523–544. Kluwer (2004)
10. Özcan, E., Bilgin, B., Korkmaz, E.E.: A comprehensive analysis of hyper-heuristics. Intelligent Data Analysis 12, 3–23 (2008)
11. Özcan, E., Uyar, Ş., Burke, E.: A greedy hyper-heuristic in dynamic environments. In: GECCO 2009 Workshop on Automated Heuristic Design: Crossing the Chasm for Search Methods, pp. 2201–2204 (2009)
12. Özcan, E., Misir, M., Ochoa, G., Burke, E.K.: A reinforcement learning - greatdeluge hyper-heuristic for examination timetabling. International Journal of Applied Metaheuristic Computing 1(1), 39–59 (2010)
13. Peng, X., Gao, X., Yang, S.: Environment identification-based memory scheme for estimation of distribution algorithms in dynamic environments. Soft Comput. 15, 311–326 (2011)
14. Wu, Y., Wang, Y., Liu, X., Ye, J.: Multi-population and diffusion umda for dynamic multimodal problems. Journal of Systems Engineering and Electronics 21(5), 777–783 (2010)
15. Yang, S.: Constructing dynamic test environments for genetic algorithms based on problem difficulty. In: Proc. of the 2004 Congress on Evolutionary Computation, pp. 1262–1269 (2004)
16. Yang, S., Richter, H.: Hyper-learning for population-based incremental learning in dynamic environments. In: Proc. 2009 Congr. Evol. Comput., pp. 682–689 (2009)
17. Yang, S., Yao, X.: Experimental study on population-based incremental learning algorithms for dynamic optimization problems. Soft Comput. 9(11), 815–834 (2005)
18. Yang, S., Yao, X.: Population-based incremental learning with associative memory for dynamic environments. IEEE Trans. on Evolutionary Comp. 12, 542–561 (2008)

Parallelization Strategies for Hybrid Metaheuristics Using a Single GPU and Multi-core Resources

Thé Van Luong[1], Eric Taillard[1], Nouredine Melab[2], and El-Ghazali Talbi[2]

[1] HEIG-VD, Yverdon-les-Bains, Switzerland
{The-Van.Luong,Eric.Taillard}@heig-vd.ch
[2] INRIA Lille Nord Europe / LIFL, Villeneuve d'Ascq, France
{Nouredine.Melab,talbi}@lifl.fr

Abstract. Hybrid metaheuristics are powerful methods for solving complex problems in science and industry. Nevertheless, the resolution time remains prohibitive when dealing with large problem instances. As a result, the use of GPU computing has been recognized as a major way to speed up the search process. However, most GPU-accelerated algorithms of the literature do not take benefits of all the available CPU cores. In this paper, we introduce a new guideline for the design and implementation of effective hybrid metaheuristics using heterogeneous resources.

1 Introduction

Metaheuristics are approximate methods that make it possible to solve in a reasonable time NP-hard complex problems. Two main categories are distinguished: population-based metaheuristics (P-metaheuristics) and solution-based metaheuristics (S-metaheuristics). Theoretical and experimental studies have shown that the hybridization between these two classes may improve the quality of provided solutions [1]. However, as it is time-consuming, there is often a compromise between the number of solutions to use and the computational complexity to explore it.

Recently, graphics processing units (GPU) have emerged as a popular support for massively parallel computing [2]. To the best of our knowledge, most GPU-accelerated metaheuristics designed in the literature only exploit a single CPU core. This is typically the case for hybrid metaheuristics on GPU [3–5]. Thus, it might be valuable to fully utilize the other remaining CPU resources. It may be particularly significant when the acceleration factors obtained by the GPU-based algorithm are relatively modest. Indeed, since all processors are nowadays multi-core, performance of GPU-based algorithms might be improved.

Nevertheless, designing optimization methods on such a heterogeneous architecture is not straightforward. Indeed, the major issues are mainly related to the distribution of tasks processing between the GPU and CPU cores. In this paper, we introduce a general guideline to deal with such issues. We propose the re-design of hybrid metaheuristics on GPU taking advantage of every available CPU cores. In this purpose, an efficient distribution of the search process

C.A. Coello Coello et al. (Eds.): PPSN 2012, Part II, LNCS 7492, pp. 368–377, 2012.
© Springer-Verlag Berlin Heidelberg 2012

between the GPU and the CPU is done. At the same time, an efficient load balancing between the GPU and the remaining CPU cores is proposed to fully utilize all the available heterogeneous resources.

As an example of application, the quadratic assignment problem (QAP) has been considered. Such a problem provides interesting irregular properties, since for optimized S-metaheuristics, most of move evaluations can be done in constant time. Thereby, speed-ups from a parallel implementation are expected to be relatively modest. Hence, the use of multi-core resources in addition with GPU-based metaheuristics is clearly meaningful.

The remainder of the paper is organized as follows: Section 2 highlights the principles of parallel models for metaheuristics on GPU. In Section 3, parallelization concepts for designing hybrid metaheuristics on GPU are described. An extension of these approaches is investigated in Section 4 for exploiting heterogeneous resources. Section 5 reports the performance results obtained for the QAP. Finally, some conclusions of this work are drawn in Section 6.

2   Parallel Metaheuristics on GPU

2.1   Parallel Models of Metaheuristics

In general, for hybrid metaheuristics, executing the iterative process of a S-metaheuristic (e.g. a local search) requires a large amount of computational resources. Consequently, parallelism arises naturally when dealing with a neighborhood. In this purpose, three major parallel models for metaheuristics can be distinguished [6]: solution-level, iteration-level and algorithmic-level.

- *Solution-level Parallel Model.* The focus is on the parallel evaluation of a single solution. Problem-dependent operations performed on solutions are parallelized. That model is particularly interesting when the evaluation function can be itself parallelized, as it is CPU time-consuming and/or IO intensive.
- *Iteration-level Parallel Model.* This model is a low-level Master-Worker model that does not alter the behavior of the heuristic. The evaluation of solutions is performed in parallel. An efficient execution is often obtained especially when the evaluation of each solution is costly.
- *Algorithmic-level Parallel Model.* Several metaheuristics are simultaneously launched for computing better and robust solutions. They may be heterogeneous or homogeneous, independent or cooperative, start from the same or different solution(s), configured with the same or different parameters.

2.2   Metaheuristics on GPU Architectures

Recently, GPU accelerators have emerged as a powerful support for massively parallel computing. Indeed, these architectures offer a substantial computational horsepower and a high memory bandwidth compared to CPU-based architectures. Due to their inherent parallel nature, P-metaheuristics such as evolutionary algorithms have been the first subject of parallelization on GPU: genetic algorithms [7], particle swarm optimization [8], ant colonies [9] and so on.

Regarding S-metaheuristics, the parallelization on GPU architectures is much harder, due to the improvement of a single solution. Therefore, only few research works have been investigated for local search algorithms [10–12]. The same goes on when dealing with hybrid metaheuristics on GPU, where there exists only few parallelization approaches [3–5].

3   Design of Parallel Hybrid Metaheuristics on GPU

3.1   Parallel Evaluation of Solutions on GPU

The parallel iteration-level model has to be designed according to the data-parallel single program multiple data model of GPUs. The CPU-GPU task partitioning is such that the CPU executes the entire sequential part of the handled metaheuristic. The GPU is in charge of the evaluation of the solutions set at each iteration. In this model, a function code called kernel is sent to the GPU to be executed by a large number of threads grouped into blocks.

This parallelization strategy has been widely used for P-metaheuristics on GPU especially for evolutionary algorithms due to their intrinsic parallel workload (e.g. in [7]). One of the major issues is to optimize the data transfer between the CPU and the GPU. Indeed, the GPU has its own memory and processing elements that are separate from the host computer.

When it comes to parallelization, the optimization of data transfers is more prominent for S-metaheuristics. As a result, when designing hybrid metaheuristics, the focus is on the embedded S-metaheuristic. In this purpose, we have contributed in [13] for the parallel evaluation of solutions (iteration-level) for local search algorithms. The key point of this approach is to generate the neighborhood of the S-metaheuristic at hand on the GPU side. Such a parallelization strategy makes it possible to minimize data transfers through the PCIe bus: the solution which generates the neighborhood and the resulting fitnesses (see Figure 1).

3.2   Parallelization Strategies for Hybrid Metaheuristics

The previous parallelization approach stands for one S-metaheuristic on GPU according to the iteration-level. For designing GPU-accelerated hybrid metaheuristics that involve a population of solutions, the algorithmic-level parallel model has to be deeply examined. In other words, multiple executions of S-metaheuristics on GPU have to be considered. For achieving this, previous approaches from the iteration-level must be adapted for the algorithmic-level. In this purpose, there are fundamentally two parallelization strategies:

- *One neighborhood evaluation on GPU.* This approach consists in evaluating one neighborhood (a set of solutions) at a time on GPU. According to Figure 1, a possible interpretation could be to repeat the whole process (i.e. the repetition of the execution of a single S-metaheuristic on GPU) to deal with as many S-metaheuristics as needed. The drawback of this approach is that the number of threads executed for one kernel on GPU might not be

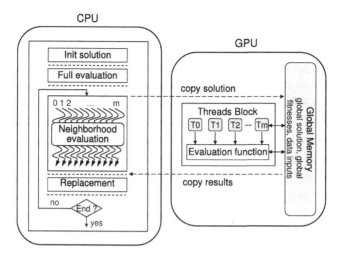

Fig. 1. For S-metaheuristics, the generation and the evaluation of the neighborhood is performed on GPU, and the CPU executes the sequential part of the search process

enough to cover the memory access latency for few optimization problems. As a result, in the rest of the paper, we will not consider this approach.

- *Many neighborhood evaluations on GPU.* In the second approach, many neighborhoods are evaluated at a time on GPU. For instance, given a certain iteration, if k embedded S-metaheuristics have to be performed on k solutions, the k associated neighborhoods will be generated and evaluated on GPU at the same time. Regarding the thread organization, a thread is associated with many neighbor calculations. For example, a thread block might represent a particular neighborhood from a given S-metaheuristic. Such a parallelization strategy deals with the issues encountered in the first approach since 1) there are enough calculations to keep the GPU multiprocessors busy; 2) the creation overhead of multiple kernel calls is reduced. However, in this second approach, homogeneous embedded S-metaheuristics are required. In such a case, the semantic of the original sequential algorithm might be altered.

4 Parallelization Strategies for Heterogenous Resources

4.1 Multiple S-Metaheuristics on Multi-core Architectures

Parallelization approaches for hybrid metaheuristics on GPU presented in the previous section only exploit one single CPU core. To exploit the remaining computational capabilities, thread-based approaches on CPU have to be examined.

In general, for a hybrid metaheuristic, a certain number of independent tasks is likely to be performed in parallel (e.g. a number of S-metaheuristic executions). Therefore, the algorithmic-level parallel model is particularly adapted to CPU architectures, since processes distributed among CPU threads do not necessary share the same instructions and the same execution context.

Algorithm 1. Template for each thread on multi-core CPUs

Require: p tasks and number_threads;
1: offset := p / number_threads;
2: tid := get_thread_id();
3: **for** k = tid * offset; k < tid * offset + offset; k++ **do**
4:     S-metaheuristic(k);
5: **end for**
6: **if** tid < (p mod number_threads) **then**
7:     k = offset * number_threads + tid;
8:     S-metaheuristic(k);
9: **end if**

Algorithm 1 provides a template for processing independent S-metaheuristics on multi-core architectures. Basically, p tasks (i.e. p S-metaheuristics) have to be equally distributed among the different threads. Each CPU thread is in charge of executing a specific number of S-metaheuristics (lines 1 and 2). Such a realization is performed in a sequential manner (lines 3 to 5). If the number of tasks is not proportional to the number of available cores, remaining tasks will be assigned to the first CPU threads (lines 6 to 9).

4.2   Hybrid Metaheuristics Using Heterogeneous Resources

As previously said, one CPU thread is actually associated with the GPU-based algorithm. The major idea for designing a hybrid metaheuristic is to manage the other CPU threads to overlap the calculations performed on GPU. Nevertheless, most of the time, in hybrid metaheuristics, the search process evolves in a synchronous manner at each iteration.

Algorithm 2. Hybrid metaheuristic template using heterogeneous resources

---

Require: m tasks, p tasks and n cores;
1: **repeat**
2:     Hybrid metaheuristic pre-treatment on host side
3:     S-metaheuristic_multi-core(p,n-1) overlap
4:     S-metaheuristic_gpu(m)
5:     Join results
6:     Hybrid metaheuristic post-treatment on host side
7: **until** a stopping criterion satisfied

For dealing with this issue, we provide in Algorithm 2 a general template for hybrid metaheuristics using heterogeneous resources. Let k be the number of tasks to assess, m the number assigned to the GPU using one CPU core, and p the number assigned to the remaining CPU cores. As quoted above, p S-metaheuristics are executed in parallel on CPU cores (number of available cores minus one) according to Algorithm 1 (line 3). Parallel techniques must be performed to obtain overlapping calculations. Meanwhile, m S-metaheuristics are evaluated on GPU as described in Section 3 (line 4). Then, a synchronization point is set to gather all the obtained results (line 5). Post-treatment operations on the hybrid metaheuristic can be applied afterwards. The process is repeated until a certain criterion is satisfied.

4.3 Load Balancing Heuristic

The remaining issue is to find an efficient load balancing between 1) the GPU using one single core; 2) the remaining CPU cores. Such a task repartition must be done in accordance with the computational capability of heterogeneous resources. We propose in Algorithm 3 a heuristic for doing this load balancing in an efficient way. The major idea of this heuristic is to automatically tune previous m and p parameters during the first iterations of the hybrid metaheuristic at hand.

Algorithm 3. Template for load balancing heuristic

Require: k tasks and n cores;
 1: m := ceil (k / 2); p := floor (k / 2);
 2: **repeat**
 3:     Hybrid metaheuristic pre-treatment on host side
 4:     gpu_time := time (S-metaheuristic_gpu(m));
 5:     cpu_time := time (S-metaheuristic_multi-core(p,n-1));
 6:     Hybrid metaheuristic post-treatment on host side
 7:     relat_speedup := cpu_time / gpu_ time;
 8:     **if** relat_speedup > 1 **then**
 9:         potential_p := p / relat_speedup; mult_coeff := k / (m + potential_p);
10:         m := round (m * mult_coeff);
11:         p := round (potential_p * mult_coeff);
12:     **else**
13:         potential_m := m * rel_speedup; mult_coeff := k / (p + potential_m);
14:         m := round (potential_m * mult_coeff);
15:         p := round (p * mult_coeff);
16:     **end if**
17: **until** a certain number of trials
Ensure: m tasks and p tasks;

At the beginning of the algorithm, tasks are equally divided between the GPU and the CPU cores (line 1). Then, the time measurement of m S-metaheuristic executions on GPU using one CPU core is accomplished (line 4). The same goes on for the creation of p S-metaheuristics on the other available CPU cores (line 5). Thereafter, the relative speed-up between the two versions is calculated (line 7). If the time to evaluate m tasks on GPU is less important than the time to compute p tasks on remaining CPU cores, then more tasks will be assigned to the GPU during the next iteration (lines 8 to 11). Otherwise, more tasks will be assigned to the remaining CPU cores (lines 12 to 16). In other words, m and p values are proportionally adjusted with the relative acceleration factor. The process is repeated until a certain number of trials.

5   Performance Evaluation

5.1   Fast Ant System

To validate the approaches presented in this paper, the fast ant system (FANT) metaheuristic [14] has been considered. Basically, the major idea of FANT is to construct each solution (active ant) in a probabilistic way from the values of the decision variables in past searches by using a memory structure. To accelerate the convergence process, a local search algorithm is performed each time a solution is built. The process is repeated until a certain number of iterations is reached. The reinforcement parameter R has an impact during the intensification phase of the FANT metaheuristic.

The embedded local search is based on the selection of the best neighbor at each iteration. Such a selection mechanism accepting non-improving neighbors, will lead to cycles during the search process. Thereby, the number of local iterations has been restricted to $\frac{n}{2}$ (n is the instance size).

5.2   Application to the Quadratic Assignment Problem

The well-known QAP arises in many applications such as facility location or data analysis. The evaluation function has a $O(n^2)$ time complexity. In the next implementations, a neighborhood based on a pair-wise exchange ($\frac{n \times (n-1)}{2}$ neighbors) has been considered. For each iteration of a local search, $\frac{(n-2) \times (n-3)}{2}$ neighbors can be evaluated in $O(1)$ and $2n - 3$ can be evaluated in $O(n)$.

From an implementation point of view, since calculations may be irregular according to the given neighbor, threads are reorganized in such a way that threads belonging to a same group of 32 threads (a.k.a. a warp) execute the same computation. In other words, groups of threads which perform $O(1)$ and $O(n)$ calculations are clearly separated. Such a mechanism allows reducing threads divergence due to conditional branches. Furthermore, to minimize the idle time due to irregular computations, $2n$ threads are associated with $O(n)$ calculations and $\frac{(n-1)}{2}$ threads execute $n \times O(1)$ calculations per local search. In this way, each thread block corresponds to one neighborhood evaluation.

5.3   Configuration

Experiments have been carried out on top of two different configurations. The first one is an Intel Core i7 930 with 4 cores cadenced at 2.8 Ghz using a NVIDIA GTX 480 graphic card (480 GPU cores). The second configuration is a bi-processor Intel Xeon E5520 with 2×4 cores cadenced at 2.26 Ghz using a Tesla C1060 (240 GPU cores). Since the first card provides on-chip memory for L1 cache memory, techniques to cache input data using the texture memory have only been applied to the second configuration. Posix threads have been considered for multi-core versions.

The average time has been measured in seconds for 30 runs, and acceleration factors are reported in comparison with a single CPU core. The standard deviation is not represented since its value is close to zero.

Table 1. Measures in terms of efficiency for the QAP using a pair-wise-exchange neighborhood. 4 FANT implementations on different architectures are considered.

Instance	Core i7 930 2.8Ghz GeForce GTX 480 4 CPU cores 480 GPU cores			Xeon E5520 2.26Ghz Tesla C1060 8 CPU cores 240 GPU cores		
	Multi-core	GPU	Heterogeneous	Multi-core	GPU	Heterogeneous
tai50a	$10.5_{\times 2.4}$	$2.1_{\times 11.8}$	$2.1_{\times 11.9 | \times 14.2}$	$7.5_{\times 3.0}$	$3.2_{\times 7.0}$	$2.3_{\times 9.8 | \times 10.0}$
tai60a	$17.8_{\times 2.4}$	$3.4_{\times 12.7}$	$3.3_{\times 13.1 | \times 15.1}$	$12.7_{\times 3.1}$	$5.4_{\times 7.3}$	$3.8_{\times 10.4 | \times 10.4}$
tai80a	$41.2_{\times 2.6}$	$7.5_{\times 14.4}$	$7.0_{\times 15.4 | \times 17.0}$	$29.4_{\times 3.3}$	$12.0_{\times 8.1}$	$8.5_{\times 11.4 | \times 11.4}$
tai100a	$81.7_{\times 2.7}$	$15.6_{\times 14.0}$	$13.9_{\times 15.8 | \times 16.7}$	$52.3_{\times 3.8}$	$22.3_{\times 9.0}$	$16.2_{\times 12.3 | \times 12.8}$
tai150b	$288.1_{\times 2.7}$	$73.7_{\times 10.6}$	$58.9_{\times 13.2 | \times 13.3}$	$138.7_{\times 5.5}$	$74.3_{\times 10.3}$	$50.3_{\times 15.2 | \times 15.8}$
tai256c	$1620.5_{\times 2.7}$	$382.7_{\times 11.3}$	$373.2_{\times 11.6 | \times 14.0}$	$610.1_{\times 6.9}$	$615.9_{\times 6.9}$	$351.9_{\times 12.1 | \times 13.8}$

5.4 Experimentation

The set of experiments consists in measuring the performance of proposed parallelization schemes. For doing this, four FANT versions have been implemented for the QAP. A CPU implementation using one single core, a multi-core version, a GPU implementation and another one using all the available heterogeneous resources. For all versions, 50 neighborhood evaluations (i.e. 50 active ants per global iteration) at a time have been considered. Regarding the semantic of the algorithms, there is no difference of the quality of solutions provided by both versions. The multi-core version does not intentionally utilize one CPU core in order to highlight the performance improvements of the heterogeneous version (since one core is associated with the GPU). The number of global iterations has been fixed to 10000, which corresponds to a realistic scenario in accordance with the algorithm convergence. Experimental results are reported in Table 1. The CPU column is not represented since the associated values can be deduced from the other columns.

Regarding the multi-core version (number of CPU cores minus one), the obtained acceleration factors grow with the instance size. For the first configuration using three cores, these speed-ups linearly vary from $\times 2.4$ to $\times 2.7$. This is not exactly the same phenomenon for the second configuration where acceleration factors alternate from $\times 3.0$ to $\times 6.9$. Indeed, for smaller instances, the overhead creation is significant in regards with the computational time. This is mainly due to the important number of threads to be created and synchronized (seven threads). But, as long as the size increases, the acceleration factor converges to the expected value.

For the GPU implementation, the obtained speed-ups are quite significant but relatively modest. They alternate from $\times 10.6$ to $\times 14.4$ for the first configuration, and from $\times 7$ to $\times 9.3$ for the second configuration. Such performance results are limited since most of move evaluations can be performed in $O(1)$. Therefore, the amount of computations is not enough to fully cover the memory access latency. Furthermore, the application is memory bound since non-coalescing accesses to the global memory drastically reduces the performance of the GPU implementation. This is due to high-misaligned accesses present in flows and distances matrices in QAP.

Regarding the heterogeneous version taking advantage of all CPU cores, the performance improvements in comparison with the GPU implementation are significant. Indeed, for the first configuration corresponding to three additional CPU cores, acceleration factors vary from ×11.6 to ×15.8, which corresponds to an improvement between 1% and 25%. For the second one with seven additional cores, better performance improvements between 39% and 75% (speed-ups varying from ×9.8 to ×15.2) can be observed.

To assess the efficiency of the heterogeneous version, potential acceleration factors are represented in italic in sub indices. These theoretical values are obtained by adding the speed-ups obtained for both multi-core and GPU versions. The performance difference, which occurs between the obtained results and the potential speed-up, is due to synchronization points between the GPU and the other CPU cores. One can clearly see that the acceleration factors obtained for the heterogeneous version are not so far from the expected theoretical ones. This is particularly the case for the second configuration containing more CPU cores. As a consequence, the heuristic for finding a parameters auto-tuning provides an efficient way to deal with load balancing for heterogeneous resources.

6   Conclusion

Hybrid metaheuristics having complementary behaviors allow improving the effectiveness and robustness in optimization. Their exploitation for solving real-world problems is possible only by using a great computational power. High-performance computing based on heterogeneous resources is recently revealed as an efficient way to use the huge amount of resources at disposal. However, the exploitation of parallel models is not trivial and many issues related to the task repartition between the GPU and multi-core architectures have to be faced.

In this paper, we have investigated on different parallelization strategies for hybrid metaheuristics on such heterogeneous resources. In the proposed parallelization approaches, the CPU manages the metaheuristic process and let the GPU be used as a coprocessor dedicated to intensive calculations. Thereafter, parts of these computations are distributed among the available CPU cores. Such a task repartition is provided by an efficient heuristic for parameters tuning.

The designed and implemented approaches have been experimentally validated on the QAP using the FANT metaheuristic. The evaluation of a neighboring solution in the QAP can be performed most of the time in constant time. As a result, for problems with modest GPU accelerations, the performance improvement provided by multi-core CPUs is particularly significant (up to 75% for eight CPU cores). In particular, we showed that our methodology enables gaining to a ×15 factor in terms of acceleration compared with a single core architecture. A perspective of this work will be to implement the proposed approaches for other combinatorial optimization problems, in which the computational complexity of move evaluations is more prominent.

With the arrival of GPU resources in clusters of workstations and grids, the next objective is to examine the conjunction of GPU computing and distributed computing to fully and efficiently exploit the hierarchy of parallel models of metaheuristics. Indeed, since all processors are currently multi-core, performance of GPU-based algorithms might be drastically improved. The challenge will be to find the best mapping in terms of efficiency of the hierarchy of parallel models on the hierarchy of CPU-GPU resources provided by multi-level architectures.

Acknowledgments. Part of this research work was supported by HES-SO RCSO-TIC research grant 25018 CPUGPU.

References

1. Talbi, E.G.: A taxonomy of hybrid metaheuristics. J. Heuristics 8(5), 541–564 (2002)
2. Ryoo, S., Rodrigues, C.I., Stone, S.S., Stratton, J.A., Ueng, S.Z., Baghsorkhi, S.S., Mei, W., Hwu, W.: Program optimization carving for gpu computing. J. Parallel Distributed Computing 68(10), 1389–1401 (2008)
3. Munawar, A., Wahib, M., Munetomo, M., Akama, K.: Hybrid of genetic algorithm and local search to solve max-sat problem using nvidia cuda framework. Genetic Programming and Evolvable Machines 10, 391–415 (2009)
4. Tsutsui, S., Fujimoto, N.: Aco with tabu search on a gpu for solving qaps using move-cost adjusted thread assignment. In: Krasnogor, N., Lanzi, P.L. (eds.) GECCO, pp. 1547–1554. ACM (2011)
5. Luong, T.V., Melab, N., Talbi, E.G.: Parallel hybrid evolutionary algorithms on gpu. In: IEEE Congress on Evolutionary Computation, pp. 1–8 (2010)
6. Talbi, E.G.: Metaheuristics: From design to implementation. Wiley (2009)
7. Wong, M.L., Wong, T.T., Fok, K.L.: Parallel evolutionary algorithms on graphics processing unit. In: IEEE Congress on Evolutionary Computation, pp. 2286–2293 (2005)
8. Mussi, L., Cagnoni, S., Daolio, F.: Gpu-based road sign detection using particle swarm optimization. In: ISDA, pp. 152–157. IEEE Computer Society (2009)
9. Bai, H., OuYang, D., Li, X., He, L., Yu, H.: Max-min ant system on gpu with cuda. In: Proceedings of the 2009 Fourth International Conference on Innovative Computing, Information and Control, ICICIC 2009, pp. 801–804. IEEE Computer Society, Washington, DC (2009)
10. Janiak, A., Janiak, W.A., Lichtenstein, M.: Tabu search on gpu. J. UCS 14(14), 2416–2426 (2008)
11. Zhu, W., Curry, J., Marquez, A.: Simd tabu search with graphics hardware acceleration on the quadratic assignment problem. International Journal of Production Research (2008)
12. Czapinski, M., Barnes, S.: Tabu search with two approaches to parallel flowshop evaluation on cuda platform. J. Parallel Distrib. Comput. 71(6), 802–811 (2011)
13. Luong, T.V., Melab, N., Talbi, E.G.: Gpu computing for parallel local search metaheuristic algorithms. IEEE Transactions on Computers 99(preprints) (2011)
14. Taillard, E.D.: Fant: Fast ant system. Technical report (1998)

Adaptive Operator Selection at the Hyper-level

Eduardo Krempser[1], Álvaro Fialho[2], and Helio J.C. Barbosa[1]

[1] Laboratório Nacional de Computação Científica, Petrópolis, Brazil
{krempser,hcbm}@lncc.br
[2] Nokia Institute of Technology (INdT), Manaus, Brazil
alvaro.fialho@indt.org.br

Abstract. Whenever a new problem needs to be tackled, one needs to decide which of the many existing metaheuristics would be the most adequate one; but it is very difficult to know their performance a priori. And then, when a metaheuristic is chosen, there are still its parameters that need to be set by the user. This parameter setting is usually very problem-dependent, significantly affecting their performance. In this work we propose the use of an Adaptive Operator Selection (AOS) mechanism to automatically control, while solving the problem, (i) which metaheuristic to use for the generation of a new solution, (exemplified here by a Genetic Algorithm (GA) and a Differential Evolution (DE) scheme); and (ii) which corresponding operator should be used, (selecting among five operators available for the GA and four operators for DE). Two AOS schemes are considered: the Adaptive Pursuit and the Fitness Area Under Curve Multi-Armed Bandit. The resulting algorithm, named as Adaptive Hyper-Heuristic (HH), is evaluated on the BBOB noiseless testbed, showing superior performance when compared to (a) the same HH without adaptation, and also (b) the adaptive DE and GA.

Keywords: Hyper-heuristics, adaptive operator selection, parameter control, multi-armed bandits, area under the curve.

1 Introduction

Metaheuristics have been used to solve a wide range of complex optimization problems. Many different algorithmic schemes can be found in the literature, each of them presenting its own specifications, resulting into different behaviors with respect to the exploration of the search space. The resulting characteristics might be more adequate or not to a given problem or class of problems. It is very difficult, however, to know *a priori* which would be the most adequate metaheuristic whenever a new problem needs to be tackled. Additionally, meta-heuristics usually have many user-defined parameters that might significantly affect their behavior and performance. In the case of evolutionary algorithms, for example, there is the population size, the selection and replacement mechanisms and their inner parameters, the choice of variation operators and their corresponding application rates, etc. As a result, once a given metaheuristic is chosen, there is still the need of correctly setting its parameters, what can be

C.A. Coello Coello et al. (Eds.): PPSN 2012, Part II, LNCS 7492, pp. 378–387, 2012.
© Springer-Verlag Berlin Heidelberg 2012

seen as a complex optimization problem *per se*. There are thus two levels of decision: (i) which algorithm should be used, and (ii) which values should be used for setting its parameters. As of today, these decisions are usually done by following the user's intuition, or by using an off-line tuning procedure aimed at identifying the best strategy for the problem at hand. Besides being computationally expensive, off-line tuning however generally delivers sub-optimal performances, as the appropriate strategy depends on the stage of the optimization process: intuitively, exploration-like strategies should be more frequently used in the early stages while priority should be given to exploitation when approaching the optimum. Regarding the first decision, a solution might come from the so-called hyper-heuristics. In [10], a hyper-heuristic is described as a heuristic scheduler that does the scheduling over a set of heuristics in a deterministic or non-deterministic way. A more comprehensive survey can be found in [2]. Concerning the second decision, more specifically in the case of setting the application rates of variation operators, a recent trend is to use methods that control, while solving the problem, which variation operator should be applied according to the recent performance of all available operators. These methods are commonly referred to as Adaptive Operator Selection (AOS) [6].

In fact, the problem of selecting which variation operator to apply can be seen as exactly the same problem of selecting which metaheuristic to use, but at a different abstraction level. Thus, in this paper we propose the use of AOS schemes at the two levels of abstraction (the hyper- and the lower level), in an independent way, while solving the problem.

We empirically analyze the use of two prominent schemes found in the literature, the probability-based Adaptive Pursuit (AP) method [13], and the most recent bandit-based method, referred to as the Fitness-based Area-Under-Curve-Bandit (AUC) [5]. Both are compared with each other at both levels, and also with what would be the choice of a Naive user, namely, the uniform selection of the operators. On the hyper-level, the AOS schemes are expected to autonomously select, while solving the problem, which metaheuristics (in our numerical experiments, Differential Evolution (DE) or Genetic Algorithm (GA)) should be applied. At the lower level, there are five operators for the GA case, and four in the case DE is chosen. DE and GA were chosen because they have been widely used in many fields and their efficiency has already been verified several times. Other heuristics as well as more than two could have been used.

A brief overview of the GA and the DE adopted in this work, as well as an introduction to AOS and to the existing schemes employed, is presented in section 2. In section 3 the proposed schemes are depicted. The computer experiments are presented in section 4, and the paper ends with some conclusions in Section 5.

2   Background

In this work, AOS schemes are used at the hyper-level in order to select between the DE and GA metaheuristics. Both of them will now be briefly described.

2.1 Genetic Algorithms

Here a real-coded GA was applied with a rank-based selection scheme. Moreover, a large number of genetic operators have already been developed and those we adopted are listed below (considering x_i, $i = 1, \ldots, N$ the variables in a chromosome, x_i^L and x_i^U respectively the lower and upper bounds for x_i):

- The one-point (1X) crossover operator which is the analogue of the standard one-point crossover for binary-coded GAs.
- The Uniform crossover (UX) [12], where each gene in the offspring is created by copying the corresponding gene from either parent according to a randomly generated crossover mask.
- The blend crossover operator (BLX-α) [3].
- A simple mutation operator (delta mutation – DM) that increments each variable with a given rate of application according to:

$$x_i = p_i + \delta\Delta_{max}$$

where p is the parent, x is the offspring, δ is a random number, and Δ_{max} is a fixed quantity, which represents the maximum permissible change in the parent.

- The non-uniform mutation (NUM) operator [9]. When applied to an individual x at generation gen and when the total number of generations allowed is $maxgen$, mutates a randomly chosen variable x_i according to

$$x_i \leftarrow \begin{cases} x_i + \Delta(gen, x_i^U - x_i) & \text{if} \quad \tau = 0 \\ x_i - \Delta(gen, x_i - x_i^L) & \text{if} \quad \tau = 1 \end{cases}$$

where τ is randomly chosen as 0 or 1 and the function $\Delta(gen, y)$ is defined as

$$\Delta(gen, y) = y(1 - \mu^{(1 - \frac{gen}{maxgen})^\eta})$$

with μ randomly chosen in $[0, 1]$ and the parameter η set to 2.

2.2 Differential Evolution

The original proposal of DE by Storm and Price [11] presents a simple and efficient algorithm for global optimization over continuous spaces. The main variants (or strategies) of the DE modify the way the individuals are selected to participate in the mutation, which in the original proposal was done randomly (called DE/rand/1/bin). The Algorithm 1 shows the pseudo-code for this variant.

The additional variants considered here basically change line 11 of Algorithm 1:

- **DE/rand/2/bin:**
  $$u_{j,i} = x_{j,r_1} + F(x_{j,r_2} - x_{j,r_3}) + F(x_{j,r_4} - x_{j,r_5})$$
- **DE/rand-to-best/2/bin:**
  $$u_{j,i} = x_{j,r_1} + F(x_{j,best} - x_{j,r_1}) + F(x_{j,r_2} - x_{j,r_3}) + F(x_{j,r_4} - x_{j,r_5})$$
- **DE/current-to-rand/1/bin:**
  $$u_{j,i} = x_{j,i} + F(x_{j,r_1} - x_{j,r_i}) + F(x_{j,r_2} - x_{j,r_3})$$

where r_1, r_2, r_3, r_4 and r_5 are randomly selected individuals and $x_{j,best}$ is the best individual in the population.

Algorithm 1. Algorithm DE/rand/1/bin.

---

input : NP (population size), GEN (# of generations), F (mutation scaling), CR (crossover rate)

1 $G \leftarrow 0$;
2 CreateRandomInitialPopulation(NP);
3 **for** $i \leftarrow 1$ to NP **do**
4   Evaluate $f(\overrightarrow{x}_{i,G})$;   /* $\overrightarrow{x}_{i,G}$ is an individual in the population */
5 **for** $G \leftarrow 1$ to GEN **do**
6   **for** $i \leftarrow 1$ to NP **do**
7   SelectRandomly(r_1, r_2, r_3) ;   /* $r_1 \neq r_2 \neq r_3 \neq i$ */
8         $jRand \leftarrow$ RandInt(1, N) ;   /* N is the number of variables */
9   **for** $j \leftarrow 1$ to N **do**
10   **if** Rand(0, 1) < CR or $j = jRand$ **then**
11              $u_{i,j,G+1} = x_{r_3,j,G} + F.(x_{r_1,j,G} - x_{r_2,j,G})$;
12   **else**
13              $u_{i,j,G+1} = x_{i,j,G}$;

14   **if** $f(\overrightarrow{u}_{i,G+1}) \leq f(\overrightarrow{x}_{i,G})$ **then**
15           $\overrightarrow{x}_{i,G+1} = \overrightarrow{u}_{i,G+1}$;
16   **else**
17           $\overrightarrow{x}_{i,G+1} = \overrightarrow{x}_{i,G}$;

2.3 Adaptive Operator Selection

The Adaptive Operator Selection aims to adjust the application of operators while the search process is performed, according to the operators performance. Thus, we need to define two aspects: how to measure the performance of the operators, usually referred to as Credit Assignment, and how to select among them after these performance measurements are made, simply called here Operator Selection.

More specifically, the Credit Assignment firstly measures the impact caused by the operator application in the optimization process, and then transforms this impact into a meaningful numerical credit that will be used for updating the empirical quality estimates of each operator. The most common impact measure is simply the fitness improvement achieved by the generated offspring w.r.t. its parent(s). Then, the credit assigned to the operator can be: (i) the Instantaneous reward, *i.e.*, received after the last application; (ii) the Average of the rewards received over a few recent applications; (iii) or the Extreme reward recently received by the operator [4]. The number of recent applications considered for the latter two is usually a user-defined parameter, referred to as W (size of the sliding window).

For the Operator Selection, we consider here two existing schemes from the literature, which were chosen for having shown superior performance in recent works. The first one is called Adaptive Pursuit (AP) [13]. It calculates an application probability for each operator, and use a roulette wheel to select the next operator to be applied. A lower bound on the probabilities is employed to preserve some level of exploration, and a winner-takes-all scheme is used to push forward the current best operator. In this work, the AP Operator Selection scheme is used in combination with the Extreme Credit Assignment.

Although alleviating the user from the need of selecting which operators should be applied to the problem at hand, and doing so in an on-line manner, the most common operator selection schemes, including the AP method, involve some hyper-parameters that need to be tuned as well. The use of credit assignment schemes based on the raw values of the fitness improvements make these hyper-parameters highly problem-dependent.

Motivated by this issue, the Fitness-based Area-Under-Curve - Bandit (AUC), a fully comparison-based adaptive operator selection was recently proposed[5]. Its robustness comes from its Credit Assignment scheme, which is based on the ranks of the fitness improvements, and not on their raw values. Briefly, it works as follows. The latest W rewards achieved by all operators are ranked, and an exponentially decaying factor is applied over the rank values, so that the top ranked rewards have a significant weight, while the very low rewards have a weight close to zero. These decayed rank values are then used to construct a curve analog to the Area Under the ROC Curve, a criterion originally used in Signal Processing and later adopted in Machine Learning to compare binary classifiers. The ROC Curve associated to a given operator s is drawn by scanning the ordered list, starting from the origin: a vertical segment is drawn when the current offspring has been generated by s, a horizontal segment is drawn otherwise, and a diagonal one is drawn in case of ties. The length of each segment is proportional to its decayed rank value. Finally, the credit associated to operator s is the area under this curve. This Credit Assignment scheme is coupled with a bandit-based Operator Selection which deterministically chooses the strategy to be applied based on (a variant of) the Upper Confidence Bound algorithm [1].

3   Adaptive Hyper-Heuristic

The adaptive algorithm proposed here combines the GA and DE techniques by choosing during the evolutionary process which metaheuristic and which operators should be used. The metaheuristics are used in an interleaved way by choosing one of them to generate each new individual in the population, *i.e.*, each new individual is generated following the choice algorithm with its operators and parent selection mechanism. The generated individual is thus compared with the target (current) individual of the population and the fittest one is maintained in the population of the next generation, following the DE replacement mechanism. The algorithms and operators are chosen by AOS methods where the impact measure is defined by the improvement in fitness between the offspring (generated individual) and its parent (target individual for DE and the best parent for GA).

A similar idea was applied to select the operators of each algorithm, *i.e.*, the four variants described in section 2.2 in the DE case, and the five operators in section 2.1 in the GA case are selected according with the response of the AOS method. The pseudo-code of the proposed algorithm is presented in Algorithm 2.

Algorithm 2. HH-AOS

---

input : NP (population size), GEN (# of generations)

1 $G \leftarrow 0$;
2 CreateRandomInitialPopulation(NP);
3 **for** $i \leftarrow 1$ **to** NP **do**
4 $\quad$ Evaluate $f(\overrightarrow{x}_{i,G})$; /* $\overrightarrow{x}_{i,G}$ is an individual in the population */
5 **for** $G \leftarrow 1$ **to** GEN **do**
6 $\quad$ **for** $i \leftarrow 1$ **to** NP **do**
7 $\quad\quad$ op $\leftarrow$ AOS-selectOperator();
8 $\quad\quad$ **if** op $== DE$ **then**
9 $\quad\quad\quad \overrightarrow{u} =$ DE-generate-one-individual(NP, $\overrightarrow{x}_G$);
10 $\quad\quad$ **else**
11 $\quad\quad\quad \overrightarrow{u} =$ GA-generate-one-individual(NP, $\overrightarrow{x}_G$, $\overrightarrow{p}$);
12 $\quad\quad$ **if** Evaluate $f(\overrightarrow{u}) <$ Evaluate $f(\overrightarrow{x}_{i,G})$ **then**
13 $\quad\quad\quad \overrightarrow{x}_{i,G+1} = \overrightarrow{u}$;
14 $\quad\quad\quad$ **if** op $== DE$ **then**
15 $\quad\quad\quad\quad$ AOS-ApplyReward($\overrightarrow{x}_{i,G}$ - $\overrightarrow{x}_{i,G+1}$);
16 $\quad\quad\quad$ **else**
17 $\quad\quad\quad\quad$ AOS-ApplyReward($\overrightarrow{p}$ - $\overrightarrow{x}_{i,G+1}$) ; /* $\overrightarrow{p}$ is the best parent
$\quad\quad\quad\quad\quad$ selected to generate the new individual */

---

4 Comparative Results

In order to evaluate the performance of our proposal, experiments were conducted using the BBOB noiseless testbed [7],which includes 24 single-objective functions from 5 different classes with very different characteristics and levels of complexity. The default guidelines were followed: 15 trials per function [8], with the maximum number of function evaluations being fixed at $10^5 \times d$. The BBOB experimental set-up uses as performance measurement the Expected Running Time (ERT), defined as follows: given a target function value, ERT is the empirical expected number of function evaluations for attaining a fitness value below the target. In other words, it is the ratio of the number of function evaluations for reaching the target value over successful trials, plus the maximum number of evaluations for unsuccessful trials, divided by the number of successful trials. Due to space constraints, the presented results are restricted to dimension $d = 20$, although similar conclusions can be taken for the other considered dimensions.

Our proposal, herein called Adaptive Operator Selection at the Hyper-Heuristic (HH) level, was compared with each algorithm (DE and GA) individually with three different selection techniques: (i) uniform selection (Naive), (ii) the adaptive pursuit selection (AP) and (iii) the fitness area under curve bandit selection (AUC).

The first analysis, depicted in Figs. 1 and 2, presents a comparison among the three operator selection techniques within HH. The uniform choice reached the target value in an least one instance, for the highest level of precision (1e-8), in only 12 of the 24 function, while the HH with any of the two AOS techniques solved 17 functions. Besides, the AOS methods require fewer function evaluations to reach the target value (Fig. 2): for 50% of the cases, HH using an AOS method is at least two times faster than HH with the naive uniform operator selection. Although no significant difference could be found between the AP and the AUC AOS schemes, considering the speed-up ratio presented in the figure 2, and also the analysis described in [6] only the latter will be used in the following. A further

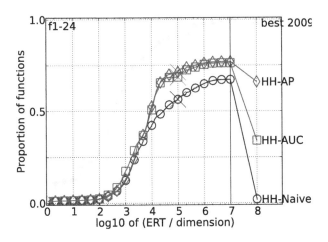

Fig. 1. Empirical cumulative distribution of the bootstrapped distribution of ERT over dimension for 50 targets in $10^{[-8..2]}$ for all functions to HH

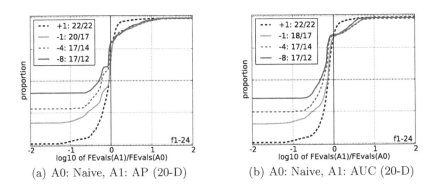

(a) A0: Naive, A1: AP (20-D)   (b) A0: Naive, A1: AUC (20-D)

Fig. 2. Empirical cumulative distributions (ECDF) speed-up ratios in 20-D to HH. ECDF of FEval ratios of Adaptive Pursuit (AP) and AUC-Bandit (AUC) divided by Naive, all trial pairs for each function. Pairs where both trials failed are disregarded, pairs where one trial failed are visible in the limits being > 0 or < 1. The legends indicate the number of functions that were solved in at least one trial (AP/AUC first)

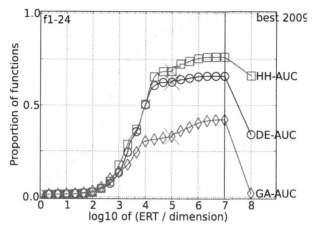

Fig. 3. Empirical cumulative distribution of the bootstrapped distribution of ERT over dimension for 50 targets in $10^{[-8..2]}$ for all functions to DE, GA and HH, all using the AUC adaptive operator selection

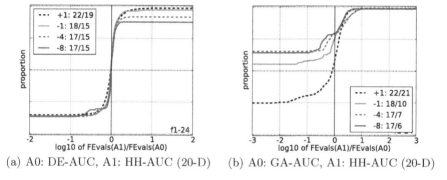

(a) A0: DE-AUC, A1: HH-AUC (20-D)   (b) A0: GA-AUC, A1: HH-AUC (20-D)

Fig. 4. Empirical cumulative distributions (ECDF) speed-up ratios in 20-D to DE-AUC, GA-AUC and HH-AUC. ECDF of FEval ratios of HH-AUC respectively divided by DE-AUC and GA-AUC, all trial pairs for each function. Pairs where both trials failed are disregarded, pairs where one trial failed are visible in the limits being > 0 or < 1. The legends indicate the number of functions that were solved in at least one trial (HH-AUC first).

analysis compares the performance of the HH-AUC with both DE and GA also using the AUC AOS mechanism. The difference is that the HH variant uses independent AOS schemes in the two levels of abstraction, while DE and GA have only one AOS instance selecting between their operators in the usual way. The results are presented in Figs. 3 and 4. As it can be seen, the autonomous selection between DE and GA done by the HH algorithm by means of the AOS methods is able to solve more instances than both DE and GA individually. This empirically confirms that the efficient mixture of DE and GA is better than each of the original methods alone.

5   Conclusions

In this paper, we propose the use of existing Adaptive Operator Selection (AOS) schemes at the Hyper-level, in order to automatically select between different metaheuristics for the generation of each new solution. The metaheuristics exemplified here were Differential Evolution (DE) and Genetic Algorithm (GA). Additionally, an independent AOS instance was also employed to automatically select between the corresponding variation operators in the usual way, selecting between four different operators whenever DE was chosen by the Hyper-AOS, and five operators otherwise. The resulting algorithm, that can be seen as an adaptive Hyper-Heuristic (HH), employs thus three independent instances of the recent Fitness Area Under Curve Multi-Armed Bandit AOS algorithm [5]: one instance controlling the choices at the Hyper-level, and the other selecting between the operators for the DE and GA algorithms. For both levels of abstraction, the impact of each AOS decision is computed by means of the fitness improvement achieved when comparing the newly generated offspring with its parent.

The proposed algorithm, tested under the light of the very comprehensive Black Box Optimization Benchmarking (BBOB) noiseless testbed [7], showed superior performance when compared to: (i)the same Hyper-Heuristic without adaptive behavior (uniformly selecting between the metaheuristics and operators), and (ii) the single-heuristic counterparts, *i.e.*, the DE and GA alone, using the same AOS mechanism to select between their corresponding variation operators. These results empirically confirm that the AOS at the Hyper-level is efficient, and that the intelligent switching between different metaheuristics is a path worth to be further investigated.

There are mainly two different paths that might be taken in the follow up of this work. One concerns its extension from the algorithmic point of view, by trying to improve and/or propose new AOS mechanisms for better efficiency at the Hyper-level. The other regards its extension from the application point of view, by analyzing the same adaptive scheme selecting among different metaheuristics and/or considering different problem domains.

Acknowledgments. The authors acknowledge the support from CNPq (grants 140785/2009-4 and 308317/2009-2).

References

1. Auer, P., Cesa-Bianchi, N., Fischer, P.: Finite-time analysis of the multi-armed bandit problem. Machine Learning 47(2-3), 235–256 (2002)
2. Burke, E.K., Hyde, M., Kendall, G., Ochoa, G., Ozcan, E., Qu, R.: Hyper-heuristics: A survey of the state of the art. Tech. rep., U. of Nottingham (2010)
3. Eshelman, L.J., Schaffer, J.D.: Real-coded genetic algorithms and interval-schemata. In: Foundation of Genetic Algorithms 2 (1993)
4. Fialho, Á., Da Costa, L., Schoenauer, M., Sebag, M.: Extreme Value Based Adaptive Operator Selection. In: Rudolph, G., Jansen, T., Lucas, S., Poloni, C., Beume, N. (eds.) PPSN 2008. LNCS, vol. 5199, pp. 175–184. Springer, Heidelberg (2008)

5. Fialho, Á., Schoenauer, M., Sebag, M.: Toward comparison-based adaptive operator selection. In: Proc. Genetic Evol. Comput. Conf. (2010)

6. Fialho, Á.: Adaptive Operator Selection for Optimization. Ph.D. thesis, Université Paris-Sud XI, Orsay, France (December 2010)

7. Hansen, N., Auger, A., Finck, S., Ros, R.: Real-parameter black-box optimization benchmarking: Noiseless functions definitions. RR-6829. Tech. rep., INRIA (2009)

8. Hansen, N., Auger, A., Finck, S., Ros, R.: Real-parameter black-box optimization benchmarking 2010: Experimental setup. RR-7215. Tech. rep., INRIA (2010)

9. Michalewicz, Z.: Genetic Algorithms + Data Structures = Evolutionary Programs. Springer (1992)

10. Özcan, E., Bilgin, B., Korkmaz, E.E.: A comprehensive analysis of hyperheuristics. Intell. Data Anal. 12(1), 3–20 (2008)

11. Storn, R., Price, K.V.: Differential evolution - a simple and efficient heuristic for global optimization over continuous spaces. J. of Global Opt. 11, 341–359 (1997)

12. Sywerda, G.: Uniform crossover in genetic algorithms. In: Proc. of the Third Int. Conf. on Genetic Algorithms, San Francisco, CA, USA, pp. 2–9 (1989)

13. Thierens, D.: An adaptive pursuit strategy for allocating operator probabilities. In: Proc. GECCO, pp. 1539–1546 (2005)

Improving Lin-Kernighan-Helsgaun with Crossover on Clustered Instances of the TSP

Doug Hains, Darrell Whitley, and Adele Howe

Colorado State University, Fort Collins CO, USA

Abstract. Multi-trial Lin-Kernighan-Helsgaun 2 (LKH-2) is widely considered to be the best Interated Local Search heuristic for the Traveling Salesman Problem (TSP) and has found the best-known solutions to a large number of benchmark problems. Although LKH-2 performs exceptionally well on most instances, it is known to have difficulty on clustered instances of the TSP. Generalized Partition Crossover (GPX) is a crossover operator for the TSP that efficiently constructs new solutions by partitioning a graph constructed from the union of two solutions. We show that GPX is especially well-suited for clustered instances and evaluate its ability to improve solutions found by LKH-2. We present two methods of combining GPX with multi-trial LKH-2. We find that combining GPX with LKH-2 dramatically improves the evaluation of solutions found by LKH-2 alone on clustered instances with sizes ranging from 3,000 to 30,000 cities.

1 Introduction

The Traveling Salesman Problem (TSP) can be stated as follows: Given n cities and an $n \times n$ cost matrix C, where entry C_{ij} is the cost of traveling between cities i and j, find a Hamiltonian circuit on the n cities which minimizes the sum of travel costs between cities on the route. We restrict our attention to symmetric instances, that is $C_{ij} = C_{ji}$.

Although the TSP is simply stated, the problem is NP-hard, necessitating the use of heuristics for larger instances. Lin-Kernighan-Helsgaun 2 (LKH-2) [2] is a state-of-the-art local search heuristic for the TSP based on the variable depth local search of Lin and Kernighan (LK-search) [4]. While LKH-2 performs exceptionally well on most instances of the TSP, its performance degrades on clustered instances [2].

Generalized Partition Crossover (GPX) [10,9] is a crossover operator for the TSP that produces offspring from a graph constructed from the union of two parent solutions. It has been shown that GPX has a high probability of producing locally optimal offspring if the parents are local optima [9]. We inspect local optima produced by LKH-2 on clustered instances and find they have a large number of common edges. As GPX relies on the common edges between two solutions to perform crossover, this indicates that the local optima produced by LKH-2 are especially well-suited for GPX.

C.A. Coello Coello et al. (Eds.): PPSN 2012, Part II, LNCS 7492, pp. 388–397, 2012.

LKH-2 can also be used in a form of iterated local search (ILS), which Helsgaun calls *multi-trial LKH-2*. Multi-trail LKH-2 incorporates a form of crossover known as Iterative Partial Transcription (IPT) based on the work of Möbius et al. [6]. Instead of using a population, multi-trial LKH-2 keeps the best-so-far solution; when it converges to a new local optimum, it applies crossover to the current candidate solution and the previous best-so-far solution. If crossover yields an improved solution, this becomes the new candidate solution. In this way, LKH-2 does not require a population, or other tuning parameters associated with population-based heuristics.

To the best of our knowledge, there are no published studies on the effectiveness of using crossover in this way. We present empirical evidence that shows incorporating a crossover operator into LKH-2 greatly improves the evaluation of solutions found by LKH-2 alone. A theoretical analysis of the differences between the GPX and IPT crossover operators proves that GPX is superior and empirical studies show that GPX produces better solutions than IPT when crossing over local optima produced by LKH-2.

Finally, there is also an advantage to doing multiple (parallel) runs of multi-trial LKH-2. This creates additional opportunities to utilize crossover. This is not like a genetic algorithm, because the runs remain independent. We present two methods of using crossover that have not previously been utilized in combination with LKH-2. Our results show that both methods improve solution quality over that of multi-trial LKH-2 with no significant increase to runtime.

2   Lin-Kernighan-Helsgaun 2

Lin-Kernighan-Helsgaun 2 (LKH-2) is a variable depth search that is based on the well known Lin-Kernighan algorithm (LK-search) [4]. LKH-2 has found the majority of best known solutions on the TSP benchmarks at the Georgia Tech TSP repository that were not solved by complete solvers[1]. At each step of the search, LKH-2 removes and replaces k edges of a given solution. This is known as a k-opt move. LKH-2 chains together a variable number of k-opt moves to find a new solution with a better evaluation than the initial solution [2].

As we are not concerned with the inner workings of LKH-2, it is sufficient to think of LKH-2 as a 'black-box' that produces a locally optimal solution when given an arbitrary initial solution. LKH-2 has a number of parameters that influence its performance; for all our experiments, we use the same settings reported by Helsgaun to produce the best results on clustered instances [2].

2.1   LKH-2 and Clustered Instances

Papadimitriou proved that LK-search solves a Polynomial Local Search (PLS) complete problem [8,3] by constructing graphs that force LK-Search to take an exponential number of steps. Papadimitriou constructed a graph containing 'bait edges' that lead LK-search into an extensive search for an improved solution.

---

[1] http://www.tsp.gatech.edu/data/index.html

It is thought that the edges between clusters of a clustered TSP instance act in a similar manner as the bait edges of Papadimitriou's proof [7]. Empirical experiments show that LKH-2 performs significantly worse on random clustered instances than uniform random instances [2]. We conjecture that its performance on clustered instances could be improved by exploiting crossover.

3   Generalized Partition Crossover

Generalized Partition Crossover (GPX) is a crossover operator for the TSP with a number of interesting properties. When given local optima as parents, GPX is highly likely to produce locally optimal children [10,9]. GPX is "respectful", meaning that any common edges in the parents are inherited by the offspring; and it "transmits alleles", meaning that *all* edges found in the offspring are directly inherited from the parents. When used in a hybrid genetic algorithm with Chained Lin-Kernighan (Chained LK) [1], GPX is able to produce higher quality solutions than Chained LK alone [10].

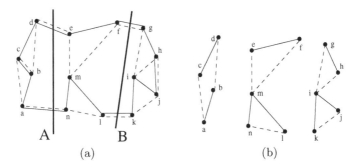

(a)   (b)

Fig. 1. An example of (a) the graph constructed by GPX from the union of two solutions, S_a (dashed edges) and S_b (solid edges), and (b) how the removal of shared edges creates subgraphs. The heavy dark lines show two partitions of cost 2, A and B. Once GPX partitions the graph, solutions are constructed from the shared edges in (a) and the dashed or solid edges from each subgraph in (b).

GPX works by first constructing a graph $G = (V, E)$ from two solutions, S_a and S_b, where V is the set of cities in the TSP instance and E is the union of edges in the two solutions (see Figure 1(a)). To partition G, the edges in G that are shared by both solutions are removed to create graph G_u as in Figure 1(b). Breadth first search is then applied to G_u to identify the connected components. Note that any connected component in G_u can be disconnected from G by removing shared edges. We refer to the connected components in G_u as *partition components*. We define the *cost* of a partition of graph G as the minimal number of shared edges that must be removed to disconnect a component from G. Whitley et al. prove that if G contains at least one partition of cost 2, it is always possible to create at least two Hamiltonian circuits distinct from the parents in $O(n)$ time [9].

After one or more partitions of cost 2 is found, GPX will recombine the parent solutions S_a and S_b to one or more offspring. First, the common edges between S_a and S_b are inherited. Then, the best possible offspring is obtained by greedily selecting the lowest cost path through each partition component of graph G. Additional offspring can be obtained by making non-greedy choices.

GPX is an ideal candidate for clustered instances of the TSP. As local optima tend to have a large number of edges in common, it is likely that some of these edges will be between clusters. This would allow GPX to partition across edges between clusters and recombine the lowest cost paths through the clusters on either side of the partition. If a partition component captures several clusters, GPX will also recombine the lowest cost edges between clusters.

To determine if this is the case, we produced one hundred different local optima with LKH-2 on a randomly generated clustered instance with 3162 cities. On average, there were 2.10±0.04 partitions of cost 2 when pairing together these solutions. Thus, GPX will likely be able to find multiple partition components of cost 2 when using two local optima produced by LKH-2 as parents.

3.1   Iterative Partial Transcription and GPX

Iterative Partial Transcription [6] (IPT) is a form of crossover used by LKH-2. When LKH-2 reaches a local optimum, it executes a random restart while retaining the best-so-far-solution. LKH-2 uses IPT to recombine the best-so-far solution with the new local optimum [2]. If IPT finds a solution better than the best-so-far, the best-so-far solution is replaced. Note that this does not require a population of solutions.

IPT constructs a graph $G = (V, E)$ from two solutions, S_1 and S_2, in the same way as GPX. IPT will attempt to partition V into two disjoint sets A and B such that the number of edges between sets is exactly 2. Let $E(A)$ be the set of edges incident only to vertices in A and $E(B)$ the set of edges incident only to vertices in B. Offspring are formed by removing the edges in S_1 that are also in $E(A)$ and replacing them with the edges found in both S_2 and $E(A)$. The same process is repeated with $E(B)$. This is identical to using Partition Crossover utilizing only a single partition of cost 2 [9].

Given G with k partition components of cost 2, GPX will return the best solution out of a possible $2^k - 2$ unique solutions [10]. As IPT uses only a single partition, it can reach only $2k - 2$ of the $2^k - 2$ solutions processed by GPX. We subtract 2 as we count only solutions different from the two parent solutions. Note that $k \geq 2$ since 1 partition break the graph into 2 partition components.

For $k > 2$, the solutions reachable by IPT are a subset of the solutions reachable by GPX. Therefore, it follows that the offspring generated by GPX is guaranteed to be equal or better than the solution generated by IPT. When k is larger, GPX will find a better solution with greater probability.

3.2 Effect of Crossover on LKH-2

To the best of our knowledge, there have been no published experiments to determine the effect of using LKH-2 with and without IPT. We therefore conduct an experiment to validate two hypotheses: 1. The addition of a crossover operator such as IPT can improve solutions over that of LKH-2, and 2. GPX can further improve solutions over IPT without a significant increase in run time.

For this experiment, we used six instances from the 8th DIMACS TSP Challenge[2]: C3k.0, C3k.1, C10k.0, C10k.1, C31k.0 and C31k.1 with sizes 3162, 3162, 10000, 10000, 31623 and 31623, respectively. For each instance, we ran three versions of LKH-2 with 10 random restarts: LKH-2 without crossover, LKH-2 with IPT and LKH-2 with GPX. When running LKH-2 with either crossover operator, the crossover operator is applied to the best-so-far solution and the local optimum produced after each restart.

As the optimal solutions are not known for all instances, we normalize the results in Table 1 by reporting the percentage above the Held-Karp bound (HKB). The HKB is guaranteed to be within 2/3 of the optimal evaluation on Euclidean instances, but in practice is often within 0.01% of optimal [3]. The HKB for each instance can be found at the DIMACS TSP Challenge homepage. We find that crossover significantly improves the solutions found by LKH-2 and that GPX can further improve over the solutions found by IPT.

Table 2 reports the average CPU time of a single call to IPT and a single call to GPX. We also report the percentage of the total CPU time per run of LKH-2 accounted for by the crossover operators. For instances of 10,000 cities and lower, IPT is slightly faster than GPX but both account for less than 0.1% of the overall running time. GPX is faster than IPT on larger problems. Therefore GPX can be used in place of IPT with no significant increase to the overall run time while increasing the potential to find improved solutions.

Table 1. The minimum percentage above the Held-Karp Bound for several clustered instances of the TSP of solutions found by ten random restarts of LKH-2 without crossover, with IPT and with GPX. Best values for each instance are in boldface. The p-value of a one-way ANOVA test are shown in the final row for each instance. A significant value is denoted with (*).

		Instance C3k.0	C3k.1	C10k.0	C10k.1	C31k.0	C31k.1
	Min	0.660	0.863	1.143	1.009	1.489	1.538
LKH-2	Avg	0.772	1.584	1.671	1.597	1.760	1.907
	Max	1.387	2.051	2.339	2.658	2.169	2.397
	Min	**0.622**	0.656	1.040	0.873	1.280	1.274
LKH-2 w/ IPT	Avg	0.665	1.329	1.160	1.022	1.433	1.595
	Max	0.786	2.051	1.396	1.419	1.663	2.397
	Min	**0.622**	**0.651**	**1.031**	**0.872**	**1.270**	**1.267**
LKH-2 w/ GPX	Avg	0.660	1.326	1.159	1.021	1.426	1.591
	Max	0.786	2.051	1.396	1.419	1.663	2.397
p-value		0.112	0.484	<0.001(*)	<0.001(*)	<0.001(*)	<0.001(*)

---

[2] http://www2.research.att.com/~dsj/chtsp/

Table 2. Average run time in seconds of IPT and GPX over the 10 runs used to produce Table 1. The numbers in parentheses represent the percentage of the total run time of LKH-2 accounted for by the crossover operators. P-values for a standard t-test are shown with significant values denoted by (*).

Instance	C3k.0	C3k.1	C10k.0	C10k.1	C31k.0	C31k.1
IPT	0.001(0.04%)	0.001(0.04%)	0.005(0.06%)	0.005(0.06%)	0.047(0.17%)	0.052(0.18%)
GPX	0.002(0.08%)	0.001(0.04%)	0.007(0.08%)	0.006(0.07%)	0.026(0.09%)	0.029(0.1%)
p-value	0.016	0.005(*)	0.036	0.014	<0.001(*)	<0.001(*)

4   Crossover and Iterated Local Search

In Iterated Local Search (ILS) [5] a perturbation operator is applied to escape local optima. The local search is restarted on the perturbed solution, and the process is repeated for a fixed number of iterations. Using ILS with LK-search based heuristics has proven to be more effective than random restarts [3,1]. Helsgaun refers to the ILS version of LKH-2 as *multi-trial LKH-2* [2].

Multi-trial LKH-2 uses a pseudo-random restart influenced by the best-so-far solution when a local optimum is reached. The next iteration of LKH-2 is then biased by ignoring any k-opt moves beginning with edges in the best-so-far solution. Given the benefits of incorporating crossover in LKH-2 shown in Section 3.2, we construct two methods for incorporating crossover with multi-trial LKH-2: *GPX across runs* and *GPX across restarts*.

4.1   GPX across Runs

GPX across runs applies crossover to improve the local optima found by independent runs of multi-trial LKH-2. At each iteration i of multi-trial LKH-2 (i.e., when it reaches a local optimum), we form a population of the local optima found at iteration i of each independent run. We then apply GPX, crossing over the best solution in the population with each other solution. The best solution found will be stored, but not returned to the multi-trial LKH-2 runs. This preserves diversity between the runs. Figure 2 depicts 10 independent runs of multi-trial LKH-2; GPX across runs will crossover the solutions with the same letters.

4.2   GPX across Restarts

Another option is to crossover solutions from the *same* run. We could crossover the best-so-far solution with the local optimum found at each iteration of multi-trial LKH-2 like IPT. However, the local optimum is discarded if it is not better than the best-so-far solution. It is possible that by doing so, low cost edges that could potentially be used to improve the best-so-far solution are discarded. To remedy this, we designed Subroutine 1 to maintain a population of local optima and to crossover the population at each iteration of multi-trial LKH-2. The best solution from the crossover becomes the starting solution for the next iteration of multi-trial LKH-2.

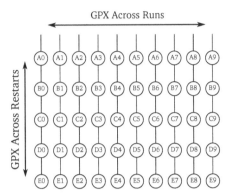

Fig. 2. A diagram depicting 10 runs of multi-trial LKH-2 run for 5 iterations per run. The circles represent local optima produced by LKH-2. GPX across runs crosses over solutions with the same letters. GPX across restarts crosses over solutions with the same numbers.

Subroutine 1. Given s^*, a local optimum passed from LKH-2; c, a cost function that sums the edge costs of a given solution, and P, a set of solutions. When the subroutine returns, control passes back to LKH-2.

1. If $P = \varnothing$, Let $P = \{s^*\}$ and return s^*.
2. Apply GPX with s^* and each solution in P.
3. Let s' be the offspring with best evaluation. If $c(s') < c(s^*)$, let $s^* = s'$.
4. If $|P| \neq$ popsize, let $P = P \cup \{s^*\}$.
5. Otherwise, let s be the solution with the poorest evaluation in P. If $c(s^*) < c(s)$, replace s in P with s^*.
6. Return s^*.

4.3   Effect of Crossover on Multi-trial LKH-2

We hypothesize that incorporating crossover with multi-trial LKH-2 should further improve solution quality, especially when GPX is the crossover operator. To test this, we ran 10 independent runs of multi-trial LKH-2 with three different methods of crossover for 50 iterations per run on the same clustered instances as before. Method one was multi-trial LKH-2 with IPT. IPT was applied at each iteration to the most recent local optimum and the best-so-far solution. If IPT produced a better solution than the best-so-far solution, it is replaced. Method two was GPX across runs. At each iteration, GPX was applied to the 10 local optima found by each run. Method three was GPX across restarts. Subroutine 1 with popsize = 10 was called at each iteration. When applying GPX across restarts, the population was set to empty at the beginning of each run.

Table 3 reports the minimum, maximum and average evaluation above the HKB after the final iteration. The data in Table 3 shows that Multi-trial LKH-2 with IPT is better than LKH-2 with IPT (see Table 1) in every problem. The

Table 3. Minimum, average, and maximum percentage of evaluation above the Held-Karp bound for solutions after 50 iterations of 10 runs of multi-trial LKH-2 using different crossovers. '*' signifies the known global optimum was found. Best solutions are in boldface. The p-value of a one-way ANOVA test are shown in the final row for each instance. All values were significant.

Instance		C3k.0	C3k.1	C10k.0	C10k.1	C31k.0	C31k.1
	Min	0.6218	0.6153	0.7184	0.7061	0.6922	0.8824
M. LKH-2 w/ IPT	Avg	0.6336	1.0822	0.9514	0.8890	0.9639	1.0237
	Max	0.6432	1.5692	1.3341	1.2489	1.2077	1.1417
	Min	**0.6180***	0.6153	**0.7037**	**0.7036**	**0.6879**	0.8660
GPX Across Runs	Avg	0.6183	0.6156	0.7144	0.7048	0.7012	0.8843
	Max	0.6190	0.6183	0.7223	0.7064	0.8053	0.9409
	Min	**0.6180***	**0.6150***	0.7151	0.7912	0.9725	**0.8131**
GPX Across Restarts	Avg	0.6188	0.7117	0.7529	0.8615	1.0199	0.8525
	Max	0.6254	1.5350	0.7691	1.1518	1.1430	0.9376
p-values		< 0.001	0.001	0.002	0.008	< 0.001	< 0.001

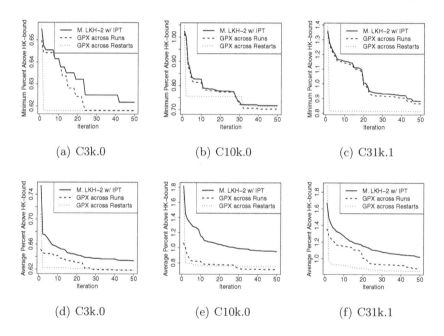

(a) C3k.0   (b) C10k.0   (c) C31k.1

(d) C3k.0   (e) C10k.0   (f) C31k.1

Fig. 3. Minimum (top row) and average (bottom row) evaluation above the Held-Karp bound at each iteration across 10 runs of multi-trial LKH-2 using various crossover methods on three instances.

same is true for the two methods using GPX. Thus, multi-trial LKH-2 with crossover does provide a significant benefit. Comparing the results for IPT to that of the two GPX based methods shows that GPX generally improves over IPT. GPX Across Restarts finds the global optimum in two cases where IPT did not. GPX Across Runs consistently improves upon IPT.

To further assess the differences between the three methods, Figure 3 shows the minimum and average evaluation at each iteration for several instances. Interestingly, GPX across restarts yields the largest gain in evaluation initially. After the second iteration, the local optima from the first iteration will be crossed over with the local optimum produced at the second iteration. As the search was biased away from investigating moves beginning with edges in the best-so-far solution, the local optimum it produces and the best-so-far solution may present ideal parents for GPX. In some cases, it finds the best solution in the second iteration and does not improve further. This may be a result of an interaction with the way multi-trial LKH-2 biases the search with edges in the best-so-far solution [2].

GPX across runs is capable of consistently improving the quality of solutions over that of IPT. The average evaluation across iterations shows larger differences of GPX over IPT. Note that the solutions produced by GPX across runs are not in any way used by LKH-2. Therefore, GPX across runs does not influence the local search. On the other hand, GPX across restarts does influence the search behavior of LKH-2. The results suggest some trade-offs in how crossover can be exploited that might offer further opportunity for improvement.

5   Conclusion

Clustered instances of the TSP are problematic for LKH-2 [2]. Examining the structure of clustered instances, it seems likely that crossover operators such as GPX and IPT will perform well on clustered instances. We examine both operators and show that they are able to significantly improve solution quality on clustered instances when combined with LKH-2.

Furthermore, GPX is a compelling replacement for IPT in LKH-2. GPX is able to find all partitions that IPT can find, but can utilize more of them when constructing offspring. This allows GPX to find higher quality solutions than IPT. GPX also has a computation cost comparable to IPT. Although IPT is slightly faster on smaller instances, both operators require less than than 0.2% of the overall run time of LKH-2. As the instance size grows, GPX scales better than IPT and GPX becomes faster than IPT on larger instances.

We introduce two methods of incorporating crossover with multi-trial LKH-2. GPX across restarts uses a subroutine to maintain a population of solutions as an alternative to applying crossover to the best-so-far solution and a local optimum produced by LKH-2. GPX across restarts produces better average solutions and finds better minimum solutions on the majority of instances tested. GPX across runs consistently improves the minimum solution quality over that of multi-trial LKH-2 w/ IPT. This method also finds the globally optimal solution for the two benchmark instances for which the global optimum is known.

Our results show that crossover offers significant benefits when incorporated with a state-of-the-art local search heuristic for the TSP. We conjecture that the benefits we observed in TSP can also be obtained in other applications using local search; crossover leverages information about good partial solutions which can be exploited in search after restarts.

Acknowledgments. This research was sponsored by the Air Force Office of Scientific Research, Air Force Materiel Command, USAF, under grant number FA9550-11-1-0088. The U.S. Government is authorized to reproduce and distribute reprints for Governmental purposes notwithstanding any copyright notation thereon.

References

1. Applegate, D., Cook, W., Rohe, A.: Chained Lin-Kernighan for large traveling salesman problems. INFORMS Journal on Computing 15(1), 82–92 (2003)
2. Helsgaun, K.: General k-opt submoves for the Lin-Kernighan TSP heuristic. Mathematical Programming Computation 1(2), 119–163 (2009)
3. Johnson, D.S., Mcgeoch, L.A.: The traveling salesman problem: A case study in local optimization. In: Local Search in Combinatorial Optimization, pp. 215–310. John Wiley and Sons (1997)
4. Lin, S., Kernighan, B.W.: An effective heuristic algorithm for the traveling-salesman problem. Operations Research 21(2), 498–516 (1973), http://dx.doi.org/10.2307/169020
5. Lourenço, H., Martin, O., Stützle, T.: Iterated local search. In: Handbook of Metaheuristics, pp. 320–353 (2003)
6. Möbius, A., Freisleben, B., Merz, P., Schreiber, M.: Combinatorial optimization by iterative partial transcription. Physical Review E 59(4), 4667 (1999)
7. Neto, D.: Efficient cluster compensation for Lin-Kernighan heuristics. Ph.D. thesis, University of Toronto (1999)
8. Papadimitriou, C.: The complexity of the Lin-Kernighan heuristic for the traveling salesman problem. SIAM Journal on Computing 21, 450 (1992)
9. Whitley, D., Hains, D., Howe, A.: Tunneling between optima: partition crossover for the traveling salesman problem. In: Proceedings of the 11th Annual Conference on Genetic and Evolutionary Computation, pp. 915–922. ACM (2009)
10. Whitley, D., Hains, D., Howe, A.: A Hybrid Genetic Algorithm for the Traveling Salesman Problem Using Generalized Partition Crossover. In: Schaefer, R., Cotta, C., Kołodziej, J., Rudolph, G. (eds.) PPSN XI. LNCS, vol. 6238, pp. 566–575. Springer, Heidelberg (2010)

A Comparative Study
of Three GPU-Based Metaheuristics

Youssef S.G. Nashed[1], Pablo Mesejo[1], Roberto Ugolotti[1],
Jérémie Dubois-Lacoste[2], and Stefano Cagnoni[1]

[1] Department of Information Engineering, University of Parma, Italy
{nashed,pmesejo,rob_ugo,cagnoni}@ce.unipr.it
[2] IRIDIA, CoDE, Université Libre de Bruxelles, Belgium
jeremie.dubois-lacoste@ulb.ac.be

Abstract. In this paper we compare GPU-based implementations of three metaheuristics: Particle Swarm Optimization, Differential Evolution, and Scatter Search. A GPU-based implementation, obviously, does not change the general properties of the algorithms. As well, we give for granted that GPU-based implementation of both algorithm and fitness function produces a significant speed-up with respect to a sequential implementation. Accordingly, the main goal of this work has been to fairly assess the efficiency of the GPU-based implementations of the three metaheuristics, based on the statistical analysis of the results they obtain in optimizing a benchmark of twenty functions within a prefixed limited time.

Keywords: Global Continuous Optimization, Particle Swarm Optimization, Differential Evolution, Scatter Search, GPGPU.

1 Introduction

Modern graphics hardware has gained an important role in the area of parallel computing, since it has been used to accelerate general computations (General Purpose Graphics Processing Unit - GPGPU - programming), in addition to playing its natural role. CUDA[TM] (Compute Unified Distributed Architecture) is a parallel computing environment by nVIDIA[TM] which exploits the massively parallel computation capabilities of its massively parallel GPUs. In particular, CUDA-C [14] is an extension of the C language that allows development of GPU routines (termed *kernels*), that can be executed in parallel by several different CUDA[TM] threads, following the Single Instruction Multiple Thread model.

Among the stochastic approaches to continuous optimization, Evolutionary Algorithms (EAs) [6] and Swarm Intelligence [1] algorithms offer a number of attractive features: robust and reliable performance, global search capability, virtually no need of specific information about the problem to solve, easy implementation, and, above all, implicit parallelism.

In this paper, we compare the GPU implementations of three real-valued population-based optimization techniques: Particle Swarm Optimization (PSO)

C.A. Coello Coello et al. (Eds.): PPSN 2012, Part II, LNCS 7492, pp. 398–407, 2012.

[8], Differential Evolution (DE) [16], and Scatter Search (SS) [7] [1]. They have been tested on a benchmark of 20 numerical problems (also implemented on GPU), comprising unimodal/multimodal and separable/non separable functions (see Table 3). The main contributions of this paper are: i. the first implementation, to the best of our knowledge, of Scatter Search in CUDA^{TM}; ii. a novel parallel version of DE that solves some of the problems of previous implementations; iii. the study of these three metaheuristics on a broader benchmark than those usually adopted to test GPU implementations; iv. an unbiased evaluation of the effectiveness of the GPU implementation, as each metaheuristic has been subjected to the same parameter optimization method and run for a pre-fixed limited amount of time.

2 Basics of the Three Metaheuristics

2.1 Particle Swarm Optimization

Particle Swarm Optimization (PSO) is a stochastic optimization algorithm which simulates the behavior of bird flocks. A set of particles (or solutions) move through a "fitness" function domain (search space) seeking the function optimum (best fitness value). Each particle's motion is described by two simple discrete-time equations which regulate the particles' velocity and position:

$$v_n(t) = w \cdot v_n(t-1)$$
$$+ c_1 \cdot rand() \cdot (BP_n - P_n(t-1))$$
$$+ c_2 \cdot rand() \cdot (BLP_n - P_n(t-1))$$
$$P_n(t) = P_n(t-1) + v_n(t)$$

where $P_n(t)$ and $v_n(t)$ are the position and velocity of the nth particle at time t; c_1, c_2 and w (inertia factor) are positive constants, $rand()$ returns random values uniformly distributed in $[0,1]$, BP_n is the best-fitness location visited so far by the particle, and BLP_n is the best-fitness location visited so far by any particle in its neighborhood, which may include a limited set of particles or even coincide with the whole swarm.

2.2 Differential Evolution

Differential Evolution has recently been shown to be one of the most successful EAs for global continuous optimization [2,18]. Unlike traditional EAs, DE perturbs the current population members with the scaled differences of randomly selected and distinct individuals. In the first iterations the elements are widely scattered in the search space and have a great exploration ability. As optimization proceeds, the individuals tend to concentrate in the regions of the search space with better fitness values, so the search automatically focuses on the most

---

[1] The code can be downloaded from http://sourceforge.net/p/libcudaoptimize [13].

promising areas. In DE, every element acts as a parent vector, for which a donor vector is created. In the original version of DE, the donor vector for the ith parent (X_i) is generated by combining three random and distinct elements X_{r1}, X_{r2} and X_{r3}. The donor vector V_i is calculated as:

$$V_i = X_{r1} + F \cdot (X_{r2} - X_{r3})$$

where F (scale factor) is a parameter that strongly influences DE's performances and typically lies in the interval $[0.4, 1]$. After mutation, every parent-donor pair generates one child (trial vector) by means of a crossover operation. Two kinds of crossover are typically used: binomial (uniform) and exponential. Crossover is applied with a certain probability Cr (crossover rate) that, like F, is one of the control parameters of DE. Then, the trial vector is evaluated and its fitness is compared to its parent's: the better survives while the other is discarded.

2.3   Scatter Search

Scatter Search is based on a systematic combination between solutions (instead of randomized, as is usual in EAs) taken from a subset of the population, named the "reference set", that is usually significantly smaller than a typical EA population. SS is composed of five structural "blocks" or methods (see Figure 1):

1. Diversification Generation: a population of solutions P is generated, having a certain degree of quality and diversity. The reference set R is then drawn from P, and includes the $|R_1|$ solutions with best fitness, and the $|R_2|$ solutions from the reference set that are farthest from P according to the Euclidean distance (hence, $|R| = |R_1| + |R_2|$); the evolution process acts only on R;
2. Solution Combination: in most problems a specific method to combine solutions is needed, which can be applied to all solutions or only to selected ones (e.g., the best solutions, and/or randomly selected ones). In many cases an existing crossover operator, borrowed from other EAs, is employed;
3. Subset Generation: the procedure deterministically generates subsets of R, to which the combination method is applied.
4. Improvement: an improvement method (typically a local search) is applied to the original solutions and/or to combined solutions;
5. Reference Set Update: once a new solution is obtained it replaces the worst solution in R only if it improves the quality of the reference set in terms of fitness and/or diversity;

3   Parallel Implementation

The first parallel versions of PSO relied on multiprocessor parallel machines or cluster of computers. With the introduction of GPUs, research shifted towards GPU-based parallel PSO (GPU PSO) to alleviate multi-processor and cluster systems inefficiencies, such as network overhead, shared memory access, etc. In 2009 and 2010, respectively, the first implementations of PSO and DE

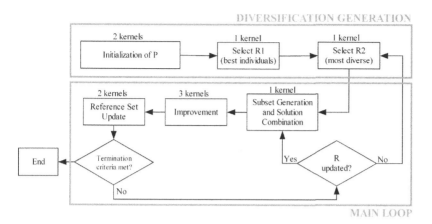

Fig. 1. Block diagram of the Scatter Search Algorithm

using nVIDIA CUDA^{TM} were developed [3,4]. Also in 2009, a hybrid between GPU PSO and pattern search, aimed at enhancing the convergence of PSO, was presented in [20]. After that, other implementations of DE have been developed [9,21], and fast versions of PSO have been implemented by relaxing the synchronicity constraints between particles' evaluation and update [12].

The early GPU PSO implementations suffered from a coarse-grained parallelization (one thread per particle), that neglected the opportunity to compute the fitness function, usually the most time-consuming process, in parallel over the problem dimensions. This aspect has been improved by the PSO implementation evaluated in this work, first introduced in [11]. In it, a thread manages a single dimension of each particle, adding a further level of parallelism. Similar inefficiencies characterized the early implementations of DE, such as a partially sequential implementation of the fitness function and random number generation. These problems were addressed by [9] using four kernels executed sequentially. Our present implementation uses only one kernel for generating the trial vectors, and another for fitness evaluation and migration. In addition, we offer three different mutations and two kinds of crossovers, while early GPU-based DE considered only one mutation strategy (DE/rand/1) and one kind of crossover. Regarding SS, to the best of our knowledge, ours is the first parallel implementation of this metaheuristic.

The performance of CUDA^{TM} code depends chiefly on the thread configuration, number of kernels, and memory access schemes used. All methods presented here aim at exploiting fast-access local and shared memory instead of slower global memory to the greatest possible extent, considering the number of kernels as the main criterion to assess how well an algorithm can be parallelized.

PSO is divided into three kernels described in [11], while DE, as mentioned earlier, can be implemented as two kernels. Each thread of the first kernel performs the following instructions:

- generate two or three distinct random numbers on the GPU, according to the mutation strategy;
- calculate an element of the donor vector from the population members randomly selected in the previous step;
- decide whether to include the donor or the parent element in the trial vector, based on the type of crossover and the crossover rate, Cr.

The second DE kernel evaluates all trial vectors simultaneously in shared memory and, if the fitness has improved, it replaces the parent with the offspring.

Clearly, SS is not as inherently parallel as the two other metaheuristics (see Figure 1). In SS a diverse population is first initialized and evaluated; diversity is simulated by generating uniform random values for each dimension over the whole search space. Then, to build the reference set R, a parallel sort operation is required to find R_1, followed by another kernel that calculates pairwise Euclidean distances between solutions in $P - R$ and R, sequentially adding the solutions that are farthest from the reference set for $|R_2|$ iterations. As for selection and crossover, a kernel selects all solution pairs in the reference set for mating, and combines them through the BLX-α crossover , generating two distinct solutions chosen with α set to $(0.5 + \lambda)$ and $(0.5 - \lambda)$, respectively. The combined solutions make up the *pool*, to which a parallel implementation of the Solis & Wets search method [19] is then applied as improvement method. For the last step, we compared two methods for updating the reference set, one of which considers both quality and diversity as in [5], while the other updates the reference set with the best $|R|$ solutions in $(R \cup pool)$. The latter yielded better results in terms of both speed and accuracy, as proven by the automatic tuning process described in the following section.

4   Experiments

We give for granted that the parallel metaheuristics we consider are faster than the corresponding sequential versions on sufficiently large problems, while their accuracy is the same for identical configurations, because they implement the same algorithm. Many comparisons between the accuracy of the sequential versions of the algorithms [5,18] have already been made, but they gave insights on their intrinsic features rather than on the computational efficiency of possible implementations. In this work, we evaluate both quality and speed of their parallel versions, analyzing the accuracy they can achieve in a limited amount of time, to assess the degree of parallelization that each of them allows to reach.

Table 1. Automatically-tuned parameter values used to test different optimization techniques

DE	PSO	SS
$Cr = 0.879$	$c_1 = 1.862$	$|P| = 140$
$F = 0.520$	$c_2 = 1.881$	$|R_1| = 9, |R_2| = 1$
Exponential Crossover	$w = 0.494$	$\lambda = 0.220$
Random Mutation	Population Size $= 125$	Solis & Wets iterations $= 85$
Size $= 48$		

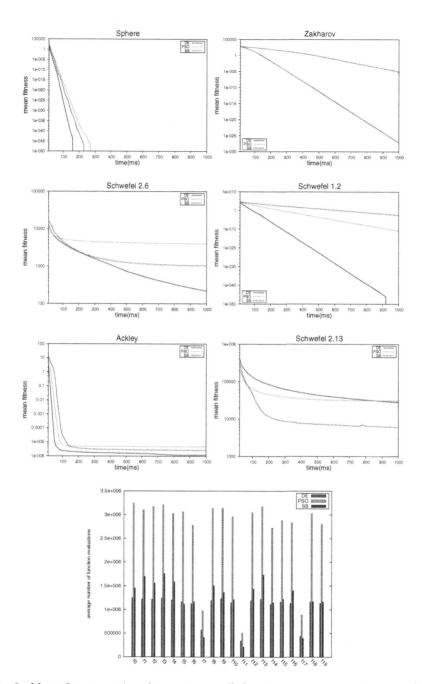

Fig. 2. Mean fitness vs time (up to 1 second) for six representative functions (unimodal separable, unimodal non-separable and multimodal non-separable), and number of function evaluations performed in 1 second by every method for each function on 30-dimensional problems.

Table 2. Results on the 20 functions

	10 dimensions						30 dimensions					
	DE		PSO		SS		DE		PSO		SS	
	Avg	Std	Avg	Std	Avg	Std	Avg	Std	Avg	Std	Avg	Std
f_0	0.0	0.0	0.0	0.0	0.0	0.0	0.0	0.0	0.0	0.0	0.0	0.0
f_1	0.0	0.0	0.0	0.0	2.5e-03	9.2e-03	0.0	0.0	0.0	0.0	2.2e-06	6.3e-06
f_2	0.0	0.0	0.0	0.0	0.0	0.0	0.0	0.0	0.0	0.0	4.1e-44	8.6e-44
f_3	0.0	0.0	0.0	0.0	7.0e-45	2.9e-44	0.0	0.0	0.0	0.0	6.7e-05	3.0e-04
f_4	0.0	0.0	0.0	0.0	1.1e-25	1.7e-25	0.0	0.0	0.0	0.0	5.1e-24	2.7e-24
f_5	0.0	0.0	0.0	0.0	0.0	0.0	2.5e-28	3.4e-28	1.8e-28	2.2e-28	1.2e-05	1.7e-05
f_6	0.0	0.0	0.0	0.0	0.0	0.0	0.0	0.0	9.7e-012	9.9e-012	2.5e-03	3.3e-03
f_7	9.8e-06	6.9e-05	1.0e-03	4.0e-04	2.7e-04	3.5e-04	2.1e+02	3.3e+02	3.8e+03	1.1e+03	9.9e+02	3.4e+02
f_8	5.0e-01	5.0e-08	3.5e-02	1.3e-01	2.9e-01	2.5e-01	5.0e-01	0.0	4.2e-01	1.9e-01	5.0e-01	2.1e-05
f_9	0.0	0.0	5.2e-01	7.8e-01	6.9e-01	9.1e-01	0.0	0.0	7.2e+01	1.6e+01	3.5e+01	9.7e+00
f_{10}	5.9e+00	3.1e+01	1.2e+02	1.2e+02	8.1e+01	1.1e+01	1.7e+01	4.8e+01	2.9e+03	4.1e+02	2.4e+03	9.0e+02
f_{11}	1.2e-03	4.9e-06	1.2e-03	2.9e-05	1.2e-03	2.4e-18	1.2e-02	5.1e-05	1.2e-02	1.8e-04	1.2e-02	8.7e-18
f_{12}	0.0	0.0	1.1e-02	1.0e-02	1.1e-03	2.9e-03	7.4e-05	7.4e-04	6.3e-10	6.0e-09	1.5e-03	4.3e-03
f_{13}	0.0	0.0	3.9e-07	4.8e-07	5.9e-01	3.3e+00	0.0	0.0	2.1e-01	7.2e-01	2.2e+01	2.6e+01
f_{14}	0.0	0.0	6.7e-07	1.1e-06	6.9e-01	9.1e-01	1.1e-06	1.2e-06	4.5e-06	9.4e-07	9.3e-03	
f_{15}	0.0	0.0	1.2e-03	6.5e-03	6.3e-31	4.4e-30	0.0	0.0	1.5e-28	1.2e-27	1.1e-27	2.7e-27
f_{16}	3.3e-02	2.9e-02	1.0e-01	2.4e-02	7.3e-01	5.4e-01	3.2e-01	3.0e-02	8.5e+00	8.6e-01	8.7e+00	1.2e+00
f_{17}	4.5e+01	2.2e+02	1.3e+00	5.1e+00	4.7e+00	8.5e+00	2.8e+04	6.1e+03	3.1e+04	1.8e+04	5.2e+03	5.6e+03
f_{18}	1.0e-01	2.8e-01	1.0e-01	2.8e-01	9.8e-02	1.4e-02	1.9e-01	3.1e-02	2.0e-01	1.7e-02	2.4e-01	5.1e-01
f_{19}	0.0	0.0	0.0	0.0	0.0	0.0	0.0	0.0	0.0	0.0	2.3e-02	1.0e-01

The algorithms we compared have a number of parameters that affect both accuracy and parallelism. "Manual" parameter tuning is time consuming and may introduce a bias in comparing an algorithm with a reference, due to better knowledge of the algorithm under consideration and to possible different time spent tuning each of them. Therefore, the automatic *tuning* of all three algorithms was performed using the `irace` software package [10], to find the configurations that yielded the best results in a given time: we set this time to one second, since it is generally short enough to avoid reaching full convergence with all three methods, allowing one to compare their short-term performances.

The tuner was run on all 20 functions with a budget of 30000 experiments, each being one run of one configuration on one function with a termination criterion of one second. Since the functions have different fitness ranges, a rank-based test is prferable to a test based on the solutions' mean values. Accordingly, the Friedman test was used to discard significantly worse configurations. We tuned the parameters for 30-dimensional problems, and assumed that such configurations are good also for lower-sized ones. Table 1 displays the parameters that have been tuned for each algorithm, and the best corresponding values.

We compared our results to the values that are most commonly used in literature. For instance, the authors in [2] suggest $F \in (0.4, 0.95)$ and $Cr \in (0.9, 1)$ for multimodal separable functions (the most common ones in our benchmark); we obtained similar results. Regarding PSO, in most papers, $c_1 = c_2 = 2.0$ [15], while our automatic tuning set them to slightly smaller values.

To evaluate both the effectiveness and the efficiency of the three parallel implementations, tests on 20 numerical benchmark functions (see Table 3) were run on a 64-bit Intel(R) Core^{TM} i7 CPU running at 2.67GHz using CUDA^{TM} v. 4.1 on a nVidia GeForce GTS450 graphics card with 1GB of DDR memory and compute capability 2.1 [14]. Table 2 reports the results obtained executing 100 runs per function (6000 independent runs) and setting 1 second as the only

Table 3. Benchmark functions. For every function, the table shows the name, the range of the search space, the formula, the multimodality (multimodal, unimodal) and the separability (separable, non separable). All minima are in $\{0\}^n$.

	Name	Range	Formula		
f_0	Sphere	$[-100, 100]^n$	$\sum_{i=0}^{n-1} x_i^2$	U	S
f_1	Elliptic	$[-100, 100]^n$	$\sum_{i=0}^{n-1} (10^6)^{\frac{i-1}{D-1}} x_i^2$	U	S
f_2	Sum of Squares	$[-1, 1]^n$	$\sum_{i=0}^{n-1} i x_i^2$	U	S
f_3	HyperEllipsoid	$[-1, 1]^n$	$\sum_{i=0}^{n-1} i^2 \cdot x_i^2$	U	S
f_4	Schwefel 2.22	$[-10, 10]^n$	$\sum_{i=0}^{n-1} \lvert x_i \rvert + \prod_{i=0}^{n-1} \lvert x_i \rvert$	U	S
f_5	Zakharov	$[-10, 10]^n$	$\left(\sum_{i=0}^{n-1} x_i^2\right) + \left(\sum_{i=0}^{n-1} 0.5 \cdot i \cdot x_i^2\right)^2 +$ $+ \left(\sum_{i=0}^{n-1} 0.5 \cdot i \cdot x_i^2\right)^4$	U	S
f_6	Schwefel 1.2	$[-100, 100]^n$	$\sum_{i=0}^{n-1} \left(\sum_{j=0}^{i} x_j\right)^2$	U	NS
f_7	Schwefel 2.6	$[-100, 100]^n$	$max\{\mathbf{A}_i\mathbf{x} - \mathbf{B}\}$, $i = 0, \ldots, n-1, \mathbf{x} = [x_0, \ldots, x_{n-1}]$, $\mathbf{A}_i, \mathbf{B}$ defined in [17].	U	NS
f_8	Dixon-Price	$[-10, 10]^n$	$(x_0 - 1)^2 + \sum_{i=1}^{n-1} \left(i \cdot \left(2x_i^2 - x_{i-1}\right)^2\right)$	U	NS
f_9	Rastrigin	$[-5.12, 5.12]^n$	$\sum_{i=0}^{n-1} \{x_i^2 - 10 \cdot \cos(2\pi x_i) + 10\}$	M	S
f_{10}	Schwefel 2.26	$[-500, 500]^n$	$418.9829 \cdot n + \sum_{i=0}^{n-1} \left(x_i \cdot \sin\sqrt{\lvert x_i \rvert}\right)$	M	S
f_{11}	Katsuura	$[-1000, 1000]^n$	$\prod_{i=0}^{n-1} \left(1 + (i+1)\sum_{k=1}^{d} round(2^k x_i)2^{-k}\right) - 1$	M	S
f_{12}	Griewank	$[-600, 600]^n$	$\sum_{i=0}^{n-1} \frac{x_i^2}{4000} - \prod_{i=0}^{n-1} \cos(\frac{x_i}{\sqrt{i}}) + 1$	M	NS
f_{13}	Rosenbrock	$[-100, 100]^n$	$\sum_{i=0}^{n-1} 100(x_i - x_{i-1}^2)^2 + (1 - x_{i-1})^2$	M	NS
f_{14}	Ackley	$[-32, 32]^n$	$-20e^{-0.2\sqrt{\frac{1}{n}\sum_{i=0}^{n-1} x_i^2}} - e^{\frac{1}{n}\sum_{i=0}^{n-1} \cos(2\pi x_i)} + 20 + e$	M	NS
f_{15}	Griewank + Rosenbrock	$[-5.12, 5.12]^n$	$f_{griewank}(f_{rosenbrock})$	M	NS
f_{16}	Scaffer	$[-100, 100]^n$	$\sum_{i=0}^{n-1} F(x_i, x_{i+1}), x_n = x_0$ where $F(x, y) = 0.5 + \frac{\sin^2\left(\sqrt{x^2+y^2}\right) - 0.5}{1 + 0.0001(x^2+y^2)}$	M	NS
f_{17}	Schwefel 2.13	$[-\pi, \pi]^n$	$\sum_{i=0}^{n-1} (\mathbf{A}_i - \mathbf{B}_i(\mathbf{x}))^2, \mathbf{x} = [x_0, \ldots, x_{n-1}]$ $\mathbf{A}_i, \mathbf{B}_i(x)$ defined as in [17].	M	NS
f_{18}	Salomon	$[-10, 10]^n$	$- \cos\left(2\pi\sqrt{\sum_{i=0}^{n-1} x_i^2}\right) + 0.1\sqrt{\sum_{i=0}^{n-1} x_i^2} + 1$	M	NS
f_{19}	Levy	$[-10, 10]^n$	$\sin^2(\pi y_0) + \sum_{i=0}^{n-2} \left[(y_i - 1)^2 \left(10\sin^2(\pi y_i + 1)\right)\right] +$ $(y_{n-1} - 1)^2 \left(1 + 10\sin^2(2\pi y_{n-1})\right)$ where $y_i = 1 + \frac{x_i - 1}{4}, i = 0, \ldots, n-1$	M	NS

termination criterion. The first column is the function under consideration. The following ones are divided into two blocks according to the number of dimensions (10 and 30). Within each block, the mean best fitness and the standard deviation over all runs are reported for each method. Results reported on a grey background highlight those case in which the median over 100 runs obtained by the method is significantly better than the other methods, according to the Kruskal-Wallis test, with a confidence level of 0.01.

5 Discussion

The results reported in Table 2 and Figure 2 allow one to draw some conclusions about the behaviour of the three parallel metaheuristics. Conforming with previous results obtained by sequential implementations, DE obtained the best results, sometimes tied with some other method, in 35 out of the 40 experiments performed, while PSO was the best method, sometimes tied with some other method, in 20 out of 40 functions, its main drawback being its tendency to stagnate and find sub-optimal solutions more often than DE, even if a higher number of function evaluations is run. Regarding SS, whose first parallel implementation is presented here, it obtained the best result in 12 out of 40 problems; however, this metaheuristic, which is not as parallelizable as the other methods, as reflected by the number of kernels, has achieved better performance over multimodal non-separable problems and time-consuming fitness functions, like Katsuura.

All tests were run with a temporal limit of one second, a short time in which all three methods can generally obtain results close to the optima without reaching full convergence. Figure 2 shows that PSO requires almost three times as many fitness function evaluations as DE to converge on 30-dimensional problems. It is important to notice that the population size in PSO is also almost three times as large as in DE, which justifies the larger number of fitness evaluations. However, this may represent a shortcoming only if applied to larger-dimensional functions than those considered in this work.

Acknowledgments. Youssef S. G. Nashed, Pablo Mesejo and Jérémie Dubois-Lacoste are funded by the European Commission (Marie Curie ITN MIBISOC, FP7 PEOPLE-ITN-2008, GA n. 238819). Roberto Ugolotti is funded by Compagnia di S.Paolo and Fondazione Cariparma. The authors want to thank Thomas Stützle for his comments and Enrico Viappiani for the help in implementing CUDA^{TM} code.

References

1. Bonabeau, E., Dorigo, M., Theraulaz, G.: Swarm Intelligence: From Natural to Artificial Systems. Oxford (1999)
2. Das, S., Suganthan, P.: Differential Evolution: A Survey of the State-of-the-Art. IEEE Transactions on Evolutionary Computation 15(1), 4–31 (2011)
3. de Veronese, L., Krohling, R.: Swarm's flight: Accelerating the particles using C-CUDA. In: Proc. IEEE Congress on Evolutionary Computation, pp. 3264–3270 (2009)
4. de Veronese, L., Krohling, R.: Differential evolution algorithm on the GPU with C-CUDA. In: Proc. IEEE Congress on Evolutionary Computation, pp. 1–7 (2010)
5. Duarte, A., Martí, R., Glover, F., Gortázar, F.: Hybrid scatter tabu search for unconstrained global optimization. Annals of Operations Research 183(1), 95–123 (2011)
6. Eiben, A.E., Smith, J.E.: Introduction to Evolutionary Computing. Springer (2003)
7. Glover, F.: Heuristics for integer programming using surrogate constraints. Decision Sciences 8(1), 156–166 (1977)
8. Kennedy, J., Eberhart, R.: Particle Swarm Optimization. In: Proc. IEEE International Conference on Neural Networks, vol. 4, pp. 1942–1948 (1995)

9. Krömer, P., Snåšel, V., Platoš, J., Abraham, A.: Many-threaded implementation of differential evolution for the CUDA platform. In: Proc. 13th Annual Conference on Genetic and Evolutionary Computation, GECCO 2011, pp. 1595–1602. ACM (2011)

10. López-Ibáñez, M., Dubois-Lacoste, J., Stützle, T., Birattari, M.: The irace package, iterated race for automatic algorithm configuration. Technical Report TR/IRIDIA/2011-004, IRIDIA, Université Libre de Bruxelles, Belgium (2011)

11. Mussi, L., Daolio, F., Cagnoni, S.: Evaluation of parallel particle swarm optimization algorithms within the CUDA architecture. Information Sciences 181(20), 4642–4657 (2011)

12. Mussi, L., Nashed, Y.S.G., Cagnoni, S.: GPU-based asynchronous particle swarm optimization. In: Proc. 13th Annual Conference on Genetic and Evolutionary Computation, GECCO 2011, pp. 1555–1562. ACM (2011)

13. Nashed, Y.S.G., Ugolotti, R., Mesejo, P., Cagnoni, S.: libCudaOptimize: an Open Source Library of GPU-based Metaheuristics. In: Proc. Genetic and Evolutionary Computation Conference, GECCO 2012 (in press, 2012)

14. nVIDIA Corporation: nVIDIA CUDA Programming Guide v. 4.0. (2011)

15. Poli, R., Kennedy, J., Blackwell, T.: Particle swarm optimization. Swarm Intelligence 1(1), 33–57 (2007)

16. Storn, R., Price, K.: Differential Evolution - a simple and efficient adaptive scheme for global optimization over continuous spaces. Technical report, International Computer Science Institute (1995)

17. Suganthan, P.N., Hansen, N., Liang, J.J., Deb, K., Chen, Y., Auger, A., Tiwari, S.: Problem definitions and evaluation criteria for the CEC 2005 special session on real-parameter optimization. Natural Computing, 1–50 (2005)

18. Vesterstrom, J., Thomsen, R.: A comparative study of differential evolution, particle swarm optimization, and evolutionary algorithms on numerical benchmark problems. In: Proc. IEEE Congress on Evolutionary Computation, pp. 1980–1987 (2004)

19. Wets, F.J., Solis, R.J.: Minimization by random search techniques. Mathematics of Operations Research 6(1), 19–30 (1981)

20. Zhou, Y., Tan, Y.: GPU-based parallel particle swarm optimization. In: Proc. IEEE Congress on Evolutionary Computation, pp. 1493–1500 (2009)

21. Zhu, W.: Massively parallel differential evolution–pattern search optimization with graphics hardware acceleration: an investigation on bound constrained optimization problems. Journal of Global Optimization 50(3), 417–437 (2011)

The Effect of the Set of Low-Level Heuristics on the Performance of Selection Hyper-heuristics

M. Mısır[1,2], K. Verbeeck[1,2], P. De Causmaecker[2], and G. Vanden Berghe[1,2]

[1] CODeS, KAHO Sint-Lieven
{mustafa.misir,katja.verbeeck,greet.vandenberghe}@kahosl.be
[2] CODeS, Department of Computer Science, KU Leuven Campus Kortrijk
patrick.decausmaecker@kuleuven-kortrijk.be

Abstract. The present study investigates the effect of heuristic sets on the performance of several selection hyper-heuristics. The performance of selection hyper-heuristics is strongly dependant on low-level heuristic sets employed for solving target problems. Therefore, the generality of hyper-heuristics should be examined across various heuristic sets. Unlike the majority of hyper-heuristics research, where the low-level heuristic set is considered given, the present study investigates the influence of the low-level heuristics on the hyper-heuristic's performance. To achieve this, a number of heuristic sets was generated for the patient admission scheduling problem by setting the parameters of a set of parametric heuristics with specific values. These values were set such that nine heuristic sets with different improvement capabilities, speed characteristics and size were generated. A group of hyper-heuristics with certain selection mechanisms and acceptance criteria having dissimilar intensification/diversification abilities were taken from the literature enabling a comprehensive analysis. The experimental results indicated that different hyper-heuristics perform superiorly on distinct heuristic sets. The results can be explained and hence result in hyper-heuristic design recommendations.

Keywords: Hyper-heuristics, Heuristic Set, Generality.

1 Introduction

Selection hyper-heuristics have been studied to effectively manage multiple algorithms, with the motivation behind their employment being to use the heuristics' strengths and eliminating their weaknesses, resulting in a better performance [1]. They take the search process to the heuristic level and perform without problem domain knowledge. Thus, hyper-heuristics are considered general algorithms capable of solving a diverse range of problems. Therefore, most of the hyper-heuristic studies in the literature deal with problem solving [2]. However, selection hyper-heuristics are not concerned with solving some problem instances, but with managing low-level heuristic sets for solving these instances as efficiently

C.A. Coello Coello et al. (Eds.): PPSN 2012, Part II, LNCS 7492, pp. 408–417, 2012.
© Springer-Verlag Berlin Heidelberg 2012

as possible. There are a limited number of studies concentrating on the heuristic set part with heuristic set reduction or heuristic elimination strategies forming the basis of these studies. In [3], heuristics were made tabu for certain iterations based on their performance. A similar tabu idea was employed for a genetic algorithm based hyper-heuristic in [4]. A heuristic subset selection mechanism after a certain number of iterations and a heuristic set reduction strategy that excludes heuristics in time to determine a good heuristic subset from a large heuristic set were studied in [5]. Another heuristic subset selection approach which temporarily eliminates poor performing heuristics was introduced in [6]. Contrastingly, in [7], different heuristic sets using multiple heuristics from the low-level heuristics for solving the DNA sequencing problem were tested with a suite of hyper-heuristics.

Due to the search level and problem-independent nature of selection hyper-heuristics, *generality* is considered their most important trait. To show their generality level, work on different heuristic sets, with differing features, is required. To the best of our knowledge, there is no study focusing on the effect of different heuristic sets on the performance of hyper-heuristics. The present contribution aims at filling the void mentioned by trying to determine what features should be considered in the design phase of a hyper-heuristic from a generality perspective. For this purpose, 11 low-level heuristics designed to solve the patient admission scheduling problem were used to generate nine heuristic sets. These heuristic sets exhibit differences regarding their improvement capabilities and the speed of the residing heuristics as well as the number of utilised heuristics. Two heuristic selection mechanisms together with seven move acceptance criteria were adopted in building 14 hyper-heuristics with distinct characteristics relying on their selection strategies and intensification/diversification capabilities. The computational results clearly indicated that the nature of the heuristics, distribution of different heuristic types, size of the heuristic sets and runtime limitations have a remarkable impact on the performance of hyper-heuristics.

In the remainder of the paper, the low-level heuristics for the patient admission scheduling problem and heuristics sets generated based on these heuristics are argued in Section 2. Followingly, Section 3 elaborates the tested hyper-heuristics. Next, the computational results are presented and discussed in Section 4. In the last section, the paper is concluded and the requirements for generality and the future research opportunities are presented.

2   Patient Admission Scheduling Problem and Heuristics

The present study focuses on patient admission scheduling (PAS) due to its combinatorial complexity as well as the existence of a set of heuristics. The PAS problem concerns assigning patients to hospital rooms or beds based on the patients' requirements [8]. The basic components of the problem are: patients, rooms, wards and time slots. Each patient is characterised by his/her gender, age, pathology, room preference, admission date, and duration of the treatment. It is assumed that every pathology can be linked to one of the hospital's specialisms. Multiple wards of the hospital can have the same specialism, but some wards are more specialized than others. We distinguish between major and minor specialisms. Every room is located on a ward of the hospital. The specialisms of the ward are inherited to a certain degree by the rooms of the ward. A room is characterized by its properties and its bed capacity. Depending on the patient's pathology, some of the room properties are mandatory or preferable.

2.1 Low-Level Heuristics

The following 11 simple low-level heuristics were used in the experiments. For this comparison study, it does not actually matter how this set of low-level heuristics is composed.

- LLH_1: Swap all the bed assignments of a randomly selected patient with the beds of randomly selected patients
- LLH_2: Transfer all the bed assignments of a randomly selected patient to randomly selected empty beds
- LLH_3: Swap all the bed assignments between two randomly selected patients
- LLH_4: Swap all the bed assignments of a randomly selected patient with randomly selected occupied beds. Transfer the remaining assignments to the randomly selected beds
- LLH_5: Transfer all the bed assignments of a randomly selected patient to a randomly selected empty bed
- LLH_6: Swap a randomly selected bed assignment with another bed while respecting room properties
- LLH_7: Swap a randomly selected bed assignment with another bed while respecting room preferences of the corresponding patient
- LLH_8: Swap a randomly selected bed assignment with another bed while respecting the room specialism
- LLH_9: Swap a randomly selected bed assignment of a randomly selected patient with another bed while respecting room properties
- LLH_{10}: Swap two randomly selected beds
- LLH_{11}: Transfer all the patients in a randomly selected room to another randomly selected room

2.2 Differentiating Heuristic Sets

The motivation here is to generate a group of heuristic sets using the aforementioned parametric low-level heuristics for PAS. Nine heuristic sets in different sizes, with different speed and improvement capabilities were generated by setting their parameters. There exist studies concerning heuristics requiring their parameters be set when applying a number of atomic steps [8]. Similarly, in the present research, each heuristic has a parameter called *sampling factor*. This parameter constitutes the number of steps to apply the same operator for different neighbouring solutions. For instance, LLH_3 with sampling factor 4 means that it should perform the corresponding swap operation 4 times at each iteration.

Heuristic Sets. Nine heuristic sets under three group headings were derived based on the 11 parametric heuristics depicted in Table 1. The first group of heuristic sets was composed of 11 heuristics with a sampling factor of four. The aim of using this first group is to measure the performance of various hyper-heuristics in a default setting. The second set involves 22 heuristics with two versions of each heuristic with sampling factors 4 and 1000. The heuristics with sampling factor 1000 is 250 times slower than the ones with sampling factor 4. It enables investigating how a hyper-heuristic behaves when the speed difference among heuristics is extremely large. The last group employs 44 heuristic with four versions of each heuristic using sampling factor values 1,4,8,16. The reasoning behind this setting is to evaluate hyper-heuristics on larger heuristic sets with relatively small speed differences.

Table 1. Heuristic sets used for the experiments

	Set size	Sampling factors	Selection type
HS_1	11	4	BEST
HS_2	11	4	FIRST_IMPROVING
HS_3	11	4	HILL_CLIMBER
HS_4	22	4, 1000	BEST
HS_5	22	4, 1000	FIRST_IMPROVING
HS_6	22	4, 1000	HILL_CLIMBER
HS_7	44	1,4,8,16	BEST
HS_8	44	1,4,8,16	FIRST_IMPROVING
HS_9	44	1,4,8,16	HILL_CLIMBER

Each of these heuristic set groups was tested under three different conditions. The first method, $BEST$, returns the best neighbouring solution after all the sample solutions were visited at each iteration. The second approach, $FIRST\_IMPROVING$, uses the first improving solution found after the sampling operations. The last technique, $HILL\_CLIMBER$, generates hill climbers based on the sampling factor value. Whenever a better or equal quality neighbouring solution is found during the sampling period, it is accepted.

Figure 1 depicts the average speed of performing one move on a PAS instance by each heuristic set. According to this metric, the heuristic sets with 22 heuristics, i.e. HS_4, HS_5, HS_6, are slower than the all others. This slowness is caused by utilising heuristics with a sampling factor of 1000. Of these, HS_4 and HS_6 are the slowest, as shown in the second graph. This severe speed difference occurs when a heuristic with the sampling factor of 1000 always checks 1000 neighbouring solutions, however, HS_5 stops looking for better neighbouring solutions whenever it finds an improving one.

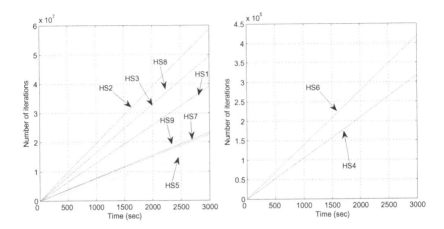

Fig. 1. Number of iterations spent over time when heuristics are randomly selected

3   Tested Hyper-heuristics

Fourteen selection hyper-heuristics (2 heuristic selection × 7 move acceptance) involving mechanisms from the literature were used for the experiments. Two selection criteria were employed for the heuristic selection process. The first approach is the simple random (SR) heuristic selection mechanism that chooses heuristics in a uniformly random manner. The second approach is the adaptive dynamic heuristic set (ADHS) strategy, determining effective heuristic subsets at runtime [9]. This strategy was also used in the winning hyper-heuristic [10] of the first international Cross-domain Heuristic Search Challenge (CHeSC 2011)[1]. The heuristic subset selection process is carried out using a performance metric involving the most relevant elements to evaluate the online behaviour of the heuristics. The details of the performance metric for heuristic i are shown in Equation 1. $C_{p,best}(i)$ represents the number of new best solutions discovered during a phase. $f_{p,imp}(i)$ shows the total amount of improvement provided during a phase. $f_{p,wrs}(i)$ indicates the total worsening caused during a phase. $f_{imp}(i)$ and $f_{wrs}(i)$ both refer to the same measurements as the last two, but during the whole search rather than a single phase. The remaining elements were used to combine the improvement capabilities of the heuristics with their speed enabling better judgement. t_{remain} denotes the remaining execution time to finalise the whole search process. $t_{p,spent}(i)$ and $t_{spent}(i)$, the former represents the spent execution time during a phase and the latter, from the start. The w_j values are set as weights to differentiate the importance of each individual performance element. It is more important to have a higher value for an earlier element.

---

[1] http://www.asap.cs.nott.ac.uk/external/chesc2011/

$$p_i = w_1 \left[\left(C_{p,best}(i) + 1 \right)^2 \left(t_{remain}/t_{p,spent}(i) \right) \right] \times b +$$
$$w_2 \left(f_{p,imp}(i)/t_{p,spent}(i) \right) - w_3 \left(f_{p,wrs}(i)/t_{p,spent}(i) \right) +$$
$$w_4 \left(f_{imp}(i)/t_{spent}(i) \right) - w_5 \left(f_{wrs}(i)/t_{spent}(i) \right) \tag{1}$$
$$b = \begin{cases} 1, & \sum_{i=0}^{n} C_{p,best}(i) > 0 \\ 0, & otw. \end{cases}$$

After a number of iterations, all the active heuristics are evaluated based on this performance metric. The length of these iterations is denoted as phase length, pl. The heuristics with comparatively poorer performance are excluded from the heuristic set for a number of phases. The duration of exclusion is referred to as tabu duration (d) and is set to $d = \sqrt{2n}$, where n refers to the number of heuristics in the heuristic set. The tabu duration of the consecutively excluded heuristics is increased by one, until $2\sqrt{2n}$. If such a heuristic survives after a phase, its tabu duration is set to its initial value. In addition, the phase length is adapted during runtime with respect to the speed of the heuristics in the current heuristic set.

The heuristic selection operation from these subsets is handled using a learning automaton [9,11]. This method accommodates a vector of heuristic selection probabilities. These values are reset at the end of each phase given the synchronously performed update operation in determining which heuristics will be excluded.

The employed move acceptance strategies are as follows: adaptive iteration limited list-based threshold accepting (AILLA) [9], great deluge (GD) [8], late acceptance (LATE) [12], simulated annealing (SA) [8], improving or equal (IE), only improving (OI) and all moves (AM). All of these acceptance mechanisms immediately accept improving solutions. The first four acceptance methods, AILLA, GD, LATE and SA, provide diversification mechanisms by accepting worsening solutions with respect to certain dynamic threshold values. IE accepts equal quality solutions to diversify the search process. OI accepts only better quality solutions, hence it has no diversification strategy. The last acceptance criterion, AM, accepts all visited solutions.

The resulting hyper-heuristics using all these sub-mechanisms, ADHS-AILLA, ADHS-GD, ADHS-SA, ADHS-LATE, ADHS-IE, ADHS-OI, ADHS-AM, SR-AILLA, SR-GD, SR-SA, SR-LATE, SR-IE, SR-OI, SR-AM, have distinct characteristics for selecting heuristics and diversifying the search process such that a comprehensive performance analysis can be performed.

4   Computational Results

14 hyper-heuristics were run 10 times on 7 PAS instances, $dataset0 \rightarrow dataset6$, attainable at http://allserv.kahosl.be/~peter/pas/ using Pentium Core 2 Duo 3 GHz PCs with 3.23 GB memory. Each hyper-heuristic was tested on 9 different heuristic sets. The time limits were taken as 10 and 50 minutes.

In Figure 2, the significantly best hyper-heuristics for each heuristic set are listed for the 10 minutes and 50 minutes experiments respectively. The significance of the performance difference is evaluated using the Wilcoxon test with a 95% confidence interval.

Regarding the 10 minutes experiments, different acceptance strategies deliver similar performances after 10 minutes of execution in most of the cases. For specific heuristic sets, even very simple acceptance mechanisms like OI and AM can find similar results to those in HS_3 and HS_6. The explanation behind this is as follows: the diversification characteristics of the selection mechanisms are no longer useful due to the heuristics' hill climbing behaviour. In addition, there is no heuristic set for which one hyper-heuristic outperforms the others except SR-LATE on HS_9. Moreover, the hyper-heuristics involving an acceptance mechanism with diversification and ADHS perform poorly on HS_9. For the majority of the test cases, AILLA and LATE perform better, yet there is no general statistically significant performance difference. For 50 minute experiments, the hyper-heuristics with GD perform best together with different hyper-heuristics on different heuristic sets. This can be considered as an effect of the execution time limit increase, from 10 minutes to 50 minutes. The hyper-heuristics using AILLA and LATE also show effective performance after running them for 50 minutes.

The hyper-heuristics generated the best results on HS_1, HS_2, HS_7 and HS_8. The heuristic sets involve heuristics using low sampling factors with a selection type of either $BEST$ or $FIRST\_IMPROVING$. This means that the fast heuristics with well balanced intensification-diversification behaviour resulted in better performance on the tested problem instances.

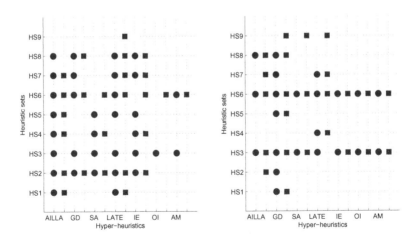

Fig. 2. The significantly best hyper-heuristics on each heuristic set after 10 minutes (left) and 50 minutes (right) (Circles refer to the hyper-heuristics with ADHS and squares refer to the hyper-heuristics with SR)

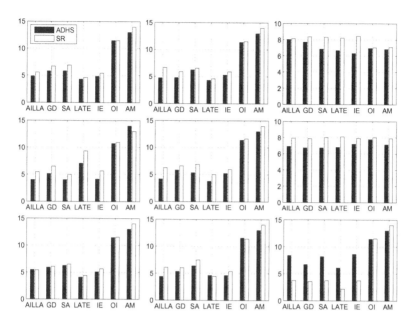

Fig. 3. Average ranking of the hyper-heuristics after 10 minutes (Each graph represents the results obtained on a heuristic set. They are ordered from left to right, top to bottom: $HS_1 \rightarrow HS_9$).

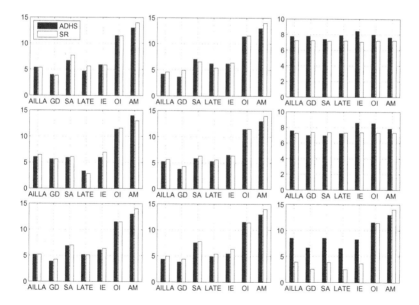

Fig. 4. Average ranking of the hyper-heuristics after 50 minutes (Each graph represents the results obtained on a heuristic set. They are ordered from left to right, top to bottom: $HS_1 \rightarrow HS_9$).

In Figure 3 and 4, the average ranking of the hyper-heuristics on each heuristic set with 10 and 50 minutes execution time limits are presented. For the 10 minute experiments, ADHS performs better than SR in the majority of the cases. However, SR provides better performance than ADHS for HS_9 that accommodates hill climbers, since ADHS mostly excludes the heuristics with a sampling factor of 1 that help to diversify the search. Consequently, the remaining heuristics may not be able to escape from certain local optima. For the 50 minute experiments, the performance difference between ADHS and SR degraded when compared to the 10 minute case. The two main reasons behind this empirical result are the low evolvability characteristic of the solution space and the longer running time. In this case, choosing wrong heuristics is not as influential when compared with the 10 minute execution time experiments.

5   Conclusion

The present study examined the performance changes of 14 selection hyper-heuristics due to different heuristic sets for the patient admission scheduling problem. The nature of the heuristic sets, size of the heuristic sets and other related limitations are all potential reasons why one hyper-heuristic delivers superior results. We reviewed these conditions in various experiments to identify the hyper-heuristics' generality levels. Nine heuristic sets were utilised in demonstrating the effect of the heuristic sets on the performance of selection hyper-heuristics. Each of these sets exhibits differences depending on the aforementioned experimental conditions. We tested, using these heuristics sets, 14 hyper-heuristics composed of an adaptive and a random selection mechanism combined with 7 move acceptance methods from the literature. The computational results on the tested heuristic sets showed that the best hyper-heuristic can change based on the heuristic set used and execution time limits employed. Particularly fast heuristic sets involving certain degree of intensification and diversification features showed better performance. These results also indicated that some of the hyper-heuristic components are more valuable than others under certain conditions. If the gap between speed and improvement capabilities of the low-level heuristics is large and the allowed execution time is short, then the heuristic selection is more important. However, if the heuristics are highly perturbative and destructive then a move acceptance strategy with effective diversification capabilities is a vital requirement. Also, a very naive acceptance mechanism like AM can deliver comparable results if the low-level heuristics have effective improvement capabilities. We have then demonstrated how by addressing different generality requirements.

In future work, the diversity of the application domains will be extended to show the performance changes with respect to the relation between heuristic search space and solution space. Additional mechanisms will be investigated to enable more general hyper-heuristics compared to traditional approaches composed of selection-acceptance pairs. Finally, a method will be devised to predict or measure the generality level of a hyper-heuristic.

References

1. Burke, E., Hart, E., Kendall, G., Newall, J., Ross, P., Schulenburg, S.: Hyper-Heuristics: An Emerging Direction in Modern Search Technology. In: Handbook of Meta-Heuristics, pp. 457–474. Kluwer Academic Publishers (2003)
2. Burke, E., Gendreau, M., Hyde, M., Kendall, G., Ochoa, G., Ozcan, E., Qu, R.: Hyper-heuristics: A survey of the state of the art. Journal of the Operational Research Society (to appear)
3. Burke, E., Kendall, G., Soubeiga, E.: A tabu-search hyper-heuristic for timetabling and rostering. Journal of Heuristics 9(3), 451–470 (2003)
4. Han, L., Kendall, G.: An investigation of a tabu assisted hyper-heuristic genetic algorithm. In: Proceedings of the IEEE Congress on Evolutionary Computation, CEC 2003, vol. 3, pp. 2230–2237 (2003)
5. Chakhlevitch, K., Cowling, P.I.: Choosing the Fittest Subset of Low Level Heuristics in a Hyperheuristic Framework. In: Raidl, G.R., Gottlieb, J. (eds.) EvoCOP 2005. LNCS, vol. 3448, pp. 23–33. Springer, Heidelberg (2005)
6. Misir, M., Verbeeck, K., De Causmaecker, P., Vanden Berghe, G.: Hyper-heuristics with a dynamic heuristic set for the home care scheduling problem. In: Proceedings of the IEEE Congress on Evolutionary Computation, CEC 2010, Barcelona, Spain, July 18-23, pp. 2875–2882 (2010)
7. Blazewicz, J., Burke, E., Kendall, G., Mruczkiewicz, W., Oguz, C., Swiercz, A.: A hyper-heuristic approach to sequencing by hybridization of DNA sequences. Annals of Operations Research, 1–15 (2011)
8. Bilgin, B., Demeester, P., Misir, M., Vancroonenburg, W., Vanden Berghe, G.: One hyperheuristic approach to two timetabling problems in health care. Journal of Heuristics 18(3), 401–434 (2012)
9. Misir, M., Verbeeck, K., De Causmaecker, P., Vanden Berghe, G.: A New Hyper-heuristic Implementation in HyFlex: a Study on Generality. In: Fowler, J., Kendall, G., McCollum, B. (eds.) Proceedings of the 5th Multidisciplinary International Scheduling Conference: Theory & Applications, MISTA 2011, Phoenix/Arizona, USA, August 10-12, pp. 374–393 (2011)
10. Misir, M., Verbeeck, K., De Causmaecker, P., Vanden Berghe, G.: An intelligent hyper-heuristic framework for CHeSC 2011. In: The 6th Learning and Intelligent Optimization Conference (LION 6). LNCS (to appear, 2012)
11. Misir, M., Wauters, T., Verbeeck, K., Vanden Berghe, G.: A hyper-heuristic with learning automata for the traveling tournament problem. In: The 8th Metaheuristics International Conference on Metaheuristics: Intelligent Decision Making, Post Conference Volume (2011)
12. Demeester, P., De Causmaecker, P., Vanden Berghe, G.: A general approach for exam timetabling: a real-world and a benchmark case. In: Proceedings of the 8th International Conference on the Practice and Theory of Automated Timetabling (PATAT 2010), Belfast, Northern Ireland, August 10-13 (2010)

Adaptive Evolutionary Algorithms and Extensions to the HyFlex Hyper-heuristic Framework

Gabriela Ochoa[1], James Walker[2], Matthew Hyde[2], and Tim Curtois[2]

[1] Department of Computing Science and Mathematics, University of Stirling, UK
[2] School of Computer Science, University of Nottingham, UK

Abstract. HyFlex is a recently proposed software framework for implementing hyper-heuristics and domain-independent heuristic optimisation algorithms [13]. Although it was originally designed to implement hyper-heuristics, it provides a population and a set of move operators of different types. This enable the implementation of adaptive versions of other heuristics such as evolutionary algorithms and iterated local search. The contributions of this article are twofold. First, a number of extensions to the HyFlex framework are proposed and implemented that enable the design of more effective adaptive heuristics. Second, it is demonstrated that adaptive evolutionary algorithms can be implemented within the framework, and that the use of crossover and a diversity metric produced improved results, including a new best-known solution, on the studied vehicle routing problem.

1 Introduction

A hyper-heuristic is *a search method or learning mechanism for selecting or generating heuristics to solve computational search problems* [6]. The main motivation is to develop automated search methodologies with higher generalisation abilities, which will potentially increase their application in practice. The HyFlex (Hyper-heuristic Flexible) framework [13] has been recently proposed to assist researchers in hyper-heuristics and autonomous search control. HyFlex consists of two parts. First, a Java programming interface for hyper-heuristics, which splits the heuristic search process into two modules. One module contains the problem-specific algorithm components and other contains the problem-independent components. Second, a library of *ready-to-use* problem domain modules covering hard combinatorial optimisation problems with a rich variety of search operators and including real-world industrial data. Two important antecedents of the HyFlex framework are the domain-barrier hyper-heuristic conceptual framework [8], and the PISA software framework [3].

Currently, six problem domain modules are implemented in HyFlex (which can be downloaded from the CHeSC 2011 website [1]). These are the original four test domains: permutation flow shop, one-dimensional bin packing, maximum satisfiability and personnel scheduling; and the two additional domains used for

C.A. Coello Coello et al. (Eds.): PPSN 2012, Part II, LNCS 7492, pp. 418–427, 2012.

the competition: traveling salesman and vehicle routing. HyFlex was used to support an international research competition: the first Cross-Domain Heuristic Search Challenge [1] that received significant international attention.

HyFlex was initially designed to support research within hyper-heuristics. However, since the framework provides search operators of different types (mutation, crossover, ruin-recreate and hill-climbing) approaches not traditionally identified as hyper-heuristics can be implemented using the framework. For example, several adaptive implementations of iterated local search (ILS) within HyFlex have been published recently [4,5,17,7]. Indeed, the algorithms ranking 2nd and 3rd in the 2011 competition can be seen as adaptive ILS methods. These approaches can be considered as hyper-heuristics as they operate in a domain independent fashion, using limited information from the search process and following a modular design. Moreover, they coordinate the effort of several move operators and local search heuristics.

The contributions of this paper are twofold. First, we describe number of extensions to the HyFlex framework that will enable the implementation of more robust and effective adaptive search heuristics. Second, we extend a previous adaptive ILS hyper-heuristic [17], which is a single-point search approach, by incorporating a population and the use of crossover heuristics. This brings hyper-heuristics close to adaptive memetic algorithms [14]. These two approaches have developed independently, but they share several features. In particular, they need to provide adaptive mechanisms to autonomously guide the choice of operators (or memes) during the search. These mechanisms have been also studied within the evolutionary computation community using the term *adaptive operator selection* [9,12].

The next section overviews the proposed extensions to the HyFlex interface, while section 3 describes their implementation within a selected problem domain: vehicle routing. Section 4 describes an empirical study illustrating that: (i) adaptive memetic algorithms can be successfully implemented within the HyFlex framework, and (ii) the distance metric incorporated in HyFlex can be used to implement state-of-the-art adaptive operator selection mechanisms. Finally, section 5 summarises our main findings and discusses routes for future research.

2   Extensions to the HyFlex Interface

Providing additional feedback information from the search process would improve the robustness and effectiveness of adaptive search heuristics. Below we discuss the proposed extensions to the HyFlex interface, including their motivation and an indication of which types of approaches may benefit from these extensions.

Distance between Solutions: An important source of feedback for population-based algorithms is an indication of the genotypic diversity in the population. Moreover, recently proposed adaptive operator selection mechanisms rely on the population diversity as a source of feedback [12]. In order to calculate the diversity of a population, a distance metric between solutions is needed. Therefore, the HyFlex interface is extended with the following two methods:

```
double getMaxDistance()
double solutionDistance(int solutionIndex1, int solutionIndex2)
```

We assume that the minimum distance between two solutions is zero, and that this occurs when they are exactly the same. Since different representations require different distance metrics and measurement ranges, the method `getMaxDistance` returns the maximum possible distance max_d between two solutions. The method `solutionDistance` returns a value between 0 and max_d representing the distance between the two solutions in the memory of solutions as indicated by the input indices.

Solution Metrics and Alternative Objective Functions: Heuristic search approaches that dynamically modify the fitness function in order to escape local optima or fitness plateaus can be found in the metaheuristics and artificial intelligence literature [2]. Moreover, a recently published hyper-heuristic approach [18], declared the winner of a computational search competition to solve the Eternity II Puzzle, employs alternative fitness functions in order to guide the search. Guided local search (GLS) proposes augmenting the objective function with a set of penalty terms on a set of *solution features* [16]. A solution feature is a non-trivial property of the solution and a cost is associated to each feature. We borrow and extend this concept in HyFlex, instead of feature, we use the term *metric* to refer to additional costs or objectives associated to a given solution. The two following two methods are included:

```
int getNumberOfMetrics()
double getMetric(int solutionIndex, int metricIndex)
```

Where the first method returns the number of solution metrics, and the second gets the value of the given metric for the indicated solution in memory. These metrics can then be used by the hyper-heuristic designer to implement their own alternative objective functions to guide the search.

Additional Instances: The version of HyFlex used in the 2011 competition contains 12 instances for the test domains (the 10 training instances and 2 additional hidden instances), and 10 instance for the new (hidden) domains. Moreover, this instance data is included within the software, and there is no flexibility for adding new instances. Having additional instances will both improve the development of robust online strategies, and facilitate the implementation of offline configuration techniques. An approach based on offline learning for algorithm selection obtained surprisingly good results in the 2011 challenge [11]. This is very promising, as the challenge was designed to encourage online approaches to heuristic selection. To enable the incorporation of additional instances the method: `void loadInstanceFromFile (String fileName)` is included in HyFlex, which loads the instance indicated in the file and set is as the current instance. The file needs to have the correct format, which will be included in the domain documentation.

Additional Utilities: Utilities for saving and retrieving solutions from files may facilitate both the reuse of previously found solutions and the analysis of previous runs. The two methods below are included:

```
void loadSolutionFromFile(String fileName, int solutionIndex)
void SolutionToFile(String fileName, int solutionIndex)
```

Where `solutionIndex` refers to the position in the memory of solutions, and `fileName` to the name of the source or destination file.

3   The Extended Vehicle Routing Domain

The vehicle routing problem with time windows involves satisfying the demand of a set of customers, using the fewest possible vehicles, and adhering to all constraints such as time windows, whereby a customer must be served between two points in time. Each vehicle starts from the same point, the depot. A route consist of a list of locations. The HyFlex VRP problem domain [17] provides 12 search operators including: 4 mutation, 2 ruin-recreate, 4 hill-climber and 2 crossover heuristics. The objective function balances the dual objectives of minimising the number of vehicles, and minimising the total distance travelled. Due to space constraints we refer the reader to [17] for a complete description. We concentrate here on the problem domain extensions.

Distance Metric: We implemented a distance metric suggested in [10], which is based on a concept formulated for the travelling salesman problem. The metric considers the number of common edges between two solutions. For the vehicle routing problem, an edge represents an undirected link between two locations. The distance metric produces a value between 0 and 1 and the formula is as follows: $distance = \frac{totalEdges - commonEdges}{totalEdges}$.

Solution Features: The solution features provided are: (1) the default objective function, which is a weighed sum of the number of routes and the distance traveled,(2) the number of routes or vehicles, (3) the total distance traveled, and (4) the distance of the shortest route.

Instance File Format: The instance format is the Solomon format. The instance file provides the number of customers and vehicle capacity. This is followed by a list of customers with he following attributes: (1) customer number, (2) X co-ordinate, (3) Y co-ordinate, (4) demand, (5)ready time, (6) due date, (7) service time.

4   Empirical Study

4.1   Algorithms

Two classes of algorithms are considered: adaptive iterated local search and adaptive memetic algorithms. These algorithms adapt the probabilities associated to the available search operators, according to the history of their performances. The operators are then selected according to these learned probabilities using a roulette wheel mechanism. Since HyFlex provides several operators belonging to different classes: mutation, ruin-recreate, crossover and hill-climbing; several adaptive mechanisms may be required for selecting different operators at different parts of an algorithm framework. This study consider two variants of each algorithm class, which differ on the feedback information used from the search process to adapt the choice of search operators. The first variant considers only the fitness function improvements or deteriorations obtained after applying the search operators, while the second is based on the *compass* mechanism [12], which considers a diversity metric and the running time of the operators in conjunction with their fitness variation as sources of feedback. The underlying idea behind the compass control mechanism is to provide an adequate exploration/exploitation balance. Thus, both diversity and quality are pertinent criteria to guide the search.

Adaptive Iterated Local Search: We consider the best performing algorithm proposed in [17], which is a multiple neighborhood ILS algorithm that includes adaptive mechanisms for both the perturbation and improvement stages. The perturbation stage selects among the set of available mutation and ruin-recreate heuristics using the *extreme value* [9] operator selection mechanism. The improvement stage considers the available hill-climbers and incorporates a simple adaptive mechanism, in which the operators are ordered according to learned propoabilities and sequentially applied using this order. We name this algorithm *AILS*. A new version of this algorithm is implemented, in which the extreme value mechanism is substituted by the by the compass mechanisms. We call this algorithm *AILS-C*.

Adaptive Memetic Algorithm: Our implementation of adaptive memetic algorithms works as follows (see Algorithm 1). A small population (of size 4) is generated and then goes through a recombination stage in which all possible recombination pairs are considered and a randomly selected crossover operator (from the available pool) is applied for each pair. From all these generated solutions the best four are kept. This is a costly stage and it is only invoked a number of times during the search process. A perturbation and improvement stage follows. For each member of the population, a mutation or ruin-recreate heuristic is selected from the pool according to operator probabilities learned using a simple reinforcement learning mechanism. The solution is thereafter improved by a hill-climbing heuristic. The improvement heuristic to apply is also selected according to learned probabilities. We call this algorithm *AMA*. A variant is also implemented in which the reinforcement learning mechanism used in the perturbation stage is substituted by the compass mechanism. We call this algorithm *AMA-C*.

4.2   Results

The experiments were conducted using the 10 VRP test instances currently available in the 2011 HyFlex software. These instances were originally taken

Algorithm 1. *Adaptive Memetic Algorithm (AMA).*

P = GenerateInitialPopulation
repeat
  P'= RecombinationStage(P)
  P'' = MutationAndImprovementStage(P')
  UpdatePerturbationOperatorProb
  UpdateImprovementOperatorProb
  P = SelectBest($P' + P'''$)
until time limit is reached

from [15] and include 5 instances from the Solomon data set and 5 from the Gehring and Homberger data set. Both data sets include three types of instances: **R**andom, **C**lustered, and **R**andom **C**lustered; according to the way in which the customers' locations are determined. Details about the instances can be found in the first three columns of Table 1. In the instance name, the first number indicates the HyFlex numbering, while the first letter whether it is a Solomom (S) or Homberger (H) instance. The second group of letters indicates the type of instance; and the final string corresponds to the identifier in the data set.

As a first test, we compared our two base adaptive algorithms AMA and $AILS$ against the best-performing algorithms for VRP in the 2011 competition and using the original HyFlex version. We considered the competition experimental setting, namely, 10 minutes per run, 31 runs per instance and the 5 competition instances. These are instances 1, 2, 5, 6 and 9. Since the instances have different objective function ranges, we selected ordinal data analysis to compare the algorithms. If m is the number of instances and n the number of competing algorithms. For each instance an ordinal value o_k is given representing the rank of the algorithm ($1 \leq o_k \leq n$). An algorithm having a rank o_k in a given instance is simply given o_k points, and the total score of an algorithm is the sum of its ranks o_k across the m instances (this metric is known as the *Borda* count). In this comparison, the number of instances $m = 5$ and the number of algorithms $n = 5$. Therefore, best possible score is 5, and the worst possible is 25. The ranks were calculated according to the median best objective function value across the 31 runs per instance. Figure 1 (a) illustrates the Borda counts for AMA, $AILS$ and the top 3 performing competitors in the 2011 challenge. Clearly, the AMA is the best performing algorithm, producing an almost perfect score.

The next set of experiments use the the new HyFlex VRP domain and the four algorithm variants described above, $AILS$, $AILS$-C, AMA and AMA-C. The whole set of 10 instances were used (see Table 1). The running time was set to 20 CPU minutes and 10 runs were conducted per instance. The machine running the tests has a 2.27 GHz Intel(R) Core(TM) i3 CPU and 4GB RAM. The Borda count is used for comparison and the median best objective function value is used for the ranking. This time we have $m = 10$ instances and $n = 4$ algorithms. Therefore, the best possible score is 10 and the worst is 40. Figure 1 illustrates the results. We can see that the two versions of the adaptive memetic algorithm have similar performance, and clearly outperform the adaptive ILS algorithms.

(a)   (b)

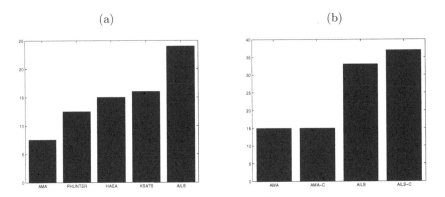

Fig. 1. Borda counts for (a) *AMA*, *AILS* and the top 3 VRP hyper-heuristics in the 2011 challenge, (b) the two variants of *AMA* and *AILS* on the full set of 10 instances and using the extended HyFlex VRP domain. Objective is minimisation.

The boxplots shown in Figure 2 illustrate the magnitude and distribution of the best objective values for 2 representative Homberger instances (instances 6 and 9). Each plot summarises the result of 10 runs from each algorithm. For both instances, the *AMA* algorithms produce the best results. The difference in performance is more noticeable for instance 6, but this behaviour is consistent across all the instances. The Borda counts in Figure 1, indicate that the two versions of *AMA* have similar performance considering the median best objective value. However, the best solutions were in most cases obtained by the *AMA − C* variant as can be seen in Figure 2 and Table 1.

Finally, Table 1 shows the best solutions found by our *AMA* algorithms together with the best-known solutions for the these instances. The adaptive memetic algorithms matched the bet-known number of vehicles for all the Solomon instances and for two of the Homberger instances. Moreover, for

(6)   (9)

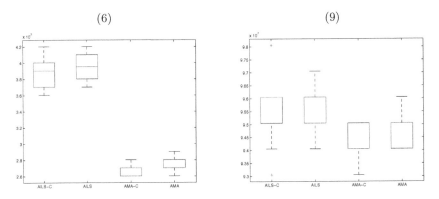

Fig. 2. Distribution of objective function values for Homberger instances 6 and 9. Objective is minimisation.

Table 1. VRP Instances. AMA best results vs. best-known results.

Instance			No. of Vehicles			Distance		
Name	Cust.	Capacity	AMA	AMA-C	$Best$-k	AMA	AMA-C	$Best$-k
0-SRC207	100	1000	4	**3**	**3**	**1047.42**	1133.83	1061.14
1-SR101	100	1000	19	19	19	1650.8	**1631.82**	1645.79
2-SRC103	100	200	11	11	11	1276.82	1263.78	**1261.67**
3-SR201	100	200	4	4	4	1261.043	1276.45	**1252.37**
4-R106	100	1000	12	12	12	1268.93	1284.23	**1251.98**
5-HC1-10-1	100	200	100	100	100	42481.26	42485.04	**42478.95**
6-HRC2-10-1	250	1000	26	26	20	33272.57	**32839.49**	63373.15
7-HR1-10-1	250	200	100	100	100	59020.74	60517.21	**53904.23**
8-HC1-10-8	250	200	101	101	93	44037.96	44120.54	**42499.59**
9-HRC1-10-5	250	200	94	93	90	52581.52	52439.09	**46631.89**

instance 1 (1-SR101) $AMA - C$ produced a shorter distance, with the same number of vehicles, which makes this a new best-known solution for this instance. Better distances were found for instances 0 and 6, but at the expense of a larger number of vehicles. These results are encouraging as HyFlex was designed to explore adaptive search heuristics that operate in a domain-independent way.

5   Conclusions

We have presented a number of extensions to the HyFlex framework that will enable the implementation of more effective adaptive heuristics, while maintaining a high degree of modularity between the problem-independent and the problem-dependent heuristic components. In particular, the new version supports the implementation of: (i) population-based approaches and mechanisms for operator selection that consider diversity metrics in the solution space, (ii) adaptive approaches that modify the fitness function or consider alternative objective functions, and (iii) offline approaches and portfolio methods that benefit from a greater number of problem instances. This article concentrated on the first of these extensions, namely using a diversity metric to implement more sophisticated adaptive operator selection mechanisms. Our results suggest that this mechanism may improve the search, in particular for locating best solutions. Indeed a new best-known solution was found for one of the studied instances. In future work we plan to further exploit this and the additional HyFlex features. Our HyFlex adaptive evolutionary algorithms also supports that using a population and crossover operators may improve the search. This is an important result, which may encourage the evolutionary computation community, as iterative hyper-heuristics have been traditionally single-point approaches.

The proposed HyFlex extensions were implemented and tested in a single domain: the vehicle routing problem. Work is in progress to incorporate these extensions in other domains such as permutation flow-shop, 1D bin packing and personnel scheduling. We envisage the incorporation of new challenging and real-world domains in HyFlex. We are also planning a second international challenge with additional features. The creativity and enthusiasm of the 2011 competitors pushed the boundary of hyper-heuristic research. We expect that the new competition will bring the interest and participation not only of hyper-heuristic researchers, but also researchers in reactive search, intelligent optimisation, adaptive operator selection, adaptive memetic algorithms, co-evolutionary memetic algorithms, guided local search, adaptive large neighborhood search, autonomous search, self-* search and automatic configuration of search heuristics to name a few.

Acknowledgements. We would like to thank the HyFlex users and CHeSC competitors for their feedback and suggestions. Our special thanks to J. Kubalic, A. Lehrbaum, K. McClymont, D. Meignan, M. Misir, A. Nunez, E. Ozcan, A. J. Parkes, K. Sim, T. Urli, T. Wauters and F. Xue.

References

1. The Cross-domain Heuristic Search Challenge (CHeSC 2011). Website (2011), http://www.asap.cs.nott.ac.uk/external/chesc2011/
2. Battiti, R., Brunato, M., Mascia, F.: Reactive Search and Intelligent Optimization. Operations research/Computer Science Interfaces, vol. 45. Springer (2008)
3. Bleuler, S., Laumanns, M., Thiele, L., Zitzler, E.: PISA – A Platform and Programming Language Independent Interface for Search Algorithms. In: Fonseca, C.M., Fleming, P.J., Zitzler, E., Deb, K., Thiele, L. (eds.) EMO 2003. LNCS, vol. 2632, pp. 494–508. Springer, Heidelberg (2003)
4. Burke, E.K., Curtois, T., Hyde, M., Kendall, G., Ochoa, G., Petrovic, S., Vazquez-Rodriguez, J.A., Gendreau, M.: Iterated local search vs. hyper-heuristics: Towards general-purpose search algorithms. In: IEEE Congress on Evolutionary Computation, CEC 2010, Barcelona, Spain, pp. 3073–3080 (July 2010)
5. Burke, E.K., Gendreau, M., Ochoa, G., Walker, J.D.: Adaptive iterated local search for cross-domain optimisation. In: Proceedings of the 13th Annual Conference on Genetic and Evolutionary Computation, GECCO 2011, pp. 1987–1994. ACM, New York (2011)
6. Burke, E.K., Hyde, M., Kendall, G., Ochoa, G., Ozcan, E., Woodward, J.: A Classification of Hyper-heuristic Approaches. In: Handbook of Metaheuristics, vol. 146, ch. 15, pp. 449–468. Springer (2010)
7. Chan, C.Y., Xue, F., Ip, W.H., Cheung, C.F.: A hyper-heuristic inspired by pearl hunting. In: International Conference on Learning and Intelligent Optimization (LION 6). LNCS, Springer (to appear, 2012)
8. Cowling, P.I., Kendall, G., Soubeiga, E.: A Hyperheuristic Approach to Scheduling a Sales Summit. In: Burke, E., Erben, W. (eds.) PATAT 2000. LNCS, vol. 2079, pp. 176–190. Springer, Heidelberg (2001)
9. Fialho, Á., Da Costa, L., Schoenauer, M., Sebag, M.: Extreme Value Based Adaptive Operator Selection. In: Rudolph, G., Jansen, T., Lucas, S., Poloni, C., Beume, N. (eds.) PPSN X. LNCS, vol. 5199, pp. 175–184. Springer, Heidelberg (2008)
10. Kubiak, M.: Distance measures and fitness-distance analysis for the capacitated vehicle routing problem. In: Metaheuristics. Operations Research/Computer Science Interfaces Series, vol. 39, pp. 345–364. Springer US (2007)
11. Mascia, F., Stutzle, T.: A non-adaptive stochastic local search algorithm for the chesc 2011 competition. In: Proceedings of Learning and Intelligent Optimization 6th International Conference, LION 6, Paris, France, January 16-20. LNCS, Springer (to appear, 2012)

12. Maturana, J., Saubion, F.: A Compass to Guide Genetic Algorithms. In: Rudolph, G., Jansen, T., Lucas, S., Poloni, C., Beume, N. (eds.) PPSN X. LNCS, vol. 5199, pp. 256–265. Springer, Heidelberg (2008)

13. Ochoa, G., Hyde, M., Curtois, T., Vazquez-Rodriguez, J.A., Walker, J., Gendreau, M., Kendall, G., McCollum, B., Parkes, A.J., Petrovic, S., Burke, E.K.: HyFlex: A Benchmark Framework for Cross-Domain Heuristic Search. In: Hao, J.-K., Middendorf, M. (eds.) EvoCOP 2012. LNCS, vol. 7245, pp. 136–147. Springer, Heidelberg (2012)

14. Ong, Y.S., Lim, M.H., Zhu, N., Wong, K.W.: Classification of adaptive memetic algorithms: a comparative study. IEEE Transactions on Systems, Man, and Cybernetics, Part B 36(1), 141–152 (2006)

15. SINTEF. VRPTW benchmark problems, on the SINTEF transport optimisation portal. Website (2011), http://www.sintef.no/Projectweb/TOP/Problems/VRPTW/

16. Voudouris, C., Tsang, E.: Guided local search and its application to the traveling salesman problem. European Journal of Operational Research 113(2), 469–499 (1999)

17. Walker, J.D., Ochoa, G., Gendreau, M., Burke, E.K.: Vehicle routing and adaptive iterated local search within the hyflex hyper-heuristic framework. In: International Conference on Learning and Intelligent Optimization (LION 6). LNCS. Springer (to appear, 2012)

18. Wauters, T., Vancroonenburg, W., Vanden-Berghe, G.: A guide-and-observe hyper-heuristic approach to the eternity ii puzzle. Journal of Mathematical Modelling and Algorithms, 1–17 (2012), doi:10.1007/s10852-012-9178-4

Applying Genetic Regulatory Networks to Index Trading

Miguel Nicolau, Michael O'Neill, and Anthony Brabazon

Natural Computing Research & Applications Group
University College Dublin
Dublin, Ireland
{Miguel.Nicolau,M.ONeill,Anthony.Brabazon}@ucd.ie

Abstract. This paper explores the computational power of genetic regulatory network models, and the practicalities of applying these to real-world problems. The specific domain of financial trading is tackled; this is a problem where time-dependent decisions are critical, and as such benefits from the differential gene expression that these networks provide. The results obtained are on par with the best found in the literature, and highlight the applicability of these models to this type of problem.

1 Introduction

The Evolutionary Computation (EC) literature tends to adapt mostly evolutionary models in a Darwinian sense: a population of individuals is created, executed, and assigned fitness scores. The most fit individuals are then more likely to survive, through a stochastic process, and the evolutionary cycle continues in this fashion, until a stopping condition is met. This model has proven to be successful throughout the years, but the knowledge of biological systems is ever increasing, and there is a growing trend in exploring more complex and realistic models [2].

One of the key aspects of genetics that is seeing increasing attention is the developmental processes that occur throughout the life of organisms. Rather than adopting a fixed, direct mapping from genotype to phenotype, developmental systems explore the lifelong, conditional expression of genes.

Genetic Regulatory Networks (GRNs) are a key element of gene expression regulation in biological organisms, and one that has seen recent attention in the EC field [1,10,13,11,5]. GRN-based algorithms explore the idea of differential gene expression through regulatory processes, and as such are potentially useful for dynamic and noisy environments.

This paper further explores the potential of GRNs for Evolutionary Computation, and exemplifies how to apply a recently introduced model [1] to a financial prediction benchmark. GRN models seem well suited to this kind of problem, where at different times of its life, an individual needs to adapt to a constantly changing environment. The results obtained further highlight the potential of GRNs as a computational device, and hopefully help to pave the future for their adoption within the EC community.

C.A. Coello Coello et al. (Eds.): PPSN 2012, Part II, LNCS 7492, pp. 428–437, 2012.

The next section introduces the biological principles behind GRNs, and details the implementation of the model used. Section 3 then introduces the problem domain, and Section 4 details the experimental setup and results achieved. Section 5 concludes this study, and highlights future work directions.

2   Gene Regulatory Networks

2.1   Background

In the cell environment, DNA segments containing genes are transcribed (i.e. expressed) into mRNA (messenger RNA) strands, which, through a translation process, are used to combine amino-acids, thus forming proteins. Some of these proteins are known as *Transcription Factors*: their role is to help regulate the expression of other genes, by binding at specific regulation sites. This process results in complex networks, with genes producing proteins regulating the expression of other genes; these are known as *Gene Regulatory Networks* (GRNs).

In the work presented here, the model originally introduced by Banzhaf [1] is used. This model consists of a binary linear genome, which is scanned for promoter regions, identifying the location of genes. It assumes that each gene is always composed of two regulatory sites (inhibiting and enhancing), and that all proteins produced are transcription factors.

This model has been used frequently in the literature. It was shown to exhibit similar dynamics to its natural counterparts, such as the appearance of specific regulatory network motifs [3] and the resulting network topologies [8], and has been evolved to optimise those topologies [12]; the resulting networks have also been extracted and used as a computational device, for a subset of Genetic Programming benchmark problems [11]. The resulting complex regulatory dynamics have also been studied, from the evolution of oscillatory dynamics [10] to actual control problems such as the pole balancing benchmark [13], and also the flag-colouring developmental problem [5].

2.2   The Model

The model used represents the genome as a binary string. This string is scanned for 32 bit long binary sequences, representing *promoter* regions; if found, these identify the location of a gene. The following 32×5 bits then represent the gene contents, and the previous 32×2 bits represent enhancing and inhibitory regions, respectively. Fig. 1 illustrates this.

In this model, a promoter site is the sequence XYZ01010101, where X, Y and Z are any 8 bit sequences. The protein produced by the gene is a 32 bit binary sequence, extracted by a majority rule between all 5 sequences of 32 bits that compose it (that is, if 3 or more equally located bits are set to 1, then the corresponding bit in the protein is set to 1).

Regulation works by matching the binary signature of transcription factors and regulating sites with the XOR operation: the result is the regulating strength.

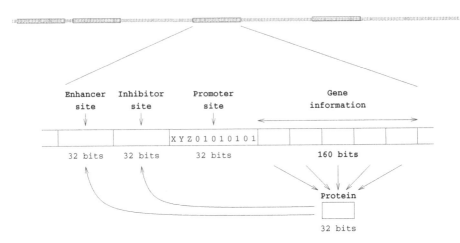

Fig. 1. Bit string encoding of a gene. If a promoter site is found, the gene information is used to create a protein, whose quantity is regulated by the attachment of proteins to the enhancer and inhibitor sites.

The enhancing and inhibiting signals regulating the production of protein p_i are then calculated as:

$$e_i, h_i = \frac{1}{N} \sum_{j=1}^{N} c_j \exp(\beta(u_j - u_{max})) \quad , \tag{1}$$

where N is the total number of proteins, c_j is the concentration of protein j, u_j is the number of complementary bits between the (enhancing or inhibitory) regulating site and protein j, u_{max} is the maximum match observed in the current genome, and β is a positive scaling factor.

The production of p_i is calculated via the following differential equation:

$$\frac{dc_i}{dt} = \delta(e_i - h_i)c_i \quad , \tag{2}$$

where δ is a positive scaling factor (representing a time unit). All the concentrations are normalised at each time step, ensuring that $\sum_i c_i = 1.0$ at all times; this results in competition for resources within the cell environment.

Input and Output. The original model is a closed world, in that there is no direct interaction with the environment. However, in most problem domains (particularly in reinforcement learning), a training set of input values are associated with a set of responses (or outputs), and the fitness of a solution is typically the difference between the responses obtained and a set of known correct outputs. To this end, a set of I/O extensions were introduced to the original model [13], which are also used in the current work.

To introduce the notion of an input signal, extra regulatory proteins (EPs) are injected into the system. These are not produced by any gene, but also contribute to the regulation of all gene expressions. They represent all the variables required to describe the state of the environment, and their concentrations reflect the (normalised) value of those inputs (and as such are unscaled).

To extract output signals from the model, genes are divided into two classes: TF-genes (i.e. genes encoding transcription factors), and P-genes (genes encoding *product* proteins). These classes of genes are established by scanning the genome for two different promoter sites: in this work, XYZ00000000 represent TF-genes, and XYZ11111111 represent P-genes. While the expression levels of TF-genes contribute to the regulatory process as before, the output of P-genes does not; the concentration of the proteins they produce is used as an output signal.

The regulation of TF-genes remains as previously stated, using Eq. 1, but they are normalised taking into account the concentration of EPs as well.

The regulation of P-genes is also determined by Eq. 1, but their expression is calculated with the following equation:

$$\frac{dc_i}{dt} = \delta(e_i - h_i) \quad .$$

(3)

Like TF-genes, all concentrations are normalised at each time step, ensuring that $\sum_i c_i = 1.0$ at all times; however, the concentration of TF-proteins and P-proteins are normalised independently.

3   Index Trading

In the financial domain, a market index is a weighted average measure of the price of individual shares that compose that market. Rather than trading single shares (or a portfolio of shares), a popular alternative is to trade on the share market index via an exchange-traded fund, which mirrors as close as possible the collective behaviour of the shares comprising the market. This type of trading has the advantage of not being tied to fluctuations of single shares, but rather to a broader market move. This also means that specific and unexpected share fluctuations are slowly absorbed by the market index [6], allowing for some degree of predictability [9].

Evolutionary algorithms have been successfully applied to financial modelling; the reasons for their applicability include their ability to efficiently explore the search space, and uncover dependencies between input variables, leading to their proper inclusion in the final models [7]. Brabazon and O'Neill [4] provide an overview of the application of evolutionary computation to financial modelling.

The work presented here follows closely the methodology of previous applications of Grammatical Evolution [15] to index trading [14,4], and uses three datasets, from the UK FTSE 100 index, the Japan Nikkei index, and the German Dax index. All data is drawn from the period between 1/1/1991 and 3/12/1997.

3.1   Technical Indicators

Rather than just observing the raw and historical market price data, it is useful to pre-process this information into technical indicators. These potentially uncover possible useful trends and other information from the raw data series, while simultaneously reducing the inherent noise in the series. Although a potentially infinite number of such indicators may exist, certain classes of indicators are regularly used by investors [9,16]. The following indicators are used in this study:

- **Moving Average.** This indicator returns the average price of the last n days; as it smooths out daily price fluctuations, it can unveil the underlying trend of the market. The parameter n controls the degree of smoothness.
- **Stochastic Oscillator.** This indicator returns the relative location of the current price in relation to its full price range over a period of of n days; it is useful in trying to predict price turning points.
- **Momentum Change.** This indicator compares the closing price with that of n days ago, and returns the rate of change. It indicates trend by remaining positive while an uptrend is sustained, or negative in the opposite case.

3.2 Datasets

Fig. 2 plots each dataset. These were divided into one training and three testing periods, of 365 days each, for the purpose of model validation. In accordance with previous studies [4], the data was pre-processed prior to evolution. Initially the raw prices were transformed into a moving average with a 75 day gap; these values were then normalised into the range of 0 to 1. This means that data from the first 75 days was not used for the purposes of trading simulation, neither was the data remaining after the four training and testing datasets.

3.3 Methodology

An evolved trader produces one of three signals for each day of the training or test periods: *buy, sell,* or *do nothing.* Starting with an initial capital of $10000, the following trading methodology is used [9,14]. If a buy signal is issued, a fixed $1000 investment is made in the market index; this position is automatically closed at the end of a ten day period. If a sell signal is issued, an investment of $1000 is sold short, and it is also closed after ten days. This means that a maximum of $10000 are invested at any given point in time. The profit or loss at the end of each trading period takes into account a one-way trading cost of 0.2%, and a further 0.3% to account for slippage. Uncommitted funds take into account a risk-free rate of return, which is approximated using the average interest rate over the entire dataset.

4   Experiments

4.1   Encoding Input and Output Variables

Four technical indicators were used in this study: a moving average of 10 days $(mAvg(10))$, momentum change of 5 $(mChange(5))$ and 10 days $(mChange(10))$, and a stochastic oscillator of 10 days $(sOsc(10))$. These were encoded using EPs, as explained in Section 2.2; the signatures for the EPs were chosen to be as different as possible, and were encoded as follows:

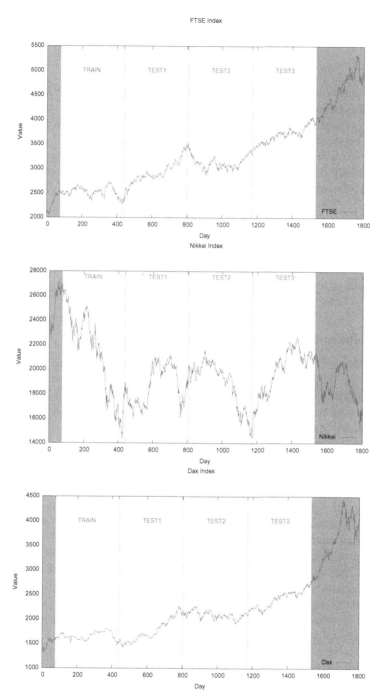

Fig. 2. Plots of the three markets analysed, and the train and test periods used. Gray shaded areas show the initial 75 day moving average gap, and the remaining data after the four year-long training and testing sets, and were not used for the simulations.

$mAvg(10)$:   0000000000000000000000000000000000
$sOsc(10)$:   0000000000000000111111111111111111
$mChange(5)$:   1111111111111110000000000000000
$mChange(10)$: 111111111111111111111111111111111

The GRN was allowed to first run for a maximum of 100000 iterations, or until all protein concentrations were stabilised; after this period, the trading session begins. To synchronise the GRN with the trading simulator, a trading signal was extracted every 2000 iterations.

To extract a trading signal from the network, the rate of change of a given P-gene is analysed: if its concentration has increased by more than 0.1%, then a *buy* signal is issued; if it has decreased by more than 0.1%, a *sell* signal is issued; otherwise, a *do nothing* signal is issued[1]. All P-genes present in the genome are tested, and the most successful one is used.

This methodology thus encodes technical indicators as regulatory proteins, which influence the internal regulatory process of the genome, and therefore influence the resulting concentration of P-genes, which can then be interpreted as a trading signal. It is a very similar process as seen in previous applications of GRNs to time-series datasets [13].

4.2 Evolutionary Setup

A $(250 + 250) - ES$ evolutionary strategy was used to evolve the binary genomes: a population of 250 individuals is used to create 250 offspring, and the best 250 of all parents and offspring are used as the new parent population (a maximum of 100 iterations were allowed). The variation operator used was a bit-flip mutation, set to 1% and adapted by the 1/5 rule of Evolution Strategies [17].

4.3 Measuring Performance

A two-set methodology was used, with the system being trained in an initial training set of one year. Once the training period was over, the system went "live", and was ran on the three test (out of sample) periods, for the purpose of a live trading simulation.

A common passive trading strategy is *Buy and Hold*, where an investor buys stocks and holds them for a long time. It is based on the idea that financial markets give a good return for investment in the long run, regardless of fluctuations and periods of volatility. In order to evaluate the performance of the evolved traders, their performance was compared to a buy and hold strategy for each of the training and test datasets.

When evolving an index trader, certain aspects required special attention. It would be inadequate to simply calculate fitness as the profit return, as this fails to consider the risk of deploying an evolved trader [14]. A measurement of this risk is provided by the maximum drawdown, that is, the maximum cumulative

---

[1] This is an entirely experimental value, and has not been optimised; it was left to the structure of the GRN to adapt to it.

loss of the system during each of the datasets. As seen in previous studies [14,4], this can be incorporated into the fitness calculation by subtracting the maximum cumulative loss from the profit of each period. This is in addition to trading costs and slippage penalties, as detailed in Section 3.3.

4.4   Results and Analysis

Table 1 presents the results obtained with the best evolved controller for the FTSE, Nikkei and Dax markets, for the train and validation periods. The best evolved trader for the FTSE and Nikkei markets outperforms the benchmark buy and hold strategy, whereas on the Dax market it slightly underperforms; this is likely linked to the fact that the Dax market is very well behaved, across the period analysed, with very rare fluctuations. These results are on par with similar EC approaches found in the literature [14,4].

Another interesting aspect of the evolved traders is their low investment risk. The buy and hold strategy keeps the full available capital of $10000 invested at all times, whereas the evolved traders kept average capital investments of only $3582.19, $2806.85 and $3225.34, for the FTSE, Nikkei and Dax markets, respectively. This is a combination of the inclusion of risk penalties in the fitness function, and the fact that the system can only trade $1000 daily.

Fig. 3 plots the best evolved trader for the FTSE market. It exhibits a very cautious approach to trading, with large periods of inactivity, resulting from a

Table 1. Best evolved traders for all datasets compared to Buy & Hold benchmark

FTSE market			
Period (days)	Buy & Hold	Best-of-run	Avg. daily inv.
Train (75 to 439)	-1269.28	3275.96	5939.73
Test 1 (440 to 804)	4886.9	1083.58	2191.78
Test 2 (805 to 1169)	-1089.8	541.806	3709.59
Test 3 (1170 to 1534)	1908.53	500.949	2487.67
Total	**4436.35**	**5402.295**	

Nikkei market			
Period (days)	Buy & Hold	Best-of-run	Avg. daily inv.
Train (75 to 439)	-6345.5	6163.38	5128.77
Test 1 (440 to 804)	1014.79	1125.6	1457.53
Test 2 (805 to 1169)	-5263.49	2144.71	3679.45
Test 3 (1170 to 1534)	4040.59	1331.56	961.644
Total	**-6553.61**	**8514.05**	

Dax market			
Period (days)	Buy & Hold	Best-of-run	Avg. daily inv.
Train (75 to 439)	-882.241	2899.86	4586.3
Test 1 (440 to 804)	4047.63	952.689	3347.95
Test 2 (805 to 1169)	-551.995	608.161	1471.23
Test 3 (1170 to 1534)	2972.24	992.868	3495.89
Total	**5585.634**	**5453.578**	

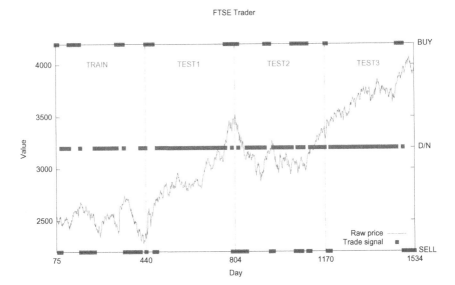

Fig. 3. Best evolved trader for the FTSE index. Blue dots indicating buy, do nothing and sell signals are plotted along with the raw market prices.

slow reaction to the regulatory proteins representing technical indicators, within the GRN. This is mostly induced by the inclusion of the risk-free rate of return and the maximum drawdown penalty into the fitness function. This is seen in Fig. 3, in the third testing period, where the trader exhibits a very cautious approach, even though the market is generally trending upwards. Further optimisation of structure and parameters of the GRN should improve this issue.

5   Conclusions

This paper explored the computational power of regulatory networks, and how to apply them to a real-world financial trading domain. The methodology required to apply GRNs was explored, and the results obtained show the potential of this approach, with results on par with the literature.

There remains much work to be done. Regarding the current problem domain, the evolved trader seems occasionally unresponsive to market changes; the use of different technical indicators could improve this issue. Also, regarding the applicability of the model, further research is required, in particular with what concerns its parameterisation.

References

1. Banzhaf, W.: Artificial regulatory networks and genetic programming. In: Riolo, R., Worzel, B. (eds.) Genetic Programming Theory and Practice, ch. 4, pp. 43–62. Kluwer Publishers, Boston (2003)

2. Banzhaf, W., Beslon, G., Christensen, S., Foster, J.A., Képès, F., Lefort, V., Miller, J.F., Radman, M., Ramsden, J.J.: From artificial evolution to computational evolution: a research agenda. Nature Reviews Genetics 7, 729–735 (2006)
3. Banzhaf, W., Kuo, P.D.: Network motifs in natural and artificial transcriptional regulatory networks. Biological Physics and Chemistry 4(2), 85–92 (2004)
4. Brabazon, A., O'Neill, M.: Biologically Inspired Algorithms for Financial Modelling. Springer (2006)
5. Cussat-Blanc, S., Bredeche, N., Luga, H., Duthen, Y., Schoenauer, M.: Artificial gene regulatory networks and spatial computation: A case study. In: Lenaerts, T., Giacobini, M., Bersini, H., Bourgine, P., Dorigo, M., Doursat, R. (eds.) Proceedings of 11th European Conference on Advances in Artificial Life, ECAL 2011, Paris, August 8-12, MIT Press (2011)
6. Hong, H., Lim, T., Stein, J.C.: Bad news travels slowly: Size, analyst coverage, and the profitability of momentum strategies. Journal of Finance 55(1), 265–295 (2000)
7. Iba, H., Nikolaev, N.: Genetic programming polynomial models of financial data series. In: Proceedings of 2000 Congress on Evolutionary Computation, CEC 2000, San Diego, CA, USA, July 16-19, vol. 2, pp. 1459–1466 (2000)
8. Kuo, P.D., Banzhaf, W.: Small world and scale-free network topologies in an artificial regulatory network model. In: Pollack, J., Bedau, M., Husbands, P., Ikegami, T., Watson, R. (eds.) Artificial Life IX: Proceedings of the Ninth International Conference on the Simulation and Synthesis of Living Systems, pp. 404–409. Bradford Books, USA (2004)
9. LeBaron, B., Lakonishok, J., Brock, W.: Simple technical trading rules and the stochastic properties of stock returns. Journal of Finance 47(5), 1731–1764 (1992)
10. Leier, A., Kuo, P.D., Banzhaf, W., Burrage, K.: Evolving Noisy Oscillatory Dynamics in Genetic Regulatory Networks. In: Collet, P., Tomassini, M., Ebner, M., Gustafson, S., Ekárt, A. (eds.) EuroGP 2006. LNCS, vol. 3905, pp. 290–299. Springer, Heidelberg (2006)
11. Lopes, R.L., Costa, E.: ReNCoDe: A Regulatory Network Computational Device. In: Silva, S., Foster, J.A., Nicolau, M., Machado, P., Giacobini, M. (eds.) EuroGP 2011. LNCS, vol. 6621, pp. 142–153. Springer, Heidelberg (2011)
12. Nicolau, M., Schoenauer, M.: On the evolution of scale-free topologies with a gene regulatory network model. BioSystems 98(3), 137–148 (2009)
13. Nicolau, M., Schoenauer, M., Banzhaf, W.: Evolving Genes to Balance a Pole. In: Esparcia-Alcázar, A.I., Ekárt, A., Silva, S., Dignum, S., Uyar, A.Ş. (eds.) EuroGP 2010. LNCS, vol. 6021, pp. 196–207. Springer, Heidelberg (2010)
14. O'Neill, M., Brabazon, A., Ryan, C., Collins, J.J.: Evolving Market Index Trading Rules Using Grammatical Evolution. In: Boers, E.J.W., Gottlieb, J., Lanzi, P.L., Smith, R.E., Cagnoni, S., Hart, E., Raidl, G.R., Tijink, H. (eds.) EvoIASP 2001, EvoWorkshops 2001, EvoFlight 2001, EvoSTIM 2001, EvoCOP 2001, and EvoLearn 2001. LNCS, vol. 2037, pp. 343–352. Springer, Heidelberg (2001)
15. O'Neill, M., Ryan, C.: Grammatical Evolution - Evolutionary Automatic Programming in an Arbitrary Language, Genetic Programming, vol. 4. Kluwer Academic (2003)
16. Pring, M.J.: Technical Analysis Explained: The Successful Investor's Guide to Spotting Investment Trends and Turning Points. McGraw-Hill (1991)
17. Rechenberg, I.: Evolutionsstrategie 1994. Frommann-Holzboog, Stuttgart (1994)

Evolutionary 3D-Shape Segmentation Using Satellite Seeds

Kai Engel and Heinrich Müller

Technische Universität Dortmund, Informatik VII (Graphische Systeme),
Otto-Hahn-Straße 16, 44221 Dortmund, Germany
{kai.engel,heinrich.mueller}@tu-dortmund.de

Abstract. The aim of 3D-shape segmentation is to divide the surface of an object into meaningful parts. We present a novel version of seed-point-based segmentation including an evolutionary optimization to obtain better segments. At first, some initial seeds are defined. Each of them generates several so-called satellite seeds which enable a more detailed control of the segment boundaries. The locations and weights of the seeds are optimized with an Evolution Strategy. The objective function takes the object's curvature at the segments' boundaries into account as well as the length of these boundaries. An extensive evaluation and comparison with important existing segmentation approaches demonstrates the great potential of our approach.

Keywords: Mesh Segmentation, Satellite Seeding, Evolution Strategy.

1 Introduction

3D-shape segmentation has become an important research topic in the field of three-dimensional computer graphics. It is used in several domains like e.g. 3D modeling, texture mapping, and collision detection [1]. The aim of 3D-shape segmentation is to create a decomposition of a 3D-model like the bunny in Fig. 1 into disjoint segments according to some criteria. We focus on models where the surface is approximated by a triangle mesh as shown in Fig. 2. In this context, shape segmentation is also known as *mesh segmentation*.

Fig. 1. Overview of our approach: normal seeds, satellite seeds, the resulting segmentation, and an optimized segmentation. The seeds are indicated by yellow circles.

C.A. Coello Coello et al. (Eds.): PPSN 2012, Part II, LNCS 7492, pp. 438–447, 2012.

This paper is concerned with part-type segmentation [10], i.e. the desired segments correspond to meaningful parts of the original mesh. A horse model, for example, should be divided into head, body, legs, and so on. For the purpose of part-type segmentation, human perception has to be transferred into criteria which can be treated algorithmically. Important and often used criteria are that patch boundaries are typically located in concave surface regions and that the patch boundaries generally have locally minimal length. These are also the major criteria which will be considered within the optimization process introduced in this paper.

Several mesh segmentation approaches use seed points. The basic idea is to determine a region around each of a finite number of properly chosen points on the surface. The regions together induce the decomposition of the surface into patches. The seed points may serve as control points for automatic optimization using an Evolutionary Algorithm. The novel concept of satellite seeds introduced in this paper extends the potential of optimizing the borders between patches without increasing the dimension of the parameter space too much.

This paper is organized as follows. It starts with a short survey on related work in Section 2. In Section 3, an overview of our approach is given. The approach is described in detail in Sections 4 and 5, where the evolutionary optimization is explained in Section 5. In Section 6, we present results including a comparison with the results of eight state-of-the-art techniques. Section 7 concludes the paper and discusses future work.

2   Related Work

Over the last decade, a large number of automatic mesh segmentation approaches has been proposed. Most of them code the objective of segmentation implicitly in an algorithm. However, some other approaches explicitly define an objective function over a set of possible partitions, which may be optimized by an adequate general-purpose solver, cf. e.g. [11]. Extensive surveys on mesh segmentation approaches can be found in [1,10,3].

Most of the solutions up to now are based on fixed heuristics with the aim of an as complete as possible characterization of segmentations. In contrast, Kalogerakis et al. [6] present a data-driven approach which learns an objective function over a feature space from a collection of labeled training meshes, independent from a concrete mesh. It offers a flexible way of learning different types of segmentations for different tasks, without requiring manual parameter tuning. A still existing disadvantage is that always adequate training models are required.

There are several possibilities of deriving a partition into segments from seed points. The potential of satellite seeding will be demonstrated on partitions induced in a way known from weighted Voronoi diagrams on the surface of the mesh. Simari et al. [11,12] have used weighted Voronoi diagrams, too, but in 3D-space using an embedding of the original mesh obtained by multi-dimensional scaling (MDS).

To our knowledge, there is only one mesh segmentation approach containing an evolutionary optimization: Simari and Singh optimize the center initialization, i.e. the positions of weighted partition centers [11]. After the initialization, they use a generalized pattern search for segmentation optimization. In contrast to their approach, we apply an Evolution Strategy for the whole process of optimization. In addition, while Simari and Singh have taken the desired forms of the parts into account by labeling the partition centers, our intention is to obtain good segments without explicit labels.

The contributions are as follows. First, we extend the seed-point-based mesh segmentation approach by the concept of satellite seeds. Second, we introduce an evolutionary optimization to obtain more natural patches. Third, we present an extensive evaluation including an automatic seeding as well as a simulated manual seeding based on ground truth segmentations by Chen et al. [3].

3   Problem Statement and Overview of Our Approach

Given a triangle mesh defining the surface of a geometric object, a decomposition into patches is desired where the boundaries are smooth and optimally located in concave regions. The first aspect is based on the assumption that the boundaries between parts of an object are generally not jagged. The latter one is based on the minima rule [5]. Our solution adopts the seed point approach and includes the following main components:

1. *Seed definition:* The seeds are defined in two steps. The first step chooses a finite set of adequate initial seeds. In the second step these seeds automatically spawn satellite seeds to get a better control of the segment boundaries (see Fig. 1(a) and (b)).
2. *Patch calculation:* Given a finite set of seeds, a decomposition of the object surface into *patches* is calculated (see Fig. 1(c)).
3. *Optimization:* Since patches usually do not match with meaningful parts, we perform a patch optimization using an Evolution Strategy. The optimized patches are often identical with meaningful parts (see Fig. 1(d)). In this article, we denote meaningful parts of an object as *components* and calculated parts, which can be regarded as being meaningful, as *segments*.

Seed definition and patch calculation are described in Section 4, the optimization process is introduced in Section 5.

4   Seed Definition and Patch Calculation

At first, some *initial seeds* are placed on the object's surface in either a manual or an automatic way. Useful heuristics of *manual seeding* are to locate one initial seed in every expected component, and to place it as central as possible within the component. One possible heuristic of automatic seeding also used in this paper is to arrange the first seed far away from the object's centroid and all following seeds one after another in such a way that their geodesic distance from each other is as large as possible. In contrast to other approaches, the seeds are presented by mesh triangles, which we call *seed triangles*.

Every initial seed defines a *patch*, which, roughly speaking, consists of all mesh triangles that are closer to it than to all other initial seeds according to an adequate distance function. The distance is calculated on the dual graph of the triangle mesh. As shown in Fig. 2, each mesh triangle corresponds to a node in the dual graph. Nodes belonging to neighbor triangles are connected with an edge.

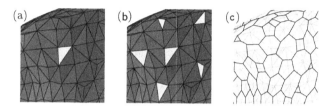

Fig. 2. Triangle mesh with a seed triangle (a), satellite seeds (b), and the dual graph (c)

The segments' boundaries should usually be surrounded by concave object areas. This condition is taken into account by using a *feature based distance function* which is defined by

$$d_{feature}^{\zeta,\eta,\nu}(t_1,t_2) := d_{geo}(t_1,t_2) + \zeta \cdot d_{ang}(t_1,t_2) + \nu \cdot \eta \cdot d_{shape}(t_1,t_2), \quad (1)$$

where t_1 and t_2 are *adjacent triangles*. It combines the geodesic distance, an angular distance and a special distance which takes into account the object's shape in a local region. The influence of the different partial distances is controlled by $\zeta, \eta \in \mathbb{R}_0^+$, while $\nu \in \{0,1\}$ decides whether or not there is a need to use d_{shape} at all. d_{geo} denotes the usual *geodesic distance* between the centers of t_1 and t_2. The distance between two neighbor nodes of the dual graph is defined as the feature based distance of the corresponding mesh triangles.

By assigning weights to the seeds, the influence of a seed on the boundary of its patch becomes adaptable. Having γ_{seed} as the weight of the seed triangle t_{seed}, the *weighted-feature-based distance* of a triangle t' to t_{seed} is defined by

$$d_{wfeature}(t',t_{seed}) := \frac{1}{\gamma_{seed}} \cdot d_{dual}(t',t_{seed}), \quad (2)$$

where d_{dual} is the length of a shortest path in the dual graph between the nodes corresponding to t' and t_{seed}. For a given seed, the weighted-feature-based distance between adjacent triangles can be estimated canonically on the dual graph by dividing the feature based distance of the dual graph edge by the seed weight. The assignment of mesh triangles to a patch is realized using a modified form of Dijkstra's shortest path algorithm on the dual graph: the seeds are processed in increasing weight order. If a dual graph node is reached which has already a lower weighted-feature-based distance to another seed, this node will not be taken into account anymore in the current run of Dijkstra's algorithm. This modification of Dijkstra's algorithm is necessary, because without the modification non-connected patches may occur when different seed weights are chosen.

The reason for combining the geodesic, the angular and our special shape-based distance is that geodesic distances are insensitive to curvature and the part boundaries. Angular distances are insensitive to part boundaries over flat regions. If the edge between two mesh triangles is concave, the angular distance is high. Thus, concave edges should be crossed less often inside a patch than edges in convex regions. The shape-based distance can be seen as an additional quality factor, which for example penalizes the transition from a cylindric region to a concave one.

The shapes of the patches can be influenced by moving seeds or changing their weights initially set on 1.0. Such a variation usually has an effect on the whole patch boundary, so that good boundary parts could be changed to the worse while worse parts are optimized. To avoid such a deterioration of good boundaries, we have developed *satellite seeding* as an extension of initial seeding. The motivation is to divide each boundary into several smaller parts. This is automatically done by substituting each patch by several smaller ones, which we call *subpatches*. To obtain subpatches of a single patch, new seeds are arranged around the initial seed like satellites, close to the initial one (see Fig. 2).

5   Optimization

In this section, we present our patch optimization approach using an enhanced *Evolution Strategy* (ES) that can also handle integer parameters and nominal-discrete ones (cf. [4]). The object variable vector consists of the seed positions in form of triangle indices (assuming that each mesh triangle can be addressed by an index), the weights of the initial and satellite seeds, the influences ζ and η, and the flag ν. All individuals of the initial population are directly derived from the given segmentation. In the majority of cases the optimization produces more natural boundaries. After optimization, every patch is regarded as a segment.

5.1   Reproduction and Selection

Recombination: An extensive survey on well-known recombination strategies is given in [2]. For our optimization problem, we have chosen different recombination strategies for the object variables. The recombination of ζ and η is realized by an intermediate recombination operator, while for the binary valued ν a discrete recombination operator is chosen. Since the seed positions are saved as indices, they cannot be recombined with an intermediate operator, so a discrete operator is used. However, a patch must not become decomposed after applying the recombination. Therefore, for each patch all seeds are randomly chosen completely from one of the individuals selected for recombination.

Mutation: All real-value parameters are mutated by adding a normally distributed random number $\chi \in \mathbb{R}$ with mean 0 and standard deviation σ. Negative values are set to a small positive value. χ is recalculated for each parameter. We have used Rechenberg's 1/5-Rule [8] to adapt the standard deviations. The mutation of seed positions has to be carried out in another way. The seeds are moved over the triangle mesh in such a way that small movements occur more often than larger ones, by using additional information stored in the mesh data structure for mutating the seed indices. The mutation operator has to ensure that each triangle contains at most one seed.

Selection: There are different kinds of Evolution Strategies like the $(\mu + \lambda)$-ES and the (μ, λ)-ES [2]. They differ from each other in the selection mechanism. We have chosen a (μ, κ, λ)-ES [9] which contains the $(\mu + \lambda)$-ES as well as the (μ, λ)-ES as special cases. The (μ, κ, λ)-ES can consider all individuals with an age smaller than the life-span κ.

5.2   Fitness Evaluation

We formulate the optimization as a minimization problem, where the fitness value corresponds to the segmentation quality. According to the knowledge about human perception presented in [5], we want to obtain segments which are mostly surrounded by concave areas. On the other hand, the component boundaries perceived by humans are usually not "jagged". Thus, a segmentation Γ with short segment boundaries is desired. In our *fitness function*

$$f(\Gamma) := \begin{cases} (1 - \Lambda) \cdot f_{concave}(\Gamma) + \Lambda \cdot f_{length}(\Gamma); \Lambda \in [0, 1] & \text{if } \Gamma \text{ is valid,} \\ \infty & \text{otherwise,} \end{cases} \quad (3)$$

which is to be minimized, the first assumption is taken into account by $f_{concave}$, the latter one by f_{length}. A segmentation is called *valid* iff each patch is connected, i.e. if no patch is decomposed into several parts. Segment boundaries are polylines, all of which have edges of the triangle mesh as line segments. The function $f_{concave}$ is defined in such a way that long line segments have more influence than shorter ones:

$$f_{concave}(\Gamma) := \frac{1}{\sum_{i=1}^{|E|} \frac{1}{l(e_i)}} \cdot \sum_{j=1}^{|E|} \left(\frac{1}{1 + \max(\alpha_j, 0)} \cdot \frac{1}{l(e_j)} \right). \quad (4)$$

E is a list of all edges belonging to the segment boundaries of Γ. The ith boundary edge is denoted by e_i and its length by $l(e_i)$. α_i denotes the signed angle between the normal vectors of the two mesh triangles adjacent to e_i. This angle is positive, if the object is concave at e_i. The first factor in equation (4) is used for scaling; it ensures that $f_{concave}(\Gamma) \in [0, 1]$. The more concave the segment boundaries of Γ are, the lower is the function value of $f_{concave}$.

The function

$$f_{length}(\Gamma) := \frac{1}{\sum_{i=1}^{|E_{mesh}|} l(e'_i)} \cdot \sum_{j=1}^{|E|} l(e_j) \quad (5)$$

evaluates how "jagged" the segment boundaries are, which is related with the length of the boundary. The smoother they are, the smaller is the function value. Once again, the list of all boundary edges is denoted by E; the list of all edges belonging to the triangle mesh is denoted by E_{mesh}. While e_j is the jth element of E, e'_i denotes the ith edge of E_{mesh}. Since composition of $f_{concave}$ and f_{length} in (3) is realized as a convex combination, the range of f is a subset of $[0, 1]$.

6   Experiments and Evaluation

6.1   Behavior of the Optimization

A drawback of state-of-the-art techniques based on seeds is that a suboptimal seeding can cause wrong segments. The optimization of seed positions and weights included in our approach compensates for this drawback. In Fig. 3, an example for the positive effects of the evolutionary optimization can be seen. The initial segmentation on the left side is taken as the source for two optimizations: one using a (μ, κ, λ)-ES and one using a $(\mu + \lambda)$-ES. Both have been applied for 150 generations. Figure 3 also shows the corresponding curves of the best fitness values per generation. In both cases, the resulting segmentations are much better. In contrast, before starting the optimization, the green and the yellow segment are not quite good; the yellow one describing the middle finger was running down the other side of the hand up to the little finger. Please also note that even optimal segmentations yield fitness values considerably larger than 0.

Other segmentations calculated by our prototype are shown in Fig. 4. The results essentially correspond to human expectations. In particular, technical models like the bearing object are almost perfectly segmented. But also the segmentations of natural models are quite good in most cases. If still necessary at all, jagged boundaries may be fixed by postprocessing [7].

6.2   Evaluation

Since we have demonstrated some results of our approach so far, we now measure the quality of our segmentation method by taking segmentations manually created by human test persons into account, that can be seen as being "optimal". Such segmentations are known as *ground truth segmentations*. We have used the *Rand Index* (RI) to evaluate the discrepancy between a calculated segmentation and a ground truth segmentation [3]. It is the relative number of all pairs of mesh triangles which either belong to the same segment in both segmentations or which belong to two different segments in both segmentations. With this information, the similarity of the segmentations is calculated. Originally, the Rand Index RI_{orig} is defined to be 1 if both segmentations are identical, and it's range is $[0, 1]$. According to [3], we use $RI = 1 - RI_{orig}$ as Rand Index in order to compare our results to those of established approaches. Thus, this Rand Index grows with increasing discrepancy. A more detailed description can be found in [3]. Since the Rand Index can also be evaluated for segmentations of other approaches, an objective comparison of our results with other ones is feasible.

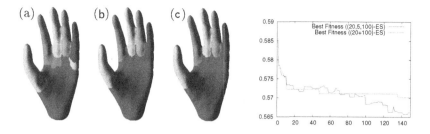

Fig. 3. The initial segmentation (a) was optimized by a (μ, κ, λ)-ES (b) and a $(\mu + \lambda)$-ES (c). The best individuals' fitness values are shown for 150 generations.

Fig. 4. Segmentations achieved by our prototype

An extensive benchmark containing 380 models as well as eleven different ground truth segmentations in the average for each model was published by Chen et al. [3]. The models are divided into 19 categories. We have used this benchmark to evaluate our segmentation approach.

In order to perform a large number of experiments, we have chosen two variants of defining a reasonable number of seeds individually for every model in an automatic way. *Variant 1* is given by our automatic seeding, where the number of initial seeds is determined by taking the average segment number of all ground truth data belonging to the current model. *Variant 2* is based on the ground truth data. The initial seeds are placed near the centroids of randomly chosen ground truth segments. This can be seen as a simulation of a manual seeding, which enables to evaluate the possible advantage of an expected optimal manual seeding against an automatic seeding. In this sense, variant 2 can also be considered as a lower bound (with respect to the Rand Index) for all automatic approaches.

We have calculated one segmentation per seeding variant for each of the 380 benchmark models mentioned above with $\mu = 3$, $\lambda = 15$, $\kappa = 5$ and an experimentally determined $\Lambda = 0.3$. The optimization was stopped already after 30 generations, which turned out to be sufficient. On the left side of Fig. 5, the influence of the optimization is shown on the basis of the Rand Index. The Rand Index values averaged over the 380 models are shown for the situations before and after optimization. *Sat1* and *Sat2* stand for our segmentation approach using seeding variant 1 and 2, respectively. Please remember that the fitness function is defined completely independent from the Rand Index and that even the best possible RI value will be positive. The latter aspect is reasoned by different subjective perceptions of a model and its components, which results in different ground truth data. For example, the wings of a high-wing plane can be interpreted as one single component or as two independent components. According to this, in [3], a RI of 0.1 was observed for manual segmentations that are regarded as being optimal. For Sat2, the optimization yields a similar value, which confirms the capabilities associated with satellite seeding.

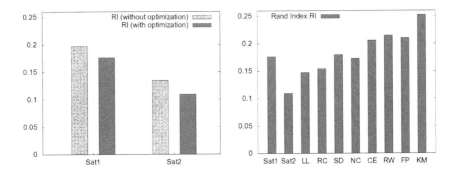

Fig. 5. Effect of the optimization (left) and averaged RI values (right)

In the right diagram of Fig. 5, the averaged RI values of our optimized segmentations and of established techniques are shown (cf. [3,6]). The established techniques are based on Labeling and Learning (LL) [6] regarding three training meshes for every evaluated segmentation, Randomized Cuts (RC), Shape Diameter Function (SD), Normalized Cuts (NC), Core Extraction (CE), Random Walks (RW), Fitting Primitives (FP), and K-Means Clustering (KM). A survey, except for the LL approach, is given in [3]. Variant 1 is superior to 5 of the 8 established techniques. A great theoretical potential of the proposed approach is demonstrated by the RI values of variant 2 which are significantly better than those of all other techniques. A trouble of our automatic seeding is that seeds may be placed nearby boundaries of meaningful components. This may be unfavorable even for satellite seeding in its current version. A further question of future research is whether the existing techniques also have a potential of improvement which can be quantitatively estimated analogously to variant 2.

We have also investigated the influence of satellite seeding on the optimization by taking only the initial seeds of Sat2 without satellite seeds. In this case, the RI after optimization is 0.146, which is clearly worse than the RI in the case of satellite seeds. Furthermore, it is also worse than in the case of Sat2 without optimization shown in Fig. 5. This behavior is caused by the fact, that the optimization algorithm tends to move "lonely" initial seeds onto parts bounded by concave areas, which often results in patches that are too small for the desired segmentation granularity. Therefore, satellite seeding as an extension of normal seeding is obviously effective for evolutionary optimization.

7   Conclusion and Future Work

We have presented a novel approach to mesh segmentation suitable for generating optimized segmentations using an Evolution Strategy. In many cases, it yields better results than important well-known techniques. Furthermore, our approach is especially well suited for semi-automatic segmentation, i.e. a manual seeding which can be done with minimal effort followed by automatic patch calculation and optimization. A major challenge of future work is to improve the automatic seeding to reduce the gap between the segmentation qualities of Sat1 and Sat2 shown in Fig. 5.

Further, the reliability of Sat2 as a simulation of a manual seeding could be investigated. First tests confirm that in most cases segmentations calculated from seedings by humans are very similar to the ones presented in this paper.

Finally, alternative fitness functions could be studied. For example, the fitness function might also force corresponding satellite seeds to stay close to each other. This could reduce the occurrence of invalid individuals in the parent population.

Acknowledgements. The bunny model is from the Stanford 3D Scanning Repository [13], all others are from the mentioned benchmark and available at [14].

References

1. Attene, M., Katz, S., Mortara, M., Patane, G., Spagnuolo, M., Tal, A.: Mesh Segmentation - A Comparative Study. In: Proceedings of the IEEE International Conference on Shape Modeling and Applications, SMI 2006 (2006)
2. Bäck, T.: Evolutionary Algorithms in Theory and Practice. Oxford University Press, Oxford (1996)
3. Chen, X., Golovinskiy, A., Funkhouser, T.: A Benchmark for 3D Mesh Segmentation. ACM Transactions on Graphics (Proc. SIGGRAPH) 28(3) (August 2009)
4. Emmerich, M., Grotzner, M., Schütz, M., Groß, B.: Mixed-Integer Evolution Strategy for Chemical Plant Optimization with Simulators. In: Parmee, I. (ed.) Evolutionary Design and Manufacture, ACDM (2000)
5. Hoffman, D.D., Richards, W.A.: Parts Of Recognition. Cognition 18, 65–96 (1984)
6. Kalogerakis, E., Hertzmann, A., Singh, K.: Learning 3D Mesh Segmentation and Labeling. ACM Transactions on Graphics 29(3) (2010)
7. Kaplansky, L., Tal, A.: Mesh Segmentation Refinement. Comput. Graph. Forum 28(7), 1995–2003 (2009)
8. Rechenberg, I.: Evolutionsstrategie: Optimierung technischer Systeme nach Prinzipien der biologischen Evolution. Frommann-Holzboog, Stuttgart (1973)
9. Schwefel, H.-P.: Evolution and Optimum Seeking. Sixth-Generation Computer Technology. Wiley Interscience, New York (1995)
10. Shamir, A.: A survey on Mesh Segmentation Techniques. Comput. Graph. Forum 27(6), 1539–1556 (2008)
11. Simari, P., Singh, K.: Multi-objective shape segmentation. Tech. rep., Departement of Computer Science, University of Toronto (2008)
12. Simari, P.D., Nowrouzezahrai, D., Kalogerakis, E., Singh, K.: Multi-objective shape segmentation and labeling. Comput. Graph. Forum 28(5), 1415–1425 (2009)
13. Stanford 3D Scanning Rep.,
    http://www.graphics.stanford.edu/data/3Dscanrep/
14. Princeton Mesh Segmentation Benchmark, http://segeval.cs.princeton.edu

Benchmarking CHC on a New Application: The Software Project Scheduling Problem

Javier Matos and Enrique Alba

Universidad de Málaga, Spain
{jmo,eat}@lcc.uma.es

Abstract. In this article we analyze the behavior and scalability of the CHC algorithm over a benchmark of instances of the software project scheduling problem. Our goal is to analyze the performance of the CHC algorithm when solving realistic NP-hard combinatorial problems and test whether its previously reported high performance on similar problems also holds on this one. We perform a preliminary study to obtain a suitable configuration of the parameters in the algorithm. After choosing the configuration, we show the results for the problem instances in the benchmark. To give a reference on how CHC performs and scales, its results are compared against those of a GA. We conclude that CHC outperforms GA in large problem instances. Moreover, CHC produces promising results for the software project scheduling problem domain, and could be used by practitioners.

Keywords: Software Project Scheduling, Metaheuristics, Evolutionary Algorithms, Comparison, Benchmark.

1 Introduction

The CHC algorithm (*Cross generational elitist selection, Heterogeneous recombination, and Cataclysmic mutation*) has been applied with success for solving hard combinatorial optimization problems. For instance, several problems in which CHC has been used include the design of robust network topologies [11], the placement of wind turbines in a wind farm [3], the scheduling of tasks to processors in an heterogeneous environment [10,12], and a multiobjective antenna placement problem [9]. Previous works have shown that CHC is a competitive algorithm for solving optimization problems, frequently obtaining results that outperform those of the algorithms that were compared with it. However, it still remains not well-known in the community, in which many theses and articles do not use this kind of GA of low complexity and high numerical benefits.

In this article we apply for the first time CHC on this software problem. We push CHC to the limit using this new problem with the purpose of studying the behavior and scalability of the algorithm. Application results themselves are competitive and help locating CHC as a state-of-the-art technique for other applications in search based software engineering [7]. For the sake of the study, and to highlight CHC benefits, we compare it with a GA.

C.A. Coello Coello et al. (Eds.): PPSN 2012, Part II, LNCS 7492, pp. 448–457, 2012.
© Springer-Verlag Berlin Heidelberg 2012

The rest of the document is organized as follows. A description of the CHC algorithm is shown in Section 2. Section 3 presents the problem instances used in the benchmark, the initial study of the parameters to tune up CHC, the discussions of the results of CHC for the benchmark, and a detailed comparison with a GA. Conclusions of the study are outlined in Section 4.

2   The CHC Algorithm

The CHC algorithm is a special type of a GA designed to promote the best individuals in the population. One of the main characteristics of CHC is that it does not use mutation, that is a way to introduce new information in the population and avoid premature convergence; instead, it uses two mechanisms to stimulate diversity: an *incest prevention*, which only allows the recombination of individuals that are different enough (in terms of the Hamming distance), and a restart of part of the population when stagnation is detected. Initially, the threshold for allowing recombination is set to $1/4$ of the chromosome length. During the recombination process, if the two randomly selected parents meet the condition to be recombined, then, the threshold is reduced by 1. As the algorithm runs, individuals become similar to each other, and eventually, the threshold to allow recombination reaches the value 0. This is how CHC detects that the population is stuck; thus, the algorithm performs a restart in part of the population: only the best p_r individuals are kept, whereas the others are restarted to increase the diversity.

The recombination operator in CHC is the *half uniform crossover* or HUX, that is a variant of the *uniform crossover* (UX), and consists in the random exchange of a half of the bits in which parents differ, as shown in Figure 1.

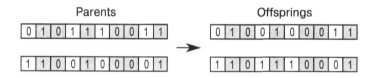

Fig. 1. The HUX recombination operator takes two parents and randomly decides on a swap for those bits at which their strings differ. Bits of the string for which the parents have the same value (highlighted in the figure) are not changed.

In Algorithm 1 we show the pseudocode of CHC as initially proposed by L. Eshelman [5]. The code reveals those features that make CHC different from traditional GAs: the elitist replacement strategy, the use of the HUX recombination operator, the absence of mutation, and the mechanism to restrict the recombination. The premature convergence of the population is reduced by the recombination policy and the diversity of individuals is ensured with the restart of a part of the population.

Algorithm 1. Pseudocode of the CHC algorithm

initialize($P(0)$)
generation ← 0
threshold ← $1/4 \cdot chromosomelength$
while not *stopcriterion* **do**
  parents ← **selection**($P(generation)$)
  if *distance(parents)* ≥ *threshold* **then**
    offspring ← **HUX**(*parents*)
    evaluate(*offspring*)
    newpop ← **replacement**(*offspring, P(generation)*)
  end if
  if *newpop* == $P(generation)$ **then**
    threshold ← *threshold* − 1
  end if
  generation ← *generation* + 1
  $P(generation)$ ← *newpop*
  if *threshold* == 0 **then**
    reinitialization($P(generation)$)
    threshold ← $1/4 \cdot chromosomelength$
  end if
end while
return best solution ever found

3 Experimental Analysis

This section presents the experimental analysis performed in this work. First, we explain the problem instances of the benchmark. Then, we tune up CHC and apply it on a set of representative instances. Finally, we compare CHC against a GA.

3.1 Problem Instances: A Wide Representative Benchmark

To carry out the analysis of CHC, we have used 250 instances of a hard combinatorial problem in our benchmark. The problem itself is the software project scheduling (SPS), that consists on the assignment of employees to tasks in a software project in order to reduce its duration and cost [1,4]. This problem belongs to the domain of search based software engineering [7]. The software project scheduling is a realistic problem with capital importance in software factories.

An instance of the SPS problem specifies a set of employees, tasks, and skills to indicate which employee can participate in which task. For every employee, it is necessary to set his/her maximum dedication, salary, and skills. For a task, it has to be known an estimation of the effort required, the skills needed to accomplish it, and a list of tasks that are prerequisite of it.

A solution to the SPS problem is an assignment matrix that represents the degree of involvement of employees to tasks (cells in this matrix have values in set [0 1]). Such a solution has to meet all the constraints imposed by the problem.

The objectives pursued in this problem are to reduce the duration and cost of the software project and to fulfil the constraints.

To solve this problem with metaheuristics like CHC and GA we have to encode the assignment matrix as a binary string. This can be done as shown in Figure 2, where cell values are discretized using four bits ($[0\ 1] \rightarrow \{0,1\}^4$). Additionally, we need a fitness function to quantify the relative quality of solutions. The fitness function that we use is presented in Equation (1), and consists on a weighted sum of the project duration (p_{dur}), the project cost (p_{cost}), the number of tasks not covered by any employee (p_{ut}), the number of skills not covered for the tasks (p_{us}), and the amount of overwork done by the employees (p_{ow}).

$$
\begin{aligned}
fitness\ function = \ &+0.1 \times p_{dur}\\
&+(5.0 \times 10^{-6}) \times p_{cost}\\
&+p_{ut} + p_{us} + p_{ow}
\end{aligned}
\tag{1}
$$

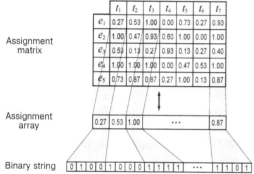

Table 1. Features of the problem instances in the benchmark. The size of the instance is the main indicator of its difficulty.

N	Size	Tasks	Employee
1	50	10	5
2	100	20	5
3	250	25	10
4	450	30	15
5	1250	50	25

Fig. 2. Representation of an assignment matrix as a binary string

In Table 1 we present the features of the 250 problem instances contained in the 5 test sets (50 instances per set). For every test set we show its identification number, the size of the contained instances, and the number of employees and tasks. All the instances share the same number of total skills, that is 10. The tasks of the instances require a random set of 4 to 6 skills. Additionally, employees have a random set of 2 to 4 skills. For a more precise understanding of the instances, we defer the interested reader to the original definition of this problem [4].

This benchmark is a large and wide set of instances since we want to actually deal with the problem class, not just with a few instances. Also, it will allow us to analyze algorithms at very different dimensions and difficulties, what will constitute a real challenge for any algorithm.

The objective of the optimization technique is to compute a solution with the lowest fitness value for every problem instance in the benchmark. The test sets have been arranged in increasing size or difficulty, where the first one has the smallest search space and the sixth the largest one. This arrangement of the test sets allows us to study trends of the algorithm with increasing size of the search space. For the benchmark, we have used a set of problem instances created by the generator described at http://mstar.lcc.uma.es.

3.2 Parameter Settings

Instead of fixing an arbitrary set of parameters, we perform an initial configuration analysis to determine the best parameter for CHC. One random problem instance of every test set is used to tune up the algorithm during the configuration analysis. The parameters studied for CHC are the population size, the recombination probability (p_c), and the percentage of population restarted (p_r). The values studied for the three parameters are:

Population size: 64, 128, 256
Recombination probability (p_c): 0.5, 0.7, 0.9
Percentage of population restarted (p_r): 40%, 60%, 80%

After the analysis of the parameters we computed Table 2 to study their impact for CHC. This table contains the average fitness, its relative standard deviation (σ), and the difference between the highest and the lowest average fitness for the test sets 1, 2, and 4 given the values of the parameters. We performed 30 independent executions for the 27 different configurations for all the 150 instances in the three test sets. If we focus on the population size, we realize that this parameter has the highest impact in the results of the algorithm. As the population is increased, the results of CHC clearly improve (the fitness reaches lower values); but there is a point at which the improvement of the fitness is nonexistent or small enough not to justify a further increase of the population. This behavior depends on the problem instance: for small instances (test set 1) large populations involve no improvement but a time penalty, whereas for large instances (test set 4) large populations produce better results.

On the other hand, if we focus on the recombination probability (p_c) and the percentage of population restarted (p_r), we conclude that the average fitness and its relative standard deviation are almost the same for the test sets. This means that whatever the value we choose for these parameters, the average result of CHC will be almost the same. There is also a second reading for these results, and it is that CHC is a robust algorithm, even if the best values for the recombination probability and the percentage of population restarted are not properly chosen. For instance, once that the population size has been fixed to 256 individuals, the differences between the best and the worst average fitness for test sets 1, 2, and 4 are 0.33%, 0.54%, and 0.44% respectively, thus, the impact of p_c and p_r in the results is upper bounded by these values.

After these initial experiments we conclude that the values for the parameters of CHC that perform the best are 256 for the population size, 0.9 for the recombination probability, and 60% for the percentage of population restarted. As a summary, the parameters used to test and study CHC with the problem instances are listed in Table 3.

Table 2. Average fitness, relative standard deviation (σ), and difference between the best and worst average fitness for a fixed value of a CHC parameter

Parameter	Value	Fitness								
		Test set 1			Test set 2			Test set 4		
		Avg.	σ	Diff.	Avg.	σ	Diff.	Avg.	σ	Diff.
	64	4.75	8.75%	0.69%	11.38	13.00%	1.18%	30.62	67.70%	2.11%
Pop. size	128	4.48	8.00%	0.38%	10.35	9.88%	0.69%	11.33	48.25%	3.37%
	256	4.35	7.52%	0.33%	9.77	9.39%	0.54%	8.09	11.19%	0.44%
	0.5	4.52	8.91%	9.37%	10.49	12.66%	15.81%	16.60	95.36%	136.33%
p_c	0.7	4.52	8.94%	9.09%	10.50	12.80%	15.69%	16.71	95.16%	136.43%
	0.9	4.53	9.0%6	9.49%	10.51	12.89%	16.14%	16.74	95.22%	136.17%
	40%	4.53	9.00%	9.25%	10.49	12.65%	15.60%	16.67	95.05%	136.69%
p_r	60%	4.53	8.96%	9.26%	10.50	12.86%	15.83%	16.69	95.87%	136.54%
	80%	4.53	8.95%	9.54%	10.51	12.85%	16.08%	16.68	94.81%	135.41%

3.3 Discussion on the Results

Here we describe the results obtained after applying the CHC algorithm to solve the problem instances of the benchmark. The two aspects in which we focus are the fitness value and the execution time. On the one side, the fitness value quantifies the performance of the algorithm to allow future comparisons. On the other side, the execution time accounts the time it takes the algorithm to compute a solution. The execution time makes it possible to study how the CHC algorithm could behave on optimization problems as complex as this, and how does it scale when the size of the problem gets increased.

In Table 4 we show the results of CHC. The table presents the number of the test set in the benchmark and for it the fitness and time values. For the fitness we show its average value for all the instances in the test set. Also, we include the maximum (σ_{max}) and the average (σ) of the relative standard deviation. The maximum is the one of the instance in the test set which has the largest deviation; the average is that of the instances in the test set. The values were computed by running 30 independent executions for the 250 instances of the 5 test sets. In total we made 7500 independent runs to get lessons on the algorithm and the problem class instead of just on one instance or small problem study.

Table 3. Parameter settings for CHC

Parameter	Value
Max. number of iterations	500
Population size	256
Offspring size	256
Recombination probability	0.9
Recombination operator	HUX
Restarted population	60%
Selection strategy	Random
Replacement strategy	Ranking

Table 4. Experimental results for CHC

N	Fitness			Time (s)	
	Avg.	σ_{max}	σ	Avg.	σ
1	4.35	4.53%	2.50%	2.98	11.70%
2	9.77	7.37%	4.06%	8.64	12.23%
3	7.91	8.15%	3.62%	21.64	13.26%
4	8.01	11.00%	3.86%	38.06	7.68%
5	25.56	23.31%	12.60%	100.93	5.32%

The analysis of the fitness indicates that CHC is a stable algorithm that produces solutions that are similar in terms of quality. This means that even with a few executions CHC is capable of finding promising solutions for the instances in a robust manner. As we studied in the previous section, the quality of the results have more to do with the population size than with the recombination probability (p_c) or the percentage of population restarted (p_r).

Now, if we turn to the execution time we find that the larger the problem size, the longer it takes to finish the computation. Generally, we observe that the execution time of the algorithm is close to linear, even when the search space grows exponentially (Figure 4). This is a great point for CHC, because it means that we can expect the algorithm to solve even larger problem instances in an acceptable amount of time.

3.4   Comparison against GA

To put the results of CHC in a wider context, we compare it with a GA. GA is a metaheuristic inspired in biological evolution [8]. It codifies problem solutions as individuals subjected to an evolutionary process [2,6]. During each iteration the algorithm selects, recombines, and mutates individuals to evolve the population. As iterations go by, new individuals are computed with better solutions codified.

We use the classic formulation of GA: it combines the *single point crossover* (SPX) recombination operator, and the mutation operator that randomly modifies selected positions in the solution. For the configuration analysis we follow the same procedure as in CHC. The parameters that we consider in the initial analysis are the populations size, the recombination probability, and the mutation probability, whose candidate values are:

Population size: 64, 128, 256
Recombination probability: 0.5, 0.7, 0.9
Mutation probability: 0.01, 0.05, 0.1

Applying the same guidelines as with CHC, we conclude that the best values for the parameters are 256 individuals for the population, 0.7 for the recombination

Table 5. Parameters used for GA

Parameter	Value
Max. number of iterations	500
Population size	256
Offspring size	256
Recombination probability	0.7
Recombination operator	SPX
Mutation probability	0.1
Bit flip probability	0.01
Selection strategy	Random
Replacement strategy	Ranking

Table 6. Results of the experiments for GA

N	Fitness			Time (s)	
	Avg.	σ_{max}	σ	Avg.	σ
1	4.65	16.12%	6.26%	6.70	5.87%
2	11.54	27.58%	13.52%	14.84	4.30%
3	21.14	39.87%	19.91%	25.06	3.33%
4	59.16	28.61%	15.25%	39.16	3.92%
5	249.89	11.23%	6.32%	106.96	4.80%

probability, and 0.1 for the mutation probability. The settings finally used for GA in the experiments are listed in Table 5.

The results obtained using the GA are listed in Table 6. The table presents the number of the test set in the benchmark and for it the fitness and time values. For the fitness we show its average value for all the instances in the test set. Also, we include the maximum (σ_{max}) and the average (σ) of the relative standard deviation. The values were computed by running 30 independent executions for the 250 instances of the 5 test sets. Compared with CHC, GA produces worse solutions and it takes more time to compute them. We also realize that GA has serious problems for test sets 3, 4, and 5: the fitness, which has to be minimized, is on average 2.67, 7.31, and 9.78 times larger than in CHC. This comparison is shown in Figure 3, where the central mark is the median, the edges of the filled box the 25th and 75th percentiles, and the whiskers extend to the most extreme values. The results show that the GA is not an efficient algorithm to solve as large and difficult instances for the software project scheduling problem.

It is interesting to study the results for each test set independently. In Figure 3 we offer a graphical comparison of the fitness value for CHC and GA. We can see that for small size instances (test sets 1 and 2) CHC beats GA by a thin margin; on the other side, when the size of the instances get increased (test sets 3, 4, and 5), then CHC overcomes GA in a notorious way. It is also important to note that CHC requires fewer fitness evaluations to reach certain fitness value. While GA carried out 128256 evaluations for every instance in all the test sets, CHC performed by average 44500, 66995, 104861, 123427, and 128256 evaluations for test sets 1 to 5. This happens because of the incest prevention mechanism in CHC, that avoids the recombination of solutions that are similar to each other. For the comparison of the fitness, the Kruskal-Wallis test has been carried out to check if the differences in the algorithms are statistically significant. All the statistical tests are performed with a confidence level of 99%, and all of them have passed this tests.

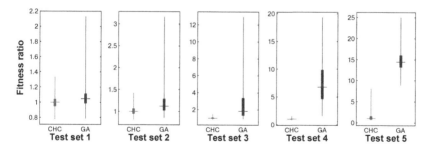

Fig. 3. Fitness comparison of CHC and GA for the test sets in the benchmark

In Figure 4 we show the execution time that it takes for CHC and GA to perform 500 iterations depending on the size of the instances in the test set. The sizes of the instances in the test sets are 50, 100, 250, 450, and 1250 respectively (Table 1). We see that the CHC algorithm always takes less time than GA

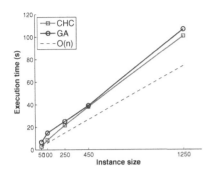
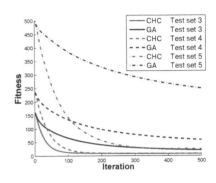

Fig. 4. Average execution time of the algorithms for increasing size problem instances in contrast with a theoretical linear time increment of complexity $O(n)$ (lower is better)

Fig. 5. Average fitness evolution through iterations (lower is better). The CHC algorithm reaches promising solutions in less iterations than GA for every test set.

to finish the computation. This is because CHC performs fewer fitness evaluations than GA, as stated previously. This means that CHC can solve the same instances than GA in less execution time.

To conclude, in Figure 5 we present the average fitness evolution of CHC and GA for the 500 iterations. We notice that the fitness value of CHC always remains below the fitness of the GA, no matter the test set. Thus, it is obvious that CHC converges faster to a promising solution. Globally, the GA needs more iterations to find promising solutions, and specifically, for test set number 5 it seems that 500 iterations are not even enough.

4   Conclusions

In this work we have presented a study on the performance of CHC when solving a benchmark of problem instances concerning software project scheduling. We focused on CHC because it has proven to be an efficient, fast, and powerful algorithm in the past, but still not well-known compared to other evolutionary algorithms. For the experiments, we faced the algorithm to a set of instances of the software project scheduling problem, that is a capital problem in software engineering. Finally, we compared the results achieved by CHC with a GA.

The analysis of the results obtained allows drawing some conclusions on the behavior of CHC. For instance, the population size was the parameter of the algorithm which had the highest impact in the results for the benchmark. Particularly, once that the population size was fixed to 256 individuals, the improvement produced by the variation of the probability of recombination and the percentage of population restarted was at most 0.54% (small, in comparison). This means that CHC is a robust algorithm that produces good results with a wide set of values for the parameters.

Regarding the comparison of CHC with GA, the CHC algorithm beats the GA in every single test set for both: in quality of the solutions and in execution time. The CHC algorithm always produce better solutions than the GA, and the larger the instance the better the result of CHC compared with the GA. Additionally, the execution time of CHC is always shorter than the execution time of the GA. In relation with this, CHC not only needs less time to find a promising solutions but also it takes less iterations to reach it. As a consequence, we can suggest that CHC is a better algorithm than the GA for this problem.

Acknowledgements. Authors acknowledge funds from the Spanish Ministry MICINN and FEDER under contracts TIN2008-06491-C04-01 (M* http://mstar.lcc.uma.es) and TIN2011-28194 (roadME) and CICE, Junta de Andalucía, under contract P07-TIC-03044 (DIRICOM http://diricom.lcc.uma.es).

References

1. Alba, E., Chicano, F.: Software project management with GAs. Information Sciences 177(11), 2380–2401 (2007) (in press)
2. Back, T., Fogel, D.B., Michalewicz, Z.: Handbook of Evolutionary Computation, 1st edn. IOP Publishing Ltd., Bristol (1997)
3. Bilbao, M., Alba, E.: CHC and SA applied to wind energy optimization using real data. In: IEEE Congress on Evolutionary Computation, pp. 1–8 (2010)
4. Chicano, F., Luna, F., Nebro, A.J., Alba, E.: Using multi-objective metaheuristics to solve the software project scheduling problem. In: Proceedings of the 13th Annual Conference on Genetic and Evolutionary Computation, GECCO 2011, pp. 1915–1922. ACM, New York (2011)
5. Eshelman, L.: The CHC adaptive search algorithm: How to have safe search when engaging in nontraditional genetic recombination. In: Rawlins, G. (ed.) Foudations of Genetic Algorithms, pp. 265–283. Morgan Kaufmann (1990)
6. Goldberg, D.E.: Genetic Algorithms in Search, Optimization and Machine Learning, 1st edn. Addison-Wesley Longman Publishing Co., Inc. (1989)
7. Harman, M., Jones, B.F.: Search-based software engineering. Information & Software Technology 43(14), 833–839 (2001)
8. Holland, J.H.: Adaptation in natural and artificial systems (1975)
9. Nebro, A.J., Alba, E., Molina, G., Chicano, F., Luna, F., Durillo, J.J.: Optimal antenna placement using a new multi-objective CHC algorithm. In: Proceedings of the 9th Annual Conference on Genetic and Evolutionary Computation, GECCO 2007, New York, NY, USA, pp. 876–883 (2007)
10. Nesmachnow, S., Cancela, H., Alba, E.: Heterogeneous computing scheduling with evolutionary algorithms. Soft Computing, 685–701 (2010)
11. Nesmachnow, S., Cancela, H., Alba, E.: Evolutionary algorithms applied to reliable communication network design. Engineering Optimization 39(7), 831–855 (2007)
12. Nesmachnow, S., Cancela, H., Alba, E.: A parallel micro evolutionary algorithm for heterogeneous computing and grid scheduling. Appl. Soft Comput. 12(2), 626–639 (2012)

Automatic Evaluation Methods in Evolutionary Music: An Example with Bossa Melodies

A.R.R. Freitas[1], F.G. Guimarães[2], and R.V. Barbosa[3]

[1] Graduate Program in Electrical Engineering, Federal University of Minas Gerais
Av. Antônio Carlos 6627, 31270-901, Belo Horizonte, MG, Brazil
alandefreitas@gmail.com
[2] Departamento de Engenharia Elétrica, Universidade Federal de Minas Gerais
Belo Horizonte, MG 31270-901, Brazil
fredericoguimaraes@ufmg.br
[3] Departamento de Teoria Geral da Música, Universidade Federal de Minas Gerais
Belo Horizonte, MG 31270-010, Brazil
rvb@musica.ufmg.br

Abstract. Evolutionary algorithms need measures of how appropriate a solution is in order to make decisions. This is always a problem for evolving art as codifying aesthetics is a complex task. In this paper we consider the problem of evaluating melodies. The evaluation of melodies in evolutionary music is an open problem that has been tackled by many authors with interactive evaluation, fitness-free genetic algorithms and even neural networks. However, all approaches based on formal analysis of databases or formal music theory have been partial, which is something to be expected for such a complex problem. Thus, we present many metrics that can be used for evaluating melodies and their practical results when applied to a Bossa Nova database of melodies coded by the authors. Although the paper is meant to extend the cycle of possible ideas for evolutionary composers, we argue that there is still much to be developed in this field and each genre of music will always need specific measures of quality.

Keywords: Evolutionary Music, Genetic Algorithms, Evaluation of Melodies.

1 Introduction

Evolutionary Algorithms and Algorithmic Composition methods need to measure how appropriate a solution is in order to make decisions. Easy ways to evaluate melodies would be comparing tunes, using only music theory or having a mentor to guide the process.

Evaluating music and art faces many challenges that we discuss in Section 2. Given the open problems for music evaluation and the methods recently proposed, we focus this paper on the definition of metrics more formally based on music theory or data extraction, as we develop the idea in Section 3.

C.A. Coello Coello et al. (Eds.): PPSN 2012, Part II, LNCS 7492, pp. 458–467, 2012.
© Springer-Verlag Berlin Heidelberg 2012

In this context, we describe a list of metrics divided in many categories from Section 4. In parallel to those metrics from a musicology research, we also show the results of those metrics on a database of Bossa Nova melodies the authors have created. We discuss each of those metrics as Information Retrieval or Computational Musicology processes.

In our discussion of the results, in Section 5, we argue that this work should be useful for scientists intending to create algorithms for generating melodies but there will always be metrics which will be more useful for genre-specific music generation.

2   Forms of Music Evaluation

Codifying aesthetics is a complex task and the biggest problem in evolutionary composition [1]. Approaches to circumvent codifying aesthetics such as interactive evolution [2,3,4], fitness-free GAs [5] and neural networks [6,7], all still present many drawbacks.

Using a human mentor usually leads to fitness bottlenecks [3,4]. Fitness-free algorithms [5,8] are bolder proposals but they also avoid studying the problem and oblige the genetic operators to be conservative. Most works based on Neural Networks do not have the ability to generalize beyond training sets [6,8,7].

Thus, an open problem is to create automatic evaluation functions [1] machine representable, capable of measuring human aesthetic properties and practically computable. They should not only define what is more likely to occur on melodies but they should also allow creativity when considering all the different aesthetic objectives to generate ideas not imagined before [9]. Computational aesthetic evaluation is a distinctly non-trivial unsolved problem [1].

3   Automatic Objective Functions

Many different metrics based on perceptions of the composer or music theory can be employed to analyze melodies in a process of algorithmic composition [10]. In this paper we present many automatic metrics and their results on a database of Bossa Nova melodies manually created by the authors.

There have been partial attempts to automate measures of fitness [1,11] and studies on which features are most important [11]. Those include four part harmonization [12] and jazz melodies [13], for instance. The influence of the genetic operators on musical features has also been partially studied [14,5]. Target values have also been used to measure fitness [15,16,17].

From an analysis over the literature, most algorithms do not examine the possible relation between all categories of metrics possible [10]. Thus, we define metrics that should be applicable to most classical, baroque or popular twentieth-century melodies and the results of their employment on a database of melodies.

With many analyses of those melodies from different points of view, we can compare the results to a potential solution from our generative algorithm. Some results indicate parameters with normal distribution, such as in the distribution of pitches, which can be tested in the candidate solution with a Jarque-Bera test [18]. Some results may show that parameters come from another distribution, such as in the distribution of rhythmic proportion, which can be compared to a candidate solution with a Two-sample Kolmogorov-Smirnov test [19]. Some parameters may only represent categorical values, which can be compared with a nominal statistical test [20]. Finally, other results indicate potential individual target values for the melodies, such as the tempo of each melody, which can be directly included in objective function values as the distance from the target.

4   Metrics and Results

In order to give a good representation of the mentioned metrics, we have created a database with 26 Bossa Nova melodies from Tom Jobim's songbook [21] and manually coded by the authors. All the data is available from the authors[1].

4.1   Tonality, Pitches, and Intervals

We first detect the key of each melody with the K-S key-finding algorithm [22], based on key profiles. As the melodies may even have key changes, it may be a simplistic approach, but all the keys detected matched the key signature in the scores and the results can give us an idea of the keys as we can see in Table 1. Thus, we transpose all songs to C (or its minor relative, Am) to make key dependent analyzes possible, such as the detection of dissonances. The correlation of the algorithm's key profiles to the pitch distribution of the pieces leads to a representation of the strength of each key, as in Figure 1(a). The correlation values are significantly higher for the C major and A minor key profiles, indicating some relevance of the method applied. The results can also be projected on a self-organizing map trained with key profiles [23], as in Figure 1(b).

Table 1. Key Profiles

C	C#	D	D#	E	F	F#	G	G#	A	A#	B	Total
12%	0%	15%	0%	4%	4%	0%	4%	0%	4%	0%	4%	**46%**

c	c#	d	d#	e	f	f#	g	g#	a	a#	b	Total
0%	0%	15%	0%	12%	4%	12%	4%	4%	4%	0%	0%	**54%**

The pitches used in all melodies are in Figure 2(a), showing that the distribution of the notes is very normal. However, by shifting all melodies to the same key, we have a large difference of occurrence between consecutive notes, as in Figure 2(b). This is due to dissonant notes, which are strange to the main scale.

Given the 12 note classes, the modulo of a pitch number by 12 is the class of this note. The occurrence of those pitch classes gives a better idea of the scales used in the melody, as in Figure 2(c). We can see a higher occurrence of notes of the diatonic

---

[1] http://www.alandefreitas.com/downloads/problem-instances.php

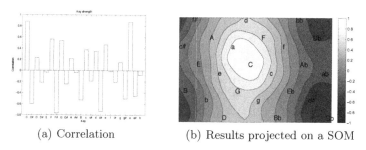

(a) Correlation   (b) Results projected on a SOM

Fig. 1. Melody keys

scale of C. The note variety is also different for each melody. A method of measuring pitch variety [11] is by dividing the number of distinct notes in a melody by 12, as in Figure 2(d), which shows that the variety of pitches is very different among the melodies but all melodies use more than 70% of the possible notes. Another aspect of pitch variety is pitch range [11], or the difference between the highest and lowest pitches. As we can see in Figure 2(e), the pitch range has a more normal distribution, centered on a range of 16 pitches.

Other useful metrics are the beginning and ending pitches, and the note distribution weighted by duration. For our database, this measure did not represent much difference, as we can see in Figure 2(f).

Perhaps, more important than the pitches themselves are the intervals between them. Figure 3(a) shows the intervals present in our melodies. In accordance with theoretical models [24], intervals of small size are more common than large ones. Figure gives a good representation of the interval sizes used in the melodies. In fact, it is a common practice to penalize very large intervals in the evaluation of the solutions [13]. However, this can be only applicable to some genres of music and an approach based on a better analysis is recommended. If

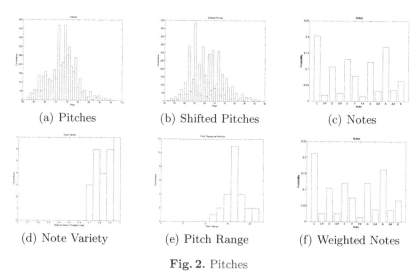

(a) Pitches   (b) Shifted Pitches   (c) Notes

(d) Note Variety   (e) Pitch Range   (f) Weighted Notes

Fig. 2. Pitches

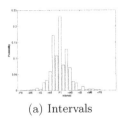
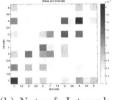

(a) Intervals   (b) Notes & Intervals   (c) Interval Variety

Fig. 3. Intervals

we combine the information of the notes to the intervals, we are going to find out that the probability of the next note depends on the current note, as shown in Figure 3(b). Also similarly to the notes, we can analyze the interval variety for each song, as in Figure 3(c).

Contour refers to the movements being performed by the melodies. There are many sorts of contour [25] and the direction of those movements may be easier to remember than the movements themselves [26]. An easy way to analyze contour is to measure how many intervals are ascenders or descenders, and the stability in relation to direction. Table 2 shows the values of ascenders and descenders in general or in relation to the last interval. The values in bold represent the contour stability, which is a criterion that has also been used in evolutionary algorithms [11], and represented for each song in Figure 4(a). Another simple form of controlling contour is through the average contour direction [13,11].

Table 2. Contour

	Ascendent	Unison	Descendent
After an ascendent	**30.17%**	18.78%	51.03%
After an unison	19.92%	**55.24%**	24.82%
After a descendent	45.36%	14.57%	**40.06%**
In general	33.52%	26.95%	39.52%

By analyzing pitches and tonality together, we can have an idea of the dissonances used in the songs. From Figure 2(c) we can see that it would be more than reasonable to analyze dissonance in terms of the proportion of notes that do not belong to the diatonic scale. Thus, the probability of a dissonant note is 30.53%, but Figure 4(b), which represents the occurrence of dissonance divided by the number of possible dissonant notes, shows how this value can vary considerably.

Attraction of dissonant notes to tonally stable notes happens to 55.25% of the dissonances. However, this measure may overlap with the measure of second order notes, as shown in Figure 3(b).

Narmour's Implication-Realization Model [27] is a study on melodic expectancy based on many principles that consider expectation of the listener after a given interval. With a quantification of the principles in model [28,29], we can either penalize melodies that disrespect the principles or measure how much the melodies follow the model.

As the model can be context-specific or inefficient to consider tonally stable intervals [30], we can also use the interval values in the melodies to infer our own model of expectancy which would be specific for our goal. Figure 4(c) shows such a model, where the rows represent implicative intervals and columns represent realized intervals. The model confirms the expectation of small intervals. Melodic attraction should also be considered by this model of expectation as we have different responses for different pitches [31]. One way of doing that would be to infer 12 different models according to the current note.

4.2 Rhythm, Patterns, and Phrases

The first feature that determines the rhythm is the duration of notes. Figure 5(a) shows a second order analysis of the proportion of note durations. From 36 possible values of duration present in the melodies, the histogram is based on the durations shorter than 4 beats and longer than 1/4 of a beat [29]. The patterns show a tendency of repetition in the duration of following notes. Another interesting pattern is that the first notes in a melody, shown in Figure 5(b), tend to have shorter duration than the last notes, shown in Figure 5(c). The rhythmic proportion in each melody is the duration of the longest note divided by the duration of the shortest note, as shown in Figure 5(d). Similarly to what we did to the pitches, we can also calculate the duration variety in the melodies, as in Figure 5(e). Part of the rhythmic analysis is not only the duration of the pitches but also how much silence we have in the melodies. In Figure 5(f) we have the amount of silence (as at most 2 beats without notes) per melody. In some cases, even more than 10% of the melody may be silent.

We have mentioned the duration of the notes but another important information is when the notes are played. A hierarchical grid of note locations may exist in the expectation of Western melodies [32]. For instance, the note positions in the musical measure are represented in Figure 5(g). Also the first notes (and last notes) may use different positions, as the example in Figure 5(h). In fact, only 6 values of note position are used for the first notes while 16 values are used for all notes. The note positions can also be weighted by the duration of those notes as it alters how listeners perceive those notes [33]. Figure 5(i) shows the relation between those two components.

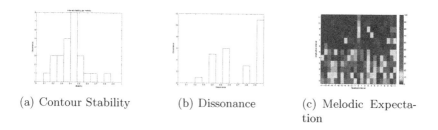

(a) Contour Stability   (b) Dissonance   (c) Melodic Expectation

Fig. 4. Expectation and Contour

Once we have information related to pitches and rhythm we can find patterns in the melodies. By autocorrelating a melody with a delayed copy of itself [34], we can identify patterns in a melody. The correlation values go from 0 to 1, and the correlation value is always 1 at point 0, when we compare a melody shape with a copy of itself, as shown in Figure 6(a). The three areas represent the maximum, mean, and minimum correlation. Similarly to the contour shape, we can apply the same technique to only pitches or duration values.

Another way of looking at the patterns is to identify the number of patterns of a specific size in a melody. We can analyze that in Figure 6(b), where each row represents a melody, each column represents a pattern size and the colors represent the amount of that pattern. Short patterns are naturally more common as longer and rare patterns may represent the repetition of phases in the melody. The same metric can be applied to notes or duration values.

We can divide melodies into musical phrases. Figure 6(c) shows the number of phrases per melody according to a rule-based approach [35]. There are also approaches based on probability [36]. The size of those phrases can also be analyzed and with those values it is possible to also study the value of the parameters for each musical phrase as well as the relation between neighbor phrases in relation to pitch and rhythm.

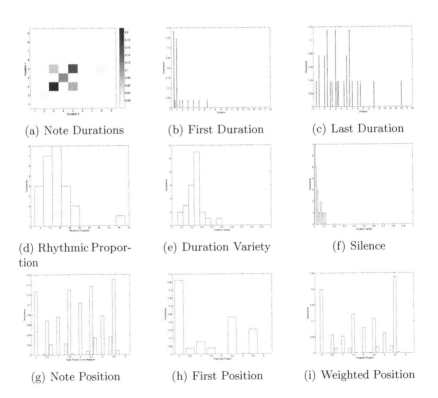

(a) Note Durations   (b) First Duration   (c) Last Duration

(d) Rhythmic Propor-   (e) Duration Variety   (f) Silence
tion

(g) Note Position   (h) First Position   (i) Weighted Position

Fig. 5. Rhythm

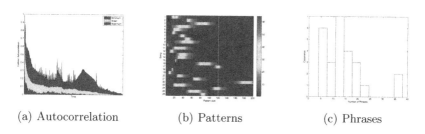

(a) Autocorrelation   (b) Patterns   (c) Phrases

Fig. 6. Patterns

5 Discussion and Future Work

All the metrics presented here can lead to different models according to the specific genres of music. Those models, in some cases, can even lead to problems which are simple to solve in polynomial time. In that case, evolutionary computation could be even unsuitable for composing. On the other hand, with all the information to be considered when generating compositions, it is unlikely to exist a good model of composition which is too trivial.

Although there are many other metrics that could be considered in the evaluation of melodies, such as contour shapes or rhythmic variation, the authors do not have the pretension to formulate all of them as it would not be feasible. However, by studying at least some of the most important metrics in relation to each category of analysis, this paper can surely give some background to scientists with intention to be evolutionary composers. Natural extensions of the ideas presented here would be to apply all the metrics on melody phrases separately and to filtrate which metrics are most important. It would be also important to perform second-order analysis on the melodies to look for potential relations between the metrics.

Once we are able to generate melodies that follow patterns of a studied database, another issue is also the diversity and originality of the solutions generated by the algorithm, as we do not want the algorithm to either return always the same "best" melody [5] or to ignore the originality needed in masterpieces [37]. Once we have considered those issues, we can focus on applying the statistical methods mentioned in Section 3 to get more formal objective values.

In regard to evolutionary computation, an important issue in the future will also be how to put all those metrics together into one or many objective functions and which genetic operators will be appropriate for those functions. So far, the formalized evaluation metrics for evolutionary music have only been partial and this paper should expand the ideas considered by evolutionary composers on their work. However, as music is a very contextual form art, specific metrics will always need to be created for specific genres of music.

Acknowledgements. This work has been supported by the Brazilian agencies CAPES, CNPq, and FAPEMIG and the Marie Curie International Research Staff Exchange Scheme Fellowship within the 7th European Community Framework Programme.

References

1. Galanter, P.: The problem with evolutionary art is... Applications of Evolutionary Computation, 321–330 (2010)
2. McCormack, J.: Open problems in evolutionary music and art. Applications of Evolutionary Computing, 428–436 (2005)
3. Todd, P., Werner, G.: Frankensteinian methods for evolutionary music. Musical Networks: Parallel Distributed Perception and Performace, 313 (1999)
4. Biles, J.: Genjam: A genetic algorithm for generating jazz solos. In: Proceedings of the International Computer Music Conference, International Computer Music Association, p. 131 (1994)
5. Freitas, A., Guimarães, F.: Originality and diversity in the artificial evolution of melodies. In: Proceedings of the 13th Annual Conference on Genetic and Evolutionary Computation, pp. 419–426. ACM (2011)
6. Biles, J., Anderson, P., Loggi, L.: Neural network fitness functions for a musical iga. In: International ICSC Symposium on Intelligent Industrial Automation, IIA 1996 and Soft Computing, SOCO 1996, International Computing Sciences Conferences (ICSC), pp. B39–B44 (1996)
7. Phon-Amnuaisuk, S., Law, E., Kuan, H.: Evolving music generation with som-fitness genetic programming. Applications of Evolutionary Computing, 557–566 (2007)
8. Biles, J.: Evolutionary computation for musical tasks. Evolutionary Computer Music (2), 28–51 (2007)
9. Freitas, A., Guimaraes, F.: Melody harmonization in evolutionary music using multiobjective genetic algorithms. In: Proceedings of the Sound and Music Computing Conference, SMC (2011)
10. Freitas, A., Guimaraes, F., Barbosa, R.: Ideas in automatic evaluation methods for melodies in algorithmic composition. In: Proceedings of the Sound and Music Computing Conference, SMC (2012)
11. Towsey, M., Brown, A., Wright, S., Diederich, J.: Towards melodic extension using genetic algorithms. Educational Technology & Society 4(2), 54–65 (2001)
12. McIntyre, R.: Bach in a box: The evolution of four part baroque harmony using the genetic algorithm. In: Proceedings of the First IEEE Conference on Evolutionary Computation, 1994. IEEE World Congress on Computational Intelligence, pp. 852–857. IEEE (1994)
13. Papadopoulos, G., Wiggins, G.: A genetic algorithm for the generation of jazz melodies. Proceedings of STeP 98 (1998)
14. Phon-Amnuaisuk, S., Wiggins, G.: The four-part harmonisation problem: a comparison between genetic algorithms and a rule-based system. In: Proceedings of the AISB 1999 Symposium on Musical Creativity, pp. 28–34 (1999)
15. McDermott, J., Griffith, N., O'Neill, M.: Toward user-directed evolution of sound synthesis parameters. Applications of Evolutionary Computing, 517–526 (2005)
16. Khalifa, Y., Foster, R.: A two-stage autonomous evolutionary music composer. Applications of Evolutionary Computing, 717–721 (2006)

17. Horner, A., Goldberg, D.: Genetic algorithms and computer-assisted music composition. Urbana 51(61801), 14 (1991)
18. Jarque, C., Bera, A.: A test for normality of observations and regression residuals. International Statistical Review/Revue Internationale de Statistique, 163–172 (1987)
19. Massey Jr., F.: The kolmogorov-smirnov test for goodness of fit. Journal of the American Statistical Association, 68–78 (1951)
20. Bashkansky, E., Gadrich, T., Kuselman, I.: Interlaboratory comparison of test results of an ordinal or nominal binary property: analysis of variation. Accreditation and Quality Assurance: Journal for Quality, Comparability and Reliability in Chemical Measurement, 1–5
21. Jobim, A., Chediak, A.: Songbook Tom Jobim, vol. 1. Irmãos Vitale (1990)
22. Krumhansl, C.: Cognitive foundations of musical pitch, vol. 17. Oxford University Press, USA (2001)
23. Toiviainen, P., Krumhansl, C., et al.: Measuring and modeling real-time responses to music: The dynamics of tonality induction. Perception-London 32(6), 741–766 (2003)
24. Dowling, W., Harwood, D.: Music cognition. Academic Press, New York (1986)
25. Nettl, B.: Music in primitive culture. Harvard University Press (1956)
26. Dowling, W.: Scale and contour: Two components of a theory of memory for melodies. Psychological Review 85(4), 341 (1978)
27. Narmour, E.: The analysis and cognition of melodic complexity: The implication-realization model. University of Chicago Press (1992)
28. Krumhansl, C.: Music psychology and music theory: Problems and prospects. Music Theory Spectrum, 53–80 (1995)
29. Eerola, T., Toiviainen, P.: MIDI Toolbox: MATLAB Tools for Music Research. University of Jyväskylä, Jyväskylä, Finland (2004)
30. Krumhansl, C., Kessler, E.: Tracing the dynamic changes in perceived tonal organization in a spatial representation of musical keys. Psychological Review 89(4), 334 (1982)
31. Lerdahl, F.: Calculating tonal tension. Music Perception, 319–363 (1996)
32. Palmer, C., Krumhansl, C.: Mental representations for musical meter. Journal of Experimental Psychology. Human Perception and Performance 16(4), 728 (1990)
33. Thompson, W.: Sensitivity to combinations of musical parameters: Pitch with duration, and pitch pattern with durational pattern. Attention, Perception, & Psychophysics 56(3), 363–374 (1994)
34. Eerola, T., Himberg, T., Toiviainen, P., Louhivuori, J.: Perceived complexity of western and african folk melodies by western and african listeners. Psychology of Music 34(3), 337–371 (2006)
35. Tenney, J., Polansky, L.: Temporal gestalt perception in music. Journal of Music Theory 24(2), 205–241 (1980)
36. Cambouropoulos, E.: Musical rhythm: A formal model for determining local boundaries, accents and metre in a melodic surface. Music, Gestalt, and Computing, 277–293 (1997)
37. Simonton, D.K.: Computer content analysis of melodic structure: Classical composers and their compositions. Psychology of Music 22(1), 31–43 (1994)

Efficient Discovery of Chromatography Equipment Sizing Strategies for Antibody Purification Processes Using Evolutionary Computing

Richard Allmendinger, Ana S. Simaria, and Suzanne S. Farid

University College London, Torrington Place, London WC1E 7JE, UK
{r.allmendinger,a.simaria,s.farid}@ucl.ac.uk

Abstract. This paper considers a real-world optimization problem involving the discovery of cost-effective equipment sizing strategies for the chromatography technique employed to purify biopharmaceuticals. Tackling this problem requires solving a combinatorial optimization problem subject to multiple constraints, uncertain parameters (and thus noise), and time-consuming fitness evaluations. After introducing this problem, an industrially-relevant case study is used to demonstrate that evolutionary algorithms perform best when infeasible solutions are repaired intelligently, the population size is set appropriately, and elitism is combined with a low number of Monte Carlo trials (needed to account for uncertainty). Adopting this setup turns out to be more important for scenarios where less time is available for the purification process.

1 Introduction

Monoclonal antibodies (mAbs) represent the fastest growing category of therapeutic biopharmaceutical drugs due to their unique binding specificity to targets. The manufacturing process for mAbs is costly and time-consuming, and can be divided into two phases (see Fig. 1): *upstream processing* (USP) and *downstream processing* (DSP). In USP, mammalian cells expressing the mAb of interest are cultured in bioreactors. Then the broth moves to DSP, where the mAb is recovered, purified and cleared from viruses using a variety of operations including a number of chromatography steps. Chromatography operations are identified as critical steps in a mAb purification process and can represent a significant proportion of the purification material costs (associated e.g. with the use of expensive affinity resins and large amounts of buffer reagents). Whilst alternatives to traditional column chromatography platforms are emerging, industry practitioners are still reluctant to perform major process changes [1]. At the same time, it is important to determine how best to use existing production facilities for mAbs [2]. This is particularly challenging given the significant improvements in USP productivities that have been accomplished over the past decade with higher mAb concentrations (*titres*) being achieved in cell culture. These improvements have not been matched in purification capacities, leading to concerns over purification bottlenecks and the desire to continuously optimize the design and operation of existing chromatography steps. Hence, to efficiently exploit these cell culture improvements, and account for the increasing demand for therapeutic mAbs, it has become critical to identify cost-effective purification processes [1].

C.A. Coello Coello et al. (Eds.): PPSN 2012, Part II, LNCS 7492, pp. 468–477, 2012.

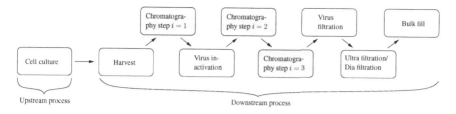

Fig. 1. Typical flowsheet for an antibody manufacturing process

An approach to realize this identification step, which is also adopted here, is to develop simulation models of mAb manufacturing processes and identify promising chromatography setups using computational methods. For example, in [3] the authors present a simulation model to identify windows of operation for the column diameter, bed height and loading flowrate of a chromatography step using productivity and cost of goods (COGs) as performance criteria. A model to find combinations of protein load and loading flowrate that meet yield and throughput constraints has been developed in [4]. The discrete-event simulation framework proposed in [5] allows the selection of optimal chromatography column diameters over a range of titres. The methodology used in [3–5] consists of selecting and evaluating specific values within the full range of variation of the critical parameters. However, such an approach may not be feasible for very large decision spaces as considered here, which drives the need for more efficient optimization methods in this domain.

This study addresses this issue by investigating the application of evolutionary optimization methods for the discovery of *chromatography column sizing strategies* — defined here by the diameter and bed height of a column, the number of columns used in parallel, and the number of cycles a column is run for — that are cost-effective in terms of COGs per gram (COG/g) of product manufactured. This discovery task can be formulated as a combinatorial (single-objective) optimization problem subject to multiple constraints and interacting decision variables, uncertain parameters and expensive fitness evaluations (represented by time-consuming computer simulations). Over the years, evolutionary algorithms (EAs) have proven to be efficient, flexible and robust optimizers for challenging optimization problems of this type — which are commonly referred to as *closed-loop optimization problems* [6, 7].

An industrially-relevant case study is used to investigate how to tune some of the simple EA configuration parameters: population size, degree of elitism, number of Monte Carlo trials (needed to cope with uncertain parameters), and constraint-handling method. The fitness landscape of different scenarios of the case study are analyzed also to observe which landscape features pose a particular challenge when optimizing equipment sizing strategies.

The rest of the paper is organized as follows. The next section describes the chromatography sizing problem considered in this work in more detail. Section 3 outlines the case study, choice of algorithms and the parameter settings considered for tackling the case study. The experimental results are presented and analyzed in Section 4, and Section 5 concludes the paper.

Fig. 2. A candidate solution (sizing strategy) with $k = 3$ chromatography steps. Each step $i = 1, ..., k$ is defined by the bed height h_i and diameter d_i of columns, number of cycles $n_{CYC,i}$ each column is used, and the number of columns $n_{COL,i}$ operating in parallel.

2 Problem Domain: Chromatography Equipment Sizing

The *chromatography equipment sizing problem* can be represented as a combinatorial optimization problem with the task of finding the most cost-effective chromatography sizing setup for a sequence of chromatography steps used in the purification process of mAbs. In the following the decision variables, objective function, constraints, and uncertain parameters to which this problem is subject to are described.

Decision Variables: Fig. 2 shows the encoding used to represent a solution x to the chromatography sizing problem. For each chromatography step $1 \leq i \leq k$ (k is the total number of steps) (e.g. affinity or ion-exchange chromatography) four discrete decision variables were defined related to the sizing and operation of chromatography columns: bed height h_i and diameter d_i of columns, number of cycles $n_{CYC,i}$ each column is used, and the number of columns $n_{COL,i}$ operating in parallel. That is, the problem is subject to $l = k \cdot 4$ discrete variables in total. For each step i, the variables define the (i) total volume of resin V_i available for the purification of a product at that chromatography step, and the (ii) processing time T_i that the chromatography step takes; both parameters are calculated according to standard mass balance equations as follows [8]:

$$V_i = \pi \cdot d_i^2/4 \cdot h_i \cdot n_{CYC,i} \cdot n_{COL,i} \tag{1}$$

$$T_i = n_{CYC,i} \cdot h_i \cdot (CV_{BUFF,i} + CV_{LOAD,i}/n_{COL,i}) \cdot u_i, \tag{2}$$

where $CV_{BUFF,i}$ and $CV_{LOAD,i}$ are the number of column volumes of buffer and product load per cycle, and u_i is the linear flowrate of the resin used at step i.

Objective Function: Our objective f is to find a chromatography sizing setup that yields minimal *cost of goods per gram (COG/g) of product manufactured*. The COGs include both direct (resource) costs (e.g. resin, buffer and labor costs) and indirect costs (e.g. facility-dependent overheads, such as maintenance costs and depreciation), and is divided by the total annual product output P to yield the metric COG/g. The COG/g values are obtained by running a detailed process economics model, which simulates the different purification steps based on mass balance and cost equations as defined in [8].

Constraints: The problem is subject to two types of constraints:

1. Each chromatography step $i = 1, ..., k$ needs to satisfy a *resin requirement constraint* to ensure that the resin volume V_i available for purification at step i is sufficient to process the mass of product M_i entering that step, given the resin's dynamic binding capacity DBC_i and the maximum utilization factor κ. Formally, this constraint can be defined as

$$V_i \geq \frac{M_i \cdot \kappa}{DBC_i} \,. \tag{3}$$

Solutions violating this constraint *are considered infeasible* and handled using one of the constraint-handling strategies introduced in Section 3.

2. There is also a *demand constraint* to ensure that the amount of product manufactured P is sufficient to satisfy the annual demand D, or $P \geq D$. This constraint may be violated for column sizing strategies with long chromatography processing times T_i. The use of COG/g as the objective function (recall that the product output P is in the denominator of this metric) was found to be sufficient to cope with this constraint. Hence, if a solution violates the demand constraint, then it is *not considered infeasible*.

Uncertainties: Uncertainty related to the product titre can have a significant impact on the annual product output P. As the equipment sizing is a function of an expected titre value for bioreactors through to chromatography columns, titre fluctuations can cause (i) failure to meet demand (if titre is lower than expected) or (ii) product waste (if titre is higher than expected and equipment capacity is insufficient to process the excess). Other sources of uncertainty (e.g. yield) may be present and are realistic but are not considered in this paper.

3 Experimental Setup

This section describes the case study, search algorithms and their parameter settings as used in the subsequent experimental analysis.

Case Study Setup: The case study considered in this work is industrially-relevant and focuses on a single-product mAb manufacturing facility that employs a process sequence as shown in Fig. 1 (with $k = 3$ chromatography steps) to satisfy a total product demand of $D = 500$kg/year with an expected titre of 3g/L. Titre variabilities were modeled using the triangular probability distribution, Tr(2.6,3.0,3.4). Three scenarios of this case study with different ratios of USP:DSP trains were investigated: 1:1, 2:1 and 4:1. The USP train refers to the number of bioreactors operating (in a staggered mode), and an increase in the USP:DSP ratio corresponds to a decrease in the *DSP window*, the time available to perform chromatography. The range of possible decision variable values is 15 cm $\leq h_i \leq$ 25 cm (11 values), 50 cm $\leq d_i \leq$ 200 cm (16 values), $1 \leq n_{\text{CYC},i} \leq 10$ (10 values), $1 \leq n_{\text{COL},i} \leq 4$ (4 values), $i = 1, 2, 3$; i.e. there are $(11 \cdot 16 \cdot 10 \cdot 4)^3 \approx 3.5 \cdot 10^{11}$ sizing strategies in total. The sizing strategy employed in industry is obtained based on empirical rules: a single column $n_{\text{COL},i} = 1$ with a fixed bed height of $h_i = 20$ cm is run for a fixed number of cycles $n_{\text{CYC},i} = 5$ with the diameter size d_i being calculated such that the resulting total resin volume V_i (Equation (1)) satisfies the resin requirement constraint (Equation (3)).

Search Algorithms: To gain insight into the behavior of evolutionary search algorithms on the chromatography sizing problem, four types of search algorithms were considered: a standard generational genetic algorithm (SGA), a genetic algorithm with generation gap (GA-GG), a genetic algorithm with a $(\mu + \lambda)$-ES reproduction scheme

(GA-ES), and a population of stochastic hill-climbers (PHC). All four algorithms began the search with the same initial population containing μ randomly generated solutions. The algorithms used also the same mutation operator, which selected a decision variable value at random from the set of possible values. SGA used uniform crossover and random flip mutation as the variation operators, and binary tournament selection (with replacement) for parental selection; for environmental selection, it replaced the entire current population with the offspring population. GA-GG and GA-ES differ from SGA in the environmental selection step only. With GA-GG, the new population was formed by selecting the fittest μ solutions from the combined pool of the offspring population and the two fittest solutions of the current population. With GA-ES, a greater degree of elitism was employed and the fittest μ solutions from the combined pool of the current population and the offspring population were selected. PHC maintained a population of stochastic hill-climbers, which, at each generation g, independently underwent mutation and replaced their parent if it was at least as fit.

Accounting of Uncertainty: To account for titre variabilities, m Monte Carlo trials (based on the probability distribution Tr(2.6,3.0,3.4)) were performed for each candidate solution. The fitness of a solution was then the average of the COG/g values across the m trials, and this average was updated if a solution happened to be evaluated multiple times during an optimization procedure.

Handling of Infeasible Solutions: Five constraint-handling strategies were analyzed to cope with infeasible solutions (violating Equation(3)). Four of them (RS1, RS2, RS3 and RS4) *repaired* infeasible solutions, i.e. modified the genotype of a solution, while strategy RS5 avoided repairing.

The four *repairing strategies* iteratively increased the values of the decision variables (associated with a particular chromatography step i), one variable at a time, until Equation (3) was satisfied or until the maximum value of a variable was reached, in which case the value of another variable was increased. The sequence in which the variables were modified affected the search. To investigate this effect, different sequences, represented by the strategies RS1 to RS4, were analyzed. The strategy RS1 applied repairing according to the decision variable sequence $d_i \rightarrow n_{CYC,i} \rightarrow h_i \rightarrow n_{COL,i}$ (where i is the chromatography step violating Equation (3)); this sequence represents typical rules applied in equipment sizing scale-up models. The strategy RS2 employed the inverse sequence of RS1. The strategies RS3 and RS4 switch between different repairing sequences during an optimization procedure. While RS3 chooses at random between the two sequences employed by RS1 and RS2, the strategy RS4 chooses at random among all possible repairing sequences (note, there are 4! sequences in total) whenever it needs to be repaired. The approach employed by RS4 is plausible e.g. if no prior knowledge about promising repairing sequences would be available. The strategy RS5 does not apply repairing but penalizes infeasible solutions by degrading their fitness by a large penalty value c.

The experimental study investigated different settings of the parameters involved in the search algorithms. The default settings used are given in Table 1. Any results shown are average results across 20 independent algorithm runs. A different seed was used for the random number generator for each EA run but the same seeds for all strategies. This allows for the application of a repeated-measures statistical test, the Friedman test, to investigate performance differences between algorithmic setups.

Table 1. Default parameter settings of search algorithms

Parameter	Setting
Parent population size μ	80
Offspring population size λ	80
Per-variable mutation probability	$1/l$
Crossover probability	0.6
Constraint-handling strategy	RS1
Number of generations G	25
Penalty value c	5000
Monte Carlo trials m	25

4   Experimental Results

Before analyzing the behavior of evolutionary search algorithms on the chromatography equipment sizing problem, an indication of the properties of the fitness landscapes spanned by three case study scenarios is given. For this, the *adaptive walks method* already used in [7] was adopted. This involved performing 1000 adaptive walks (using a fixed titre of 3g/L) on the landscape of each scenario, and recording the length and final fitness of each walk. Figure 3 shows the distribution of both measurements in the form of boxplots. From Figure 3(a) it can be observed that increasing the USP:DSP ratio decreases the average length of an adaptive walk. That is, the landscape becomes more rugged, or, equivalently, the number of local optima increases. This pattern is due to tighter DSP windows, which cause more solutions to violate the demand constraint and thus makes the problem harder to solve. This also causes an increase in the COG/g values as indicated in Figure 3(b). The next section presents an analysis of how the search algorithms fared, for both the deterministic (using a fixed titre of 3g/L) and stochastic scenario.

Deterministic Product Titre: Figure 4(a) analyzes the performance of the search algorithms as a function of the population size μ. The aim of this experiment was to understand whether a large population should be evolved for few generations, or a small population for many generations. This understanding is important when optimizing subject to limited resources, such as limited computational power and time constraints. The figure illustrates that: (i) a population size of around $40 \leq \mu \leq 80$ yielded the best performance for the GA-based algorithms, (ii) GA-ES found the most cost-effective strategies, and (iii) random search outperforms PHC. Small population sizes, or search algorithms employing no elitism, such as SGA, did not perform well due to the high probability of getting trapped in one of the many local optima of the fitness landscape. Large population sizes converged slowly due to the low number of generations available for optimization. PHC was inferior to random search because the hill-climbers could get trapped in local optima, in which case further improvements were unlikely, while random search kept on generating (at random) new and potentially fitter solutions. (The performance of random search is constant for varying μ as it depends only on the number of function evaluations performed.)

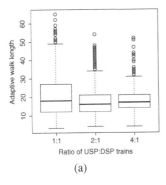

(a)

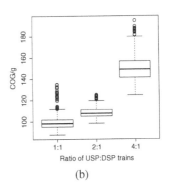

(b)

Fig. 3. Boxplots showing the distribution of the (a) length and (b) final fitness (COG/g) of 1000 adaptive walks for different USP:DSP ratios. The box represents the 25th and 75th percentile with the median indicated by the dark horizontal lines. The whiskers represent the observations with the lowest and highest value still within $1.5 \cdot$ IQR of the 25th and 75th percentile, respectively; solutions outside this range are indicated as dots.

Figure 4(b) investigates the performance impact of the constraint-handling strategies RS1 to RS5 when augmented on GA-ES (a similar performance impact was present for the other search algorithms). It demonstrated that the constraint-handling strategy employed had an effect on the convergence speed and the final solution quality. It also indicated that a repairing strategy (RS1, RS2, RS3 and RS4) should be preferred over a non-repairing one (RS5). The superior performance of RS1 is due to the fact that the variable d_i is modified (increased) first to repair a solution. Unlike to the other variables, an increase in d_i is often sufficient to just satisfy the resin requirement constraint without increasing the processing time. From the performance obtained with RS2, RS3 and RS4 it can be concluded that if d_i cannot be changed, then either the variable $n_{CYC,i}$ or h_i should be modified to meet the resin requirement constraint.

Stochastic Product Titre: The performance of the algorithms was then investigated in the presence of uncertain product titres. Figure 5 indicates that uncertainty impacts negatively the convergence speed and under certain circumstances also the final solution quality. This impact tends to be less severe as the degree of elitism employed by an algorithm increases (i.e. the performance of GA-ES is less affected than the one of SGA). Elitism can help circumventing this issue as it causes a population to converge (quickly) to a (local) optimal region and then exploit this region. However, on the other hand, too much elitism (Figure 5(a)) may disturb and prevent the discovery of innovative solutions; here, optimization in a stochastic environment using relatively small values of m can yield better performance than optimization in a deterministic environment due to the greater randomness in the search. When the optimizer does not employ elitism (Figure 5(b)), however, any additional randomness in the search may be a burden (as it can cause a population to oscillate between different search space regions, preventing or slowing down convergence towards promising regions).

Figure 6 shows the sizing strategies for the most expensive chromatography step ($i = 1$) found by GA-ES for the USP:DSP ratios 1:1 (Figure 6(a)) and 4:1 (Figure 6(b)) at the end of the search across 20 independent algorithmic runs. For both scenarios,

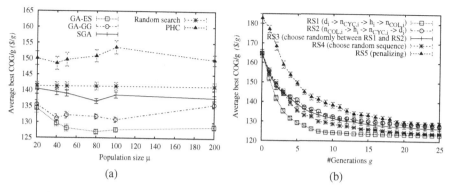

(a)   (b)

Fig. 4. (a) Average best COG/g (and its standard error) obtained by different search algorithms as a function of the population size μ; the total number of fitness evaluations was fixed to 2000, i.e. the number of generations is $G = \lfloor 2000/\mu \rfloor$. (b) Average best COG/g, as a function of the generation counter g, obtained by GA-ES using different repairing strategies. Both experiments were conducted on a chromatography equipment sizing problem featuring a ratio of USP:DSP trains of 4:1. For each setting shown on the abscissa, a Friedman test (significance level of 5%) has been carried out: In (a), GA-ES performs best for $\mu > 40$, and in (b), RS1 performs best in the range $1 < g < 15$.

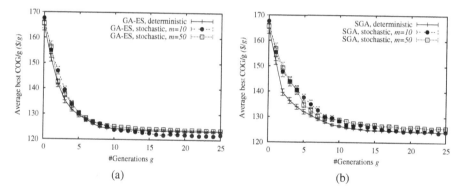

(a)   (b)

Fig. 5. Average best COG/g (and its standard error) obtained by (a) GA-ES and (b) SGA in a deterministic and stochastic environment (using different values for the number of Monte Carlo trials m) as a function of the generation counter g. For each setting shown on the abscissa, a Friedman test (significance level of 5%) has been carried out: In (a), GA-ES with $m = 10$ performs best for $g > 15$, and in (b), SGA, deterministic, performs best in the range $1 < g < 6$.

the solutions shown have COG/g values that do not differ by more than 3% of each other. Comparing the most cost-effective sizing strategy found by GA-ES (filled bubble) with the strategy used in industry (filled diamond), GA-ES is able to reduce the COG/g for 1USP:1DSP and 4USP:1DSP by up to 5% (mainly through sizing strategies featuring smaller h_1 and/or d_1 in combination with more cycles $n_{CYC,1}$) and 20% (through sizing strategies exhibiting fewer cycles $n_{CYC,1}$ and larger d_1), respectively. Another advantage of EAs is that the result of an optimization procedure is a set of

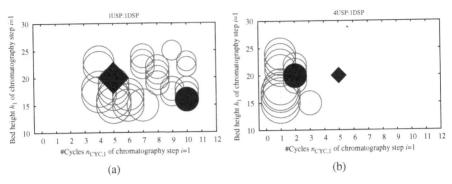

Fig. 6. Column sizing strategies for the most expensive chromatography step ($i = 1$) found by GA-ES at the end of the search across 20 independent algorithm runs (within an uncertain optimization environment) (bubbles) for the scenarios (a) 1USP:1DSP and (b) 4USP:1DSP. The size of a bubble is proportional to the variable d_1; all solutions feature the setup $n_{COL,1} = 1$. The fitness values of all solutions found by the EA for a particular scenario are within 3% of each other. For each scenario, the filled bubble represents the optimal setup found by the EA. The setup used by industry is indicated with a filled diamond and was not part of the solution set found by the EA.

cost-efficient sizing strategies (rather than a single strategy), providing flexibility and freedom to account for facility space restrictions and user preferences when it comes to selecting a final sizing strategy. Note, the EA finds more similar solutions for the scenario 4USP:1DSP than for 1USP:1DSP because the problem is harder to solve, as already indicated in the landscape analysis conducted previously.

5   Conclusion and Future Work

This paper has considered a real-world problem concerned with the discovery of cost-effective equipment sizing strategies for purification processes (with focus on chromatography steps) of biopharmaceuticals. This application can be formulated as a combinatorial closed-loop optimization problem subject to (i) expensive fitness evaluations, (ii) multiple dependent decision variables, (iii) constraints, and (iv) uncertain parameters.

The study revealed that EAs can identify a diverse set of equipment sizing strategies that are more cost-efficient than the strategies used in industry. In particular, the analysis demonstrated that an EA performs best when elitism is used in combination with a small number of Monte Carlo trials (to cope with uncertain parameters), infeasible solutions are repaired using a non-trivial strategy, and (when resources are limited) a medium-sized population (a size between $30 \leq \mu \leq 80$) is evolved for a relatively large number of generations.

Future research will look at extending the equipment sizing problem considered here with decision variables related to the sequence of a purification process employed. This will make the optimization tool developed more versatile, and also help gain more insights into the working of EAs.

Acknowledgement. Financial support from the EPSRC Centre for Innovative Manufacturing in Emergent Macromolecular Therapies with a consortium of industrial and government users is gratefully acknowledged.

References

1. Langer, E.: Downstream factors that will continue to constrain manufacturing through 2013. BioProcessing Journal 8(4), 22–26 (2009)
2. Kelley, B.: Industrialization of mAb production technology: The bioprocessing industry at a crossroads. MAbs 1(5), 443–452 (2009)
3. Joseph, J.R., Sinclair, A., Tichener-Hooker, N.J., Zhou, Y.: A framework for assessing the solutions in chromatographic process design and operation for large-scale manufacture. Journal of Chemical Technology and Biotechnology 81, 1009–1020 (2006)
4. Chhatre, S., Thillaivinayagalingam, P., Francis, R., Titchener-Hooker, N.J., Newcombe, A., Keshavarz-Moore, E.: Decision-Support Software for the Industrial-Scale Chromatographic Purification of Antibodies. Biotechnology Progress 23, 88–894 (2007)
5. Stonier, A., Smith, M., Hutchinson, N., Farid, S.S.: Dynamic simulation framework for design of lean biopharmaceutical manufacturing operations. Computer Aided Chemical Engineering 26, 1069–1073 (2009)
6. Knowles, J.: Closed-Loop Evolutionary Multiobjective Optimization. IEEE Computational Intelligence Magazine 4(3), 77–91 (2009)
7. Allmendinger, R.: Tuning Evolutionary Search for Closed-Loop Optimization. PhD thesis, University of Manchester, Manchester, UK (2012)
8. Farid, S.S., Washbrook, J., Titchener-Hooker, N.J.: Modelling biopharmaceutical manufacture: Design and implementation of SimBiopharma. Computers & Chemical Engineering 31, 1141–1158 (2007)

Beware the Parameters: Estimation of Distribution Algorithms Applied to Circles in a Square Packing

Marcus Gallagher

School of Information Technology and Electrical Engineering,
University of Queensland, Q. 4702. Australia
marcusg@itee.uq.edu.au
http://www.itee.uq.edu.au/~marcusg

Abstract. Simple continuous estimation of distribution algorithms are applied to a benchmark real-world set of problems: packing circles in a square. Although the algorithms tested are very simple and contain minimal parameters, it is found that performance varies surprisingly with parameter settings, specifically the population size. Furthermore, the population size that produced the best performance is an order of magnitude larger that the values typically used in the literature. The best results in the study improve on previous results with EDAs on this benchmark, but the main conclusion of the paper is that algorithm parameter settings need to be carefully considered when applying metaheuristic algorithms to different problems and when evaluating and comparing algorithm performance.

Keywords: Estimation of Distribution Algorithms, Circles in a Square Packing Problems, Parameter Settings.

1   Introduction

Some of the strengths of metaheuristic optimization algorithms are their general applicability and ease of implementation to produce good results. When an algorithm is proposed and evaluated in the literature, parameter values are required to be specified. While it is recognized that parameters need to be adjusted for different problems (or, with self-adaptive parameter tuning techniques, dynamically during execution), the hope is that performance should be relatively insensitive to parameter values and that the default values will be a reasonable starting point. These assertions are not frequently checked in experimental studies in the literature.

In this paper simple continuous Estimation of Distribution Algorithms (continuous Univariate Marginal Distribution Algorithm (UMDA$_c^G$) and the Estimation of Multivariate Normal Algorithm (EMNA)) are applied to a benchmark real-world set of geometric packing problems (circles in a square). The problems are representative of real-world packing problems and are known to be challenging for optimization algorithms. In Sec. 2 we describe the EDAs used and the

C.A. Coello Coello et al. (Eds.): PPSN 2012, Part II, LNCS 7492, pp. 478–487, 2012.

circles in a square packing problems. The methodology used is a sequence of experiments and results which are described in Sec. 3. In Sec. 4 a discussion of the results is given and the paper is concluded.

2   Background

2.1   Continuous EDAs: UMDA$_c^G$ and EMNA

Estimation of Distribution Algorithms are a class of population-based meta-heuristics that utilize a probability distribution to direct the search process. One of the simplest EDAs is the continuous Univariate Marginal Distribution Algorithm UMDA$_c^G$ [10]. In UMDA$_c^G$, new individuals/candidate solutions are generated using a factorized product of univariate Gaussian distributions. Following truncation selection (retaining the fraction τ of the fittest individuals), the parameters of these distributions are updated via their maximum likelihood estimates over the selected individuals. The algorithm is summarized in Table 1.

Table 1. General pseudocode for UMDA$_c^G$

Given: population size p, selection parameter τ
BEGIN (set $t = 0$) Generate p individuals uniformly random in the search space
REPEAT for $t = 1, 2, \ldots$ until stopping criterion is met
    Select $p_{sel} < p$ individuals via truncation selection
    Estimate model parameters μ_t, σ_t^2 via Max. likelihood
    Sample p individuals from $\mathcal{N}(\mu_t, \sigma_t^2)$
    $t = t + 1$
ENDREPEAT

EMNA is a similar algorithm but uses a full multivariate Gaussian distribution instead of a factorized product of univariate distributions. The only change to the pseudocode is the estimation of a sample covariance matrix Σ_t rather than individual σ_t^2 values. Many of the continuous EDAs proposed in the literature use a Gaussian distribution and extend on the basic UMDA$_c^G$ and EMNA algorithms.

For implementation, EMNA and UMDA$_c^G$ have two parameters: p and τ. For τ, a range of values have been used in the literature (e.g $\tau = 0.8$ [7]) though there seems to be a preference for smaller values (e.g. τ=0.2 or 0.3 [3,8,14], τ=0.5 [5,6,10]). For p, various values have been used from 200 up to 2000 for test problems typically of dimensionality up to 50-D. p has also been scaled with dimensionality, e.g. in [2] it is suggested that for UMDA$_c^G$, $p \geq 15d^{0.5} + 5$ and for EMNA, $p \geq 4d^{1.5} + 16$ (note that these guidelines are for modified versions of the algorithms, incorporating adaptive variance scaling and anticipated mean shift). In the results below problems up to 60-D are used; these rules would recommend $p \geq 121$ and $p \geq 1875$ respectively.

2.2   Circles in a Square Packing Problems

Circles in a square (CiaS) is a class of well-studied geometric packing problems. Given the unit square defined in a 2D Euclidean space and a pre-specified number of circles, n_c, constrained to be of equal size, the problem is to find an optimal packing; i.e. to position the circles and compute the radius length of the circles such that the circles occupy the maximum possible area within the square. All circles must remain fully enclosed within the square, and cannot overlap. Mathematically, the problem can be stated as follows [1]. Let $C(\mathbf{z}^i, r)$ be the circle with radius r and center $\mathbf{z}^i = (y_1^i, y_2^i) \in I\!\!R^2$. Then the optimization problem is:

$$r_n = \max r \tag{1}$$

$$C(\mathbf{z}^i, r) \subseteq [0,1]^2, i = 1, \ldots, n_c \tag{2}$$

$$C^{int}(\mathbf{z}^i, r) \cap C^{int}(\mathbf{z}^j, r) = \emptyset \; \forall \; i \neq j \tag{3}$$

where C^{int} is the interior of a circle. Alternatively, the problem can be reformulated as finding the positions of n_c points inside the unit square such that their minimum pairwise distance is maximized. In this case the problem (and constraint) can be restated as:

$$d_n = \max \min_{i \neq j} \| \mathbf{w}^i - \mathbf{w}^j \|_2 \tag{4}$$

$$\mathbf{w}^i \in [0,1]^2, i = 1, \ldots, n_c \tag{5}$$

It is known that a solution to (4) can be transformed into a solution to (1) using the following relation:

$$r_n = \frac{d_n}{2(d_n + 1)}.$$

From the point of view of evaluating metaheuristic optimization algorithms, the problem given by (4) is convenient because generating a feasible candidate solution simply requires placing a set of n points within the unit square. Note that the optimization problem is over $2n_c$ continuous variables (the coordinates of each point $\mathbf{w}^i$ in the unit square).

CiaS packing problems represent a challenging class of optimization problems. In general, they cannot be solved using analytical approaches or via gradient-based mathematical optimization. These problems are also believed to generally contain an extremely large number of local optima. For the related problem of packing equal circles into a larger circular region, Grosso et al. use a computational approach to estimating the number of local optima by repeatedly running a local optimization algorithm over a large number of trials [9]. Although a conservative estimate, this indicates that the number of local optima grows unevenly but steadily, with at least 4000 local optima for $n_c = 25$ and more than 16000 local optima for $n_c = 40$.

Castillo et al.[4] present a survey of industrial problems and application areas that involve circle packing: including cutting, container loading, cylinder packing, facility dispersion, communication networks and facility and dashboard layout problems. CiaS packing problems can be considered as a simplified version of such real-world problems and have received a large amount of attention in the mathematical, optimization and operations research literature (see [4] for a recent overview). For most values of n_c below 60 and for certain other values, provably optimal packings have been found using either theoretical or computational approaches (see [12] and the references therein). For larger values of n_c, finding provably optimal packings in general becomes increasingly difficult and time-consuming. The Packomania website [11] maintains an large list of the optimal (or best known) packings for many values of n_c from 2 up to 10000, along with references and other related resources.

Previous work has also applied heuristics and global optimization algorithms to CiaS problems with the aim of finding good solutions but without a guarantee of global optimality. Some general-purpose metaheuristics such as simulated annealing have been applied [13] as well as special-purpose metaheuristics designed for the problem (see [1,4,12]). There have been few applications of evolutionary or population-based metaheuristics to CiaS problems. $UMDA_c^G$ was previously applied to CiaS with little attempt to optimize the algorithm parameters, and performance was relatively poor.

2.3   Formulating EDAs for CiaS Packing

Using the formulation given in (4), the feasible search space of a CiaS problem is defined by the unit hypercube $[0, 1]^{2n_c} \subset \mathbb{R}^{2n_c}$. The $UMDA_c^G$ algorithm starts with an initial population generated uniformly across the search space. It is clearly easy to achieve this for CiaS problems and this is the approach taken here. Note however that the use of heuristics exploiting problem knowledge may lead to improved starting populations.

The feasible search space for CiaS problems is similar to the simple (often symmetric) box boundary constraint assumed on many commonly used mathematical test functions for benchmarking continuous metaheuristics. However for CiaS problems, any candidate solution with one or more coordinate values outside this region will be infeasible (with an undefined objective function value), in contrast to an artificial test function (e.g. mathematical equation) where the objective function can still be evaluated outside the feasible search space. A naive application of $UMDA_c^G$ will thus result in large numbers of infeasible solutions being generated, since all univariate Gaussian distributions within the model are capable of generating component solution values in the range $[-\infty, \infty]$. While it is possible to employ a general constraint handling technique (e.g. creating a penalty function), a simple approach is taken here by repairing infeasible solutions utilizing a small amount of prior knowledge about the problems. Given that the objective of the problem (in Eqn. 4) is to maximize the minimum pairwise distance of the n_c points to be positioned in the unit square, it is to be expected that optimal solutions for any size problem will involve positioning a subset of points on the boundary of the square[1]. Therefore, to facilitate the generation of

---

[1] Equivalently, an optimal packing of circles will always contain circles that touch the boundary of the square.

such candidate solutions, any value in a solution vector generated during a run of UMDA$_c^G$ that lies outside the feasible region is reset to the (nearest) boundary. That is, $\forall \mathbf{w}^i = (w_1, w_2), i = 1, \ldots, n_c$, if $w_1 < 0$, then set $w_1 = 0$ or if $w_1 > 1$ then set $w_1 = 1$, with identical conditions for w_2. This simple check is performed on every generated individual and guarantees the feasibility of every candidate solution.

3   Experimental Methodology and Results

In this Section a series of experiments are described and results presented for UMDA$_c^G$ and EMNA on CiaS problems. Experiment (a) is a screening experiment to determine reasonable algorithm parameter values which are then applied to a range of CiaS problems (Experiments (b) and (c)). The first objective is to provide a set of results on these problems that can be used in comparison for future research. However the experiments are sequential and dependent, hence the results obtained in Experiment (a) lead to further investigation of the value of p (with interesting results) in Experiments (b) and (c). The intention is to be as transparent as possible about the experiments in contrast to reporting only the best results obtained after an unreported amount of parameter tuning has taken place.

The maximum number of function evaluations for each experimental trial needs to be sufficiently large to ensure that the algorithm has sufficient time to converge. From the point of view of solving the problem, we anticipate that the algorithm can obtain good performance in a reasonable amount of function evaluations (e.g. polynomial in the problem dimensionality). From the experimental point of view, we have a finite amount of computing resources and want to conduct repeated trials of experiments to examine the statistics of the experiments. For all the experiments in this paper, the maximum number of functions evaluations was set to $MAXF = 10^7$ and each experiment was run for 30 repeated trials (random initializations).

3.1   Experiment (a): Screening for τ and p

The aim of the first experiment was to determine values for τ and p that provide good performance. A single problem size ($n_c = 5$) was selected for the experiment, towards the smaller end of the range (for computational speed) but not at the end of the range (in the hope that higher-dimensional problem features are present in this problem). The assumption is that parameter values determined on this problem size will also work well for other CiaS problem sizes. Note however that it is an assumption made for practical considerations, to find a reasonable starting point for further experiments. It is unlikely to be optimal (indeed the experiments further below confirm this with respect to p). Since $0 < \tau \leq 1$, a linear grid of 10 values $(0.1, 0.2, \ldots, 1.0)$ are selected for experimentation. This assumes that varying τ on a linear scale in this range will provide a reasonable indication of the sensitivity of performance to τ. For population size,

$1 \leq p < P_{MAXF}$. In practice, $p = P_{MAXF}$ only allows for a single generation of the algorithm (which amounts to uniform random search). In addition, p_{sel} must be large enough to provide a valid estimate of the UMDA$_c^G$ model parameters. Given that $MAXF = 10^7$, the possible range for p spans orders of magnitude, hence the value chosen here were 20, 200, 2000 and 20000.

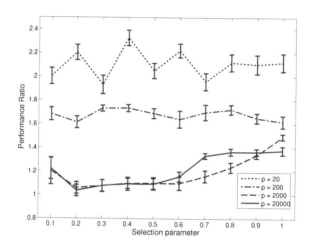

Fig. 1. Results from Experiment (a): performance of UMDA$_c^G$ on the $n_c = 5$ problem. Lines show averages and error bars standard deviations of performance over 30 trials.

Fig. 1 shows the average and standard deviation performance results for this experiment. Performance is reported as a ratio given by $\frac{d_n}{f(\mathbf{x}^b)}$ where d_n is the objective function value of the known global optimum (or best known solution) and $f(\mathbf{x}^b)$ is the objective function value of a solution found by the algorithm (or a statistic over such values). This means that the optimal performance value is 1 and e.g. a value of 1.5 indicates a solution that is 1.5 times (or 50% worse than) the value of the global optimum (smaller values are better). It is clear that the larger population sizes ($p = 2000$ and $p = 20000$) provide much better performance. In these cases, a small selection parameter value gives the best performance and the performance seems relatively insensitive to τ in the range $[0.2, 0.5]$. Outside this range, performance deteriorates significantly. Note that $\tau = 1$ corresponds to no selection pressure (i.e. UMDA$_c^G$ repeatedly samples and estimates its distribution in a form of random search). For $p = 20$ the value of p_{sel} is very small for small values of τ, which is a likely cause of the high variability of results in the $p = 20$ curve when $\tau < 0.8$.

The best performing values of τ are in close agreement with the values used previously in most of the literature. A population size of 2000 has also been used previously although it is a relatively large value. However the largest value used ($p = 20000$) is an order of magnitude larger and provides good performance (including the best average value at $\tau = 0.2$).

3.2   Experiment (b): Performance across Problem Size and p

On the basis of the results from Experiment (a), $\tau = 0.2$ was chosen as a suitable value for the selection parameter. In Experiment (b), UMDA$_c^G$ was run with $\tau = 0.2$ for problem sizes $n_c = 2, 3, \ldots, 30$ (i.e. 4D - 60D problems). Different population sizes were also run again at the same orders of magnitude except the smallest (i.e, $p = 100, 1000, 10000$).

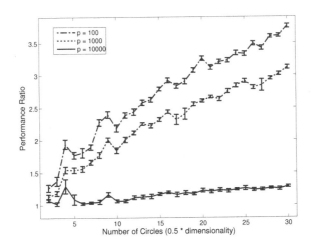

Fig. 2. Results from Experiment (b): Performance of UMDA$_c^G$ across problem sizes $n_c = 2, 3, \ldots, 30$

These results (Fig. 3) show that the performance with $p = 10000$ is dramatically better than the smaller sizes across all problem sizes tested and the difference increases steadily with problem dimensionality. For $p = 10000$, the average performance remains within approximately 20% of the global optimum. This improves on previous results on these problems with UMDA$_c^G$ [7], highlighting the importance of algorithm parameter settings even with a simple algorithm such as UMDA$_c^G$. Although for the most part performance worsens gradually with problem size, there are a few exceptions that were more difficult for UMDA$_c^G$ (e.g. $n_c = 4, 9, 16$) than the next largest problem in each case (regardless of p). This is surprising because these problems have an intuitively simple solution (corresponding to a uniform grid layout of circles in the square). Further experiments are required to understand why these problems are more difficult for UMDA$_c^G$.

3.3   Experiment (c): Further Evaluation of p

To investigate more closely the performance of UMDA$_c^G$ for large populations, the next experiment tested additional population sizes of $p = 10000, 12000, 13000,$

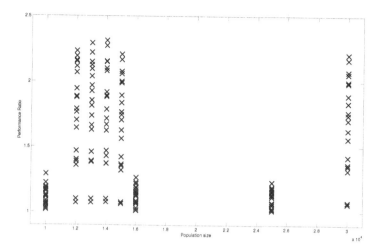

Fig. 3. Results from Experiment (c): mean performance results for all problem sizes $n_c = 2, 3, \ldots, 30$ (crosses) across population sizes tested

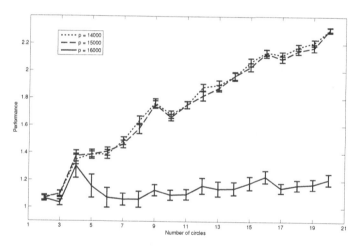

Fig. 4. Results from Experiment (d): mean and standard deviation of performance for EMNA across problem sizes

$14000, 15000, 16000, 25000, 30000$ for all problem sizes $n_c = 2, 3, \ldots, 30$. The average performance results are shown in Fig. 3. The results show that performance has unexpected sensitivity to the population size: performance is relatively good for $p = 10000$, deteriorates rapidly at $p = 12000, 13000, 14000$ and 15000, becomes good again at $p = 16000$ and 25000 before finally becoming worse again at 30000. This behaviour is unusual and does not seem explainable purely by properties of the algorithm. One possibility is that there is some property of the CiaS problems that has a complex interaction with the estimation of the EDA Gaussian model, but this requires further investigation.

3.4 Experiment (d): Evaluating EMNA with the Same Parameter Settings

In the final experiment, the performance of UMDA$_c^G$ is compared with EMNA. It is assumed that parameter values that worked well for UMDA$_c^G$ could work well with EMNA because the algorithm are somewhat similar, however this cannot be guaranteed. EMNA was run with $\tau = 0.2$ and $p = 14000, 15000, 16000$ across all problem sizes. The results are shown in Fig. 4.

The performance of EMNA for these parameter settings was very similar to that obtained for UMDA$_c^G$. The large variability in performance between the population sizes used is also true for EMNA.

4 Discussion and Conclusions

The experimental results above are an example of the impact that algorithm parameters can have on the performance of a metaheuristic optimization algorithm. UMDA$_c^G$ and EMNA have a minimal number of parameters compared to most other algorithms and performance was not expected to be highly sensitive to their settings. Nevertheless, the results show that it is important to devote some effort to choosing these values to obtain good performance. The values obtained and used here may not provide optimal performance for other problems and some experimentation with parameter values is necessary. For the best parameter settings tested, UMDA$_c^G$ and EMNA were able to obtain reasonable performance (i.e. within 20% of the global optimum value) for the range of CiaS problems considered, which is a significant improvement over previous results with UMDA$_c^G$ [7]. These results were obtained with a typical value for τ (0.2) but with a much larger population size ($p = 10000, 25000$) than has been previously used.

The results show surprising performance variability with p between 10000 and 30000 which demands further investigation. To our knowledge, the literature does not provide an immediate explanation for this. It would be very interesting if this performance difference can be attributed to some specific characteristics of the CiaS problems themselves. An alternative is some kind of implementation or numerical error (64-bit Matlab R2011b was used for all the experiments on a Linux PC). The non-monotonic behaviour of performance with problem size (Fig. 3) also needs to be investigated and explained. One possibility is that the repair mechanism used to handle the problem constraints has had a significant impact on these particular problem sizes.

References

1. Addis, B., Locatelli, M., Schoen, F.: Disk packing in a square: A new global optimization approach. INFORMS Journal on Computing 20(4), 516–524 (2008)
2. Bosman, P.A.N., Grahl, J., Thierens, D.: Enhancing the Performance of Maximum–Likelihood Gaussian EDAs Using Anticipated Mean Shift. In: Rudolph, G., Jansen, T., Lucas, S., Poloni, C., Beume, N. (eds.) PPSN X. LNCS, vol. 5199, pp. 133–143. Springer, Heidelberg (2008)

3. Bosman, P.A.N., Thierens, D.: Expanding from Discrete to Continuous Estimation of Distribution Algorithms: The IDEA. In: Deb, K., Rudolph, G., Lutton, E., Merelo, J.J., Schoenauer, M., Schwefel, H.-P., Yao, X. (eds.) PPSN VI. LNCS, vol. 1917, pp. 767–776. Springer, Heidelberg (2000)
4. Castillo, I., Kampas, F.J., Pintér, J.D.: Solving circle packing problems by global optimization: Numerical results and industrial applications. European Journal of Operational Research 191, 786–802 (2008)
5. Dong, W., Yao, X.: Covariance matrix repairing in gaussian based edas. In: IEEE Congress on Evolutionary Computation, CEC 2007, pp. 415–422 (2007)
6. Dong, W., Yao, X.: Unified eigen analysis on multivariate gaussian based estimation of distribution algorithms. Information Sciences 178(15), 3000–3023 (2008)
7. Gallagher, M.: Investigating circles in a square packing problems as a realistic benchmark for continuous metaheuristic optimization algorithms. In: The VIII Metaheuristics International Conference, MIC 2009, Hamburg, Germany (July 2009)
8. Grahl, J., Bosman, P., Rothlauf, F.: The correlation-triggered adaptive variance scaling idea. In: Proceedings of the 8th Annual Conference on Genetic and Evolutionary Computation, pp. 397–404 (2006)
9. Grosso, A., Jamali, J.U.A., Locatelli, M., Schoen, F.: Solving the problem of packing equal and unequal circles in a circular container. Tech. rep., Eprint - Optimization Online (March 2008),
http://www.optimization-online.org/DB_HTML/2008/06/1999.html
10. Larrañaga, P., Lozano, J.A. (eds.): Estimation of Distribution Algorithms: A New Tool for Evolutionary Computation. Kluwer (2002)
11. Specht, E.: Packomania (2012), http://www.packomania.com/ (retrieved)
12. Szabó, P.G., Markót, M.C., Csendes, T.: Global optimization in geometry - circle packing into the square. In: Audet, P., Hansen, P., Savard, P. (eds.) Essays and Surveys in Global Optimization. Kluwer (2005)
13. Theodoracatos, V.E., Grimsley, J.L.: The optimal packing of arbitrarily-shaped polygons using simulated annealing and polynomial-time cooling schedules. Computer Methods in Applied Mechanics and Engineering 125, 53–70 (1995)
14. Yuan, B., Gallagher, M.: Experimental results for the special session on real-parameter optimization at CEC 2005: A simple, continuous EDA. In: The 2005 IEEE Congress on Evolutionary Computation, 2005, vol. 2, pp. 1792–1799 (2005)

Block Diagonal
Natural Evolution Strategies

Giuseppe Cuccu and Faustino Gomez

IDSIA
USI-SUPSI
6928 Manno-Lugano, Switzerland
{giuse,tino}@idsia.ch
http://www.idsia.ch/~giuse,~tino

Abstract. The *Natural Evolution Strategies* (NES) family of search algorithms have been shown to be efficient black-box optimizers, but the most powerful version xNES does not scale to problems with more than a few hundred dimensions. And the scalable variant, SNES, potentially ignores important correlations between parameters. This paper introduces Block Diagonal NES (BD-NES), a variant of NES which uses a block diagonal covariance matrix. The resulting update equations are computationally effective on problems with much higher dimensionality than their full-covariance counterparts, while retaining faster convergence speed than methods that ignore covariance information altogether. The algorithm has been tested on the *Octopus-arm benchmark*, and the experiments section presents performance statistics showing that BD-NES achieves better performance than SNES on networks that are too large to be optimized by xNES.

1   Introduction

Natural Evolution Strategies (NES; [11]) have been shown to efficiently optimize neural network controllers for reinforcement learning tasks [2; 7; 10]. This family of algorithms searches the space of network weights by adapting a parameterized distribution (usually Gaussian) in order to optimize expected fitness by means of the natural gradient. The two main variants of NES, xNES [3] and SNES [7], make a trade-off between generality and efficiency: xNES (like CMA-ES [5]) uses a full covariance matrix, capturing all possible correlations between the weights but at a cost of $\mathcal{O}(w^3)$, where w is the number of weights. Unfortunately, xNES does not scale to the space of even modest size neural networks, with hundreds of weights. At the other extreme, SNES ignores weight correlations altogether in exchange for $\mathcal{O}(w)$ complexity, by using a diagonal covariance matrix. Even though it cannot solve non-separable problems, it seems to work well for neuroevolution, arguably because of the high number of possible solutions for any given network structure.

SNES updates its search distribution two orders of magnitude faster than xNES, but, by not taking into account epistatic linkages between network weights

C.A. Coello Coello et al. (Eds.): PPSN 2012, Part II, LNCS 7492, pp. 488–497, 2012.

(e.g. arising from correlated inputs), does not make full use of strong regularity inherent in many control problems. For example, sensors positioned near each other on a robot body will likely generate correlated readings, and therefore the corresponding weights processing sensory information will probably be correlated as well.

In this paper, we introduce a new NES variant that is intermediate between SNES and xNES in that it allows for correlations between subsets of search dimensions (e.g. weights), by using a search distribution with a block-diagonal covariance matrix. By allowing correlations only between some weights, the computational complexity can be reduced significantly vis-a-vis xNES, but this requires first identifying which weights should be grouped together. In a general, unconstrained optimization setting such properties of the objective (fitness) function are not known *a priori*. However, in neuroevolution, the phenotypic structure provides a natural way to decompose the search space by grouping together those weights which belong to the same neuron (i.e. network sub-function).

This Block Diagonal NES (BD-NES) uses one full covariance matrix for each neuron, allowing correlations between all weights of a given neuron, but ignoring correlation between weights of different neurons. This approach is similar to *cooperative coevolution* [4; 6], where each neuron is represented by a separate sub-genotype, and the complete individual is constructed by concatenating the sub-genotypes.

The next section derives the new algorithm from the NES family. Section 3, presents comparative results against SNES. Section 4, discusses the results and provides some ideas on how to further improve this approach.

2   Block Diagonal Natural Evolution Strategies

BD-NES can be viewed as multiple xNES [3] algorithms running in parallel, one for each block in the covariance matrix of the search distribution. Of course, the blocks can be of different size if the relationship between problem dimensions is known in advance (i.e. whether any two dimension are separable). Here, in the context of neuroevolution and in the absence of this kind of knowledge, the division of the network weights into blocks is determined by the number of neurons in the network architecture, Ψ.

Figure 1 describes the block-diagonal covariance matrix used by the search distribution. Each neuron, i has its own block, Σ_i, that captures all of the covariances between its incoming connections. Algorithm 1 presents the code for BD-NES. First, the mean vectors, $\mu_i \in \mathbb{R}^c$, and $c \times c$ covariance matrices, Σ_i, $i = 1..n$, are initialized, where n is the number of neurons, and c is the number of incoming connections per neuron. Each generation (the while loop), λ networks are constructed by sampling from each Gaussian sub-distribution to obtain ψ neuron chromosomes, z^i, $i = 1..\psi$, (line 5) which are then concatenated into a complete genome, z, (line 7). The genomes are then transformed into networks, and evaluated. The fitness achieved by a networks is passed to its constituent

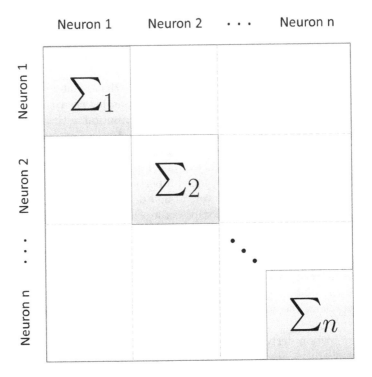

Fig. 1. Block Diagonal covariance matrix: The search distribution has a separate block in its covariance matrix for the each neuron (i.e. the covariance between neurons is zero) in the network architecture being evolved. The block size for a given neuron is the number of connections entering that neuron. To evaluate the gradient from the distribution, samples are drawn from the blocks, and then concatenated to construct the full genotype.

neuron chromosomes (line 10) and used to update the corresponding mean, and dedicated covariance block using xNES (line 14), described next.

Let $p(z \,|\, \theta)$ denote the density of the Gaussian with parameters $\theta = (\mu, \Sigma)$. Then, the expected fitness under the search distribution is

$$J(\theta) = \mathbb{E}_\theta[f(z)] = \int f(z) \, p(z \,|\, \theta) \, dz \ .$$

The gradient w.r.t. the parameters can be rewritten as

$$\nabla_\theta J(\theta) = \nabla_\theta \int f(z) \, p(z \,|\, \theta) \, dz$$
$$= \mathbb{E}_\theta \left[f(z) \, \nabla_\theta \log \left(p(z \,|\, \theta) \right) \right] \ ,$$

(see [11] for the full derivation) from which we obtain the Monte Carlo estimate

$$\nabla_\theta J(\theta) \approx \frac{1}{\lambda} \sum_{k=1}^{\lambda} f(\mathbf{z}_k) \, \nabla_\theta \log\left(p(\mathbf{z}_k \mid \theta)\right) \tag{1}$$

of the search gradient. The key step then consists of replacing this gradient, pointing into the direction of (locally) steepest descent w.r.t. the given parameterization, by the natural gradient

$$\widetilde{\nabla}_\theta J = \mathbf{F}^{-1} \nabla_\theta J(\theta) \ ,$$

where $\mathbf{F} = \mathbb{E}\left[\nabla_\theta \log\left(p\left(\mathbf{z}|\theta\right)\right) \nabla_\theta \log\left(p\left(\mathbf{z}|\theta\right)\right)^\top\right]$ is the Fisher information matrix; leading to a straightforward scheme of natural gradient descent for iteratively updating the search distribution

$$\theta \leftarrow \theta - \eta\widetilde{\nabla}_\theta J = \theta - \eta\mathbf{F}^{-1}\nabla_\theta J(\theta) \ ,$$

with learning rate parameter η. The sequence of (1) sampling an offspring population, (2) computing the corresponding Monte Carlo estimate of the fitness gradient, (3) transforming it into the natural gradient, and (4) updating the search distribution, constitutes one generation of NES.

In order to render the algorithm invariant under monotonic (rank preserving) transformations of the fitness values, *fitness shaping* [11] is used to normalize the fitness into rank-based *utilities* $u_k \in \mathbb{R}$, $k \in \{1, \ldots, \lambda\}$. The individuals are ordered by fitness, with $\mathbf{z}_{1:\lambda}$ and $\mathbf{z}_{\lambda:\lambda}$ denoting the most and least fit offspring, respectively. The distribution parameters are then updated using the "fitness-shaped" gradient:

$$\nabla_\theta J = \sum_{k=1}^{\lambda} u_k \cdot \nabla_{(\theta)} \log\left(p(\mathbf{z}_{k:\lambda} \mid \theta)\right) \ . \tag{2}$$

Typically, the utility values are either non-negative numbers that sum to one, or a shifted variant with zero mean.

Using the same exponential local coordinates as in [3], the update equations for the sub-distributions are:

$$\boldsymbol{\mu}_{new}^i \leftarrow \boldsymbol{\mu}^i + \eta_\mu \cdot \sum_{k=1}^{\lambda} u_k \cdot \mathbf{z}_k^i$$

$$A_{new}^i \leftarrow A^i \cdot \exp\left(\frac{\eta_A}{2} \cdot \sum_{k=1}^{\lambda} u_k \cdot \left(\mathbf{z}_k^i \mathbf{z}_k^{i^\top} - I\right)\right)$$

where A^i is the upper triangular matrix resulting from the Cholesky decomposition of covariance block Σ^i, $\Sigma^i = A^{i\top} A^i$.

This approach assumes weights of different neurons not to be correlated, but given the high number of feasible solutions in continuous control problems such constraint does not usually limit the search.

Algorithm 1. BD-NES(Ψ)

---

1 INITIALIZE $(\mu_1, \Sigma_1) \ldots (\mu_n, \Sigma_n)$
2 **while** *not solved* **do**
3   **for** $k \leftarrow 1$ *to* λ **do**
4   **for** $i \leftarrow 1$ *to* ψ **do**
5 | $z_k^i \sim \mathcal{N}(\mu_i, \Sigma_i)$
6   **end**
7          $\mathbf{z}_k \leftarrow$ CONCATENATE($z_k^1 \ldots z_k^\psi$)
8          $fit_k \leftarrow$ EVALUATE($\mathbf{z}_k$)
9   **for** $j \leftarrow 1$ *to* ψ **do**
10 | $fit_k^i \leftarrow fit_k$
11   **end**
12   **end**
13   **for** $i \leftarrow 1$ *to* ψ **do**
14 | $(\mu_i, \Sigma_i) \leftarrow$ UPDATEXNES($\mu_i, \Sigma_i, (z_1^i \ldots z_\lambda^i)$)
15   **end**
16 **end**

Computational Complexity

SNES and xNES can be thought of as special cases of BD-NES. Let P be a partition of the weights consisting of b blocks of the same size s, and w be the total number of weights. SNES considers all weights to be uncorrelated, so $s = 1$, and $b = w$, whereas in xNES all of the weights are considered to be correlated: $b = 1$ and $s = w$, producing the full covariance matrix.

The dominant operation in the NES update step in terms of computational complexity is the covariance matrix inversion. The computational cost under this framework is proportional to the cost of inverting each matrix block, times the number of blocks being inverted, $\mathcal{O}(bs^3)$. For BD-NES, *with neurons defining the covariance blocks*, $b = \psi$ and $s = c$, where ψ is the number of neurons in the network, and c is the number of connections per neuron (i.e. the node degree).

For single layer feed-forward networks, c depends only on the number of input/output units specified by the problem domain, and not on the number of neurons, so the complexity becomes $\mathcal{O}(\psi)$; the same as SNES, with an hidden constant depending on the number of input units in the network.

For fully-connected recurrent neural networks, c grows with the number of neurons $c \sim \psi$ (each additional neuron adds a connection to every other neuron), thus the complexity becomes $\mathcal{O}(\psi \times \psi^3) = \mathcal{O}(\psi^4)$, which is between SNES and xNES, as in such networks $w \sim \psi^2$, making the complexity of SNES $\mathcal{O}(\psi^2)$ and that of xNES $\mathcal{O}(\psi^6)$. The complexity improves further if we assume that for large networks the connectivity is sparse, with neurons having a fixed average number of recurrent connections, k, as is the case in real-world complex networks, which exhibit the small world property [1]. In this case, BD-NES reduces to $\mathcal{O}(\psi)$, since $c = k$ is constant.

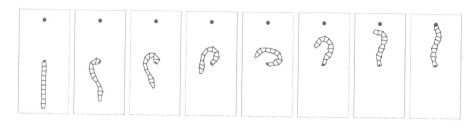

Fig. 2. Octopus-Arm Acrobot Task. A flexible arm consisting of n compartments, each with 3 controllable muscles, must be lifted from its initial downward-pointing position (left), against gravity, to touch a goal location (black dot) with its tip. The behavior shown was evolved through BD-NES.

3 Experiments

BD-NES was tested on a version of the octopus arm control benchmark. This environment was chosen because it requires networks with thousands of weights and therefore cannot be solved using modern evolutionary strategies like xNES and CMA-ES that use a full covariance matrix for the search distribution.

3.1 Octopus-Arm Acrobot Task

The Octopus Arm [12; 13] (see figure 2) consists of p compartments floating in a 2D water environment. Each compartment has a constant volume and contains three controllable muscles (dorsal, transverse and ventral). The state of a compartment is described by the x, y-coordinates of two of its corners plus their corresponding x and y velocities. Together with the arm base rotation, the arm has $8p + 2$ state variables and $3p + 2$ control variables.

In the standard setup, the goal of the task is to reach a target position with the tip of the arm, starting from three different initial positions, by contracting the appropriate muscles at each 1sec step of simulated time. It turns out that it is very easy to get close to the target from two of the initial positions. Therefore, we devised a version, shown in figure 2, where the arm initially hangs down (in stable equilibrium due to gravity), and must be lifted to touch the target above, on the opposite side of the environment, with its tip. The task is termed the *octopus arm acrobot* due to its similarity with the classic acrobot swing-up task [9].

Also, instead of the standard 8 "meta"–actions that simplify control by contracting groups of muscles simultaneously (e.g. all dorsal, all ventral, etc.), the controllers must instead contract each individual muscle independently.

3.2 Network Architecture

Networks were evolved to control a $n = 10$ compartment arm using fully-connected recurrent neural networks having 32 neurons, one for each muscle (see figure 3).

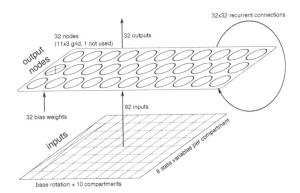

Fig. 3. Network architecture. The octopus arm is controlled by a single layer recurrent neural network with 82 inputs and 32 neurons (outputs), one for each muscle in the arm.

The networks have a 32×82 input weight matrix, a 32×32 recurrent weight matrix and bias vector of length 32, for a total of $3,680$ weights.

The size of a full covariance matrix to search in $3,680$ dimensions is $3,680^2 = 13,542,400$ entries. Yet each of the 32 neurons has only $82+32+1 = 115$ incoming connections, so the covariance blocks in BD-NES that have to be inverted have only $115^2 = 13,225$ entries, three orders of magnitude fewer than for the full covariance matrix.

3.3   Setup

BD-NES was compared with SNES. To provide a baseline, random weight guessing (RWG; [8]) was also used, where the network weights are chosen at random (i.d.d.) from a uniform distribution. This approach gives us an idea of how difficult the task is to solve by simply guessing a good set of weights.

The population size λ is proportional to the number of weights, w, being evolved; set here to $\lambda = 50$. The learning rates are $\eta_\mu = \frac{\log(w)+3}{5\sqrt{w}}$ and $\eta_\sigma = \frac{\eta_\mu}{2}$. Each run was limited to $10,000$ fitness evaluations. The fitness was computed as:

$$\left[1 - \frac{t}{T}\frac{d}{D}, 0\right], \tag{3}$$

where t is the number of time steps before the arm touches the goal, T (set to 100) is the maximum number of time steps in a trial, d is the final distance of the arm tip to the goal, and D is the initial distance of the arm tip to the goal. This fitness measure is different to the one used in [12], because minimizing the integrated distance of the arm tip to the goal causes greedy behaviors. In the viscous fluid environment of the octopus arm, a greedy strategy using the shortest length trajectory does not lead to the fastest movement: the arm needs to be contracted first, then rotated, and finally stretched upwards towards the

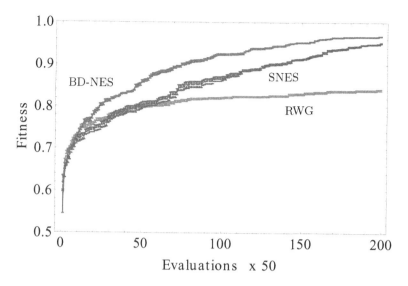

Fig. 4. Performance on octopus-arm acrobot. BD-NES (top, blue) and SNES (bottom, red) performance on the octopus arm benchmark. Curves are averages over 20 runs, with error bars showing the corresponding variance.

goal. The fitness function favors behaviors that reach the goal within a small number of time steps.

3.4   Results

Figure 4 shows the fitness of the best network found so far for the three methods, averaged over 20 runs (bars indicate variance). BD-NES reaches a fitness equal to the final fitness of SNES at around 7000 evaluations (30% fewer evaluations), and does so in 15% less cpu-time[1] (55.4min for BD-NES, 65.4min for SNES).

Figure 5 shows the neuron chromosomes at the end of a typical run of (a) SNES and (b) BD-NES, projected into 3D via principal component analysis. Each point denotes a neuron sampled from one of final sub-distributions. In all SNES runs the neuron distributions overlap (note scale), suggesting that the neurons are functionally more similar than the neurons comprising a network BD-NES, where neuron clusters are more distinct. While similarity between neurons is to be expected given the similar function that muscles in adjacent compartments must perform, different parts of the arm must perform slightly different tasks (e.g. the muscles controlling the rotation at the base), so that the specialization occurring in BD-NES could explain the better performance.

---

[1] Reference machine: intel i7 640M at 3.33GHz and 4GB of ram DDR3 at 1066MHz. Mathematica implementation of search algorithm using the Java implementation of the octopus arm available at: http://www.cs.mcgill.ca/~idprecup/ workshops/ICML06/octopus.html.

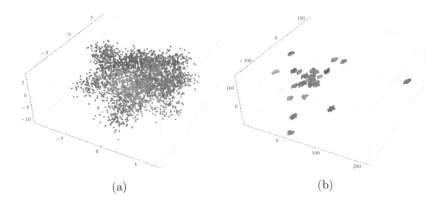

(a)   (b)

Fig. 5. Neuron specialization. The plots show the 115-dimensional neuron chromosomes in the final population of a typical run, projected into 3D by PCA, for (a) SNES and (b) BD-NES. For SNES, The neuron clusters overlap and are concentrated in a small region of the search space. For BD-NES, neuron distributions form distinct clusters, that are more spread out.

4   Discussion

BD-NES is a novel algorithm of the NES family allowing for partial correlation information to be retained while limiting the computational overhead. The experiments show that block diagonal covariance matrix adaptation can scale up to over 3000 dimensions, and search more efficiently than its diagonal-covariance counterpart, SNES.

For problems where the number of inputs is very large (e.g. video input), decomposing the network at the level of neurons will not work. In this case, neurons can use receptive fields that receive only part of the full input, as is done in convolutional networks. Or, blocks can be built based on inputs rather than neurons, each block representing the covariance matrix for the weights of all connections from a particular input to all neurons. Future work will start by applying the method to a vision version of the task used here, where the network does not receive the state of the arm, but instead sees the arm configuration from a 3rd-person perspective, and must solve the task using high-dimensional images as input.

Acknowledgments. This research was supported by Swiss National Science Foundation grant #137736: "Advanced Cooperative NeuroEvolution for Autonomous Control". Thanks to Jan Koutník for inspiration and support.

References

[1] Albert, R., Barabási, A.-L.: Statistical mechanics of complex networks. Reviews of Modern Physics 74 (January 2002)

[2] Cuccu, G., Luciw, M., Schmidhuber, J., Gomez, F.: Intrinsically motivated evolutionary search for vision-based reinforcement learning. In: Proceedings of the 2011 IEEE Conference on Development and Learning and Epigenetic Robotics IEEE-ICDL-EPIROB. IEEE (2011)

[3] Glasmachers, T., Schaul, T., Sun, Y., Wierstra, D., Schmidhuber, J.: Exponential natural evolution strategies. In: Genetic and Evolutionary Computation Conference, GECCO, Portland, OR (2010)

[4] Gomez, F., Miikkulainen, R.: Incremental evolution of complex general behavior. Adaptive Behavior 5(3-4), 317 (1997)

[5] Hansen, N., Ostermeier, A.: Completely derandomized self-adaptation in evolution strategies. Evolutionary Computation 9(2), 159–195 (2001)

[6] Potter, M.A., De Jong, K.A.: Evolving neural networks with collaborative species. In: Proceedings of the 1995 Summer Computer Simulation Conference (1995)

[7] Schaul, T., Glasmachers, T., Schmidhuber, J.: High dimensions and heavy tails for natural evolution strategies. In: Genetic and Evolutionary Computation Conference, GECCO (2011)

[8] Schmidhuber, J., Hochreiter, S., Bengio, Y.: Evaluating benchmark problems by random guessing. In: Kremer, S.C., Kolen, J.F. (eds.) A Field Guide to Dynamical Recurrent Neural Networks. IEEE Press (2001)

[9] Spong, M.W.: Swing up control of the acrobot. In: Proceedings of the 1994 IEEE Conference on Robotics and Automation, San Diego, CA, vol. 46, pp. 2356–2361 (1994)

[10] Sun, Y., Wierstra, D., Schaul, T., Schmidhuber, J.: Stochastic search using the natural gradient. In: International Conference on Machine Learning, ICML (2009)

[11] Wierstra, D., Schaul, T., Peters, J., Schmidhuber, J.: Natural evolution strategies. In: Proceedings of the Congress on Evolutionary Computation, CEC 2008, Hongkong. IEEE Press (2008)

[12] Woolley, B.G., Stanley, K.O.: Evolving a Single Scalable Controller for an Octopus Arm with a Variable Number of Segments. In: Schaefer, R., Cotta, C., Kołodziej, J., Rudolph, G. (eds.) PPSN XI. LNCS, vol. 6239, pp. 270–279. Springer, Heidelberg (2010)

[13] Yekutieli, Y., Sagiv-Zohar, R., Aharonov, R., Engel, Y., Hochner, B., Flash, T.: A dynamic model of the octopus arm. I. Biomechanics of the octopus reaching movement. Journal of Neurophysiology 94(2), 1443–1458 (2005)

Finding Good Affinity Patterns
for Matchmaking Parties Assignment
through Evolutionary Computation

Sho Kuroiwa[1,2], Keiichi Yasumoto[1], Yoshihiro Murata[3], and Minoru Ito[1]

[1] Nara Institute of Science and Technology, 8916-5, Takayama, Ikoma, Nara, Japan
{sho-k,yasumoto,ito}@is.naist.jp
[2] Hopeful Monster Corporation, 8916-5, Takayama, Ikoma, Nara, Japan
sho-k@hopefulmonster.jp
[3] Hiroshima City University, 3-4-1 Ozuka-Higashi, Asa-Minami, Hiroshima, Japan
yosihi-m@hiroshima-cu.ac.jp

Abstract. There is a demand to maximize the number of successful couples in matchmaking parties called "Gokon" in Japanese. In this paper, we propose a method to find good affinity patterns between men and women from resulting Gokon matches by encoding their attribute information into solutions and using an evolutionary computation scheme. We also propose a system to assign the best members to Gokons based on the method. To derive good affinity patterns, a specified number of solutions as chromosomes of evolutionary computation (EC) are initially prepared in the system. By feeding back the results of Gokon to the solutions as fitness value of EC, semi-optimal solutions are derived. To realize the proposed system, we need simultaneous search of multiple different good affinity patterns and efficient evaluation of solutions through as small number of Gokons as possible with various attribute members. To meet these challenges, we devise new methods for efficient selection operation inspired by Multi-niches Crowding method and reuse of past Gokon results to evaluate new solutions. To evaluate the system, we used the *NMax* problem assuming that there would be N good affinity patterns between men and women as a benchmark test. Through computer simulations for $N = 12$, we confirmed that the proposed system achieves almost twice as many good matches as a conventional method with about half the evaluation times.

Keywords: Evolutionary Computation, Matchmaking Party, Multi-niches Crowding.

1 Introduction

Recently, the low birthrate has become a serious problem in Japan. One of the reasons for the problem is the lack of opportunities to find a marriage partner. For this reason, several local governments and enterprises have provided opportunities for unmarried people to meet potential marriage partners. In particular,

C.A. Coello Coello et al. (Eds.): PPSN 2012, Part II, LNCS 7492, pp. 498–507, 2012.

matchmaking parties called "Gokon" are now attracting considerable attention in Japan. It is important for such a matchmaking party to assign participants so that the number of man and woman pairs likely to begin relationships is maximized. We regard this pair as a "good match". However, assigning members to a Gokon so as to maximize the number of good matches is difficult since affinity between men and women is not yet well-understood (prediction problem), and determining the best Gokon members is also difficult (combinatorial optimization problem). In this paper, we propose a system to solve these two problems.

To resolve the prediction problem, Evolutionary Computation (EC) [1] is used. EC is a well-established method for solution search of the target system and has a wide range of applications [2]. The system has concatenations of man and woman attribute information (called the *attribute*, hereafter) as solutions of EC (chromosomes or individuals) to find semi-optimal solutions representing a good affinity by feeding back the number of good matches in a Gokon as fitness values. To resolve the combinatorial optimization problem, we set Gokons as many as possible and assign, to each Gokon, a specified number of men and women who have attribute similar to good affinity obtained as a solution. It is perhaps not the best way, but we focus mainly on resolving the prediction problem in this paper.

Since our target prediction problem is a multimodal problem, EC has to find many peaks (good affinity patterns) in the domain of affinity patterns. However, most of the existing application studies treat how to find the peak of a unimodal problem [2]. Our previous research also did not devise a method for solving multimodal problem [3]. So, the method inspired by Multi-niches Crowding [4] selection which has a reputation in the EC domain as a technique to calculate the multi-maximum of multimodal function is adopted in the proposed system. To find optimal solution of the target problem, an incredibly large number of Gokons are needed. Thus, the method using archival records to evaluate the solution instead of doing new Gokons is adopted.

We compared performance of the proposed system with the greedy approach using computer simulations. The volume of attribute and number of peaks in affinity domain is twice as large as in previous research [3], and the function of good match is defined taking into account the uncertainty in the real world at this time. Through computer simulations, we confirmed that the proposed system can generate almost twice good matches in half real Gokons compared to the greedy approach.

2   Gokon Problem

The Gokon problem is to divide a large population (system users) into small groups (Gokon) with the same number of men and women with good affinity where each user can be included in different Gokons until he/she makes a match.

Input: The input of the problem is the sets of male participants and female participants denoted by $B = \{b_1, b_2, \cdots\}$ and $G = \{g_1, g_2, \cdots\}$, respectively.

Each man $b_i \in B$ and each woman $g_j \in G$ have k attributes $(b_{i1}, b_{i2}, \cdots, b_{ik})$ and l attributes $(g_{j1}, g_{j2}, \cdots, g_{jl})$ respectively, where each attribute represents his/her feature and/or personality.

Output: The output of the problem is to make h groups (Gokons) that can be overlapped, where each Gokon has L men and L women from B and G.

Number of Gokons is h, and the i-th Gokon is represented by $[B_i, G_i]$. The output is Gokon assignment $[B_1, G_2], \cdots, [B_h, G_h]$ to optimize objective function and unknown evaluation function F.

Objective Function: Evaluation function F, which returns the number of good matches in the Gokon $[B_i, G_i]$, is defined. Objective function is to maximize the sum of F for all h Gokons, and is given by

$$\text{maximize} \sum_{i \in \{1,\ldots,h\}} F(B_i, G_i) \tag{1}$$

3   Proposed System

We propose a system to evolve the evaluation function of the Gokon problem using EC and compute Gokon participant lists as shown in Fig. 1. The system is composed of **solutions** representing good affinity between men and women and the following three operators. The **EC operator** executes EC calculation (Selection, Crossover and Mutation) to the solutions and generates **candidate solutions** (indicated in Fig. 1 as Step.1). The **Assignment Operator** generates a participant list corresponding to each candidate solution (indicated in Fig. 1 as Step.2). The **Evaluation Operator** gives fitness value to each candidate solution (indicated in Fig. 1 as Step.3). Finally, we update the solutions using candidate solutions. Running through these steps, solutions will improve gradually toward optimal solutions. Here, we define initial solutions as first generation and the solutions updated $t-1$ times as t-th generation. At the evaluation of candidate solution, we want to reduce the number of Gokon times because the optimal participant lists cannot be obtained until the system finds good affinity. Until this point, the system inflicts a time and money loss on participants to make good matches. So, it is desirable to be able to evolve candidate solutions without doing actual Gokons. Then, we devise a method using archival records of attributes of good matches indicated in Fig. 1 as **Stock**.

3.1   Solution

Each solution of EC is coded as a concatenation of attribute of a man and a woman who are likely to make a good match. The first half of each solution shows the attribute of a man, and the remaining half shows that of a woman as in Fig. 2.

Input of the system, man and woman attribute correspond to each half of the solution. Those attributes are obtained by the system from a user questionnaire and personality test, the Temperament and Character Inventory (TCI) [5].

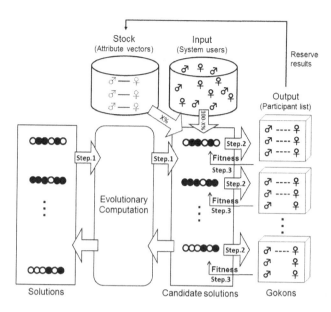

Fig. 1. Outline

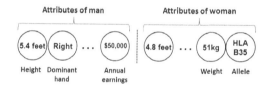

Fig. 2. Example of solution representation

However, what information is important for determining an affinity is not known well academically, and matchmaker companies use slightly different information on their own. For example, TCI has many question items but the result of the test is categorized into 12 patterns (it takes only 4 bits). To determine the element factor of affinity, we estimate that the necessary information size to encode the solution is down to 20 bits each for man and woman (totally 40 bits).

3.2 EC Operator

The EC Operator performs three operations in solutions: selection, crossover and mutation. We adopted one-point crossover and mutation that are commonly used as the EC operations.

However, the Gokon problem is a multimodal function with multiple good affinities, and solutions need to have diversity to search for multi-optimal so-

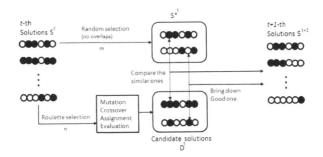

Fig. 3. Scheme of Multi-niches Crowding Factor

lutions. Therefore, we adopted a selection method considering Multi-Niches Crowding [4](hereafter, we call this the MNC method), which is one of the representative methods for keeping diversity in solutions. The MNC method prevents the increase of similar solutions in subsequent generations.

In the proposed method, first we select randomly m ($m \leq |S^t|$) solutions as $S*^t$ without overlapping from S^t as shown in Fig. 3. Second, we select n ($n \leq |S^t|$) solutions by roulette selection from t-th solutions S^t. The we apply EC operations, crossover and mutation, assignment and evaluation to those, and get D^t ($|D^t| = n$). Finally, we find the most similar $s* \in S*^t$ for each $d \in D$ and bring down d or $s*$, which has the higher fitness value, into next generation S^{t+1}.

3.3 Assignment Operator

This operator assigns members into the Gokon corresponding to each candidate solution (generate participant list of Gokon). If a man and woman have the same attributes as a candidate solution and make a good match, it is natural and preferable to use this result in calculating a fitness value of this candidate solution. However, there are very few men and women who have exactly the same attributes as the candidate solution. So, this operator assigns a specified number of men and women in the order of **similarities**. After this operation, participants are ready to start the Gokon.

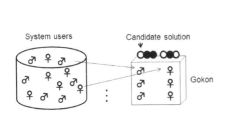

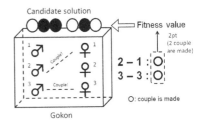

Fig. 4. Operation of Assignment   **Fig. 5.** Operation of Evaluation

3.4   Evaluation Operator

The evaluation operator gives results of the Gokon as a fitness value of the corresponding candidate solution. These results are obtained by the participants joining each Gokon through a questionnaire. After a Gokon, each participant answers the questionnaire about the participants of the opposite sex. We use the number of good matches in the Gokon as fitness value of the solution, with adjustment considering similarities (described in Sect. 3.3) between their attributes and candidate solution. That means if a candidate solution is not similar to the attributes of man and woman who make a good match, the match should not contribute to the evaluation of the solution.

Then, we prepare the new index of the similarities called **matchrate**. Candidate solution is denoted by vector $\mathbf{s} = (s_1, \cdots, s_{k+l})$. We define a pair of man and woman as (b, g), where attributes of b are $(b_1, b_2, \cdots, b_k)$ and g are $(g_1, g_2, \cdots, g_l)$, a vector concatenating those attributes is $\mathbf{x} = (b_1, b_2, \cdots, b_k, g_1, g_2, \cdots, g_l)$ denoted by $(x_1, x_2, \cdots, x_{k+l})$. Hereafter, we call this $\mathbf{x}$ the attribute vector. At this time, matchrate C is given by

$$C(\mathbf{s}, \mathbf{x}) = \frac{\sum_{i=1}^{k+l} match(s_i, x_i)}{k + l} \tag{2}$$

Here, $match$ is defined as follows.

$$match(s_i, x_i) = \begin{cases} 1 & when\ (s_i = x_i) \\ 0 & (otherwise) \end{cases} \tag{3}$$

Thus, the sum of the matchrate value in the Gokon is given as the fitness value of the candidate solution as shown in Fig. 6. The fitness value is going to be 0 when there are no good matches.

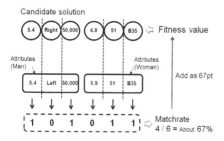

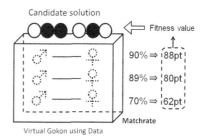

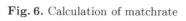

Fig. 6. Calculation of matchrate   **Fig. 7.** Evaluation using stock

Furthermore, using this matchrate, we can give fitness value using archival records of good matches without doing a real Gokon. As shown in Fig. 1, we reserve attribute vectors of good matches, called stock, in the early phase. When evaluating the candidate solution, we apply the assignment operator to the stock and real users in the ratio of $X\%$ to $100 - X\%$. We regard data from Stock as attribute vectors of real users for evaluating the candidate solutions as shown in Fig. 7.

4   Experiments

We conducted a computer simulation to evaluate how efficiently the proposed system can solve the Gokon problem. For evaluation, we defined a benchmark test called the *NMax* problem [3].

In the experiment, first we want to confirm if the MNC method works well. Thus, we compared the solutions obtained with and without MNC. We define **the optimum achievement rate** as a metric to evaluate the solutions for this purpose.

We also measured the total number of good matches for all Gokons using solutions at 1, 000-th generation. We compared the results between the proposed method and the greedy method, which is the conventional way to greedily search for good affinity patterns (i.e., peaks).

4.1   The Benchmark Test (*NMax* Problem)

The *NMax* Problem is the problem where N arbitrary bit strings represent the solutions [3].

Input: Input of *NMax* Problem is attribute vector $\mathbf{x}$ defined in Sect. 3.4 and N peak vector $P = (\mathbf{p}^1, \cdots, \mathbf{p}^N)$ where $\mathbf{p}^i = (p_1^i, \cdots, p_{k+l}^i)$. In the experiment, $b = (b_1, b_2, \cdots, b_k)$ and $g = (g_1, g_2, \cdots, g_l)$ are given randomely like $(1\ 0 \cdots 1)$.

Output: Output of *NMax* is given as follows.

$$f_{NMax}(\mathbf{x}) = max_{1 \leq j \leq N}(match(\mathbf{x}, \mathbf{p}^j)) \tag{4}$$

Here, j is the index of N which indicates the peak of *NMax* and $match(\mathbf{x}, \mathbf{p})$ is defined as Eq.(3). The output of the *NMax* Problem is the degree of how close the input vector $\mathbf{x}$ is to one of the peaks of the *NMax* problem. When $N = 1$, $\mathbf{p}^1 = \{1, 1, \cdots, 1\}$, the *NMax* Problem is identical to the *OneMax* Problem. We make a strong assumption that the *NMax* Problem can represent the existence of N good affinity patterns (attribute vectors) in real-world.

In this experiment, we improve the *NMax* Problem to be more realistic. In real-world Gokon, human decision is often affected by the situation or the state of mind. So, we add the randomness into the *NMax* function. The function G, output of *NMax* Problem, returns the probability of making a good match by using f_{NMax} as follows.

$$G = \begin{cases} 1 - \frac{1-f_{NMax}}{1-\beta} & (\beta \leq f_{NMax}) \\ 0 & (otherwise) \end{cases} \tag{5}$$

Here, β is the threshold to make a good match. When output of f_{NMax} is smaller than β, probability of a good match G equals 0, while for f_{NMax} more than β, P increases linearly.

4.2   Optimum Achievement Rate

To evaluate solution $\mathbf{s}$, using optimal solutions $\mathbf{p}^i$ indicated in Fig. 9 and *match* function defined as Eq.(3), the optimum achievement rate R is given by

$$R = max_{1 \leq i \leq n}(match(\mathbf{s}, \mathbf{p}^i)) \tag{6}$$

This is the degree of how close the solution is to one of the optimal solutions of the *NMax* problem. So, the average value of the solutions described as bellow is key to understand how close the solutions are to good affinities.

$$\overline{R} = \frac{\sum_{i=0}^{|S^t|} R_i}{|S^t|} \tag{7}$$

4.3   Comparative Method

In the greedy method, the candidate solution is updated to its most similar attribute vector of man and woman who make a good match as shown in Fig. 8 and bring down solutions which has the higher fitness value into next generation, instead of EC operation. This method intends to improve the solutions gradually as generations progress.

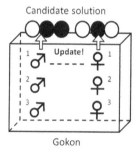

Candidate solution

Gokon

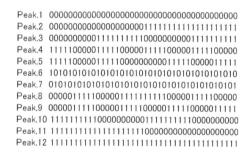

Peak.1 0000000000000000000000000000000000000000
Peak.2 0000000000000000000000011111111111111111111
Peak.3 0000000000111111111110000000000111111111111
Peak.4 1111100000111110000011111000001111100000
Peak.5 1111100000111110000000000011110000011111
Peak.6 1010101010101010101010101010101010101010
Peak.7 0101010101010101010101010101010101010101
Peak.8 0000011111000001111111111000001111100000
Peak.9 0000011111000001111100000111110000011111
Peak.10 1111111110000000000011111111110000000000
Peak.11 1111111111111111111100000000000000000000
Peak.12 1111111111111111111111111111111111111111

Fig. 8. Greedy Method   **Fig. 9.** *NMax* Problem ($N = 12$)

4.4   Experimental Setup

Input Data: The number of participants is $6{,}000$ ($3{,}000$ men and $3{,}000$ women), length of attribute of participant ($|b_{ik}|$, $|g_{jl}|$) is 20 bits (so that attribute vector and solution length are 40 bits each), the number of attribute vectors in stock is 500, the threshold β to make a good match is 0.5 and Gokon size is 30 ($|B_i| = 15$, $|G_i| = 15$). Each user can only participate in Gokon five times. The system replaces the user who made good match or participated Gokon five times with a new user in every generation.

The EC Parameters: The number of solutions is 30, the selection rate is 0.5, the crossover rate is 0.95, the mutation rate is 0.2. The solutions are initially generated by uniform random number.

In this experiment, if there exist no man and woman who have attribute similar to the peak of *NMax* indicated in Fig. 9, the number of good matches will dramatically decrease. This prohibits the progress in evolving solutions. Thus, we prepare new attribute randomly which has 1 to 100% similarity to each peak.

In general, it is important to consider what attributes are needed and how the attributes are coded as we described in Sect. 3.1. However, in this experiment, we focus on investigating whether our proposed algorithm can effectively find optimal solutions (i.e., *N* good affinities).

4.5 Results

We show the optimum achievement rate for $12Max$ Problem at 1,000-th generation in the cases using MNC and not using MNC, respectively in Figs. 10 and 11. The optimum achievement rate $\overline{R}$ is shown as an average value of solutions.

```
00000000100110000010010001000100000010001 Peak1 77%
00000111001000101000000001100100100001100 Peak1 72%
00100110100010000000000001000110000000000 Peak1 80%
00000001000000100000000001010010000001000 Peak1 82%
00110001000100111000111101111111111110111 Peak2 77%
00001000000000000001001111111111111111011111 Peak2 92%
00001000010000000000001100111101010101011011111 Peak2 80%
00100000010000000000011010111111111111111 Peak2 90%
00000111000000000000110111011011110111111 Peak2 80%
00001000101111111011010000100001111110 Peak3 80%
00000000001111001111001000000001011010111 Peak3 85%
11011000010111000000101101001111110000 Peak4 80%
01111000001111010100111110110011011010000 Peak4 82%
01111000010111111010011110000001111101000 Peak4 82%
11111010101111101110000000110010010101101 Peak5 80%
10111001000011000000000001011100010111 Peak5 77%
01111000011110100000000001100101011011111 Peak5 80%
10101011101010010010101110101001010000000 Peak6 87%
01010111110001010101010111110010010101011 Peak7 80%
01011101010100010010100011101000101010101 Peak7 80%
10000111101000010100101111000000111100000 Peak8 77%
00000111101000111010010000001101110000000 Peak8 82%
00001111010110011111111110000010111000000 Peak8 85%
00000001100010011110100110000100011111 Peak9 75%
110000001000001101110010011110000011111 Peak9 77%
00000110111000011111000101101000001111110 Peak9 82%
00001111100010001110100100111110000010111 Peak9 85%
110011110101011111110000000100000000001 Peak11 82%
111011111101100110100100010001000100000 Peak11 77%
111101011110110101111101111011101111111 Peak12 80%
```

Fig. 10. Using MNC ($\overline{R} = 81\%$)

```
11100010100000100011001000110000110000111 Peak1 60%
00001010000101000111101101010010101000111 Peak2 60%
00110010000100101111111011010101111001011 Peak2 62%
110000001010011001000100011100011110011 Peak3 62%
11010010001001011110011010100001110000010 Peak3 60%
111000101010111110011000010000001111100111 Peak3 75%
111010000011011000110011001110011110001100 Peak4 70%
11111101000100111000100100010001010101101 Peak4 65%
100100110011001010000111111001100100100 Peak4 62%
11110001100001011100110100010011100011011 Peak4 57%
101011000111011001001010111000111100110 Peak5 60%
1011001100110110101111101010101000011111 Peak5 62%
101000111010111000110011011110010110010100 Peak6 62%
1110011010001110001111110011001100100110 Peak6 60%
10100010101111110011001000110010011000010 Peak6 65%
11111101001010101111010100110101011010 Peak7 62%
11000111100001110110111111010101011100111 Peak8 62%
1010011101100001111011000010101010000110 Peak9 65%
1100001110100010110010110110101010101111 Peak9 60%
00001110111010000001010011011110100001110 Peak9 65%
1011100111100000011101010110010010101011 Peak9 65%
100011100111100101111000010111011010010111 Peak9 67%
0001001110101011010111010011001110111101 Peak9 60%
11011000111000000111010110100000110001 Peak10 60%
0110110101000010000010011110010100011001 Peak10 65%
110110111100001101110101010010011001110 Peak10 60%
01110110100101111110011010101010010101000 Peak11 60%
01010101011110111111101010000101010101 Peak12 65%
01110110111110000111111110110000110000110 Peak12 62%
111011010000111001110110010111110101010 Peak12 62%
```

Fig. 11. Not using MNC ($\overline{R} = 63\%$)

As a result, the method using MNC can obtain about 18% higher value than the method not using MNC.

As shown in Table 1, optimum achievement rate is obtained after 1,000 EC generations, and number of couples (good matches) is calculated by applying an assignment operator to the solutions at the 1,000-th generation. Here, $\overline{R}$ is shown as an average value of 30 trials, and plus-minus means the standard deviation which started from different initial solutions generated randomly. Proposed (Stock use: 50%) in Table 1 is the case using stock. In this case, the total times of real Gokon is 5,250, almost half the times as without stock (stock use: 0%).

Table 1. Comparison of proposed and greedy method

	$\bar{R} \pm S.D.$	Number of couples	Number of real Gokon
Optimal*	$100 \pm 0\%$	6302 ± 24 pairs	0 Times
Proposed (Stock use: 0%)	$82 \pm 1\%$	5003 ± 102 pairs	10,000 Times
Proposed (Stock use: 50%)	$80 \pm 1\%$	4698 ± 132 pairs	5,250 Times
Greedy	$64 \pm 7\%$	2530 ± 334 pairs	10,000 Times

* At the case when giving optimal solutions (Fig. 9) into solutions.

5   Conclusions and Future Work

In this paper, we defined the Gokon problem, which assigns men and women who are likely to make good matches in the same Gokon. We also proposed the Evolutionary System to solve the problem and evaluated the system using computer simulation. For the $12Max$ problem, which is the problem of finding 12 unknown peaks (i.e., good affinity patterns), we confirmed that the proposed system achieves 82% similarities to optimal solutions and almost twice as many good matches as the conventional method with about half the evaluation times.

As part of future research, we plan to fill the gap for applying the proposed system to actual Gokons. To obtain the results of this experiment, we needed Gokon data for 5 to 10 thousand times. Major companies in this domain in Japan perform Gokons about 2 to 15 thousand times a year. Thus, if we can use such data, the system will be available within a more practicable timeframe.

References

1. Foster, J.A.: Evolutionary Computation. Nature Rev. Genet. 2, 428–436 (2001)
2. Cagnoni, S., Poli, R., Smith, G.D., Corne, D., Oates, M., Hart, E., Lanzi, P.L., Willem, E.J., Li, Y., Paechter, B., Fogarty, T.C. (eds.): Real-World Applications of Evolutionary Computing. LNCS, vol. 1803. Springer, Heidelberg (2000)
3. Kuroiwa, S., Murata, Y., Kitani, T., Yasumoto, K., Ito, M.: A Method for Assigning Men and Women with Good Affinity to Matchmaking Parties through Interactive Evolutionary Computation. In: Proc. of Simulated Evolution and Learning, pp. 645–655 (2008)
4. Cedeno, W., Vemuri, V., Slezak, T.: Multi-Niche Crowding in Genetic Algorithms and its Application to the Assembly of DNA Restriction-Fragments. Evolutionary Computation 2(4), 321–345 (1995)
5. Cloninger, C.R., Svrakic, D.M., Przybeck, T.R.: A Psychobiological Model of Temperament and Character. Archives of General Psychiatry 50, 975–990 (1993)

A Benchmark Generator
for Dynamic Permutation-Encoded Problems

Michalis Mavrovouniotis[1], Shengxiang Yang[2], and Xin Yao[3]

[1] Department of Computer Science, University of Leicester
University Road, Leicester LE1 7RH, United Kingdom
mm251@mcs.le.ac.uk
[2] Department of Information Systems and Computing, Brunel University
Uxbridge, Middlesex UB8 3PH, United Kingdom
shengxiang.yang@brunel.ac.uk
[3] CERCIA, School of Computer Science, University of Birmingham, UK
Birmingham B15 2TT, UK
x.yao@bham.cs.ac.uk

Abstract. Several general benchmark generators (BGs) are available for the dynamic continuous optimization domain, in which generators use functions with adjustable parameters to simulate shifting landscapes. In the combinatorial domain the work is still on early stages. Many attempts of dynamic BGs are limited to the range of algorithms and combinatorial optimization problems (COPs) they are compatible with, and usually the optimum is not known during the dynamic changes of the environment. In this paper, we propose a BG that can address the aforementioned limitations of existing BGs. The proposed generator allows full control over some important aspects of the dynamics, in which several test environments with different properties can be generated where the optimum is known, without re-optimization.

1 Introduction

Over the years, there has been a growing interest for dynamic optimization problems (DOPs). However, after years of research, the field of dynamic optimization still has many open issues. One of them is the development of suitable dynamic benchmark generators (BGs) that can be easily adapted by researchers. A benchmark can be defined as standard test problems designed for the development of new algorithms and the comparison with existing ones.

A DOP can be otherwise defined as a series of several static instances. Hence, a straightforward, but not efficient, method to construct a dynamic test problem is to switch between different static instances that will cause dynamic changes. However, several general dynamic BGs have been proposed that re-shape the fitness landscape, including: 1) Moving Peaks [1]; 2) DF1 [10]; and 3) exclusive-or (XOR) [16]. The first two benchmark problems work for the continuous domain where they use functions with adjustable parameters to simulate shifting landscapes. The continuous space can be modelled as a "field of cones" [10], where

C.A. Coello Coello et al. (Eds.): PPSN 2012, Part II, LNCS 7492, pp. 508–517, 2012.

each cone is adjusted individually to represent different dynamics. A similar approach is not feasible for the combinatorial space because the landscape is indistinct and cannot be defined without reference to the optimization algorithm.

The XOR DOP generator [16] is the only dynamic BG for the combinatorial space that can convert any static binary-encoded problem (with known optimum) to a dynamic environment without affecting the global optimum value. In this way, one can see how close to the optimum each algorithm performs. A similar approach does not exist for permutation-encoded problems, such as the travelling salesman problem (TSP) and the vehicle routing problem (VRP).

In this paper, a general dynamic BG for permutation-encoded (DBGP) problems is developed which can convert a static problem instance to a dynamic one, by modifying its encoding. The experiments on some static benchmarks with known optimum show that the dynamic changes affect the algorithm but the optimum remains the same. Furthermore, the experiments show that DBGP can be directly applied to other variants of the fundamental TSP, e.g., the capacitated VRP [9].

The rest of the paper is organized as follows. Section 2 gives a summary of dynamic optimization, including the performance measurements, algorithmic methods, and BGs currently used. Section 3 describes the proposed DBGP. Section 4 presents the experimental study, in which DBGP is used to generate several dynamic test problems from static TSPs and VRPs. Finally, Section 5 concludes this paper with concluding remarks and directions for future work.

2 Background

2.1 Dynamic Combinatorial Optimization

The objective for static optimization problems is to find the optimum solution efficiently. For DOPs, the environment changes as a function of time t, which causes the global optimum to move. Hence, the objective for DOPs is to track the moving optimum efficiently. Formally, a combinatorial DOP can be defined as $\Pi = (X, \Omega, f, t)$, where Π is the optimization problem, X is a set of feasible solutions, Ω is a set of constraints, f is the objective function which assigns an objective value to a solution $x(t)$, where $x(t) = \{x_1, \ldots, x_n\}$ is a vector of n discrete optimization variables that satisfy the constraints Ω, and t is the time.

The main aspects of "dynamism" are the frequency and the magnitude of environmental changes. The former corresponds to the speed and the latter to the degree of an environmental change, respectively. An environmental change may involve factors like the objective function, input variables, problem instance, constraints, and so on, that cause the optimum to change.

The environmental changes are classified into two types: dimensional and non-dimensional changes. Dimensional changes correspond to adding/removing variables from the problem. Such environmental changes affect the representation of the solutions and alter a feasible solution to an infeasible one. A repair operator may address this problem, but requires a prior knowledge of the problem and the dynamic changes. Non-dimensional changes correspond to the change of the variables of the problem. Such environmental changes do not affect the representation of the solutions, and, thus, are easier to address.

2.2   Nature-Inspired Approaches in Dynamic Environments

There is a growing interest in the evolutionary computation (EC) community to apply nature-inspired algorithms to address combinatorial DOPs due to their inspiration from nature, which is a continuous adaptation process [5]. Popular examples of such algorithms are evolutionary algorithms (EAs) [6] and ant colony optimization (ACO) algorithms [2].

One common characteristic of these nature-inspired algorithms is that they are iterative. Hence, they are able to transfer knowledge from previous iterations and adapt to the new environment [1]. EAs have search operators to exchange information between a population of individuals. In ACO algorithms, ants share their pheromone trails with other ants to communicate.

The difference between an EA and an ACO algorithm lies in that the former maintains an actual population of μ solutions, whereas the latter consists of a "virtual" population. More precisely, ACO is a constructive heuristic in which all the ants deposit pheromone to mark their solution every iteration. Therefore, the information of the solutions is only kept to the pheromone trails. The constructive procedure of ACO is biased by the existing pheromone trails and some heuristic information available a priori [2].

2.3   Performance Measurements

For DOPs, it is difficult to analyze the adaptation and searching capabilities of an algorithm. This is because there is no agreed measurement to evaluate algorithms and researchers view their algorithms from different perspectives. Different measurements have been proposed to compare algorithms such as "accuracy, stability, and reactivity" [14], "collective mean fitness" [11], "accuracy and adaptability" [15], and "performance and robustness" [13].

A common method used to compare algorithms for DOPs is the best offline performance [5], which is defined as:

$$P_B = \frac{1}{G} \sum_{i=1}^{G} \left(\frac{1}{R} \sum_{j=1}^{R} P_{ij}^* \right), \tag{1}$$

where G is the number of iterations, R is the number of independent runs, and P_{ij}^* is the fitness of the best solution of iteration i in run j after the last dynamic change. Apart from the performances which describe the best the system can do, other researchers are concerned for measurements which can characterize the population as a whole [13]. Traditionally, the target in DOPs is to track the moving optimum over time [5]. Recently, a new perspective on DOPs has been established, known as robust optimization over time (ROOT), where the target is to find the sequence of solutions which are robust over time [17]. More precisely, a solution is robust over time when its quality is acceptable to the environmental changes during a given time interval.

Other researchers want to observe "how close to the moving optimum a solution, either robust or not, is?". Probably it is the best way to evaluate the effectiveness of an algorithm for DOPs, in addition to the time needed to converge to that optimum. However, the global optimum is needed for every changing environment and this is very challenging due to the $\mathcal{NP}$-Hardness of most combinatorial optimization problems (COPs). Since a DOP can be considered as several static instances, a direct way is to solve each one to optimality, which may be non-trivial or even impossible, especially for large problem instances. It may be possible on small problem instances, but then it will reduce the usefulness of benchmarking. Hence, the need for a BG to address the challenges of comparison is increased, but it is even harder to develop a BG for DOPs with known optimum in COPs, without re-optimization.

2.4 Benchmark Generators for DOPs

The field of dynamic optimization is related to the applications of nature-inspired algorithms [5]. The area is rapidly growing on strategies to enhance the performance of algorithms, but still there is limited theoretical work, due to the complexity of nature-inspired algorithms and the difficulty to analyze them in the dynamic domain. Therefore, the development of BGs to evaluate the algorithms is appreciated by the EC community. Such tools are not only useful to evaluate algorithms but also essential for the development of new ones.

The XOR DOP generator [16] is the only benchmark for the combinatorial space that constructs a dynamic environment from any static binary-encoded function $f(x(t))$, where $x(t) \in \{0,1\}^n$, by a bitwise XOR operator. It simply shifts the population of individuals into a different location in the fitness landscape. Hence, the global optimum is known during the environmental changes.

In the case of permutation-encoded problems, where $x(t)$ is a set of numbers that represent a position in a sequence, researchers prefer their own benchmark instances to address different real-world applications, e.g., the dynamic TSP (DTSP) with exchangeable cities [4], DTSP with traffic factors [8], and dynamic VRP (DVRP) with dynamic demands.

3 Proposed Dynamic Benchmark Generator

3.1 General Framework

Most research on dynamic optimization has been done with EAs on binary-encoded COPs. Recently, ACO has been found effective on permutation-encoded DOPs, e.g., DTSP [8]. Due to the high number of specialized BGs for permutation-encoded COPs the establishment of a generalized one that converts the base of a static COP to a dynamic one is vital. Most of the existing BGs are not easily available and they are difficult to be adapted. Moreover, on each environmental change, the fitness landscape is modified no matter whether dimensional or non-dimensional changes are applied. Hence, it is impossible to know how close to the optimum an algorithm performs on each environmental change.

Younes *et al.* [18] introduced a general benchmark framework that applies a mapping function on each permutation-encoded individual. The mapping function swaps the labels, i.e., the city index, between two objects and all the individuals in the population are treated in the same way. This way, the individuals represent different solutions after a dynamic change but the fitness landscape of the problem instance does not change. However, this generator is restricted to the range of algorithms and COPs that they are compatible with, and it is limited to the accuracy regarding the magnitude of change.

The proposed DBGP is designed to allow full control over the important aspects of dynamics and convert the base of any benchmark static COP with known optimum to a dynamic one without causing the optimum to change. Such static instances can be obtained from the TSPLIB[1] and VRPLIB[2], where most of the instances have been solved to optimality.

The basic idea of the proposed DBGP is to modify the encoding of the problem instance, instead of the encoding of each individual, i.e., the distance matrix, without affecting its fitness landscape. To illustrate such a dynamic change, let $G = (V, E)$ be a weighted graph where V set of n nodes and E is a set of links. Each node u_i has a location defined by (x, y) and each link (u_i, u_j) is associated with a non-negative distance d_{ij}. Usually, the distance matrix of a problem instance is defined as $D = (d_{ij})_{n \times n}$. Then, an environmental change may occur at any time by swapping the location of some node i with the location of some node j. In this way, the values in the distance matrix are re-allocated but the optimum remains the same; see Fig. 1.

The dynamic environments constructed by DBGP[3] may not reflect a full real-world situation but achieve the main goal of a benchmark in which the optimum is known during all the environmental changes. In other words, DBGP sacrifices the realistic modelling of application problems for the sake of benchmarking. Moreover, it is simple and can be adapted to any TSP and its variants to compare algorithms in dynamic environments.

3.2   Frequency and Magnitude of Change

Every f iterations a random vector $r(T)$ is generated that contains all the objects of a problem instance of size n, where $T = \lceil t/f \rceil$ is the index of the period of change, and t is the iteration count of the algorithm. For example, for the TSP the objects are the cities which have a location (x, y). The magnitude m of change depends on the number of swapped locations of objects.

More precisely, $m \in [0.0, 1.0]$ defines the degree of change, in which only the first $m \times n$ of $r(T)$ object locations are swapped. In order to restrict the swaps to the first cities, a randomly re-ordered vector $r'(T)$ is generated that contains only the first $m \times n$ objects of $r(T)$. Therefore, exactly $m \times n$ pairwise swaps are

---

[1] Available at http://comopt.ifi.uni-heidelberg.de/software/TSPLIB95/

[2] Available at: http://neo.lcc.uma.es/radi-aeb/WebVRP/ or
   http://www.or.deis.unibo.it/research_pages/ORinstances/
   VRPLIB/VRPLIB.html

[3] Available at: http://www.cs.le.ac.uk/~mm251

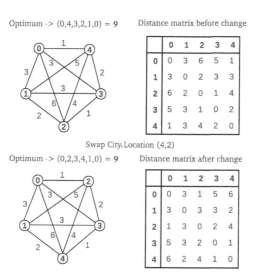

Fig. 1. Illustration of the distance matrix with the optimum solution of the problem instance before and after a dynamic change

performed using the two random vectors starting from the first pairs. In Younes' generator [18], the magnitude of change is expressed as the number of swaps imposed on the mapping function. In this way, the objects affected from the dynamic change may not correspond to the predefined magnitude parameter. For example, if $m = 0.5$, half of the objects may be swapped with the remaining half of the objects of the optimization problem. Hence, the change affects all the objects and may be considered as $m = 1.0$.

The frequency of change is defined by the constant parameter f which is usually defined by the algorithmic iterations. However, before each environmental change, the previous pairwise swaps are reversed, starting from the end of $r(T-1)$ and $r'(T-1)$. In this way, the environmental changes are always applied to the encoding of the initial static problem instance.

3.3 Effect on Algorithms

DBPG can be applied to algorithms that either maintain an actual population or not, because the dynamic changes occur to the encoding of the actual problem instance. In this way, the solutions of EAs with the same encoding as before a dynamic change have a different cost after a dynamic change. On the other hand, the constructive procedure of ACO is affected since different heuristic information is generated whereas the pheromone matrix remains unchanged.

In general, DBGP shifts the population of EAs and biases the population of ACO algorithms to a new location in the fitness landscape. Younes' generator assumes that the solver has a population of solutions since the mapping function is applied on the encoding of each individual, e.g., EAs. Hence, it cannot be applied to algorithms that do not maintain an actual population, e.g., ACO.

Another advantage of the DBGP against Younes' generator [18] is that in the VRP the solutions after a dynamic change may represent an infeasible solution when EAs are used. This is because when the label of the customer changes then its demand changes, and the capacity constraint is possible to be violated. Hence, a repair operator or a penalty function has to be applied. The proposed DBGP overcomes this problem since only the location of the customer changes whereas the label and the demand remain unchanged.

3.4 Cyclic Dynamic Environments

The default dynamic environments generated by DBGP do not guarantee that any of the previously generated environment will re-appear. Such environments are called *random dynamic environments* in this paper. In fact, some algorithms that are enhanced with memory are expected to work better on dynamic environments that re-appear in the future [1]. Such environments are called *cyclic dynamic environments* in this paper, and they can be generated as follows. First, we generate K random vectors $(\boldsymbol{r}(0), \ldots, \boldsymbol{r}(K-1))$ with their corresponding re-ordered vectors as the base states in the search space. Initially, the first base state is applied. Then, every f iterations the previous dynamic changes are reversed, and then the new ones are applied from the next base state. In this way, it is guaranteed that the environments generated from the base states will re-appear.

DBGP has two options for cyclic dynamic environments regarding the way the base states are selected: 1) cyclic, where the base states are selected as in a fixed logical ring; and 2) randomly, where the base states are selected randomly.

From the above cyclic environment generator, we can further construct cyclic dynamic environments with noise as follows. Each time a new base state is to be selected, swaps are performed from the objects that are not in the $\boldsymbol{r}(T)$ with a small probability, i.e., p_{noise}. Note that the swaps occurring from the noise are reversed in the same way as with the dynamic changes above.

3.5 Varying f and m Parameters

In the random and cyclic environments, the f and m parameters remain fixed during the execution of the algorithm. An additional feature of DBGP is to vary the values of f and m with a randomly generated number with a uniform distribution in $[1, 100]$ and $[0.0, 1.0]$, respectively, for each environmental change.

4   Experimental Study

In this section, we perform some preliminary experiments based on two of the best performing ACO algorithms, i.e., $\mathcal{MAX} - \mathcal{MIN}$ Ant System ($\mathcal{MMAS}$) and Ant Colony System (ACS) [2], on two well-known COPs, i.e., TSP and VRP. From some static benchmark instances of these problems a set of dynamic test cases is generated using the proposed DBGP to test if the parameter m corresponds to the degree of a dynamic change.

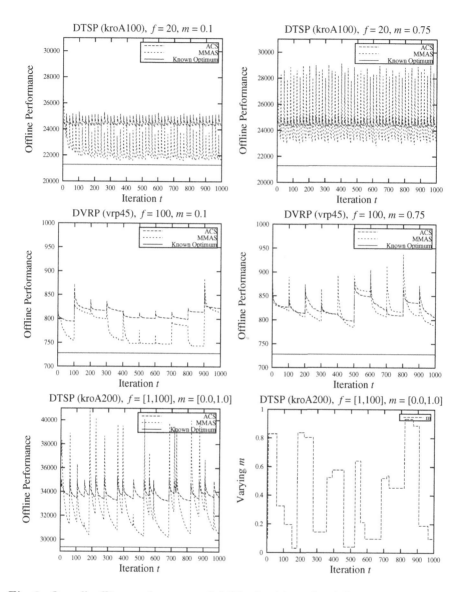

Fig. 2. Overall offline performance of ACO algorithms for different problems with different dynamic properties against the known global optimum

In Fig. 2 we illustrate our preliminary results on dynamic test environments with different properties. The algorithms perform 1000 iterations for 30 runs and $\boldsymbol{P}_B$, defined in Eq. (1), is chosen. On each iteration the algorithms perform the same number of function evaluations for a fair comparison. The first two graphs represent the performance of the aforementioned algorithms to a DTSP with a cyclic environment (of four base states), the middle graphs represent a DVRP with a random environment and the last ones a DTSP with varying m and f (left) and the corresponding values of m for each iteration (right).

From the experiments we can observe that $\mathcal{MMAS}$ performs much closer to the optimum when $m = 0.1$, since the changing environments are similar and the knowledge transferred is useful. On the other hand, the performance of ACS is inferior to $\mathcal{MMAS}$ in all dynamic test cases. The most important observation from the experiment is the reaction of algorithms to different values of m. In Fig. 2, the algorithms have a small drop in the offline performance when a change occurs with $m = 0.1$, and a large one when $m = 0.75$. This shows that DBGP defines and controls the degree of change with parameter m appropriately.

5   Conclusions and Future Work

The construction of benchmark DOP generators is important for the empirical comparison of algorithms in the EC community, due to the limited theoretical work available. This paper proposes a general BG for dynamic permutation-encoded COPs that modifies the encoding of the problem instances, to introduce dynamic changes. The proposed DBGP converts any static benchmark instance to a dynamic test environment with different properties.

In order to test DBGP, some preliminary experiments with dynamic test environments generated with DBGP for the DTSP and DVRP are carried out using ACO algorithms. From the experiments, it can be investigated how close to the optimum each algorithm converges on each environmental change. Although DBGP lacks a real-world application model, it is a simple method to empirically analyze algorithms on permutation-encoded DOPs, and can be easily adapted.

Therefore, an interesting future work is to add to the DBGP more real-world related models, such as the time-linkage property where the future behaviour of the problem depends on the current or a previous solution found by the algorithm [12]. Furthermore, it will be interesting to integrate DBGP with the ROOT framework [3,17]. Another future work, is to test DBGP on more permutation-encoded problems, such as the capacitated arc routing problem, which is the arc counterpart of the VRP [9].

Acknowledgement. This work was partially supported by three EPSRC grants, EP/E058884/1 and EP/E060722/2 on "Evolutionary Algorithms for Dynamic Optimisation Problems: Design, Analysis and Applications", and EP/I010297/1 on "Evolutionary Approximation Algorithms for Optimisation: Algorithm Design and Complexity Analysis".

References

1. Branke, J.: Memory enhanced evolutionary algorithms for changing optimization problems. In: Proc. of the 1999 IEEE Congr. on Evol. Comput., pp. 1875–1882 (1999)

2. Dorigo, M., Stützle, T.: Ant Colony Optimization. The MIT Press, London (2004)
3. Fu, H., Sendhoff, B., Tang, K., Yao, X.: Characterising environmental changes in robust optimisation over time. In: Proc. of the 2012 IEEE Congr. on Evol. Comput., pp. 551–558 (2012)
4. Guntsch, M., Middendorf, M.: Applying Population Based ACO to Dynamic Optimization Problems. In: Dorigo, M., Di Caro, G.A., Sampels, M. (eds.) ANTS 2002. LNCS, vol. 2463, pp. 111–122. Springer, Heidelberg (2002)
5. Jin, Y., Branke, J.: Evolutionary optimization in uncertain environments - a survey. IEEE Trans. Evol. Comput. 9(3), 303–317 (2005)
6. Holland, J.: Adaption in Natural and artificial systems. University of Michigan Press, Ann Arbor (1975)
7. Kilby, P., Prosser, P., Shaw, P.: Dynamic VRPs: A study of scenarios, Tech. Rep. APES-06-1998, University of Strathclyde, U.K. (1998)
8. Mavrovouniotis, M., Yang, S.: Memory-Based Immigrants for Ant Colony Optimization in Changing Environments. In: Di Chio, C., Cagnoni, S., Cotta, C., Ebner, M., Ekárt, A., Esparcia-Alcázar, A.I., Merelo, J.J., Neri, F., Preuss, M., Richter, H., Togelius, J., Yannakakis, G.N. (eds.) EvoApplications 2011, Part I. LNCS, vol. 6624, pp. 324–333. Springer, Heidelberg (2011)
9. Mei, Y., Tang, K., Yao, X.: A memetic algorithm for periodic capacitated arc routing problem. IEEE Trans. on Syst. Man and Cybern., Part B: Cybern. 41(6), 1654–1667 (2011)
10. Morrison, R.W., De Jong, K.A.: A test problem generator for non-stationary environments. In: Proc. of the 1999 IEEE Congr. on Evol. Comput., pp. 2047–2053 (1999)
11. Morrison, R.W.: Performance measurement in dynamic environments. In: Proc. of the 2003 Genetic and Evol. Comput. Conf., pp. 5–8 (2003)
12. Nguyen, T.T., Yao, X.: Dynamic Time-Linkage Problems Revisited. In: Giacobini, M., Brabazon, A., Cagnoni, S., Di Caro, G.A., Ekárt, A., Esparcia-Alcázar, A.I., Farooq, M., Fink, A., Machado, P. (eds.) EvoWorkshops 2009. LNCS, vol. 5484, pp. 735–744. Springer, Heidelberg (2009)
13. Rand, W., Riolo, R.: Measurements for understanding the behavior of the genetic algorithm in dynamic environments: A case study using the shaky ladder hyperplane-defined functions. In: Proc. of the 2005 Genetic and Evol. Comput. Conf., pp. 32–38 (2005)
14. Weicker, K.: Performance Measures for Dynamic Environments. In: Guervós, J.J.M., Adamidis, P.A., Beyer, H.-G., Fernández-Villacañas, J.-L., Schwefel, H.-P. (eds.) PPSN VII. LNCS, vol. 2439, pp. 64–73. Springer, Heidelberg (2002)
15. Trojanowski, K., Michalewicz, Z.: Evolutionary algorithms for non-stationary environments. In: Proc. of 8th Workshop on Intelligent Information Systems, pp. 229–240 (1999)
16. Yang, S.: Non-stationary problem optimization using the primal-dual genetic algorithm. In: Proc. of the 2003 IEEE Congr. on Evol. Comput., pp. 2246–2253 (2003)
17. Yu, X., Jin, Y., Tang, K., Yao, X.: Robust optimization over Time – A new perspective on dynamic optimization problems. In: Proc. of the 2010 IEEE Congr. on Evol Comput., pp. 3998–4003 (2010)
18. Younes, A., Calamai, P., Basir, O.: Generalized benchmark generation for dynamic combinatorial problems. In: Proc. of the 2005 Genetic and Evol. Comput. Conf., pp. 25–31 (2005)

Evolving Femtocell Algorithms with Dynamic and Stationary Training Scenarios

Erik Hemberg[1], Lester Ho[2], Michael O'Neill[1], and Holger Claussen[2]

[1] Natural Computing Research & Applications Group
Complex & Adaptive Systems Laboratory
School of Computer Science & Informatics
University College Dublin
erik.hemberg@ucd.ie, m.oneill@ucd.ie
[2] Bell Laboratories
Alcatel-Lucent
Dublin
{lester.ho,holger.claussen}@alcatel-lucent.com

Abstract. We analyse the impact of dynamic training scenarios when evolving algorithms for femtocells, which are low power, low-cost, user-deployed cellular base stations. Performance is benchmarked against an alternative stationary training strategy where all scenarios are presented to each individual in the evolving population during each fitness evaluation. In the dynamic setup, different training scenarios are gradually exposed to the population over successive generations. The results show that the solutions evolved using the stationary training scenarios have the best out-of-sample performance. Moreover, the use of a grammar which produces discrete changes to the pilot power generate better solutions on the training and out-of-sample scenarios.

1 Introduction

Femtocells are low power, low-cost, user-deployed cellular base stations (BS), which operate in dynamic environments. A significant issue facing the developers of the algorithms which control the behaviour of femtocells is how best to design the algorithms to handle these unforeseen, dynamic environments. In previous studies [12, 11] we have successfully examined the suitability of Genetic Programming (GP), and a grammar-based form of GP [14], Grammatical Evolution (GE) [6], to generate control algorithms for these devices. In these earlier studies a predefined, static set of scenarios are exposed to the evolving population to determine the quality of the evolving solutions.

Our aim in this study is to examine the impact of the training scenarios employed on the quality of the evolved solutions. More specifically we ask:

– *Is there a difference in the robustness of solutions (out-of-sample) based on the use of stationary versus dynamic training scenarios?*

The remainder of the paper is structured as follows. In Sect. 2 the femtocell problem is described. Experiments and results are in Sect. 3 and 4. Finally, the conclusion and future work is presented in Sect. 5.

C.A. Coello Coello et al. (Eds.): PPSN 2012, Part II, LNCS 7492, pp. 518–527, 2012.

2   The Femtocell Problem

As is the case with other physical network infrastructure such as base stations, there are a number of issues surrounding the optimal placement of the hardware in addition to the design of algorithms which manage the performance of hardware networked in this manner. Femtocells are low power, low-cost, **user-deployed** cellular base stations. Therefore, in the case of femtocells the designer of the software does not know *a-priori* where (and how many) femtocells might be deployed in a site.

If we consider an intended area of coverage, e.g. an office environment as shown in Fig. 1(c), where a group of femtocells is deployed to jointly provide end-user services, we focus on the problem of distributed coverage optimisation by adjusting the pilot power of the BS in order to alter the coverage of the femtocells. The objectives are:

Mobility: To minimise mobility events (handovers) between femtocells and macrocells within the femtocell group's intended area of coverage.

Load: To balance the load amongst the femtocells in the group to prevent overloading or under-utilisation.

Leakage: To minimise the leakage of the femtocell group's coverage outside its intended area of coverage.

There have been previous studies of applying EC to telecommunication problems [1]. But only two specifically regarding femtocell coverage algorithms and EC, one using GP [12] and another using GE [11]. Most related work in the literature regarding cellular coverage optimisation deals with centralised computation methods [16, 8], e.g. the calculation of parameters such as the number and locations of BS, pilot channel transmit powers, or antenna configurations using a central server running an optimisation algorithm. Many studies also focus on determining the optimal BS numbers or placements to achieve the operator's quality of service or coverage target. This approach is not always practical because network design is restricted by BS placements, and in the case of femtocells these are physically deployed by the end-user.

3   Experimental Setup

In the femtocell problem we face a number of challenges, the most pressing of which are (i) fitness evaluations are computationally expensive, and (ii) it is not clear how best to design the fitness evaluations in terms of the type and number of training scenarios presented to the evolving population. In this study we focus on understanding how to best design a fitness function by examining the robustness of solutions evolved using dynamic and stationary training scenarios. In terms of computational expense, the dynamic training scenarios are potentially attractive as less scenarios are presented to each individual of the population, thereby reducing the evolutionary algorithms run time. In addition, there are potentially performance gains to be achieved by adopting dynamic environments during evolutionary runs, for example, see [15]. We therefore study the robustness of solutions depending on how they have been evolved. The two approaches we use to drive evolution are:

Stationary training scenario: The fitness function employs multiple training scenarios at each generation. The training fitness is calculated as the average fitness across each training scenario presented to the individual.

Dynamic training scenario: The fitness function is comprised of a single training scenario at each generation and as the generations progress the training scenario changes.

The guidance of the search towards a solution is different for each setup. The stationary setup is comprised of multiple training scenarios, where evolution is trying to find a general solution by averaging the fitness over several scenarios. The reasoning is that solutions that are too specialised will be avoided since a solution must have good fitness on all the scenarios. In contrast, the dynamic setup exposes the evolving population to a single scenario which changes over time, attempting to guide the population towards solutions which can behave well on scenarios presented over different environmental conditions. There are more assumptions and uncertainties in the dynamic scenario regarding the robustness of the solution. First, the search must be given enough evaluations in each scenario to find good solutions. Then, the next scenario needs to be similar enough to allow the search to gain advantage from existing parts in the solutions. Thus, this is a potentially powerful approach for stimulating generic parts in solutions. Although it requires the capability of the search method and setup to represent and identify general components which are preserved during the search.

3.1   Simulation Model

A user mobility and traffic model is employed, where users are initially randomly placed at way points on the map and moving at a speed of $1ms^{-1}$, spending some time at a way point before moving to another. In total 50, 200 and 400 users are modelled, in *low* (l), *medium* (m) and *high* (h) load scenarios. Each user's voice traffic model produces 0.2 Erlangs of traffic during 24 hours of simulated operation time, with the algorithm adjusting the femtocell pilot power after collecting statistics for 30 minutes. The algorithm start time for each femtocell is randomly dithered with, and the initial pilot channel power $\rho = -30$dBm, $\rho \in [-50, -49, \dots, 11]$. Femtocell to macrocell handovers are triggered when a user terminal's pilot channel receive power from the best femtocell goes below -100dBm. Outside cell users move east-west and west-east on the north and south edges of the map. When the signal leakage is strong enough the outside user request a handover to the femtocell and a rejection is recorded. The outside user tries to connect once to each leaking femtocell when moving through the femtocell coverage.

Office (O12, O8, O4). The number of BS in the office environment are 12, 8 and 4, shown in Fig. 1. The scenario with 12 BS is denoted **O12**. In **O4** the coordinates have been slightly altered compared to the **O8** and **O12** scenarios, by moving the BSs closer to the walls.

The building is an office with cubicles, closed meeting rooms, and toilets. The exterior of the building is mainly glass and the interior is mostly light interior walls and cubicle partitions. This is a realistic plug-and-play femtocell deployment, which can be

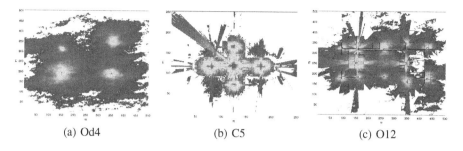

| (a) Od4 | (b) C5 | (c) O12 |

Fig. 1. Received pilot power for the **Outdoor(Od4)** 1(a), **Cross(C5)** 1(b) and **Office(O12)** 1(c)

sub-optimal due to the lack of exhaustive cell planning. In the simulation each femto-cell has a maximum capacity of 8 voice calls, a macrocell underlay coverage is also assumed. A path loss map is generated for the 450m x 500m area for each femto-cell. For shorter distances the PL, path loss (dB) at d (meters) from a BS is mod-elled as $38.5 + 20log_{10}(d) + PL_{walls}$, with a smooth transition to $28 + 35log_{10}(d) + PL_{walls}$ otherwise. A correlated shadow fading with a standard deviation of 8 dB and spatial correlation of $r(d) = e^{d/20}$. The assumed transmission losses for the explicit building model are a function of the incident angle, this model is taken from Ho et al. [12].

Outdoor (Od4). There are no walls and the BS placement is the same as in **O4**.

Cross (C5). There are walls and 5 BS. All the way points and hot-spots are different from **O12** and set to explicitly model the need for load balancing by overloading some cells and under utilizing others. Moreover, a different path loss model is used.

The training scenarios are **Od4, C5, O12** with medium load, the validation scenario is **O4l**, and the test scenarios are **O8, O4** at low, medium and high load. The dynamic setup starts with **Od4**, see Fig. 1(a). The next scenario is **C5** Fig. 1(b). The last scenario is **O12** Fig. 1(c). Thus, there is an increase in number of BS and the walls between each scenario. The stationary scenarios evaluate on all the scenarios at every iteration.

3.2 Evolutionary Algorithm

In this study we use a Matlab implementation of GE, GEM [1]. Two different grammars are tested (denoted CG and SRCG). A conditional grammar that changes the pilot power with discrete values and a conditional equation grammar changing the pilot power with continuous values calculated from generated equations. The search space is very different for the grammars. A difference in performance is to be expected with the number of fitness evaluations used. In addition, the expected result from the different scenarios would be that stationary scenarios should perform well since it is always the same underlying simulation model. The setup is the same as Hemberg et al. [11].

---

[1] http://ncra.ucd.ie/GEM/GEM.tgz

```
<CODE> ::= if gt(my_handover, MT)
              if gt(my_load, LT)
                if gt(my_macro_requests, LeT)
                  <function>
                else
                  <function>
              else if gt(my_macro_requests, LeT)
                  <function>
                else
                  <function>
            else if gt(my_load, LT)
              if gt(my_macro_requests, LeT)
                  <function>
                else
                  <function>
              else if gt(my_macro_requests, LeT)
                  <function>
                else
                  <function>
<function> ::= <terminal><function> | <terminal>
<terminal> ::= my_power = increase_power(my_power);
             | my_power = decrease_power(my_power);
             | my_power = do_nothing(my_power);
```

Fig. 2. Conditional statement grammar (CG)

```
<function> ::= my_power = <expr_0>;
<expr_0> ::= (<expr><op><expr>) | <pre-op>
<expr> ::= (<expr><op><expr>) | <var> | <var> | <var>
         | <pre-op> | <pre-op_step> | <pre-op_monotone>
<pre-op> ::= sin(real(<expr>)) | cos(real(<expr>))
           | log(real(<expr>)) | tan(real(<expr>))
<pre-op_monotone> ::= exp(real(<expr>)) | uminus(<expr>)
<pre-op_step> ::= atan(<expr>) | tanh(<expr>) | sigmoid(<expr>)
<var> ::= my_power | my_load | my_handover | my_macro_requests
        | <cnst>
<cnst> ::= <nr><nr> | <nr> | 0.<nr><nr> | 0.<nr>
<nr> ::= 1 | 2 | 3 | 4 | 5 | 6 | 7 | 8 | 9
```

Fig. 3. Symbolic Regression and Conditional Statement Grammar (SRCG). Only the differences between the CG (Fig. 2) is shown

Conditional Statement Grammar (CG). We construct a grammar using conditional statements. The thresholds and the size of the increase and decrease of power needs to be predetermined. Here the change is 1dBm and the thresholds are mobility ($MT = 0$), leakage ($LeT = 0$) and load ($LT = 7$).

Symbolic Regression and Conditional Statement Grammar (SRCG). Creates equations and uses thresholds as in CG. Only the differences in CG and SRCG are shown in Fig. 3. To create the SRCG we combine the grammar in Fig. 2. The multiple <var> productions keeps the grammar from "exploding", see Harper [10]. The grammar adopted in this study is in MATLAB syntax. A wide range of functions were used and only the real valued part of the function values was used. The unary minus is uminus.

Fitness Function. Statistics of mobility, load and leakage are collected over a specified update period. These statistics are then used as inputs into the algorithm, and for calculating the fitness. The duration of the simulation is T, the number of femtocells is

N, and $\mathbf{x}$ is a vector of femtocells. The fitness is a vector comprised of the fitness for each function, $\mathbf{f} = [f_M(\overline{M(h,r)}), f_L(\overline{L(\mathbf{x})}), f_{Le}(\overline{Le(\mathbf{x})})]$. The mobility objective is conflicting with load and leakage, leakage can also conflict with load. All the objectives are normalized and equally important.

Mobility fitness is the number of handovers and relocations of users. The mobility events between femtocells and macrocells are recorded for each period. The number of femtocell handovers is h, macrocell handovers is h^M, femtocell relocations is r, and macrocell relocations is r^M. Mobility, M, is the ratio of update periods where a mobility event occurs divided by the total number of update periods.

$$M_b^M(h,r) = \sum_{t=0}^{T}\sum_{i=1}^{N} h_{it}^M + \sum_{t=0}^{T}\sum_{i=1}^{N} r_{it}^M$$

$$M_b(h,r) = M_b^M(h,r) + \sum_{t=0}^{T}\sum_{i=1}^{N} h_{it} + \sum_{t=0}^{T}\sum_{i=1}^{N} r_{it}$$

The mobility fitness is maximised when there are no handovers or relocation to the macrocell underlay, and is 0 when all femtocell user handovers are to or from macrocells, otherwise

$$\overline{M(h,r)} = \begin{cases} M_b^M(h,r)/M_b(h,r) & \text{if } M_b(h,r) > 0 \\ 1 & \text{if } M_b(h,r) = 0 \end{cases}$$

Load fitness has the objective that the femtocells should serve enough users. It is based on the ratio of average number of times the load has been greater than a defined maximum load threshold, LT, and the total load, including the macrocell. If the mean cell load during an update period exceeds LT then L is equal to one, else it is equal to zero. Cell load is $0 \leq \mathbf{x} \leq 8$ in this scenario, $LT = 7$, below the capacity of the femtocell, to prevent operation at full capacity. Total load is the sum of the femtocells and the macrocell, L_M.

$$L(\mathbf{x}) = \begin{cases} LT & \text{if } \mathbf{x} > LT \\ \mathbf{x} & \text{if } \mathbf{x} \leq LT \end{cases}$$

Average load is $\overline{L(\mathbf{x})} = \sum_{t=0}^{T}\sum_{i=1}^{N} L(\mathbf{x}_{it})/L_M(\mathbf{x}_t)$.

Leakage fitness is the number of outside users trying to use the femtocell. Leakage increases the number of unwanted users captured, which increases the signalling load to the core network. The leakage, Le is the ratio of blocked calls, y, to the maximum number of macrocell users, C_{MU}, $0 \leq y \leq C_{MU}$ with $Le(y) = 1 - y/C_{MU}$.

GE Parameters. The evolutionary parameter settings for the GE algorithm are presented in Table 1.

Nodal mutation [4] is used instead of the standard integer mutation. The multiple objectives are tackled with the NSGA-II, see Deb et al. [5]. When reinitializing individuals the max derivation tree depth is picked from the distribution of derivation tree depths in the first front. This is both an attempt to restrict bloat and search at derivation

Table 1. Parameter settings for the experiments (dynamic setup (DTS) stationary setup (STS))

Parameter	Value
Max wraps	2
Codon size	128
Population size	20
Initialisation	Ramped half-and-half
Initialisation depth	8
Generations	STS:10; DTS:30
Tournament size	2
Crossover probability	0.5
Mutation	1 event per individual
Parsimony pressure	True
Runs	28

tree depths where good solutions have been found. All evaluated solutions are added to a tabu list and if a solution is in the tabu list the solution will be reinitialized [11]. Furthermore, monotone solutions are not allowed, i.e. only static, increasing or decreasing power.

To find solutions which maximizes one objective and those which have uniform fitness components we use the method from Jain et al. [13] to modify the fitness, where a score of one is the components are uniform and zero they are non-uniform, $\phi(\mathbf{x}) = \frac{(\sum_{i=0}^{n} x_i)^2}{n \sum_{i=0}^{n} x_i^2}$. We penalise the fitness function vector, $\mathbf{f}(\mathbf{x})$ to get $\mathbf{f}'(\mathbf{x})$ by modifying it with its score, $\mathbf{h}(\mathbf{x})$, where $\mathbf{h}(\mathbf{x}) = 1 - \phi \circ \mathbf{f}(\mathbf{x})$ and $\mathbf{f}'(\mathbf{x}) = e^{-\mathbf{h}(\mathbf{x})}(1 - \mathbf{h}(\mathbf{x})^{1/4})$.

4   Results

The grammars and setups are run independently 28 times with different seeds for the pseudo-random number generator. To simplify the presentation the average of the fitness function vector is shown. Figure 4(b) outlines the results, in terms of the run time of the dynamic (DTS) versus stationary (STS) setups. We can observe substantially lower run times for the DTS. Both setups use the same number of fitness evaluations and the total run-time was also significantly different, from a t-test at a 0.05-level, for all comparisons except SRCG in DTS and STS.

With respect to the quality of solutions evolved using the different training scenarios, training results are presented in Figure 5, validation performance in Fig. 4(a), and out-of-sample performance outlined in Table 2.

The mean fitness of all objectives from the training fitness of all solutions in the population, excluding extreme solutions, progresses towards higher fitness, shown in Fig. 5. Since there are changes in the fitness function and the values are the average of the front it is possible for the fitness value to drop. The values decrease when more solutions with lower average are added to the first front. The difference between the methods is significant as can be seen by the non-overlapping error-bars. The graphs show that the representation in CG finds good solutions very fast in comparison to SRCG. The SRCG also has a larger standard deviation. Note that the graphs only show

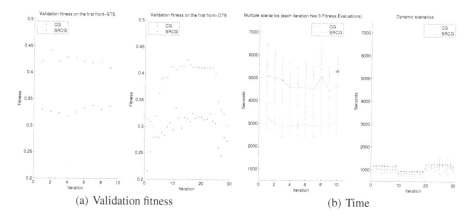

(a) Validation fitness   (b) Time

Fig. 4. In Fig. 4(a) the average validation fitness of the first front is shown, which is slightly increasing. The values decrease when more solutions with lower average are added to the first front. In Fig. 4(b) comparison of evolutionary run times for the dynamic and stationary training scenarios. On average run times are considerably lower for the dynamic training scenarios, and on comparison of the grammars adopted the SRCG form provides additional gains.

Table 2. Fitness on test data for the non-extreme solutions on the first front. The columns show total number of solutions on the fronts in the runs (Total), average fitness of the solutions on the first front (Avg Fit), standard deviation (Std), median (Med), minimum (Min) and maximum (Max).

Version	Total	Avg Fit	Std	Med	Min	Max
CG STS	59	0.467	0.033	0.471	0.374	0.521
SRCG STS	97	0.301	0.077	0.316	0.130	0.458
CG DTS	84	0.349	0.172	0.421	0.031	0.506
SRCG DTS	110	0.244	0.120	0.265	0.000	0.451

the training fitness during the runs and it is not possible to compare the values between DTS and STS since the fitness in the dynamic scenario is only for the current scenario. Thus, a validation scenario was used and in Fig. 4(a) the average validation fitness of the first front is shown, which is slightly increasing.

The non-extreme solutions from the first front for each run are evaluated on the test scenarios. The average of the fitnesses and the average of the first front is chosen in order to allow simple comparisons. This approach was chosen since there can be multiple solutions with the same fitness but different phenotypes and the out-of-sample quality of the solutions is unknown. There is a significant difference in fitness according to the non-parametric Wilcoxon rank sum test for equal medians at a 0.05-level for all values. Thus, we can conclude that for the femtocell scenarios examined here the test performance was best when using the STS setup. It is worth noting that with the SRCG and the DTS some solutions generated invalid values in the test scenarios. As expected the DTS have a higher standard deviation compared to the STS.

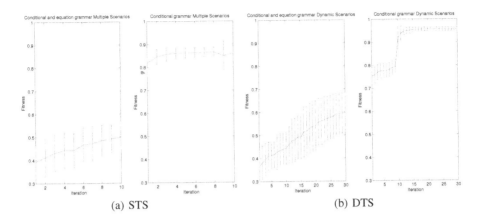

(a) STS   (b) DTS

Fig. 5. Average training fitness of non-extreme solutions

5   Conclusions and Future work

A significant issue facing the developers of the algorithms which control the behaviour of femtocells is how best to design the algorithms to handle these unforeseen, dynamic environments. In this study we examined this issue with respect to the design of fitness functions for an evolutionary algorithm which evolves algorithms to control femtocell behaviour. More specifically we asked *"Is there a difference in the robustness of solutions (out-of-sample) based on the use of stationary versus dynamic training scenarios?"*. Given the experimental setup adopted in this study it was found that, while the dynamic training scenarios result in more efficient run times, the stationary training scenarios produce more robust solutions. In future work we will examine different approaches to the dynamic environment setup, and adopt a wider range of scenarios in each case. There are also potentially many lessons to be learned from, for example, the statistical machine learning literature on best to design training to achieve solutions which generalise beyond training data (e.g., [7, 3, 9, 2]). We will examine if these methods can complement the evolutionary search adopted here.

Acknowledgement. This research is based upon works supported by the Science Foundation Ireland under Grant No. 08/IN.1/I1868.

References

[1] Alba, E., Chicano, J.F.: Evolutionary algorithms in telecommunications. In: MELECON 2006, pp. 795–798. IEEE (2006)

[2] Bersano-Begey, T.F., Daida, J.M.: A discussion on generality and robustness and a framework for fitness set construction in genetic programming to promote robustness. In: Genetic Programming Conference, pp. 11–18 (1997)

[3] Breiman, L.: Bagging predictors. Machine Learning 24(2), 23–140 (1996)

[4] Byrne, J., O'Neill, M., McDermott, J., Brabazon, A.: An Analysis of the Behaviour of Mutation in Grammatical Evolution. In: Esparcia-Alcázar, A.I., Ekárt, A., Silva, S., Dignum, S., Uyar, A.Ş. (eds.) EuroGP 2010. LNCS, vol. 6021, pp. 14–25. Springer, Heidelberg (2010)

[5] Deb, K., Pratap, A., Agarwal, S., Meyarivan, T.: A fast and elitist multiobjective genetic algorithm: NSGA-II. IEEE TEC 6(2), 182–197 (2002)

[6] Dempsey, I., O'Neill, M., Brabazon, A.: Foundations in Grammatical Evolution for Dynamic Environments. Springer (April 2009)

[7] Efron, B.: Bootstrap methods: another look at the jackknife. The Annals of Statistics 7(1), 1–26 (1979)

[8] Fagen, D., Vicharelli, P.A., Weitzen, J.: Automated wireless coverage optimization with controlled overlap. IEEE Transactions on Vehicular Technology 57(4), 2395–2403 (2008)

[9] Freund, Y., Schapire, R.: A Desicion-Theoretic Generalization of On-line Learning and an Application to Boosting. In: Vitányi, P. (ed.) EuroCOLT 1995. LNCS, vol. 904, pp. 23–37. Springer, Heidelberg (1995)

[10] Harper, R.: Ge, explosive grammars and the lasting legacy of bad initialisation. In: WCCI 2010 (July 2010)

[11] Hemberg, E., Ho, L., O'Neill, M., Claussen, H.: A symbolic regression approach to manage femtocell coverage using grammatical genetic programming. In: GECCO, pp. 639–646. ACM (2011)

[12] Ho, L., Ashraf, I., Claussen, H.: Evolving femtocell coverage optimization algorithms using genetic programming. In: 2009 IEEE Personal, Indoor and Mobile Radio Communications, pp. 2132–2136. IEEE (2010)

[13] Jain, R., Chiu, D.M., Hawe, W.R.: A quantitative measure of fairness and discrimination for resource allocation in shared computer system. Eastern Research Laboratory, Digital Equipment Corp. (1984)

[14] Mckay, R.I., Hoai, N.X., Whigham, P.A., Shan, Y., O'Neill, M.: Grammar-based Genetic Programming: a survey. Genetic Programming and Evolvable Machines 11(3), 365–396 (2010)

[15] O'Neill, M., Nicolau, M., Brabazon, A.: Dynamic environments can speed up evolution with genetic programming. In: GECCO, pp. 191–192. ACM (2011)

[16] Siomina, I., Varbrand, P.: Automated optimization of service coverage and base station antenna configuration in umts networks. IEEE Wireless Communications 13(6), 16–25 (2006)

Author Index